Molten Salts: From Fundamentals to Applications

NATO Science Series

A Series presenting the results of scientific meetings supported under the NATO Science Programme.

The Series is published by IOS Press, Amsterdam, and Kluwer Academic Publishers in conjunction with the NATO Scientific Affairs Division

Sub-Series

I. Life and Behavioural Sciences — IOS Press
II. Mathematics, Physics and Chemistry — Kluwer Academic Publishers
III. Computer and Systems Science — IOS Press
IV. Earth and Environmental Sciences — Kluwer Academic Publishers
V. Science and Technology Policy — IOS Press

The NATO Science Series continues the series of books published formerly as the NATO ASI Series.

The NATO Science Programme offers support for collaboration in civil science between scientists of countries of the Euro-Atlantic Partnership Council. The types of scientific meeting generally supported are "Advanced Study Institutes" and "Advanced Research Workshops", although other types of meeting are supported from time to time. The NATO Science Series collects together the results of these meetings. The meetings are co-organized bij scientists from NATO countries and scientists from NATO's Partner countries – countries of the CIS and Central and Eastern Europe.

Advanced Study Institutes are high-level tutorial courses offering in-depth study of latest advances in a field.
Advanced Research Workshops are expert meetings aimed at critical assessment of a field, and identification of directions for future action.

As a consequence of the restructuring of the NATO Science Programme in 1999, the NATO Science Series has been re-organised and there are currently Five Sub-series as noted above. Please consult the following web sites for information on previous volumes published in the Series, as well as details of earlier Sub-series.

http://www.nato.int/science
http://www.wkap.nl
http://www.iospress.nl
http://www.wtv-books.de/nato-pco.htm

Series II: Mathematics, Physics and Chemistry – Vol. 52

Molten Salts: From Fundamentals to Applications

edited by

Marcelle Gaune-Escard

Institut Universitaire des Systèmes Thermiques Industriels,
Marseille, France

Kluwer Academic Publishers

Dordrecht / Boston / London

Published in cooperation with NATO Scientific Affairs Division

Proceedings of the NATO Advanced Study Institute on
Molten Salts: From Fundamentals to Applications
Kas, Turkey
4–14 May 2001

A C.I.P. Catalogue record for this book is available from the Library of Congress.

ISBN 1-4020-0458-3 (HB)
ISBN 1-4020-0459-1 (PB)

Published by Kluwer Academic Publishers,
P.O. Box 17, 3300 AA Dordrecht, The Netherlands.

Sold and distributed in North, Central and South America
by Kluwer Academic Publishers,
101 Philip Drive, Norwell, MA 02061, U.S.A.

In all other countries, sold and distributed
by Kluwer Academic Publishers,
P.O. Box 322, 3300 AH Dordrecht, The Netherlands.

Printed on acid-free paper

All Rights Reserved
© 2002 Kluwer Academic Publishers
No part of the material protected by this copyright notice may be reproduced or utilized in any form or by any means, electronic or mechanical, including photocopying, recording or by any information storage and retrieval system, without written permission from the copyright owner.

Printed in the Netherlands.

TABLE OF CONTENTS

List of Contributors VII
Preface IX

Molten Salts: Fundamentals
Interionic Forces and Relevant Statistical Mechanics 1
M.P. Tosi

Electronic Properties Of Metal/Molten Salt Solutions 23
W.W. Warren, Jr.

Light Scattering from Molten Salts: Structure and Dynamics 47
G. N. Papatheodorou and S. N. Yannopoulos

Neutron Scattering:
Technique And Applications To Molten Salts 107
A.K. Adya

Metal-Molten Salt Interfaces: Wetting Transitions and
Electrocrystallization 149
W. Freyland

Modelling of Thermodynamic Data 179
H.A. Øye

Thermodynamic Model of Molten Silicate Systems 213
V. Daněk, Z. Pánek

Data Mining and Multivariate Analysis in Materials Science:
Informatics Strategies for Materials Databases 241
K. Rajan, A. Rajagopalan and C. Suh

Pyrochemistry in Nuclear Industry 249
T. Inoue and Y. Sakamura

Molten Salts for Safe, Low Waste and Proliferation Resistant
Treatment of Radwaste in Accelerator Driven and Critical Systems 263
V.V. Ignatiev

Electrochemical Techniques
Some Aspects of Electrochemical Behaviour of Refractory
Metal Complexes 283
S.A. Kuznetsov

Origin and Control of Low Melting Behavior in Salts
Polysalts, Salt Solvates and Glassformers 305
C.A. Angell

Electrodes and Electrolytes for Molten Salt Batteries:
Expanding the Temperature Regimes 321
R.T. Carlin and J. Fuller

Synthesis and Catalysis in Room-Temperature Ionic Liquids 345
P.J. Smith, A. Sethi and T. Welton

COORDINATION COMPOUNDS IN MELTS 357
S.V. Volkov

Calorimetric Methods 375
M. Gaune-Escard

SUBJECT INDEX 401

LIST OF CONTRIBUTORS

A. K. ADYA
University of Abertay Dundee,
School of Molecular of Life Sciences
Division of Applied Chemistry, Bell Street
Dundee DD1 1HG, U.K

C.A. ANGELL
Arizona State University
Tempe AZ 85287, USA

R.T. CARLIN
Office of Naval Research, code 33
Ballston Centre Tower One,
800 North Quincy Street
Arlington - VA 22217-5660, USA

V. DANEK
Institute of Inorganic Chemistry,
Slovak Academy of Sciences
Dubravska cesta 9, Bratislava
SLOVAK REPUBLIC

W. FREYLAND
University of Karlsruhe, kaiserstr.12
D-76128 Karlsruhe
GERMANY

M. GAUNE-ESCARD
I.U.S.T.I.
Technopole de Château Gombert
5 Rue Enrico Fermi
13453 Marseille Cedex 13
FRANCE

V. IGNATIEV
RRC Kurchatov Institute,
Kurchatov sq. 1
Moscow,
RUSSIA

T. INOUE
Central Res. Inst. of Electric Power Ind., Komae
Research Laboratory
2-11-1 Iwado Kita, Komae-shi, Tokyo
201, JAPAN

S. KUZNETSOV	Institute of Chemistry, Kola Science Centre RAS, 14 Fersman Str. Apatity, RUSSIA
H. ØYE	Dept. of Chemistry Norwegian University of Science and Technology Trondheim N 7491 NORWAY
G. PAPATHEODOROU	ICE/HT FORTH P.O. Box 1414 University Campus GR 26500-Patras GREECE
K. RAJAN	Dept. of Material Science & Engineering Faculty of Information Technology Rensselaer Polytechnic Institute Troy NY 12180-3590 USA
M. TOSI	Scuola Normale Superiore di Pisa Piazza dei Cavalieri 7 Pisa ITALY
S. VOLKOV	Inst. of General and Inorg. Chem. 32-34 Prospect Palladina 142 Kiev GSP, UKRAINE
W.W. WARREN	Oregon State University Department of Physics Weniger Hall 301 Corvallis, OR 97331-6507 USA
T. WELTON	Imperial College Department of Chemistry London, UK

Preface

This **NATO Advanced Study Institute** was devoted to **MOLTEN SALTS** and several aspects ranging **FROM FUNDAMENTAL TO APPLICATIONS** were highlighted by the lectures reproduced in this volume. Other valuable inputs to the training of NATO-ASI students were given as tutorials.

Finally, some short research oral and poster presentations were given as illustrations of the basic principles detailed in the courses. Several will be published separately (Z. Naturforsch A., 2001) since the objective of the present volume is to produce a textbook rather than ordinary proceedings.

The NATO-ASI, was held in Kas, Turkey, May 4-14, 2001. Financial support came from the NATO Science Committee but also from several generous sources including: US Naval Research Office, US Aerospace and Commissariat à l'Energie Atomique (C.E.A.). Also specific support was granted to nationals of Greece (NATO), Portugal (NATO), Turkey (TUBITAK) and U.S.A. (National Science Foundation). The practical help offered by the region and the town of Kas, Turkey gave the participants the opportunity to discover the nice surroundings of the NATO-ASI location.

All of the above have to be thanked.

The organizing committee comprised of Marcelle GAUNE-ESCARD (France, NATO country director), Sergey VOLKOV (Ukraine, NATO partner country co-director) together with:Z. AKDENIZ (Turkey), V. DANEK (Slovakia), V. KHOKLHLOV (Russia), H. ØYE (Norway) and M. TOSI (Italy) members of the scientific committee.

In addition to lecturers, the participants included 76 students from 23 countries with 36,8 % from NATO partner countries.

Special thanks also to Ms Joyce BARTOLINI for her untiring help and dedication, not only before and after the NATO-ASI, but also in the preparation of proceedings.

M. GAUNE-ESCARD
September 2001

MOLTEN SALTS: FUNDAMENTALS
Interionic Forces and Relevant Statistical Mechanics

M.P. TOSI
*INFM and Classe di Scienze, Scuola Normale Superiore
Piazza dei Cavalieri 7, I-56126 Pisa, Italy*

Abstract. General background on molten metal halides: chemical coordinates, melting parameters, transport coefficients. Overview on liquid structure. Primitive model for structure and thermodynamics. Background on interionic forces: cohesion and lattice vibrations in alkali halide crystals, ionic binding in alkali halide molecules. Structure of alkali halide melts and chemical short-range order. Liquid-solid and liquid-gas coexistence. Cluster formation in trichloride melts and their mixtures with alkali chlorides. Clusters in aluminium-alkali fluoride mixtures and solutions of sodium metal in molten cryolite. Ionic transport, viscosity and dynamics of density fluctuations.

1. General background

The crystal structures of metal halide compounds arise from electronic charge transfer and local compensation of positive and negative ionic charges by formation of chemical order. This charge compensation is achieved by nature in condensed ionic phases in two qualitatively distinct ways. The first involves halogen sharing and high coordination for the metal ions, as for example in alkali, alkaline-earth and lanthanide metal halides. In the second type charge compensation takes place within well defined molecular units, either monomeric as for example in $HgCl_2$ and $SbCl_3$ or dimeric as in $AlBr_3$.

Studies of metal halide melts by neutron and X-ray diffraction have shown that melting usually preserves the type of chemical order which is found in the crystal. For example, the melting of NaCl, $MgCl_2$ or YCl_3 can be viewed as a transition from an ionic crystal to an ionic liquid (ionic-to-ionic, in short) and that of $SbCl_3$ or $AlBr_3$ as a molecular-to-molecular transition. However, $AlCl_3$ and $FeCl_3$ are known instances of ionic-to-molecular melting (see § 6.1). These two instances tell us that there is no major fundamental difference in the character of the bonding between ionic-type and molecular-type halides. Just as these two types of compound are amenable to basically the same quantum-chemical approaches at the Hartree-Fock level and beyond, so essentially the same interionic-force model should be useful for both.

In addition to chemical short-range order, diffraction studies have revealed that intermediate-range order can exist in both network-type and molecular-type ionic melts. This type of order is well known in glassy materials and reflects structural correlations extending over distances of about 1 nm, i.e. larger than atomic bond lengths (≈ 0.3 nm) but smaller than the scale of textural inhomogeneities (> 5nm).

Macroscopic data on melting parameters and transport coefficients already reflect the melting mechanism and the type of liquid structure. Table 1 refers to representative cases.

TABLE 1. Macroscopic properties of metal chlorides (from [1])

Salt	χ_M	Crystal	T_m (K)	ΔS_m (e.u.)	$\Delta V_m/V_l$	σ (Ω^{-1}cm^{-1})	η (cp)
NaCl	0.40	NaCl	1074	6.30	0.28	3.6	1.0
CuCl	1.20	Wurtzite	696	2.43	0.16	3.7	4.1
SrCl$_2$	0.55	CaF$_2$	1146	3.44	0.11	2.0	3.7
CaCl$_2$	0.60	CaCl$_2$	1045	6.44	0.043	2.0	3.4
MgCl$_2$	1.28	CdCl$_2$	980	9.74	0.28	1.0	2.2
HgCl$_2$	1.32	HgCl$_2$	554	9.11	0.21	3x10^{-5}	1.6
ZnCl$_2$	1.44	ZnCl$_2$	570	4.09	0.14	1x10^{-3}	4x10^3
LaCl$_3$	0.705	UCl$_3$	1131	11.49	0.16	1.3	6.7
YCl$_3$	0.66	AlCl$_3$	994	7.56	0.0045	0.39	--
FeCl$_3$	0.99	FeCl$_3$	573	17.80	0.39	0.04	--
AlCl$_3$	1.66	AlCl$_3$	466	18.14	0.47	5x10^{-7}	0.36
GaCl$_3$	1.68	GaCl$_3$	351	7.84	0.17	2x10^{-6}	1.8
SbCl$_3$	2.08	SbCl$_3$	346	8.96	0.17	2x10^{-4}	--

for various types of ordering in chlorides (complete references can be found in specialized reviews [1, 2]). Each compound is labelled by a parameter χ_M as an indicator of the chemical bond, which has been taken for the metal atom from the chemical scale of the elements proposed by Pettifor [3]. In essence, this chemical coordinate sets these compounds in order of increasing covalency against ionicity as one moves downwards in Table 1 for each value of the nominal valence.

The data in the Table are (i) the structure of the hot crystalline phase, (ii) the melting temperature T_m and the corresponding values of the melting parameters (entropy of melting ΔS_m and relative volume change $\Delta V_m/V_{liquid}$), and (iii) transport coefficients near freezing (ionic conductivity σ and shear viscosity η). The melting parameters give useful indications on the melting mechanism. Tallon [4] has proposed the empirical relation

$$\Delta S_m = \nu R \ln 2 + \gamma C_V \Delta V_m / V \quad (1)$$

where ν is the number of atoms in a formula unit, R the gas constant, γ the Gruneisen parameter and C_V the specific heat. Insofar as γ and C_V take similar values for similar systems in corresponding states, Eq. (1) implies an approximately linear relationship between ΔS_m and $\Delta V_m/V$, extrapolating for $\Delta V_m/V \to 0$ to a constant $\Delta S_m/\nu \to R\ln 2$.

Such a linear relationship is approximately verified by many halides of mono-, di- and trivalent metals [5]. However, some systems show a deficit in ΔS_m relative to the 'norm'. These exceptions are associated with special melting mechanisms, i.e. melting from a disordered solid (e.g. CuCl and SrCl$_2$) and melting into a network-forming liquid (e.g. ZnCl$_2$) or into a molecular liquid (e.g. GaCl$_3$ and SbCl$_3$). The deficit in ΔS_m is associated in the first case with disordering of the hot crystal into a fast-ion conducting state before melting, and in the other cases with residual order in the melt. The very large values of ΔS_m and $\Delta V_m/V_l$ for FeCl$_3$ and AlCl$_3$ reflect drastic changes in the state of local order on melting through an ionic-to-molecular transition.

The values of transport coefficients of the melt near freezing also reflect the type of liquid structure. As is illustrated in Table 1, the ionic conductivity σ is very low both in molecular-type melts (HgCl$_2$, AlCl$_3$, GaCl$_3$ and SbCl$_3$) and in network-type melts (ZnCl$_2$). In the latter system the viscosity of the melt is very high, allowing this pure chloride to be brought into a glassy state by normal quenching. A fast-ion conducting state in the hot solid, as in noble-metal halides, is instead associated with disorder in one crystalline sublattice and is reflected in the melt by a marked difference between the diffusion coefficients of the two ionic species (see § 8.1 below).

2. Overview on liquid structure

Resolution of chemical short-range order out of total diffraction patterns from a multi-component liquid can be achieved experimentally by various methods. Levy *et al.* first used in 1960 a combination of neutron and X-ray diffraction on LiCl. Most revealing are the multi-pattern neutron diffraction experiments based on isotopic enrichment, as first applied to CuCl by Page and Mika in 1971. Useful analyses are also allowed by the combined study of presumed isomorphous melts, *e.g.* the DyBr$_3$ and YBr$_3$ pair [6].

The intensity $I(k)$ of radiation coherently scattered through an angle 2θ is

$$I(k) = \sum_{\alpha,\beta}(n_\alpha n_\beta)^{1/2}b_\alpha b_\beta S_{\alpha\beta}(k): \tag{2}$$

here $k = (4\pi/\lambda)\sin\theta$ is the scattering wave number, n_α is the number density of ions of chemical species α, b_α is their coherent scattering amplitude and $S_{\alpha\beta}(k)$ are the partial structure factors of the liquid. These are defined through the Fourier transform of the partial radial distribution functions $g_{\alpha\beta}(r)$,

$$S_{\alpha\beta}(k) = \delta_{\alpha\beta} + 4\pi(n_\alpha n_\beta)^{1/2}\int_0^\infty (\sin kr / kr)[g_{\alpha\beta}(r)-1]r^2 dr \tag{3}$$

where the quantity $4\pi r^2 n_\beta g_{\alpha\beta}(r)dr$ is the mean number of β-type ions in a spherical shell of radius r and thickness dr centred on an α-type ion. The b_α's for neutron scattering in Eq. (2) are isotope-averaged amplitudes at given isotopic composition of the liquid. In the multi-pattern method melts of different isotopic composition are examined - most often chlorides, where the scattering amplitudes of ^{35}Cl and ^{37}Cl differ by a factor of about five.

Special linear combinations of the partial structure factors emphasize certain aspects of the short-range order, as Bhatia and Thornton [7] first pointed out for liquid metal alloys. Thus, if in Eq. (2) we take the b_α's to be identical for all species, we obtain the number-number (NN) structure factor

$$S_{NN}(k) = \sum_{\alpha,\beta}(n_\alpha n_\beta / n^2)^{1/2}S_{\alpha\beta}(k) \tag{4}$$

where $n = \sum_\alpha n_\alpha$ is the total particle-number density. This function, with the corresponding pair function $g_{NN}(r)$, describes the correlations between fluctuations in the total number density and reflects the topological order from excluded volume associated to finite ionic sizes. The position r_N of the main peak in $g_{NN}(r)$ is close to the first-neighbor bond length and is only slightly greater than the distance of closest approach between first neighbors, as measured by the sum of their ionic radii. In molten alkali halides the main peak in $S_{NN}(k)$ at k_N is very broad and $g_{NN}(r)$ is almost flat beyond its first peak. From this viewpoint these melts look like hot and dense hard-sphere fluids.

On the other hand, if we consider a binary liquid with $n_1b_1 = -n_2b_2$ we obtain from Eq. (2) the concentration-concentration (or charge-charge) structure factor

$$S_{QQ}(k) = [n_2 S_{11}(k) - 2(n_1 n_2)^{1/2} S_{12}(k) + n_1 S_{22}(k)] / n. \tag{5}$$

This function and the corresponding $g_{QQ}(r)$ describe the correlations between fluctuations in composition, *i.e.* the chemical short-range order. In molten alkali halides $g_{QQ}(r)$ has marked oscillations extending over distances of about 1 nm: starting from a given ion, one meets regions that are alternatively enriched and depleted in the other species. The period r_Q of these oscillations is close to the second-neighbor distance. $S_{QQ}(k)$ shows a strong and rather narrow peak at $k_Q \approx 0.7 k_N$: this is known as the Coulomb prepeak. The important notion here is that of Debye screening, though extended to encompass oscillatory rather than merely monotonic behavior of the screening charge density.

In molten halides of polyvalent metals, topological order and Coulomb order are not as simply related as they are in the alkali halides. In strongly ionic di- and tri-halides the dominant role of the Coulomb repulsions between the metal ions lends special relevance to the metal-metal structure factor $S_{MM}(k)$. In $SrCl_2$ $S_{MM}(k)$ shows a strong peak near 1.5 Å^{-1}, so that the main peak in $g_{MM}(r)$ lies as far out as 0.5 nm [8]. Within the structure imposed by the polyvalent metal ions, the halogen component of the melt is more weakly ordered. A similar picture is useful for molten $LaCl_3$, with a progressive reduction in metal-ion coordination and a progressive growth in network features being driven across the rare-earth chloride series by the diminishing size of the metal ion [9].

In a compound such as $ZnCl_2$, on the other hand, $g_{MM}(r)$ peaks at the sharply reduced distance of 0.38 nm, only slightly greater than the Cl-Cl distance of 0.37 nm [10]. Its liquid structure is best described as a dense random packing of halogens accommodating the metal ions in tetrahedral sites. This structure results from the constraints imposed by a network of partially covalent bonds, arising from hybridization of the d states on Zn^{2+} with the p states on Cl^-. Network formation in the melt is accompanied by intermediate-range order, which is most prominently evidenced by a sharp peak near 1 Å^{-1} in both the total and the metal-metal structure factors. From later structural studies of glasses [11] this feature is referred to as the first sharp diffraction peak (FSDP). As already noted in § 1, the same feature is present in the diffraction patterns of molecular-type trihalide melts such as $AlCl_3$ and $AlBr_3$. Here an appropriate picture is that of a random close packing of halogens accommodating the trivalent metal ions in tetrahedral sites to form a liquid of strongly correlated molecular dimers.

While some workers prefer to view the FSDP as arising from holes in the liquid structure, it can be formally demonstrated that the distribution of structural holes and the hole-hole pair correlations are wholly determined by the order in the pair distributions of the atoms.

3. Primitive model for structure and thermodynamics

The two-component fluid of oppositely charged hard spheres (CHS) affords a primitive model for strongly ionic liquids. The CHS model can be solved analytically [12] in the so-called mean spherical approximation (MSA) and is especially useful in regard to Coulomb short-range order and its consequences in molten alkali halides.

The CHS model has the following exact properties:

(i) the pair distribution functions obey the hard-sphere boundary conditions
$$g_{\alpha\beta}(r) = 0 \qquad (r < \sigma_{\alpha\beta}) \qquad (6)$$
enforcing sharply defined regions of excluded volume with distances $\sigma_{\alpha\beta}$ of closest approach for the various ion pairs (the additivity property $\sigma_{12} = (\sigma_{11} + \sigma_{22})/2$ is usually assumed, thus identifying $\sigma_{\alpha\alpha}/2$ with ionic radii R_α); and

(ii) the direct correlation functions $\hat{c}_{\alpha\beta}(k)$ obey the asymptotic behaviors
$$\lim_{k \to 0} \hat{c}_{\alpha\beta}(k) = -4\pi z_\alpha z_\beta e^2 / k^2 k_B T \qquad (7)$$
where z_α are the ionic valences. As usual, these functions are related to the inverse of the matrix of partial structure factors by the Ornstein-Zernike relation
$$\hat{c}_{\alpha\beta}(k) = (n_\alpha n_\beta)^{-1/2} [\delta_{\alpha\beta} - S^{-1}_{\alpha\beta}(k)]. \qquad (8)$$
Equation (7) embodies the property of complete screening of any ionic charge by the ionic conducting fluid.

Indeed, the linearized Debye-Hückel theory of dilute electrolyte solutions is recovered when the form of the direct correlation functions in Eq. (7) is taken to apply at all wave numbers. This yields upon Fourier transform the Debye-Hückel result
$$c^{DH}_{\alpha\beta}(r) = -z_\alpha z_\beta e^2 / r k_B T. \qquad (9)$$
The MSA corrects the Debye-Hückel theory for finite ionic sizes by combining the exact property in Eq. (6) with the approximate expression
$$c^{MSA}_{\alpha\beta}(r) = -z_\alpha z_\beta e^2 / r k_B T \qquad (r > \sigma_{\alpha\beta}), \qquad (10)$$
again satisfying Eq. (7). The liquid structure equations (6), (8) and (10) can still be solved analytically within this approximation. Let us consider for simplicity the case of a symmetric CHS model, with $z_+ = -z_- = 1$ and $R_+ = R_- = \sigma/2$ (the so-called restricted primitive model or RPM). From the symmetry property $g_{++}(r) = g_{--}(r)$ in the RPM it is easily seen that the total number density $N(r)$ and the charge density $Q(r)$ become uncorrelated, leading to especially transparent analytic results as reported below.

For the number-density correlations the Wertheim-Thiele solution of the Percus-Yevick theory of the one-component fluid of uncharged hard spheres is recovered, i.e.
$$c^{MSA}_{NN}(r) = \begin{cases} a + b(r/\sigma) + c(r/\sigma)^3 & (r < \sigma) \\ 0 & (r > \sigma) \end{cases} \qquad (11)$$
where $a = -(1+2\eta)^2(1-\eta)^{-4}$, $b = 6\eta(1+\eta/2)^2(1-\eta)^{-4}$ and $c = \eta a/2$ with $\eta = \pi n \sigma^3/6$ being the packing fraction. The charge-charge correlations are instead given by
$$c^{MSA}_{QQ}(r) = \begin{cases} (k_D^2 \sigma^2 / 12\eta)(2B + B^2 r/\sigma) & (r < \sigma) \\ -e^2 / r k_B T & (r > \sigma) \end{cases} \qquad (12)$$
where $k_D = (4\pi n e^2/k_B T)^{1/2}$ is the Debye inverse screening length and $B = -1 + (Lk_D)^{-1}$ with L being a length given by
$$L = \sigma[(1 + 2k_D\sigma)^{1/2} - 1]^{-1}. \qquad (13)$$
Notice that $L \to 1/k_D$ in the limit $\sigma \to 0$.

The above structural results in the MSA lead to physically transparent results for the main thermodynamic properties of the RPM. The Coulomb internal energy U_C per unit volume can be calculated with the help of Eq. (12) from the general expression

$$U_C = \frac{1}{2}\int d\mathbf{r}\,(e^2/r)\sum_{\alpha,\beta} n_\alpha n_\beta z_\alpha z_\beta g_{\alpha\beta}(r)$$
$$= -\frac{1}{2}nz_+z_-(2\pi)^{-3}\int d\mathbf{k}\,(4\pi e^2/k^2)[S_{QQ}(k)-1]. \quad (14)$$

The MSA result is

$$U_C^{MSA} = -(ne^2/2L)(1+\sigma/2L)^{-1}. \quad (15)$$

The p-n-T equation of state is found from Eq. (15) as

$$p = p_0 + (k_BT/4\pi\sigma^3)\left[x + x(1+2x)^{1/2} - \frac{2}{3}(1+2x)^{3/2} + \frac{2}{3}\right], \quad (16)$$

with p_0 the pressure for neutral hard spheres and $x = k_D\sigma$. From Eq. (15) one also finds that the potential induced on an ion charge e by the surrounding charge distribution is

$$\phi_{ind} = (2/ne)U_C = -(e/L)/(1+\sigma/2L) \quad (17)$$

Thus, the screening charge $-e$ can be viewed as located at a distance L_s from the ion, with

$$L_s = L + \sigma/2. \quad (18)$$

This is the *screening length* of the symmetric CHS fluid in the MSA.

Finally, the total potential drop from the surface of the ion to infinity is $\phi = \phi_{ind} + 2e/\sigma$, the latter being the self-potential at the ion surface. The capacitance of the ion (thought of as a microscopic spherical electrode) is $C_{ion} = e/(\phi\pi\sigma^2)$ per unit area and hence, taking the limit $\sigma \to \infty$ we find the electrical capacitance of a flat impenetrable electrode (in the absence of image effects):

$$C^{MSA} = 1/4\pi L. \quad (19)$$

The result in Eq. (19) does not depend on the charge carried by the electrode, and this exposes the main limitation of the MSA: the Coulomb interactions outside the ionic cores are being treated in a linear-response approximation, which may be valid only at weak coupling. Nevertheless, contact with electrochemical differential-capacitance data on molten salts facing metallic electrodes near the point of zero charge has been made with the help of these results [13].

4. Background on interionic forces: the alkali halides

4.1. COHESION AND LATTICE VIBRATIONS IN ALKALI HALIDE CRYSTALS

A primary source of information on interionic forces is the binding energy of the crystal in a given structure as a function of the lattice parameter, the ions being in an ideal frozen state at their lattice sites (*static crystal*). The theory of binding in these large-gap materials is conveniently developed starting from Hartree-Fock free-ion wave functions [14]. Orthogonalization of overlapping parallel-spin orbitals centred at near sites leads to repulsive contributions to the crystal energy, which increase rapidly and approximately exponentially as the lattice spacing decreases.

Good results for the binding and the equation of state have also been obtained by Kohn-Sham density functional methods, in which one deals directly with the electron density in the crystal. These calculations have drawn attention not only to the role of orbital orthogonalization [15], but also to subtle differential deformations of the ionic electron densities relative to the free-ion state [16]. These deformations may be described as a relative contraction of the anions and a relative expansion of the cations in the crystal and emerge from detailed comparisons between theory and experiments of X-ray and γ-ray diffraction. Ionic deformations are crucial in determining how the various contributions to the total energy vary with lattice spacing: this variation would be much weaker if the electron density in the crystal were taken as the simple superposition of the free-ion densities [15]. Conversely, absurdly strong repulsions would arise in such quantal calculations if overlap orthogonality were replaced by volume exclusion.

In the models of cohesion stemming from the early work of Born and Mayer (see *e.g.* [17, 18]), integral valences are assumed for the Coulomb interactions and the overlap repulsive energy is represented as the sum of two-body central terms, given by

$$\Phi_{\alpha\beta}(r) = A_{\alpha\beta} \exp[(R_\alpha + R_\beta - r)/\rho_{\alpha\beta}]. \tag{20}$$

Here, r is the distance between the two ions and $\rho_{\alpha\beta}$ are empirical parameters of order one tenth of the first-neighbor distance. In an accurate redetermination of the repulsive parameters from cohesive properties of alkali halide crystals Fumi and Tosi [19] stressed that the ionic radii R_α in Eq. (20) reflect in their *relative* magnitudes the sizes of ions in the crystalline state rather than in the free state. The sum of ionic radii for cations and anions, $R_+ + R_-$, can be adjusted to the lattice spacing by a rescaling of the factors $A_{\alpha\beta}$ in Eq. (20), but the differences $R_+ - R_-$ exceed those obtained from Pauling's radii by several tenths of an Ångström. This fact is crucial in giving the correct relative weight to the overlap repulsive interactions between the various types of ion pairs.

Including van der Waals attractive interactions at the dipole-dipole level, the potential energy U_S of the static crystal as a function of the interionic bond vectors $\{\mathbf{r}_{ij}\}$ is

$$U_S(\{\mathbf{r}_{ij}\}) = \sum_{i>j} \left[\frac{z_i z_j e^2}{r_{ij}} + \Phi_{ij}(r_{ij}) - \frac{C_{ij}}{r_{ij}^6} \right]. \tag{21}$$

The electric fields generated by the ionic charges vanish at the ideal lattice sites in highly symmetric structures such as those of the alkali halides and are not included in Eq. (21). They play a crucial role, however, when we turn to consider distortions of the perfect crystal lattice. Thus Jost and later Mott and Littleton (see *e.g.* [20]), in dealing with the point defects which are responsible for charge and mass transport in the crystal, attributed to ionic displacements and to ion-core polarization the large difference between the crystalline cohesive energy (≈ 8 eV per ion pair) and the energy of formation of a Schottky defect pair (≈ 2 eV). Later studies of lattice vibrations recognized the role of ion-core polarization in response not only to the prevailing electric field, but also to relative displacements of neighboring ions leading to changes in the state of overlap.

Let us consider with Born and Huang [17] a longitudinal optical vibration of long wavelength, in which the ion pair inside each lattice cell takes a relative displacement $\mathbf{u}_+ - \mathbf{u}_-$. The macroscopic polarization $\mathbf{P}$ of the crystal is given by

$$\mathbf{P} = v_c^{-1}[(ze-q)(\mathbf{u}_+ - \mathbf{u}_-) + (\alpha_+ + \alpha_-)\mathbf{E}_{eff}] \tag{22}$$

where v_c is the volume of the unit cell, α_+ and α_- are the electric polarizabilities and $\mathbf{E}_{eff} = \mathbf{E} + 4\pi\mathbf{P}/3$, $\mathbf{E}$ being the macroscopic electric field set up in the crystal. The quantity q, which in the linear response regime enters the theory merely as a correction to the bare ionic charge ze, arises from the mutual distortion of the electron clouds of neighboring ions and can be written as

$$q = \frac{1}{3} M[m'(r_0) + 2m(r_0)/r_0]. \tag{23}$$

Here, M is the first-neighbor coordination number and $m(r_0)$ - a function of the first neighbor distance r_0 with derivative $m'(r_0)$ - is the dipole due to overlap. By making the reasonable assumption that $m(r_0)$ varies with distance as the overlap-repulsive interaction in Eq. (20), i.e. $m(r_0) \propto \exp(-r_0/\rho_{+-})$, Born and Huang show from the analysis of dielectric-constant data for many halides and oxides that in the equilibrium configuration the electron cloud of the negative ion is more strongly pushed back than that of the positive ion (so that the overlap deformation dipole points towards the positive ion) and that the magnitude of this dipole increases with increasing overlap.

Further studies of all vibrational modes in ionic and semiconducting crystals have led to the so-called shell model (see e.g. Cochran [21]). In brief, the model envisages each ion as having an outer shell of valence electrons which is elastically tied to a rigid inner core. Thus, all relative displacements of neighboring ions are accompanied by the formation of dipoles due to changes in the state of close-shell overlap, in supplement to the dipoles induced by the prevailing electric field. The overlap deformation dipole may be thought to lie in the cation-anion overlap region: it appears, however, that its precise location is not important in practice, so that it may be taken to lie at the centre of the negative ion where it counterbalances the dipole due to induction by the electric field.

The last step in developing such a phenomenological model of interionic forces is to transcend the linear regime of lattice vibrations. We again assume that the overlap dipole is proportional to the overlap repulsive force. We thus can write the potential energy U of an ionic assembly, taken as a function of the interionic bond vectors $\{\mathbf{r}_{ij}\}$ and of electronic coordinates described by the dipole moments $\{\mathbf{p}_i\}$ carried by the ions, as

$$U(\{\mathbf{r}_{ij}\},\{\mathbf{p}_i\}) = U_S(\{\mathbf{r}_{ij}\}) + U_{pol}^{cl}(\{\mathbf{r}_{ij}\},\{\mathbf{p}_i\}) + U_{shell}(\{\mathbf{r}_{ij}\},\{\mathbf{p}_i\}). \tag{24}$$

Here, U_S is given by Eq. (21), U_{pol}^{cl} is the classical polarization energy given by

$$U_{pol}^{cl} = \sum_{i \neq j}\left[-z_i e \frac{\mathbf{p}_i \cdot \mathbf{r}_{ij}}{r_{ij}^3} + \frac{\mathbf{p}_i \cdot \mathbf{p}_j}{2r_{ij}^3} - 3\frac{(\mathbf{p}_i \cdot \mathbf{r}_{ij})(\mathbf{p}_j \cdot \mathbf{r}_{ij})}{2r_{ij}^5}\right] + \sum_i \frac{p_i^2}{2\alpha_i}, \tag{25}$$

and U_{shell} is the shell-deformation energy,

$$U_{shell} = \frac{\alpha_s}{\alpha_-}\sum_{i_j,j_-}\mathbf{p}_j \cdot \hat{\mathbf{r}}_{ij}\left|\frac{d\Phi_{ij}(r_{ij})}{dr_{ij}}\right|. \tag{26}$$

Here, the quantity α_s is a short-range polarizability and the sum in Eq. (26) runs over $j_- =$ negative ions and over $i_j =$ positive ions which are first neighbors of the j-th negative ion. Minimization of U with respect to the dipoles yields the dipole $\mathbf{p}_h$ on a halogen as

$$\mathbf{p}_h = \alpha_- \mathbf{E}_h(\{\mathbf{r}_{ij}\},\{\mathbf{p}_i\}) + \alpha_s \sum_{i_j} \hat{\mathbf{r}}_{ih}\left|\frac{d\Phi_{ih}(r_{ih})}{dr_{ih}}\right|, \tag{27}$$

where $\mathbf{E}_h$ is the self-consistent electric field on the halogen. The overlap dipole is the second term on the RHS of Eq. (27) and each of its components lies along a cation-anion bond, pointing to the cation. Therefore, the net dipoles saturate as the ions deform in reaching their equilibrium positions.

The model of ionic interactions given in Eqs. (24) - (27) seems to have been first proposed by Tosi and Doyama [22] in connection with the alkali halide molecular monomers. We turn to discuss this application as a test of the usefulness of the model and to present further evidence for its essential correctness.

4.2. IONIC BINDING IN ALKALI HALIDE MOLECULES

The two lowest adiabatic potential energy curves of an alkali halide molecule display an avoided crossing representing a rapid change from atomic to ionic constituents as the bond length R decreases. Large-scale configuration-interaction calculations on the LiF molecule have placed great attention on elucidating the nature of the avoided crossing [23]. *Ab initio* evidence for a strongly ionic character near equilibrium is also available for alkali halide dimers: in particular, extraction of an electron from Na_2Cl_2 leaves a $(Na_2Cl)^+$ residue to which a chlorine *atom* is very weakly bound [24].

TABLE 2. Equilibrium bond length, dipole moment, binding energy relative to free ions, and vibrational frequency of alkali halide molecular monomers in the ionic model (from [22]) and from configuration interaction calculations on NaCl (CI).

	R_0 (Å) Theory	R_0 (Å) Expt	d (Debye) Theory	d (Debye) Expt	$-U(r_0)$ (eV) Theory	$-U(r_0)$ (eV) Expt	ω_e (cm^{-1}) Theory	ω_e (cm^{-1}) Expt
NaCl	2.383	2.3606	9.58	9.0	5.71	5.64	402	380
CI:	2.382		10.		5.43		378	
NaBr	2.527	2.5020	9.97	---	5.42	5.41	325	315
KCl	2.635	2.6666	10.24	10.48	5.21	5.01	312	305
KBr	2.823	2.8207	10.86	10.41	4.94	4.79	253	230

Following early rigid-ion model calculations of the Born-Mayer type, Rittner [25] stressed the need to include ion-core polarization in dealing with the ground-state potential energy curves of alkali halide monomers near equilibrium. The main argument for this is that the measured dipole moment is about one-half of what one would calculate by placing positive and negative unitary charges at a distance equal to the measured bond length. Tosi and Doyama [22] supplemented the Rittner model by including the overlap deformation dipole, with the very remarkable consequence that *a fit of interionic force parameters to molecular data was no longer necessary to obtain a good quantitative account of the molecular properties*! Table 2 illustrates for a few of these molecules the agreement between their results, as obtained with interaction parameters taken from alkali halide *crystal properties*, and experiment.

The main message from Table 2 is that for alkali halides the same model handles to useful quantitative accuracy both crystals and molecular monomers near equilibrium. The model should thus be applicable to alkali halide melts - although, as it turns out, near freezing these are much closer to the crystalline state. Transferability of interionic-force parameters has important implications for those polyvalent-metal halide melts in which neutral and charged molecular clusters are believed to be present. Approximate transferability between different compounds is both a justification for the correctness of the model and a very useful simplification in constructing its various realizations.

Table 2 includes the results of configuration interaction calculations on the NaCl molecule based on the use of Gaussian basis sets for the valence electrons moving in *ab initio* effective core potentials (Andreoni, Galli and Tosi, unpublished). At the equilibrium value R_0 of the bond length the ground state is characterized by a highly ionic bond, a transfer of $0.85e$ from Na to Cl being estimated from generalized-valence-bond calculations of Mulliken populations. The magnitude of the charge transfer calculated in this way increases as the molecule is stretched, to reach a value of $0.99e$ just before the avoided crossing. This lies at $R_x \approx 1$ nm, in agreement with the simple estimate $R_x \approx e^2/$(ionization potential - electron affinity) obtained by balancing the Coulomb energy gain against the electronic energy loss in the electron transfer between isolated atoms. An effective R-dependent polarizability can also be extracted from the calculated dipole moment of the molecule as a function of internuclear distance: this function takes the value ≈ 3 Å^3 near equilibrium and rapidly decreases at lower values of R, undoubtedly because of quenching of ionic deformability from overlap effects as are included in the deformation dipole model.

Similar configuration-interaction calculations on the NaCl$^+$ ion show that its ground state is obtained by removing an electron from the highest π orbital of the NaCl molecule. In a vertical ionization process the removed electron is highly localized on the chlorine site, but even after relaxation the hole remains localized on the chlorine and the bonding arises primarily from polarization of the chlorine atom in the field of the sodium ion. The inclusion of correlations is crucial to obtain good agreement with experiment for the binding energy (0.44 eV) and for the vertical and adiabatic ionization potentials, the latter being underestimated by about 1 eV in an Hartree-Fock calculation.

A detailed theoretical account of the deformation dipole model for alkali halide monomers has been given in a quantum-mechanical exchange perturbation approach by Brumer and Karplus [26]. They choose the ground states of M$^+$ and X$^-$ as the zero-order basis in dealing with the MX molecule and expand the molecular energy E to second order in their interaction potential V, i.e. $E = E^{(0)} + E^{(1)} + E^{(2)}$. Since exchange gives rise to quantities which are of second or higher order in the overlap integrals, $E^{(1)}$ and $E^{(2)}$ are expanded up to second-order terms in the overlap, i.e. $E^{(1)} = E^{(1,0)} + E^{(1,2)}$ and $E^{(2)} = E^{(2,0)} + E^{(2,2)}$. Finally, V is treated by the multipole expansion, yielding $E^{(1,0)} = -e^2/r$ and $E^{(2,0)} = -(e^2/2r^4)(\alpha_+ + \alpha_-) - C/r^6$. The $E^{(1,2)}$ term is of second order in the overlap S and can be approximated by S^2/r, so that near equilibrium $E^{(1,2)} \approx A \exp(-r/\rho)$. Finally, the $E^{(2,2)}$ term, which was ignored in the Rittner treatment, has the major effect of modifying the polarizabilities through corrections which decay exponentially as the internuclear separation increases. Evidently, this is the origin of the overlap deformation dipoles: they first arise in energy terms which correspond to dipoles in the multipolar expansion and to second-order effects in the exchange perturbation expansion.

5. Liquid structure and phase diagram of the alkali halides

A general feature of simulation results on molten alkali halides near freezing is that the experimental data on thermodynamic properties and liquid structure are described rather well by a rigid-ion model such as that obtained from crystal data by Fumi and Tosi [19]. Detailed simulation studies of polarizable-ion models [27] have evidenced rather minor changes in the results. The availability of a simple, yet realistic model has had some important consequences for statistical mechanical theories, on which we briefly pause.

5.1. THEORIES OF LIQUID STRUCTURE AND THERMODYNAMICS

The evaluation of liquid structure from given interaction pair potentials $\phi_{\alpha\beta}(r)$ combines the Ornstein-Zernike relations in Eq. (8) with the relations

$$g_{\alpha\beta}(r) = \exp[-\phi_{\alpha\beta}(r)/k_B T + g_{\alpha\beta}(r) - 1 - c_{\alpha\beta}(r) - b_{\alpha\beta}(r)], \qquad (28)$$

where the functions $b_{\alpha\beta}(r)$ are exactly defined by an infinite series of so-called bridge diagrams. The hypernetted-chain approximation (HNC) sets to zero the bridge functions $b_{\alpha\beta}(r)$ and as a consequence suffers from two main faults: (i) it oversoftens the pair distribution functions in the approach to the excluded-volume regions, and (ii) it yields inconsistent results for the compressibility of the liquid from the virial theorem and from thermodynamic density fluctuations. For one-component fluids Rosenfeld and Ashcroft [28] corrected for the inadequacy of the HNC by taking over the bridge function from the fluid of hard spheres in the Percus-Yevick approximation, with a free parameter being used to enforce consistency between liquid structure and thermodynamics.

The availability of three different pair distribution functions in a molten alkali halide allows one to examine the theory of liquid structure over a more extended region in space, with main regard to Coulomb ordering in addition to topological order. Specific attention must be given to the excluded volumes from Coulombic repulsions in the like-ion distribution functions and to the detailed shapes of these functions beyond their main peak. Ballone *et al.* [29] have shown that these structural features reveal correlations within four-particle clusters, which determine the leading set of bridge diagrams. Table 3 shows how the theory fares in a rather severe test, *i.e.* the calculation of thermodynamic properties (excess internal energy, pressure and compressibility at the measured atmospheric-pressure density) for molten NaCl and KCl near freezing.

In connection with shock-wave studies of CsI Ross and Rogers [30] have shown that in the liquid along the melting curve a gradual pressure-induced change occurs from an open to a closer-packed structure, as short-range interactions progressively grow with increasing pressure in relative importance over the Coulomb interactions. Ultimately the N-N distribution function comes to resemble the pair function of a Xenon-like fluid.

TABLE 3. Thermodynamic properties of molten NaCl and KCl near freezing (from [1])

	NaCl (1148 K) Theory	Expt	KCl (1043 K) Theory	Expt
$-U_{exc}$ (kJ/mol)	713.8	714.3	649.4	650.5
p (kbar)	1.3	0.0	1.3	0.0
K_T (10^{-12} cm^2/dyn)	31.7	32.4	39.0	36.7

5.2. FREEZING OF ALKALI HALIDES AT STANDARD PRESSURE

I have emphasized earlier in § 2 that in an alkali halide melt near freezing at standard pressure the main peak in the charge-charge structure factor is strong and sharp, acting as a marker of chemical short-range order. According to linear response theory, this features implies a high compliance of the melt against modulation by charge density waves at the corresponding wave vectors.

To connect this property to equilibrium freezing, let us write the periodic single-particle densities $n_\alpha(\mathbf{r})$ for the two species in a NaCl-type crystal in the form

$$n_\alpha(\mathbf{r}) = \frac{1}{2} n(1+\eta) + \sum_{\mathbf{G}(\neq 0)} n_{\mathbf{G}(\alpha)} \exp(i\mathbf{G} \cdot \mathbf{r}) \tag{29}$$

where n is the mean liquid density, η is the fractional change of density in freezing and $n_{\mathbf{G}(\alpha)}$ are Fourier components at the reciprocal lattice vectors, which are indicated by $\mathbf{G}$. The order parameters of the phase transition are η and $n_{\mathbf{G}(\alpha)}$: the latter determine the Debye-Waller factors of the various Bragg reflections from the crystal. In fact, the position of the (111) star of reciprocal lattice vectors can be put into correspondence with the main peak of $S_{QQ}(k)$, and it is then found that the first three even stars of reciprocal lattice vectors underlie the broad peak in $S_{NN}(k)$ [1]. Crystal and liquid structure are therefore compatible: (i) the even-indexed Bragg reflections, which describe the simple-cubic crystal to which the NaCl crystal structure would reduce if the difference between the two species could be ignored, are weighted by the average form factor as is $S_{NN}(k)$ in the melt; and (ii) the odd-indexed Bragg reflections, which distinguish the NaCl crystal structure from a simple-cubic crystal, are weighted by the difference in form factors of the two ionic species as is $S_{QQ}(k)$.

Calculations of the free-energy difference between the crystalline and liquid phases at equilibrium under standard pressure by an HNC-type approach [31] show indeed that the dominant order parameters of the phase transition are the relative density change η and the Fourier components of the crystalline density at the (111) star. In essence, the volume contraction in freezing is accompanied by a free energy release as needed to balance the free energy loss involved in the modulation of the particle densities in the liquid through the locking-in of charge density waves with wave vectors at the (111) star.

5.3. CRITICAL BEHAVIOR

Although not specific to alkali halides, the issue of the critical behavior of ionic fluids is a topic of high interest (for a review see Fisher [32]). While continuum fluids with short range or van der Waals interactions exhibit Ising-type critical exponents, various observations on ionic solutions show either classical criticality or a crossover to Ising behavior occurring only very close to criticality. A persistence of classical behavior was attributed by Mott [33] to the presence of long-range Coulomb interactions.

It is well established especially from work on the RPM that inclusion of ion pairing and of the associated free-ion depletion is essential at the relatively low temperatures where criticality occurs in this model. The results of this work still predict only classical criticality, but are in quite satisfactory agreement with Monte Carlo data. However, at higher temperatures all pairing theories violate thermal convexity, owing to problems in implementing a chemical picture of ion pairing and in specifying appropriately the association constant. To overcome these problems an exact thermodynamic formulation of chemical association has been developed [34].

A number of novel results have been obtained in this context for both density and charge correlations [32]: some important points being (i) the use of the density correlation length to evaluate the crossover scale for departure from classical/mean field behavior with the help of Ginzburg's criterion; and (ii) the role of charge density oscillations as markers of incipient charge ordering and its competition with density fluctuations in the critical region.

6. Trihalides of Al and their mixtures with salts and metals

In the preceding sections I have focused on the alkali halides both as paradigms of interionic force models and as objects of attention for fundamental statistical mechanics. The rapid overview of liquid structures for other ionic compounds that was given in § 2 will certainly be expanded and enriched in other contributions to this volume. In this section I should like to give a summary of our own work on clusters in Aluminium trihalide melts and on their mixtures with alkali halides and with Sodium metal.

6.1. ALUMINIUM CHLORIDE CLUSTERS

Aluminium trichloride crystallizes in a layer-type structure, which is formed from a slightly distorted cubic close packing of chlorines accommodating the metal ions inside every second (111) plane of octahedral sites. On melting the coordination of the metal ions drops to essentially fourfold. Diffraction, Raman light scattering, thermodynamic

TABLE 4. Structure of the Al_2Cl_6 molecular dimer (bond lengths in Å, bond angles in degrees; from [35]). Cl^B and Cl^T denote bridging and terminal halogens, respectively.

	Al-Cl^B	Al-Cl^T	Al-Al	Cl^B-Cl^B	Cl^T-Cl^T	$\angle Cl^B AlCl^B$	$\angle Cl^T AlCl^T$
IM	2.276	<u>2.065</u>	3.20	3.23	3.59	89.5	120.8
ED	2.252	2.065	3.21	3.16	3.64	91.0	123.4
QC	2.289	2.083	3.26	3.21	3.64	89.1	121.8

and transport data are consistent with the melt being made of strongly correlated Al_2Cl_6 molecules, each dimer being formed by two $AlCl_4$ tetrahedra sharing an edge. The related $AlBr_3$ compound already possesses such a dimer-based structure in the crystalline state.

In adjusting the force model presented in § 4 we have found that one can combine a good account of the structure of ionic clusters with better quantitative results for their vibrational frequencies by allowing for a reduction of the ionic valences below their classical integral values, subject to overall charge neutrality [35]. This reduction is by 18 % in $AlCl_3$, compared with 19 % in $AlBr_3$ and with 5 % in AlF_3. The model whose results are reported below thus accounts for electron-shell deformability of the halogens through (i) effective valences and (ii) electrical and overlap polarizabilities. As noted in § 4, the number of adjustable parameters is kept at a minimum by assuming transferability between different compounds whenever reasonable. Halogen-ion parameters are commonly taken to be transferable, so that typically two or at most three parameters need to be adjusted when the halides of a new metallic element are tackled.

Table 4 compares our results for the structure of the Al_2Cl_6 molecule in an ionic model (IM) with those obtained from electron diffraction on the vapour phase (ED) and from quantum chemical calculations at the Hartree-Fock level (QC). Here and in the following we underline the results that have been adjusted to experiment in determining the model parameters. The agreement between model and experiment evidently is of high quality and is especially remarkable in regard to the molecular bond angles. These specifically reflect the description of the electrical dipoles induced on the halogens.

Model calculations on vibrational motions and on fluctuations of the dimer into ionized states allow a better assessment of the usefulness of the model in studies of the melt. Of special interest is the ion-transfer reaction

$$2(M_2X_6) \Leftrightarrow (M_2X_7)^- + (M_2X_5)^+. \tag{30}$$

The Al_2Cl_7 anion is formed by two corner-bridged $AlCl_4$ tetrahedra and has four inequivalent zero-force structures corresponding to relative rotations of the two tetrahedra. Quantum-chemical calculations have yielded the ground-state structure as being of type C_s or of type C_2. We find it to be of type C_2 and almost degenerate with type C_s. A comparison of our results [35] with those of QC calculations based on two different basis sets is shown in Table 5. In fact, we find in our calculations that only the C_2 structure is mechanically stable: the others have at least one imaginary mode frequency and therefore correspond to a multiplicity of saddle points separating several true minima in a very complex free-energy landscape. An alkali counterion finds a suitable location near four chlorines, at a distance of 3.60 Å in KAl_2Cl_7 [36].

TABLE 5. Structure of the $(Al_2Cl_7)^-$ molecular ion (from [35]). The ranges of values shown span those appropriate to non-equivalent terminal chlorines.

Cluster/structure		Al-Cl^B	Al-Cl^T	$\angle$AlCl^{B}Al	$\angle$Cl^{T}AlCl^T
$(Al_2Cl_7)^-/C_2$	IM	2.35	2.10-2.12	111.	100.-108.
$(Al_2Cl_7)^-/C_s$	IM	2.34	2.10-2.12	112.	98.-110.
$(Al_2Cl_7)^-/C_s$	QC (STO-3G)	2.24	2.086-2.089	133.6	103.9-105.6
$(Al_2Cl_7)^-/C_s$	QC (3-21G)	2.37	2.200-2.203	131.6	103.3-105.2

TABLE 6. Activation energies for the indicated reactions (in eV; from [35]).

	IM	DF
$Al_2Cl_6 \rightarrow 2AlCl_3$	0.85	1.1
$[(Al_2Cl_5)^+]_{3b} \rightarrow [(Al_2Cl_5)^+]_{2b}$	0.34	0.39
$2(Al_2Cl_6) \rightarrow (Al_2Cl_7)^- + [(Al_2Cl_5)^+]_{3b}$	3.3	1.66

For the Al_2Cl_5 cation we have found two mechanically stable structures: a symmetric triple-bonded one built from two face-sharing tetrahedra and an asymmetric double-bonded one at a somewhat higher energy. The latter is obtained by stripping a terminal halogen from Al_2Cl_6 and therefore contains a tetrahedrally coordinated Al and a trigonally coordinated one. These results have been confirmed in as yet unpublished calculations by P. Ballone using a density functional (DF) approach based on an LDA supplemented by gradient corrections. More generally, for all clusters examined so far the first-principles DF calculations of Ballone almost exactly reproduce the molecular shapes that we have obtained in an ionic model. A comparison with Ballone's results for several reaction energies, with the reaction products being at infinite separation, is reported in Table 6.

Ionization reactions are evidently in need of further study. Electron correlations beyond the Hartree-Fock approximation are included in these DF calculations by the same basic methods as are used nowadays in evaluating binding and structure of condensed disordered phases by the Car-Parrinello method.

A large body of evidence on liquid mixtures of AlCl$_3$ with alkali chlorides supports the presence of complex anions such as Al$_2$Cl$_7$. In essence, the added salt acts as a chlorine donor, allowing the Al ions to maintain their preferred fourfold coordination as the composition of the mixture is changed. Thus, in the (AlCl$_3$)$_x$.(NaCl)$_{1-x}$ system the stoichiometric mixture ($x = 1/2$) is best viewed as a Na$^+$.(AlCl$_4$)$^-$ melt and, on increasing x, the (AlCl$_4$)$^-$ tetrahedra tie up *via* chlorine sharing into units such as (Al$_2$Cl$_7$)$^-$ and (Al$_3$Cl$_{10}$)$^-$. The end result of this evolution with composition is the edge-sharing bitetrahedron in the pure AlCl$_3$ melt. Face sharing as in Al$_2$Cl$_5$ would arise if the concentration of chlorines could be further reduced.

The same flexibility under rotations around the Al-Cl^B bonds that is shown by Al$_2$Cl$_7$ is displayed by the chain-like structures of higher oligomeric anions. Thus, for Al$_3$Cl$_{10}$ we find four inequivalent structures, which are all mechanically stable and differ in binding energy by only hundredths of an eV [36]. Starting from a stretched configuration in which the metal ions and bonding chlorines lie all in the same plane, the ground-state structure is obtained by rotating the terminal groups out of this plane. Two further stable states are a winged structure, in which the terminal Al-Cl^B bonds are twisted out of the plane, and a cart structure, in which the planar skeleton of the molecule is preserved but the central and terminal chlorines go into a staggered configuration.

By extending these calculations to the Al$_4$Cl$_{13}$ tetramer we could see convergence in the binding energy of the polymeric series to a value of ≈ 0.5 eV per monomer [36].

6.2. ALUMINIUM-SODIUM FLUORIDE CLUSTERS

Molten cryolite (Na$_3$AlF$_6$ or AlF$_3$.3NaF) is a stoichiometric compound in a continuous range of liquid mixtures, which enters the industrial Hall-Héroult process for the electrodeposition of Al metal from alumina. Raman scattering and other evidence has shown that, as the concentration of NaF is increased starting from the NaAlF$_4$ composition, there is a gradual shift in cluster population from (AlF$_4$)$^-$ to higher species identified as (AlF$_5$)$^{2-}$ and (AlF$_6$)$^{3-}$ [37 - 39]. The evidence shows that the equilibrium between these charged clusters is strongly shifted towards the (AlF$_5$)$^{2-}$ anion species in cryolite.

Table 7 reports our results for the structure of the NaAlF$_4$ cluster in comparison with those obtained in *ab initio* QC calculations [40]. Three zero-force states for this cluster correspond to a corner-bridged (I), an edge-bridged (II) and a face-bridged (III) Na bound to a distorted AlF$_4$ tetrahedron. The NaAlF$_4$(I) structure is mechanically unstable and the other structures are competitive in the QC calculations, their energy difference being at the level of hundredths of an eV and being sensitive to the choice of the basis set and to the inclusion of correlations by semi-empirical methods. We find that the NaAlF$_4$(II) structure is stable by 0.1 eV relative to NaAlF$_4$(III), this result being consistent with the C$_{2v}$ structure indicated by the Raman spectrum of matrix-isolated NaAlF$_4$ molecules [41]. The mean Al-F bond length in the NaAlF$_4$ molecule is reported from experiment as 1.69

Å and the ionic model also yields a reasonably accurate description of the data on the vibrational spectrum of the NaAlF$_4$ molecule [40].

TABLE 7. Equilibrium structure of NaAlF$_4$(II) and NaAlF$_4$(III) (from [40]; F* denotes a fluorine bonding the Na ion).

Cluster	Na-F*	Al-F*	Al-F	∠F*AlF*	∠FAlF	∠FAlF*
NaAlF$_4$(II):						
IM	<u>2.11</u>	1.73	1.65	93.6	116.7	--
QC	2.12	1.721	1.646	93.6	116.4	--
NaAlF$_4$(III):						
IM	2.34	1.71	1.64	--	--	119.2
QC	2.282	1.698	1.634	--	--	119.7

We have next applied the model without further adjustments to evaluate the isolated Na$_2$AlF$_5$ and Na$_3$AlF$_6$ clusters in various structures allowing for fourfold, fivefold or sixfold coordination of the Al ion [40]. In all cases stable structures can be obtained by attaching one or two NaF molecules to the NaAlF$_4$(II) cluster without changing the coordination of the Al ion. This result agrees with observations showing that the fourfold coordinated state for Al is present in these mixtures from the equimolar melt down to the cryolite composition and beyond.

With regard to the relative stability of the fivefold and sixfold-coordinated structures for the Al ion at the cryolite composition, our calculations yield a rather complex potential energy landscape for the Na$_3$AlF$_6$ cluster, involving a number of saddle points and relative minima. However, we find that the deepest energy minimum corresponds to a fivefold coordinated Al in combination with three double-bridged Na ions and a fluorine coion. This configuration is deeper in energy by about 0.6 eV than the deepest sixfold-coordinated configuration, the latter being a distorted octahedron with three triple-bridged Na ions. The variety of configurations that we find for each basic structure and the relatively small separations in energy suggest (i) further stabilization of the fivefold-coordinated (AlF$_5$)$^{2-}$ complex anion by entropy gains at finite temperature and (ii) high diffusivity of counterions and coions in the melt. For further details the reader is referred to the discussion given in the original paper [40].

6.3. SOLUTIONS OF SODIUM METAL IN MOLTEN CRYOLITE

Macroscopic experiments on the addition of metals to molten cryolite have aimed to clarify the relative roles of Al and Na in the cathodic processes occurring during the production of Al metal from electrolytic baths containing cryolite [42]. Representative examples of these observations are (i) the deposition of specks of Al metal in cryolite after exposure to Na metal and freezing, and (ii) the emission of hydrogen bubbles containing small partial pressures of AlF, NaAlF$_4$ and Na from addition of Al metal to molten cryolite in the presence of moisture.

We have studied the process of addition of sodium metal to molten cryolite [43] on the assumptions that (i) Na enters the liquid in the form of monovalent ions and electrons, and (ii) the electrons may partly be localized on free Al ions (if any) and partly reside in a uniform neutralizing background. The binding of electrons to metal ions is governed by the respective ionization potentials. The equilibria between the various species is then treated by a series of laws of mass action, involving activation free energies which self-consistently depend on the concentrations of all species.

We find that with increasing concentration of sodium metal the higher complex anions in the cryolite bath dissolve, while the concentrations of both $(AlF_4)^-$ anions and of Al ions in various valence states increase. At Na concentration above a few percent, the most stable species becomes that of Al^+ ions in the monovalent state. Thus, consistently with the above-mentioned macroscopic evidence, valence electrons are transferred from Na to Al in the melt and the species Al^+ and $(AlF_4)^-$ are stabilized.

7. Stability of local structures in molten salt mixtures

Whereas in the preceding section I have focused on microscopic models for Al-alkali halide mixtures, it should be recalled that there is a large body of evidence on the stability of different local structures regarding some 140 liquid systems from mixing polyvalent and alkali metal halides. The role of the alkali halide component is primarily to donate halogens to the polyvalent metal ions, while the alkalis act as counterions to provide charge compensation and screening. Long-lived fourfold coordinations (*e.g.* for Zn^{2+}, Mg^{2+} and Al^{3+} in chlorides) and long-lived sixfold coordinations (*e.g.* for Y^{3+}) have been reported. In other systems (*e.g.* the Sr- and Ba-alkali chlorides) the characteristic features of complex anions are not observed in the Raman spectrum. However, for $CaCl_2.2KCl$ the coexistence of fourfold and sixfold coordinations has been proposed from Raman scattering data, suggesting that this may be an interesting border case.

The ample evidence on local coordinations in molten-salt mixtures has been classified [44] on the basis of Pettifor's chemical scale of the elements, as introduced in § 1 for pure molten salts. In essence, this is a phenomenological scale incorporating the progress made in understanding the cohesion of solids from first-principle calculations and can be used to build structural diagrams assembling compounds with similar structure in certain regions of a plane whose coordinates are constructed from the parameters of the component elements. In a pseudopotential viewpoint the elemental parameters are radii associated with valence-electron orbitals from atomic structure calculations, which correlate with classical ionic radii to a considerable extent but also reflect the basic character of the binding.

The structural diagram of molten salt mixtures [44] uses as coordinates the quantities $X_{M-A(X)} = \chi_M - \chi_A$ and $Y_{M-A(X)} = \chi_M + \chi_A - 2\chi_X$ for each M-A(X) mixture from the Pettifor coordinates of the polyvalent metal M, the alkali A and the halogen X. The former coordinate separates out the three classes of mixtures in which the stability of fourfold coordination is strong, marginal or absent at compositions that would favour it. The second coordinate evidences the role of the halogen partner (chlorides, bromides and iodides have broadly similar behavior, but fluorides behave differently) and that of the alkali counterion (Li counterions tend to lower the stability of complex anions).

The length L introduced in Eq. (13) may also be used as a classification parameter for local coordinations in molten-salt mixtures [45]. In analogy with Mott's criterion for the stability of a bound state in a conducting fluid, a complex anion should be stable when the ratio L/d is larger than some critical value, d being the metal-halogen bond length. Table 8 shows as an example how this criterion works (with a critical ratio of about 1.6) to separate chloride mixtures into fourfold-complex forming (Al, Be and Mg) and non-forming (Sr and Ba), with Ca being an intermediate case as remarked above.

TABLE 8. Range of ratio L/d for mixtures of MCl_n and ACl with A from Li to Cs (from [45]).

M	Al	Be	Mg	Ca	Sr	Ba
L/d	1.74-1.80	1.79-1.85	1.66-1.72	1.56-1.61	1.52-1.57	1.48-1.53

8. Transport and dynamics in molten salts

The transport properties and the microscopic dynamical behavior of molten salts are correlated with the types of liquid structure that have been outlined in § 2. Some examples of these correlations will be given in this section.

8.1 IONIC TRANSPORT

The persistence times of local structure in molten alkali halides near freezing are of the order of several picoseconds. These correspond to values of ionic conductivity and shear

TABLE 9. Calculated diffusion coefficients in superionic-conductor melts (in units of 10^{-5} cm^2 s^{-1}, from [47]), compared with the results of molecular dynamics (MD).

Salt	T(K)	D_+	$(D_+)_{MD}$	D_-	$(D_-)_{MD}$	D_+/D_-	$(D_+/D_-)_{MD}$
AgI	873	3.6	3.8	0.34	0.3	11.	12.7
CuCl	773	20.	10.	4.3	2.5	4.8	4.0
CuBr	880	13.	10.	2.3	2.7	5.6	3.9
CuI	923	18.	8.8	2.2	1.3	8.1	6.8

viscosity as given for NaCl in Table 1. The self-diffusion coefficients of the two ions in molten NaCl have similar values (D_{Na} = 1.7 and D_{Cl} = 1.3x10^{-4} cm^2 s^{-1}), in spite of the large differences in atomic masses and ionic radii. This is consistent with the similarity of $g_{NaNa}(r)$ and $g_{ClCl}(r)$ seen in diffraction experiments, which implies similarity in restoring forces and residence times for the two ionic species.

On the other hand, for the melts of noble-metal halides computer simulation studies [46] and theoretical calculations embodying sum rules on velocity autocorrelations [47] indicate a large difference in the mobilities of cations and anions, by up to a factor of order ten in molten AgI (see Table 9). The observed liquid structure, showing that the metal-ion component of the melt is more strongly disordered, implies major differences in the structural back-scattering of cations by cations and of halogens by halogens. It is also clear from the Table that the calculation of transport coefficients is delicately sensitive both to the model of the interionic forces and to the theoretical approach for given interionic force model.

A further example of connections between structure and transport is the Chemla effect. This was first discovered in experiments on molten (Li,K)Br mixtures, showing that the mobility of the K ion overtakes that of the Li ion with increasing content of KBr. The transport behavior of the two cations is related to the structural features of the cation-anion pair distribution functions, combined with the volume dilation which accompanies the increase of KBr content [48]. A Li ion spends a relatively long residence time in oscillations at a fourfold-type site before being able to diffuse out, and this 'trapping' is strengthened with decreasing density and temperature of the melt.

With regard to the connection between ionic diffusivities and ionic conductivity, deviations from the Nernst-Einstein relation of up to 20% have been reported for alkali halides both from experiment and from computer simulations [49]. Including a correction for correlations, this relation is

$$\sigma = (ne^2 / k_B T)(D_+ + D_-)(1 - \Delta) \qquad (30)$$

where n is the number density of ion pairs and the factor $(1 - \Delta)$ arises from cross-correlations between the motions of a single ion and those of other ions. Positive values of Δ are observed and show a tendency of oppositely charged ions to diffuse together.

In molten $ZnCl_2$ near freezing Sjöblom and Behn [50] report diffusion coefficients of about 1.5×10^{-7} cm^2/s for both ions. From the Nernst-Einstein relation and the measured ionic conductivity (see Table 1), one would predict diffusivities lower than this by several orders of magnitude [51]. It is evident that motions of neutral units are mainly responsible for diffusion.

8.2 VISCOSITY

As discussed by Hirschfelder *et al.* [52], dimensional analysis suffices to suggest scaling laws for transport coefficients through microscopic interaction parameters. Abe and Nagashima [53] have shown that the correlation

$$1/\eta^* = -5.960 + 23.37 V^* T^{*1/2} \qquad (32)$$

holds in 18 molten alkali halides for the reduced shear viscosity $\eta^* = \eta\sigma^2/(m\varepsilon)^{1/2}$ as a function of the reduced volume $V^* = V/N\sigma^3$ and of the reduced temperature $T^* = k_B T/\varepsilon$, with σ and ε being suitably chosen parameters of the interionic potential and m the molecular weight. Other systems discussed by the same authors include alkali nitrates, liquid metals, and hydrocarbons.

Precise viscosity measurements on pure and mixed ionic melts (alkali halides and nitrates) including glass-formers have revealed a universal behavior in the form of a modified Arrhenius law [54],

$$\eta \propto (T/T_0)\exp(qT_0/T) \qquad (33)$$

where $q = 5.9 \pm 0.1$ and T_0 is a suitably chosen characteristic temperature. A fractional-power relation exists between conductivity and viscosity of pure and mixed ionic melts,

$$\sigma T \propto (T/\eta)^m \qquad (34)$$

with $m = 0.8 \pm 0.1$ [55]. This relation holds over a range of nine orders of magnitude for η in the glass-forming $Ca_2K_3(NO_3)_7$ compound. In a range of temperature where both σ and $1/\eta$ can be represented by an Arrhenius law, the relation (34) implies a systematic difference in their respective activation energies.

8.3 DYNAMICS OF DENSITY FLUCTUATIONS

In the simplest model of a molten alkali halide, one would expect that the dynamics of density fluctuations be described by a 'superposition' of oscillations of the total mass density, leading to sound waves at long wavelengths as for a monatomic fluid, and of oscillations of the charge density as a counterpart of the propagating plasma mode in a plasma. This viewpoint, though quantitatively oversimplified, is essentially correct. In particular, Coulomb ordering is responsible for a propagating collective excitation of charge fluctuations. I refer to a specialized review [1] for a discussion of detailed analyses of infrared reflectivity, neutron inelastic scattering and Raman scattering data.

Raman scattering is especially useful in network-like melts such as $ZnCl_2$ and in molecular-like liquids such as $AlCl_3$. Experiment shows that the density of vibrational states in these liquids is strongly structured: it mainly reflects the vibrational motions of the basic structural unit and these appear to be only moderately affected by coupling induced by intermediate-range order. An extensive collection of Raman frequency data for ionic systems has been given by Brooker and Papatheodorou [41].

9. Concluding remarks

Alkali halide systems, including diatomics, dimers, melts and crystals, have been the subject of very many experimental and theoretical investigations since the early part of the past century. A unified model of ionic interactions allows a theoretical description of their static and dynamic structures, to useful quantitative accuracy in comparison with much of the available evidence. Their condensed phases have been the main testing ground for a number of ideas in the statistical mechanics of dense ionic systems.

In more recent years these models and ideas have been extended to help in understanding more complex ionic materials, such as polyvalent metal halides and oxides, mixtures of salts, and metal-salt solutions. Only a few examples of work in this vast field of great physical richness and technical importance could be covered in this brief review.

References

1. Rovere, M. and Tosi, M.P. (1986) Structure and dynamics of molten salts, *Rep. Progr. Phys.* **49**, 1001-1081.
2. Tosi, M.P., Price, D.L., and Saboungi, M.-L. (1993) Ordering in metal halide melts, *Annu. Rev. Phys. Chem.* **44**, 173-211.
3. Pettifor, D.G. (1986) The structure of binary compounds: I. Phenomenological structure maps, *J. Phys.* C **19**, 285-313.
4. Tallon, J.L. (1982) The entropy change on melting of simple substances, *Phys. Lett.* A **87**, 139-143.
5. Akdeniz, A. and Tosi, M.P. (1992) Correlations between entropy and volume of melting in halide salts, *Proc. R. Soc. London* A **437**, 85-96.
6. Wasse, J.C., Salmon, P.S., and Delaplane, R.G. (2000) Structure of molten trivalent metal bromides studied by using neutron diffraction: the systems $DyBr_3$, YBr_3, $HoBr_3$ and $ErBr_3$, *J. Phys.: Condens. Matter* **12**, 9539-9550.
7. Bhatia, A.B. and Thornton, D.E. (1970) Structural aspects of the electrical resistivity of binary alloys, *Phys. Rev.* B **2**, 3004-3013.
8. McGreevy, R.L. and Mitchell, E.W.J. (1982) The determination of the partial pair distribution functions for molten strontium chloride, *J. Phys.* C **15**, 5537-5550.
9. Tatlipinar, H., Akdeniz, Z., Pastore, G., and Tosi, M.P. (1992) Atomic size effects on local coordination and medium-range order in molten trivalent metal halides, *J. Phys.: Condens. Matter* **4**, 8933-8944.
10. Biggin, S. and Enderby, J.E. (1981) The structure of molten zinc chloride, *J. Phys.* C **14**, 3129-3136.

11. Price, D.L., Moss, S.C., Reijers, R., Saboungi, M.-L., and Susman, S. (1989) Intermediate-range order in glasses and liquids, *J. Phys.: Condens. Matter* **1**, 1005-1008.
12. Waisman, A. and Lebowitz, J.L. (1972) Mean spherical model integral equation for charged hard spheres, *J. Chem. Phys.* **56**, 3086-3099.
13. Painter, K.R., Ballone, P., Tosi, M.P., Grout, P.J., and March, N.H. (1983) Capacitance of metal-molten salt interfaces, *Surf. Sci.* **133**, 89-100.
14. Löwdin, P.O. (1956) Quantum theory of cohesive properties of solids, *Phil. Mag. Suppl.* **5**, 1-172.
15. Gygi, F., Maschke, K., and Andreoni, W. (1984) Electron charge density of alkali halides beyond the rigid-ion approximation, *Solid State Commun.* **49**, 437-439.
16. Böbel, G., Cortona, P., Sommers, C., and Fumi, F.G. (1983) Electron density in NaF and KCl crystals in the self-consistent local-density-functional approximation, *Acta Crystallogr.* A **39**, 400-407.
17. Born, M. and Huang, K. (1954) *Dynamical Theory of Crystal Lattices*, Oxford University Press, Oxford.
18. Tosi, M.P. (1964) Cohesion of ionic solids in the Born model, *Solid State Phys.* **16**, 1-120.
19. Fumi, F.G. and Tosi, M.P. (1964) Ionic sizes and Born repulsive parameters in the NaCl-type alkali halides: the Huggins-Mayer and Pauling forms, *J. Phys. Chem. Solids* **25**, 31-44.
20. Fumi, F.G. and Tosi, M.P. (1957) Lattice calculations on point imperfections in the alkali halides, *Disc. Faraday. Soc.* **23**, 91-98.
21. Cochran, W. (1971) Lattice dynamics of ionic and covalent crystals, *Crit. Rev. Solid State Sci.* **2**, 1-44.
22. Tosi, M.P. and Doyama, M. (1967) Ionic-model theory of polar molecules, *Phys. Rev.* **160**, 716-718.
23. Kahn, L.R., Hay, P.J., and Shavitt, I. (1974) Theoretical study of curve crossing: ab initio calculations on the four lowest $^1\Sigma^+$ states of LiF, *J. Chem. Phys.* **61**, 3530-3546.
24. Galli, G., Andreoni, W., and Tosi, M.P. (1986) Stability and ionization-induced structural transitions of sodium chloride microclusters from Hartree-Fock calculations: $Na_2Cl_2^{(+)}$ and $Na_2Cl^{(+)}$, *Phys. Rev.* A **34**, 3580-3586.
25. Jordan, K.D. (1979) Structure of alkali halides: theoretical methods, in *Alkali Halide Vapors: Structure, Spectra and Reaction Dynamics*, Academic, New York, 479-534.
26. Brumer, P. and Karplus, M. (1973) Perturbation theory and ionic models for alkali halide systems: diatomics, *J. Chem. Phys.* **58**, 3903-3918.
27. Sangster, M.J.L. and Dixon, M. (1976) Interionic potentials in alkali halides and their use in simulations of the molten salts, *Adv. Phys.* **25**, 247-342.
28. Rosenfeld, Y. and Ashcroft, N.W. (1979) Theory of simple classical fluids: universality in the short-range structure, *Phys. Rev.* A **20**, 1208-1235.
29. Ballone, P., Pastore, G., and Tosi, M.P. (1984) Structure and thermodynamic properties of molten alkali chlorides, *J. Chem. Phys.* **81**, 3174-3180.
30. Ross, M. and Rogers, F.J. (1985) Structure of dense shock-melted alkali halides: evidence for a continuous pressure-induced structural transition in the melt, *Phys. Rev.* B **31**, 1463-1468.
31. D'Aguanno, B., Rovere, M., Tosi, M.P., and March, N.H. (1983) Freezing of ionic melts into normal and superionic phases, *Phys. Chem. Liquids* **13**, 113-122.
32. Fisher, M.E. (1999) Understanding criticality: simple fluids and ionic fluids, in *New Approaches to Problems in Liquid State Theory*, Kluwer, Dordrecht, 3-8.
33. Mott, N.F. (1974) *Metal-Insulator Transitions*, Taylor and Francis, London.
34. Fisher, M.E. and Zuckerman, D.M. (1998) Exact thermodynamic formulation of chemical association, *J. Chem. Phys.* **109**, 7961-7981.
35. Akdeniz, Z. and Tosi, M.P. (1999) A refined ionic model for clusters relevant to molten chloroaluminates, *Z. Naturforsch.* **54 a**, 180-186.
36. Akdeniz, Z., Çaliskan, M., Çiçek, Z., and Tosi, M.P. (2000) Polymeric structures in aluminium and gallium halides, *Z. Naturforsch.* **55 a**, 575-580.
37. Gilbert, B., Mamantov, G., and Begun, G.N. (1975) Raman spectra of aluminum fluoride containing melts and the ionic equilibrium in molten cryolite type mixtures, *J. Chem. Phys.* **62**, 950-955.
38. Gilbert, B. and Materne, T. (1990) Reinvestigation of molten fluoroaluminates Raman spectra: the question of the existence of AlF_5^{2-} ions, *Appl. Spectrosc.* **44**, 299-305.
39. Robert, E., Olsen, J.E., Danek, V., Tixhon, E., Østvold, T., and Gilbert, B. (1997) Structure and thermodynamics of alkali fluorides-aluminum fluoride-alumina melts. Vapour pressue, solubility, and Raman spectroscopic studies, *J. Phys. Chem.* B **101**, 9447-9457.
40. Akdeniz, Z., Çiçek, Z., and Tosi, M.P. (1999) Theoretical evidence for the stability of the $(AlF_5)^{2-}$ complex anion, *Chem. Phys. Lett.* **308**, 479-485.
41. Brooker, M.H., and Papatheodorou, G.N. (1983) Vibrational spectroscopy of molten salts and related glasses and vapors, *Adv. Molten Salt Chem.* **5**, 26-184.
42. Grjotheim, K., Krohn, C., Malinovsky, M., Matiasovsky, K., and Thonstad, J. (1982) *Aluminium Electrolysis - Fundamentals of the Hall-Héroult Process*, Aluminium-Verlag, Dusseldorf.
43. Akdeniz, Z. and Tosi, M.P. (1991) Structure breaking and electron localization in liquid cryolite-sodium solutions, *Phil. Mag.* B **64**, 167-179.
44. Akdeniz, Z. and Tosi, M.P. (1989) Stability diagrams for fourfold coordination of polyvalent metal ions in molten mixtures of halide salts, *J. Phys.: Condens. Matter* **1**, 2381-2394.

45. Akdeniz, Z., Li, W., and Tosi, M.P. (1988) Classification of stability for tetrahedral halocomplexes in molten-salt mixtures, *Europhys. Lett.* **5**, 613-617.
46. Trullas, J., Girò, A., and Silbert, M. (1990) Potentials and correlation functions for the copper-halide and silver-halide melts. II: Time correlation functions and ionic transport properties, *J. Phys.: Condens. Matter* **2**, 6643-6650.
47. Tankeshwar, K. and Tosi, M.P. (1991) Ionic diffusion in superionic-conductor melts, *J. Phys.: Condens. Matter* **3**, 7511-7518.
48. Tankeshwar, K. and Tosi, M.P. (1992) Theory of the Chemla effect in molten (Li,K)Cl, *Solid State Commun.* **84**, 245-247.
49. Ciccotti, G., Jacucci, G., and McDonald, I.R. (1976) Transport properties of molten alkali halides, *Phys. Rev. A* **13**, 426-436.
50. Sjöblom, C.-A. and Behn, A. (1968) Self-diffusion in molten zinc chloride, *Z. Naturforsch.* **23 a**, 495-497.
51. Tatlipinar, H., Amoruso, M., and Tosi, M.P. (2000) Ionic charge transport in strongly structured molten salts, *Physica B* **275**, 281-284.
52. Hirschfelder, J.O., Curtiss, C.E., and Bird, R.B. (1964) *Molecular Theory of Gases and Liquids*, Wiley, New York.
53. Abe, Y. and Nagashima, A. (1981) The principle of corresponding states for alkali halides viscosity, *J. Chem. Phys.* **75**, 3977-3985.
54. Voronel, A., Veliyulin, E., Grande, T., and Øye, H.A. (1997) Universal viscosity behaviour of regular and glassforming ionic melts, *J. Phys.: Condens. Matter* **9**, L247-L249.
55. Voronel, A., Veliyulin, E., Machvariani, V.Sh., Kisliuk, A., and Quitmann, D. (1998) Fractional Stokes-Einstein law for ionic transport in liquids, *Phys. Rev. Lett.* **80**, 2630-2633.

ELECTRONIC PROPERTIES OF METAL/MOLTEN SALT SOLUTIONS

WILLIAM W. WARREN, JR.
Oregon State University
Department of Physics
Weniger Hall 301
Corvallis, OR 97331-6507, USA

1. Introduction

1.1 HISTORICAL BACKGROUND

Investigation of solutions of metals in molten salts has a long history extending back nearly 200 years to the work of Sir Humphrey Davy. For a long time, however, it was unclear whether these systems were true solutions or simply colloidal suspensions of metal particles in more of less pure molten salts. The pioneering experimental studies of Aten[1] in the early 20th century provided the first indications that, in fact, metals can form true, microscopically homogeneous solutions with their salts. The issue was finally put to rest in the 1950's and early 1960's by Bredig and coworkers who carried out comprehensive investigations of the phase diagrams and fundamental physical properties of a large number of metal/molten salt systems. Much of the work of this period was motivated by the molten salt reactor program in the United States and is summarized in Bredig's classic 1964 review[2].

The 1970's and 1980's saw a renewal of research activity on metal/molten salt solutions, driven this time by fundamental interest in the metal-nonmetal transition. Solutions of metals with their molten salts comprise one of a number of classes of physical system that undergo continuous transformations from a metallic state to a nonmetal under change of one or more thermodynamic variables, i.e. temperature, pressure, or composition[3]. Other metal-nonmetal systems include heavily doped crystalline semiconductors, metal-ammonia solutions, expanded fluid metals, and liquid semiconducting alloys. The latter are alloys of metallic liquid elements that develop non-metallic electronic properties at alloy compositions close to specific stoichiometries. Perhaps the most dramatic example, and the one directly responsible for stimulating new work on metal/molten salt solutions, is the liquid cesium-gold system. These alloys are essentially molten salts at the equiatomic composition CsAu and they transform continuously to the metallic state on addition of excess cesium or excess gold[4]. During this period, the earlier studies of fundamental physical properties (phase diagrams, electrical conductivity, magnetic susceptibility) were extended by the application of the spectroscopic tools of solid state physics and physical chemistry (electron spin resonance, nuclear magnetic resonance, picosecond optical techniques, neutron diffraction, Raman spectroscopy).

Research on metal/molten salt solutions has been led through the 1990's by Freyland and coworkers. These workers have refined the use of spectroscopic techniques, especially for study of alkali metal/alkali halide solutions, and have extended their studies to heterogeneous systems. The latter include investigations of molten salt/solid interfaces and the wetting transition. Study of metal/molten salt interfaces is discussed by Freyland elsewhere in this volume. The present chapter will be devoted to the electronic properties of bulk metal/molten salt solutions.

1.2 THEORETICAL ISSUES

1.2.1 *The Metal-nonmetal Transition*

As stated above, metal/molten salt solutions are examples of material systems that undergo a transition from a metallic to a non-metallic state. In this case, the driving thermodynamic parameter is the composition, often expressed as the mole fraction of metal, x_M. Virtually all such systems, including the "textbook example" of heavily doped crystalline silicon, are structurally disordered. In the case of doped silicon, the disorder is associated with the random distribution of donor or acceptor atoms in the crystalline matrix. Disorder in metal/molten salt solutions is, of course, represented by the lack of long-range periodicity in the liquid state.

The fundamental metal-nonmetal dilemma is illustrated schematically for a molten alkali metal/alkali halide solution in Fig. 1. The pure metal is essentially a disordered assembly of positive ions embedded in a "sea" of delocalized conduction electrons. Because the effective electron-ion interaction is so weak, as we discuss below, the

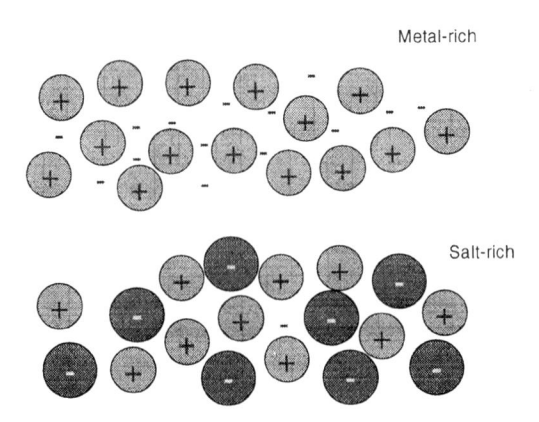

Figure 1. Schematic representation of the metallic and non-metallic limits of a molten alkali metal/alkali halide solution.

electrons form a nearly free electron gas. In contrast, a dilute solution of metal in a molten salt represents a very low density of "excess" electrons in the molten salt matrix. The obvious question is, "What is the state the excess electrons in the dilute metal limit?" Do these electrons form a highly dilute electron gas, or are they localized in some form? If the dilute electrons are localized, the question becomes, "What is the nature of the localized states?" And finally, in this case, "How does the system transform continuously from a non-metallic solution, with a low density of localized electrons, to the metallic limit with a high density of nearly free electrons?"

The data reproduced in Figure 2 show that a macroscopic property such as the DC conductivity is not especially helpful in answering the kinds of questions posed above. In the case of the alkali metal/alkali halide solutions K-KBr and Cs-CsI, the conductivity increases continuously from the ionic conductivity values (~ 1 $(\Omega cm)^{-1}$) of the pure melts to the high electronic conductivities ($\sim 10^4$ $(\Omega cm)^{-1}$) characteristic of liquid metals. Although there is a kind of "knee" in the data at about 15 at. % metal, there is no clear indication of a "transition." Nor is there in these data a clue as to the nature of electronic transport in the salt-rich limit.

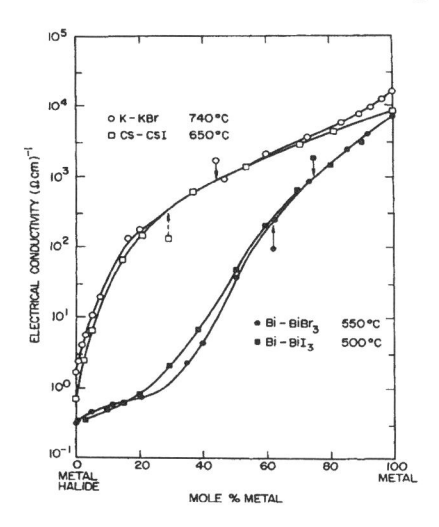

Figure 2. DC electrical conductivity at constant temperature versus mole % metal for K-KBr[5], Cs-CsI[6], Bi-BiBr$_3$[7], and Bi-BiI$_3$[8].

It is further shown in Figure 2 that not all metal/molten salt solutions behave the same. In the case of the Bi-BiBr$_3$ and Bi-BiI$_3$ solutions, there is little initial change in the DC conductivity on adding Bi metal. The eventual metal-nonmetal transition is shifted to much higher metal concentrations than in the case of the alkali metal solutions. Again, the conductivity data do not tell us how this transition occurs or what is the nature of the localized states in the non-metallic limit. In order to make progress, it is necessary to turn to spectroscopic techniques that provide a more microscopic perspective.

1.2.2 *State-dependent interactions*

One of the central theoretical concepts associated with the metal-nonmetal transition is that of *state-dependent interactions*[9]. To a theorist, a material is not so much defined by its chemical constituents as by the interactions among those constituents. Thus, one writes a Hamiltonian

$$\mathbf{H} = (K.E.)_{electrons} + (K.E.)_{ions} + V_{ion\text{-}ion} + V_{electron\text{-}ion} + V_{electron\text{-}electron} \qquad (1)$$

in which the first two terms describe the kinetic energies of the electrons and ions, respectively, and the last three potential energy term essentially define the material. In the presence of a metal-nonmetal transition, the interaction potentials are profoundly sensitive to the thermodynamic state of the system. Elemental mercury provides a simple example of this effect. In the liquid state under ordinary conditions of room temperature and ambient atmospheric pressure, mercury is a typical liquid metal. The atoms are held together by the metallic bonding provided by a high density of delocalized electrons. The vapor, on the other hand, is a gas of atoms interacting primarily through the van der Waals interactions. Thus, there is no single set of interatomic potentials can be used to describe both the liquid state and the vapor with which it is in equilibrium.

The idea of state-dependent interactions in metal/molten salt solutions can be illustrated explicitly by considering the screening of coulomb potentials. An isolated electron in a highly salt-rich solution is subject to the bare coulomb interaction with the charge q of a neighboring ion:

$$V_{electron-ion}(r) = \frac{q}{4\pi\varepsilon_0 r}. \tag{2}$$

In a metal, on the other hand, the ionic potentials are strongly screened by the high density of conduction electrons. The Thomas-Fermi model of screening[10] gives a simple expression for the screened potential:

$$V_{electron-ion}(r) = \frac{q}{4\pi\varepsilon_0 r} e^{-k_{TF} r} \tag{3}$$

in which the screening length k_{TF}^{-1} depends on the electron density n through

$$k_{TF} = 4\pi\varepsilon_0 n / k_B T . \tag{4}$$

At metallic densities, the screening length is very short; in cesium, for example, $k_{TF}^{-1} = 0.8$ Å. Thus, as a function of decreasing metal content in a metal/molten salt solution, the electron-ion interaction changes dramatically from the strongly screened potential of Eq. (3) to the "bare" coulomb interaction described by Eq. (2). This has profound implications for the stability of localized electronic states. The ion-ion and electron-electron interactions are similarly dependent on the state of the system represented, in these solutions, by the density of nearly free electrons.

The electron-ion interactions are further weakened in the metallic state by the effect of "core orthogonalization." The plane-wave eigenfunctions representing conduction electrons in a metal must be orthogonal to the electronic states of electrons in the ion cores. Under given conditions, the effect of this orthogonalization can be represented as a contribution to an effective electron-ion potential. This is not a true potential since it depends on the electron density. This point is, however, that the effect of orthogonalization is to partially cancel the bare ionic potential leading to a relatively weak total "pseudopotential[11]."

The nearly-free-electron characteristics of elemental liquid metals are, in large measure, a result of the screening of the long-range coulomb interactions and reduction of the local electron-ion interactions by core orthogonalization. The lack of long-range crystalline order is therefore relatively unimportant for electron transport in liquid metals. For example, the mean-free-path for electron scattering in pure liquid sodium is approximately 10 interatomic spacings. In the extreme salt-rich limit, on the other hand, excess electrons are at the mercy of the bare coulomb potentials of the ions. One can reasonably ask whether a highly dilute, nearly-free-electron gas is even possible under these conditions.

1.2.3 *Dilute Metals?*

The question of whether low densities of valence electrons localize or form a dilute metal was first addressed by Mott[12]. His intuitive answer was that given an array of, say, sodium atoms each separated from the others by 1000 m, common sense tells us that such a system would not form a metal. Rather, the system should be viewed as a highly dispersed collection of atoms on each of which the valence electron is localized in an atomic orbital. Under what conditions, then does a metal form as the atoms are brought closer together? Mott proposed that the transition from non-metal to metal should occur when the screening length k_{FT}^{-1} becomes comparable with the effective Bohr radius a^* of the localized atomic state. At higher electron densities, the screened electron-ion potential becomes too weak to bind an electron to its ion.

The Mott criterion can be related to the conduction electron density by the condition

$$a^* n^{1/3} \geq \sim 0.25. \tag{5}$$

This condition can also be seen from simple geometrical arguments to define the approximate density at which spheres of radius a^* will begin touch. Overlap of the atomic orbitals on neighboring atoms is, of course, essential for formation of a metallic state.

Assuming the radius of a localized state, a^*, to be roughly the cation-ion distance in the pure salt, we can apply the Mott criterion molten Cs-CsI solutions. In this case, the Mott criterion predicts a metal-nonmetal transition at an electron concentration of about 2×10^{20} cm^{-3}, or a metal concentration of approximately three atomic percent. These considerations thus lead us to expect that these solutions will be metallic over most of the concentration range. At metal concentrations below a few atomic percent, on the other hand, screening becomes ineffective, and the low density metal transforms to an assembly of localized electronic states. We shall see that these expectations are confirmed by the spectroscopic experiments on alkali metal/alkali halide solutions at low metal concentrations.

1.3 TECHNOLOGICAL CONSIDERATIONS

Technological applications of molten salts are discussed elsewhere in this volume. It should be clear, that an understanding of metal/molten salt reactions and solutions is important for any application in which molten salts are in contact with metals. These include use of molten salts as thermodynamic working fluids in which corrosion of containers and conduits is an issue and, in fact, any industrial process in which molten salts play a role. Metal/molten salt reactions are also of importance in battery applications in which molten salt electrolytes are in contact with metal electrodes and containment cells. It should be noted however, that most of these technological situations involve contact between molten salts and metals other than those of their own cation. Most fundamental research on metal/molten salt solutions, on the other hand, involves common cation solutions, i.e. a metal with one of its halides.

The other technological role for metal/molten salt solutions concerns their value as model systems for the study of electrons in disordered media and metal-nonmetal transitions in doped ionic semiconductors, amorphous semiconductors, conducting glasses, etc. Thus while specific metal/molten salt solutions may not have found commercial application, their study can provide insight into the behavior of analog materials of technological importance.

2. **Fundamental Physical Properties**

The phase diagrams and fundamental thermodynamic, electrical and magnetic data for a large number of metal/molten salt solutions have been collected in various review articles[2, 13, 14].

2.1 PHASE DIAGRAMS

Figure 3 shows some examples of phase diagrams that illustrate the diversity of phase behavior found in metal/molten salt solutions. Consider, for example the two alkali metal/halide solutions shown in Fig. 3(a) and 3(b). The system Na-NaCl exhibits a wide range of liquid state immiscibility with a consolute temperature nearly 300 °C above the monotectic. At the melting temperature of NaCl, sodium is soluble in NaCl only up to about 2 at. %. The solubility of the salt in liquid sodium at this temperature is similarly low. In contrast, liquid cesium and its halides are mutually soluble over the full range of composition from pure salt to pure metal. The relatively flat liquidus curves at intermediate concentrations, however, suggests an unfulfilled tendency to phase-separate. Thermodynamic studies have, in fact, revealed the presence of concentrations fluctuations in this range as might be expected for a solution at

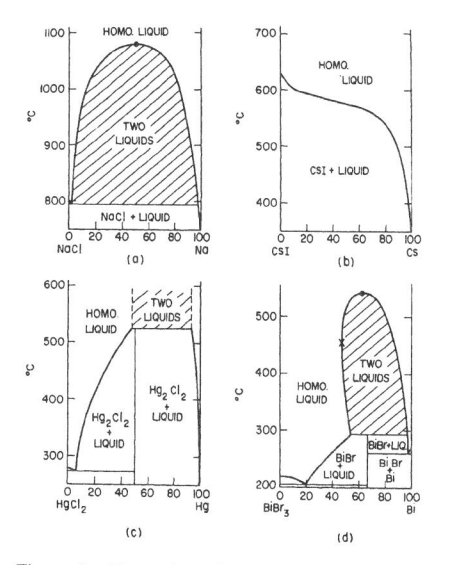

Figure 3. Examples of phase diagrams[2] for metal/molten salt solutions: (a) Na-NaCl, (b) Cs-CsI, (c) Hg-HgCl$_2$, and (d) Bi-BiBr$_3$.

temperatures not too far above its consolute point. The potassium and rubidium metal/halide solutions exhibit phase behavior intermediate between that of the sodium and cesium solutions illustrated in Figs. 3(a) and 3(b) i.e. regions of liquid immiscibility that are greatly reduced compared to those of the sodium solutions.

The phase diagrams of the polyvalent metal solutions Hg-HgCl$_2$ and Bi-BiBr$_3$ shown in Figs. 3(c) and 3(d), respectively, are quite different from those of the alkali metal solutions. Liquid-liquid phase separation also occurs in these solutions, but the "miscibility gap" is shifted to metal-rich compositions. The melts of the pure salts in these cases are essentially molecular in structure. The salt-rich solutions are thus solutions of the molecular salts with the dissociated melts found around the 50 at. % regions. The latter range is, of course, only accessible at temperatures well above the melting temperatures of the pure salts.

2.2 ELECTRICAL PROPERTIES

The initial variation of the electrical conductivity on adding metal provides a clue as to the nature of transport of the "excess electrons" associated with the metal. In general we can consider the conductivity σ_{DC} to be the sum of contributions σ_{elec} and σ_{ion} from the electrons and the ions, respectively. The electronic term can be expressed in the form

$$\sigma_{elec} = \overline{n}_e \times e \times \overline{\mu}_e \tag{6}$$

where $\overline{n}_e = x_M z N_0 / \Omega_{mole}$ is the average electron concentration obtained on adding a mole fraction x_M of metal of valence z, and $\overline{\mu}_e$ is the average mobility of these electrons. The average mobility, like the equivalent molar conductance Λ_M, is a measure of the average effectiveness of added metal in increasing the electrical conductivity. It is easy to show

$$\Lambda_M \cong z N_0 e \overline{\mu}_e. \tag{7}$$

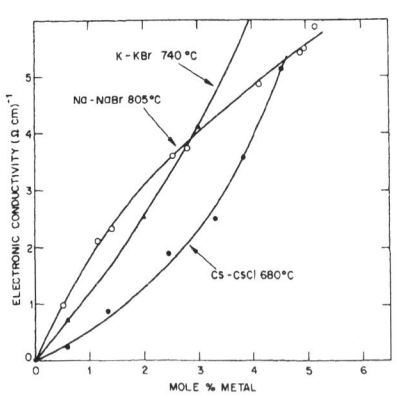

Figure 4. Initial increase of electronic conductivity $\sigma_{elec} = \sigma_{DC} - \sigma_{salt}$ with added metal for Na-NaBr[5], K-KBr[5], and Cs-CsCl[15].

If the mobility of added electrons is initially independent of metal concentration, and if each atom of added metal contributes z electrons to $\bar{n}_e$, then we should expect σ_{DC} to increase linearly with x_M. In fact, as illustrated in Fig. 4, this is rarely the case. For cesium/cesium halide solutions such as Cs-CsCl, the electronic conductivity increases more rapidly than linearly indicating that the average mobility of the excess electrons must increase with increasing metal concentration. For sodium solutions, on the other hand, the dependence is weaker than linear. In this case, either the effective number of charge carriers is less than $\bar{n}_e$ as defined above, i.e. electrons are trapped, or the electronic mobility is reduced at higher metal concentrations. We caution, however, that it is difficult to distinguish these effects from the conductivity data alone since localized (trapped) electrons could be in dynamic equilibrium with delocalized electrons. To address this question and to understand better the difference between the various alkali metal solutions, we need more detailed microscopic information.

The behavior of the electrical conductivity of the bismuth/bismuth halide solutions[7, 8] is quite different from that of the alkali metal systems. As metal is added to the pure halides at temperatures above the consolute temperatures, the initial variation of the conductivity is rather weak so that the transition to metallic values is shifted to higher metal concentrations than is the case for the alkalis. At lower temperatures, below the consolute temperature, the total conductivity remains close to the values of the pure salts up to the limits of metal solubility. That is, there is essentially no electronic contribution to the conductivity in the salt-rich solutions at low temperatures.

The electrical behavior of the divalent alkaline earth/halide solutions[16, 17] is similar in some respects to that of the alkali/alkali metal halide solutions. The conductivity of calcium/calcium halide solutions has been measured well below the very high consolute temperatures and is essentially independent of metal concentration. In contrast, σ_{DC} rises immediately on adding barium metal to the barium halides and, because of reduced ranges of liquid-liquid phase separation, the data can be extended to higher metal concentrations. The conductivity data for strontium/strontium halides solutions is intermediate between that of the calcium and barium solutions. Like the sodium/sodium halide solutions, there is a tendency for electron trapping in the lighter alkaline earth solutions under conditions well below the consolute temperatures.

2.3 MAGNETIC PROPERTIES

The fundamental magnetic property of a material is the total magnetic susceptibility, the quantity that describes the magnetization developed by a volume V of material in reponse to imposition of a uniform, static magnetic field $\vec{B}$:

$$\vec{M} = \chi_V V \vec{B}. \qquad (8)$$

The total susceptibility includes contributions from the diamagnetism of the ion cores, the paramagnetism of electron spins, and the paramagnetic and diamagnetic effects of the orbital motions of valence electrons:

$$\chi_{total} = \chi_{ion}^{dia} + \chi_{elec\ spin}^{para} + \chi_{elec\ orbital}^{para} + \chi_{elec\ orbital}^{dia} \qquad (9)$$

The paramagnetic electron spin contribution is particularly informative in salt-rich solutions because it reveals whether the states assumed by excess electrons are singly-occupied (paramagnetic) or doubly-occupied (spin-paired, diamagnetic). Quantitative determination of the spin-paramagnetic contribution is difficult, however, because one must somehow separate it from all the other contributions to the total measured susceptibility. Nevertheless, magnetic susceptibilities have been widely measured on metal/molten salt solutions because of the fundamental character of this property and the fact that the measurement can be made without physical contact to samples in sealed cells. This is an obvious technical advantage when dealing with highly reactive materials at high temperatures.

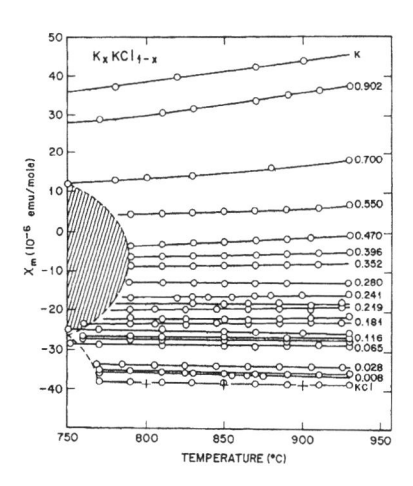

Figure 5. Total magnetic susceptibility versus temperature for K-KCl solutions[18]. The shaded region represents the domain of liquid-liquid phase separation.

The magnetic properties of a alkali metal/alkali halide solution are illustrated in Fig.5 for the case of K-KCl[18]. The susceptibility of the pure salt is diamagnetic and essentially temperature-independent. As potassium metal is added to the salt, the susceptibility immediate increases and with further metal addition, rises to the paramagnetic and slightly temperature dependent susceptibility of the pure metal. It is evident that the addition of metal introduces paramagnetic electronic states to the solutions, but for quantitative analysis is it is necessary to turn to the spectroscopic techniques of electron and nuclear spin resonance.

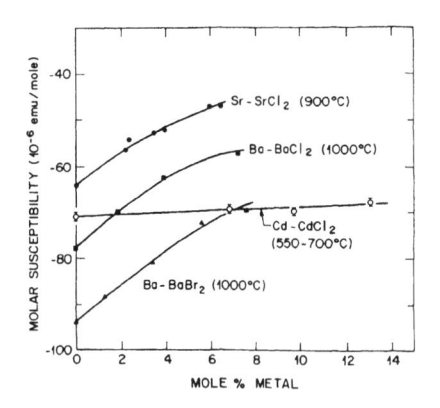

Figure 6. Total magnetic susceptibility versus metal concentration for Sr-SrCl₂, Ba-BaCl₂, Ba-BaBr₂[19], and Cd-CdCl₂[20].

The value of the magnetic susceptibility as a means of characterizing the state of the excess electron is particularly evident from the data for some divalent metal/halide solutions shown in Fig. 6. The total susceptibility increases in the paramagnetic direction immediately on addition of metal to $SrCl_2$, $BaCl_2$, and $BaBr_2$. This is again a clear indication of the formation of singly-occupied electronic states on addition of metal. In contrast, the susceptibility of Cd-$CdCl_2$ solutions scarcely changes with up to 14 at. % added metal. This is strong evidence for formation of diamagnetic traps. Recent results[21] from Raman spectroscopy reveal, in fact, that the diamagnetic species Cd_2^{2+} is formed in this system.

3. Useful Spectroscopic Techniques

3.1 NEUTRON DIFFRACTION

Elastic and inelastic scattering of neutrons has proven to be a powerful tool for investigating the structure of liquids, including molten salts and their mixtures. The basic principles of the method are described in the chapter by Adya and need not be repeated here. It is sufficient for the present purposes to summarize the kind of information that can be obtained for metal/molten salt mixtures.

The fundamental structural information for a binary liquid MX_n is contained in the three partial structure factors $S_{M-M}(q)$, $S_{X-X}(q)$, and $S_{M-X}(q)$. These functions of the scattering wave number q describe, respectively, correlations of metal cations (M-M), anions (X-X), and the correlations of cations with anions (M-X). The Fourier transforms of these functions yield the corresponding pair correlations, $g_{M-M}(r)$ etc. A single scattering experiment can measure only the total structure $S(q)$ which is essentially the sum of the three partial functions, weighted by the respective scattering cross-sections of the scattering nuclei. However, Enderby and coworkers[22] showed that it is possible to measure the separate partial structure factors by carrying out three scattering experiments on samples of different isotopic composition. The studies established, for example, that the molten alkali halides exhibit a high degree of local compositional order in which cations are surrounded by a first-neighbor shell of anions and vice-versa. Moreover, the charge correlation extends over many shells of neighbors. As we discuss later, the local compositional order is of great importance for understanding the states of excess electrons in dilute alkali metal/alkali halide solutions.

Bhatia[23] showed that the partial structure factors described above can be transformed into appropriate linear combinations that describe, respectively, density correlations, concentration correlations, and the cross-correlations between density and concentration. The concentration-concentration correlations are particular important in metal/molten salt solutions near the consolute point for liquid-liquid phase separation. In this range, $S_{CC}(q)$ exhibits critical behavior associated with the divergence of the correlation length ξ for concentration fluctuations;

$$S_{CC}(q) = S_{CC}(0)/(1+q^2\xi^2) \qquad (10)$$

Pioneering studies of concentration fluctuations in metal/molten salt solutions were carried out by Chieux and coworkers[24].

3.2 MAGNETIC RESONANCE

The problem of separating the contribution of spin paramagnetism from the total magnetic response of a metal/molten salt solution can be solved in large measure through the application of nuclear magnetic resonance (NMR) and electron spin resonance (ESR) techniques. For a discussion of the technical aspects of these methods as applied to studies of molten salt systems, see Reference 25.

3.2.1 *Nuclear Magnetic Resonance (NMR)*

Nuclear magnetic resonance (NMR) is the spectroscopy of the nuclear Zeeman levels associated with the interaction of nuclear dipole moments with an externally applied magnetic field of strength B_0[26]. The frequency corresponding to the fundamental splitting of the energy levels, the nuclear Larmor frequency, is $\omega_0 = \gamma_n B_0$ where γ_n is the gyromagnetic ratio of a given nuclear isotope. The details of the resonant absorption of radio-frequency energy at, or near, ω_0 are highly sensitive to the local environment of the nucleus. Further, because the value of γ_n is different for each isotope, the method is chemically selective, yielding atomic scale information about the environment of specific chemical constituents in the material studied. The information comes in essentially two forms. Static (time-averaged) interactions of the nuclei with their surroundings yield shifts of the resonance frequency, whilee dynamic interactions are responsible for linewidths and spin-relaxation phenomena.

In the particular case of metal/molten salt solutions, the most important interactions are the magnetic hyperfine interactions with the spins of unpaired electrons in the system. These are expressed as static local field contributions that shift the NMR frequency from the nominal value ω_0, and fluctuating local fields that dominate the relaxation behavior of the nuclei in these systems. The spins responsible for the hyperfine fields may be delocalized conduction electrons in metallic solutions, or singly-occupied localized states in salt-rich solutions. The principal advantages of NMR are its local, chemically selective nature, general applicability to a wide variety of materials and, for high temperature studies, the fact that NMR is a "contactless" method. Examples of the use of NMR methods to investigate specific classes of metal/molten salt solutions are described later in this chapter.

3.2.2 Electron Spin Resonance (ESR)

Electron spin resonance (ESR) is analogous to NMR in that it is the spectroscopy of electron spins in an applied magnetic field. Because of the much larger dipole moment of electrons, the resonance frequencies for ESR fall in the microwave range, rather than the radio-frequency range, for typical laboratory magnetic fields. As a probe of the electron spin susceptibility, ESR has the advantage that the integrated intensity of the resonance is exactly the spin susceptibility:

$$I_{ESR} = \chi_{elec\ spin}^{para}. \qquad (11)$$

Thus, the spin-susceptibility can be determined directly from a properly calibrated ESR experiment. In contrast, the NMR shift due to electron paramagnetism is only proportional to $\chi_{elec\ spin}^{para}$, and the constant of proportionality (hyperfine coupling) is not known *a priori*. The disadvantages of ESR for metal/molten salt solutions are the need for microwave techniques, somewhat cumbersome for high temperature studies, and the fact that ESR lines are often immeasureably broad as a result of the short lifetimes of electron spin states in high temperature liquids.

3.3 OPTICAL TECHNIQUES

Optical absorption was the first spectroscopic technique to be applied to metal/molten salt solutions[27, 28]. In the 1930's Mollwo[27] obtained the spectra of salt-rich alkali metal/alkali halide solutions and observed bands similar to the F-bands of alkali halide crystals that had been exposed to metal vapors. These are examples of photo-excitation of the electronic levels of localized electrons. In metallic solutions, optical absorption and reflectivity are due to free carrier (Drude) absorption. The recent papers of Freyland and coworkers[29–31] provide a good introduction to the current state of optical absorption techniques.

The availability of picosecond and sub-picosecond pulsed light sources has introduced new possibilities for studying the dynamics of electronic states in molten salts and metal/molten salt solutions. Photo-generated electronic states can be observed in transient absorption experiments and their subsequent reactions with other species in the melt can be followed as a function of time[32]. An example of the use of this technique will be discussed in § 4.3.

4. Alkali Metal/Alkali Halide Solutions

Because of their relative simplicity, the alkali metal/alkali halide solutions have been the most intensively investigated metal/molten salt solutions in recent years. This is despite the experimental complications introduced by the high chemical reactivity of the alkali metals. The electronic properties of these solutions raise a number of basic questions:

- What is the state of the valence electrons associated with dilute concentrations of excess metal in alkali halides?
- What is the mechanism of electronic conduction in dilute solutions?
- What is the nature of the metal-nonmetal transition in these solutions?
- How does the electronic (metal-nonmetal) transition relate to the phase separations occurring in some alkali metal/alkali halide solutions?
- What accounts for the differing behavior among the various alkali metal/alkali halide solutions?

As we discuss below, application of spectroscopic techniques has led to considerable progress in finding answers to questions such as these.

4.1 DILUTE ELECTRONS IN MOLTEN HALIDES

4.1.1 Electronic localization: "F-centers"

Optical absorption spectroscopy provides an important clue as to the states occupied by low concentrations of electrons in molten alkali halides. All such solutions exhibit strong absorption bands in the visible or near infra-red; the intensities of the bands increase progressively with increasing metal concentration. The Cs-CsI data of von Blanckenhagen et al.[31] reproduced in Fig. 7 provide a good example of this behavior. The similarity of these bands to those associated with F-centers in alkali

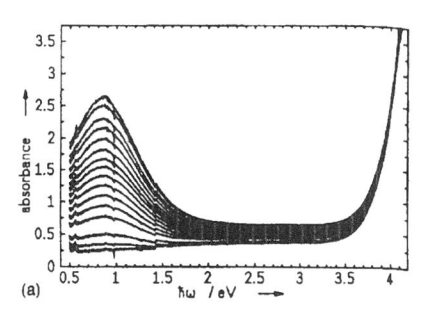

Figure 7. Optical absorption spectra[31] of Cs-CsI solutions for molar metal concentrations ranging from pure CsI (lowest curve) to 3.91 % Cs (uppermost curve).

halide crystals led Pitzer[33] to suggest a liquid-state analogy, i.e. that the excess electrons are bound in a localized state coordinated by cations as illustrated in Fig. 8.

What properties does one expect F-center analogs to exhibit? First, since such states are occupied by single electrons, the localized electrons should introduce spin paramagnetism to the otherwise diamagnetic molten halide. This is clearly the case as

CRYSTAL   LIQUID

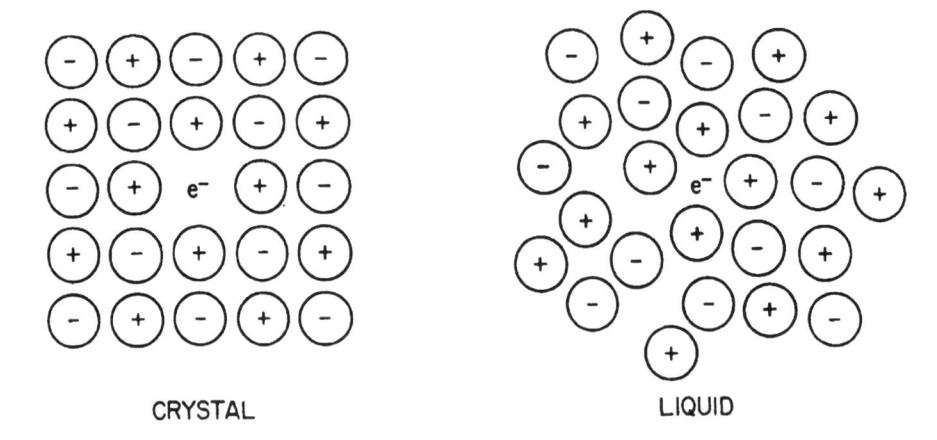

Figure 8. Schematic diagram of electron solvated by cations (F-center) in crystalline and molten alkali halide.

is shown by the magnetic susceptibility data for K-KCl shown in Fig. 5. However, the contributions of spin paramagnetism can be isolated more effectively by means of resonance techniques. To explain this, it is necessary to digress briefly to consider the origin of paramagnetic frequency shifts in nuclear magnetic resonance (NMR).

The actual frequency at which NMR is observed depends on the combined effects of an external applied magnetic field (B_0) and a local magnetic field (B_{local}) produced at the nuclear site by various interactions within the material:

$$v_{NMR} = \gamma_n (B_0 + B_{local})/2\pi \tag{12}$$

where γ_n is the nuclear gyromagnetic ratio introduced in §3.2.1. In materials containing unpaired electrons, the internal field is very often produced by the magnetic hyperfine interactions between the nuclear and electronic spins, $\mathsf{H}_{hf} = A\vec{I} \cdot \vec{S}$. The local field is proportional to the external field through the electron spin paramagnetic susceptibility:

$$B_{local} = (N_0 \gamma_n \gamma_e \hbar^2)^{-1} \langle A \rangle \chi^{para}_{elec\ spin} B_0 \tag{13}$$

where N_0 is Avogadro's number and the average hyperfine coupling $\langle A \rangle$ is related to tthe average probability density $\langle |\psi(R_n)|^2 \rangle$ of the electron at the resonant nucleus by

$$\langle A \rangle = \hbar \langle \omega_{hf} \rangle = \frac{8\pi}{3} \gamma_e \gamma_n \hbar^2 \langle |\psi(R_n)|^2 \rangle. \tag{14}$$

There are two important limits. In metals, the spin susceptibility is the Pauli paramagnetic susceptibility of conduction electrons obeying Fermi-Dirac statistics:

$$\chi_{elec\ spin}^{para} = 2\mu_B^2 N(E_F) \tag{15}$$

where $N(E_F)$ is the density of electronic states at the Fermi energy (E_F). The NMR shifts associated with the susceptibility of Eq. (15) are the characteristic "Knight shifts" of metals. The other limit is represented by a low density of localized electrons such as occurs in alkali metal/alkali halides solutions at low metal concentrations. In this case, the electrons obey classical Boltzmann statistics and the susceptibility is given by the Curie susceptibility for a molar fraction of spins, c_s at temperature T:

$$\chi_{elec\ spin}^{para} = \frac{c_s N_0 \gamma_e^2 \hbar^2}{4 k_B T}. \tag{16}$$

It is important to recognize that these shifts due to electron spin effects are distinct from the usual "chemical shifts" which are associated with electronic *orbital* currents.

The paramagnetic NMR shifts in an alkali halide/alkali metal solutions are illustrated in Fig. 9 for the case of Cs-CsI solutions[34]. These data show the large ^{133}Cs Knight shifts (greater than 1%) in metal-rich solutions and paramagnetic shifts proportional to the excess metal concentration in the salt-rich limit. The ^{127}I shifts observed in salt-rich solutions show that the electronic states have substantially higher density at the Cs cation sites than at the anions. The values of $\langle|\psi(Cs)|^2\rangle$ and $\langle|\psi(I)|^2\rangle$ are known for the F-centers in CsI from electron spin resonance experiments[35, 36]. Using Eq. (16) with c_s equal to the concentration of excess metal, we can test the F-center model of the solutions by comparing the values of $\langle|\psi(Cs)|^2\rangle$

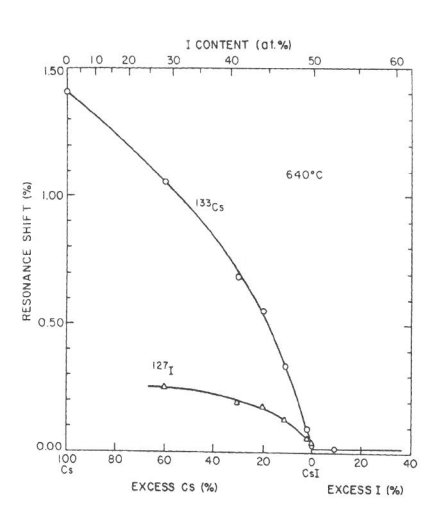

Figure 9. NMR resonance shifts versus composition for ^{133}Cs and ^{127}I in molten Cs-CsI solutions[34].

and $\langle|\psi(I)|^2\rangle$ with the solid state values (Table 1). To account for the slightly different volumes of the states in liquid and solid, we compare the respective values of $d_I^3\langle|\psi|^2\rangle$ where d_I is the average cation-anion distance. The agreement for Cs is excellent while that for I is acceptable, the lower density at the anions in the molten solutions probably reflecting the greater disorder in the second neighbor shell compared with the crystal.

Table 1. Comparison of electronic charge density at Cs and I nuclei in molten Cs-CsI solutions and F-centers in crystalline CsI.

	Liquid	Solid
$d_I^3 \langle |\psi(Cs)|^2 \rangle$	71	72
$d_I^3 \langle |\psi(I)|^2 \rangle$	12	17

Overall, the NMR shift data are good evidence for the F-center model of excess electrons in molten CsI with the concentration of F-centers being roughly that of the excess metal.

Electron spin resonance data[37], however, show that we cannot generalize the above conclusions to all alkali metal/alkali halide solutions. The integrated ESR intensities shown in Fig. 10 for K-KCl show that the spin susceptibility is substantially smaller than expected if all excess electrons are in singly-occupied F-center states. This is true even for concentrations well below 1 at. % metal. It is clear that in K-KCl, a large fraction of the excess electrons are in diamagnetic ("spin-paired")

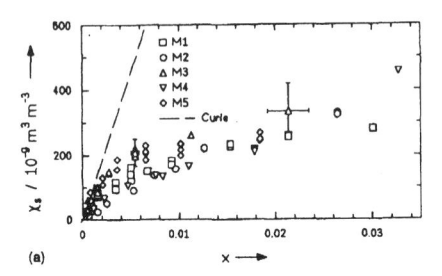

Figure 10. Spin susceptibility of salt-rich K-KCl solutions obtained from the integrated electron spin resonance intensities[37]. Dashed line shows the Curie law, Eq. (16), expected if all excess electrons are in singly-occupied localized states.

configurations. Bredig[2] suggested early on that dimers, e.g. K_2, might form in these solutions. A more recent alternative is suggested by the quantum molecular dynamics simulations of Fois et al.[38] These calculations, show that doubly-occupied F-centers with anti-parallel spins, so-called "bipolarons," are more stable in K-KCl than the singly-occupied states. While these simulations are highly suggestive, there is as yet no experimental evidence for any particular structure of the spin-paired species.

4.1.2 Electron Transport and Dynamics

Let us return to the question raised earlier concerning the properties we might expect of low concentrations of excess electrons localized in a molten alkali halide. In particular, we can inquire as to the dynamic properties of such electronic states. It is clear from the DC electrical conductivity (see, for example, Fig. 4) that the electrons are mobile even at low concentrations of excess metal. Nuclear spin-lattice relaxation measurements provide a means of probing the electronic dynamics on the atomic scale.

Spin-lattice relaxation describes the process by which a system of nuclear magnetic moments achieves thermal equilibrium with the thermal reservoir ("lattice") of the material in which they are located. A spin system out of equilibrium must exchange energy with its surroundings in order to restore equilibrium populations of the various spin energy levels. The transitions ("spin-flips") necessary for this process are induced by time-dependent local fields. In the case of alkali metal/alkali halide solutions, these are the fluctuating components of the same magnetic hyperfine fields whose time-averages generate the resonance shifts discussed above. The local field fluctuations can be characterized by a correlation time τ which is essentially the time that a given electron and nucleus interact. Clearly τ is related to the microscopic mobility of the electrons and, in particular, is quite different for localized and delocalized (metallic) electrons.

Time-dependent quantum perturbation theory[39] shows that the spin-lattice relaxation rate, $1/T_1$, is related to τ and the mean-square hyperfine field (in frequency units) by

$$1/T_1 = \langle \omega_{hf}^2 \rangle \tau \qquad (17)$$

provided that the fluctuations are fast relative to the NMR time scale, i.e. $\omega_0 \tau \ll 1$. The mean-square hyperfine field is

$$\langle \omega_{hf}^2 \rangle = c_s (32\pi/9) \gamma_e^2 \gamma_n^2 \hbar^2 \langle |\psi(R_n)|^2 \rangle^2 . \qquad (18)$$

The spin-lattice relaxation rates[34] for ^{133}Cs and ^{127}I in Cs-CsI are shown in Fig. 11. These data show that even as the concentration of excess electrons decreases on passing from the metallic solutions to the salt-rich solutions, the relaxation rate increases to a strong peak. Eventually the rate goes to a very low value in the pure molten salt. The peak is the result of the competition between a decreasing concentration of unpaired spins, c_s, and a lengthening τ as the electrons become more localized. Analysis of these data showed that at metal concentrations below a few at. %, the correlation time τ approaches a limit of a few ps. This is the time scale of individual ionic motions in the liquid state and this result suggests that the mobility of F-center electrons is determined by the lifetimes of particular solvation configurations.

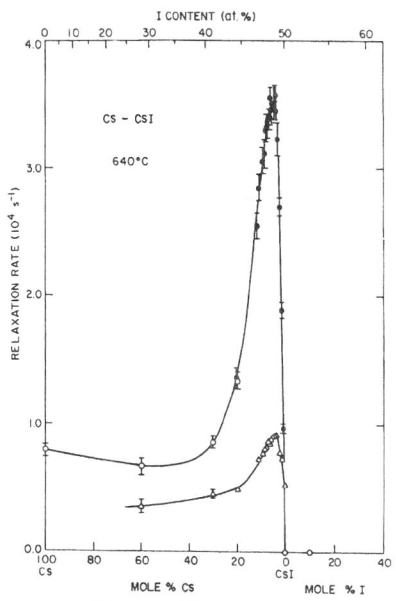

Figure 11. ^{113}Cs and ^{127}I spin-lattice relaxation rates versus composition in Cs-CsI solutions[34].

4.2 THE METAL NON-METAL TRANSITION AND PHASE SEPARATION

NMR data also help resolve the dilemma posed by the differing DC conductivity data of different alkali metal/alkali halide solutions (Fig. 4) as well as the differing importance of spin-paired species in, say, Cs-CsI and K-KCl. For localized electrons, the resonance shift can be written

$$\frac{\Delta v}{v} = \frac{\hbar}{4k_B T} \frac{\gamma_e}{\gamma_n} \langle \omega_{hf} \rangle. \qquad (19)$$

When combined with Eq. (17), this yields the following relation between the NMR shift and the relaxation rate

$$\frac{1}{T_1} = \frac{4k_B T}{\hbar} \frac{\gamma_e}{\gamma_n} \frac{\langle \omega_{hf}^2 \rangle}{\langle \omega_{hf} \rangle} \tau \frac{\Delta v}{v}. \qquad (20)$$

As the metal concentration is varied, Eq. (20) implies a linear dependence of the relaxation rate on shift if the correlation time is independent of concentration. The plot in Fig. 12 of $1/T_1$ versus $\Delta v/v$ reveals very different behavior for Cs-CsI and Na-NaBr. In the case of Cs-CsI, the dependence is highly non-linear due to the strong reduction of the correlation time with increasing metal concentration. This is a clear precursor of the metal-nonmetal transition. In Na-NaBr, on the other hand, the essentially linear dependence indicates that τ does not depend on concentration – there is no indication of an impending metal-nonmetal transition.

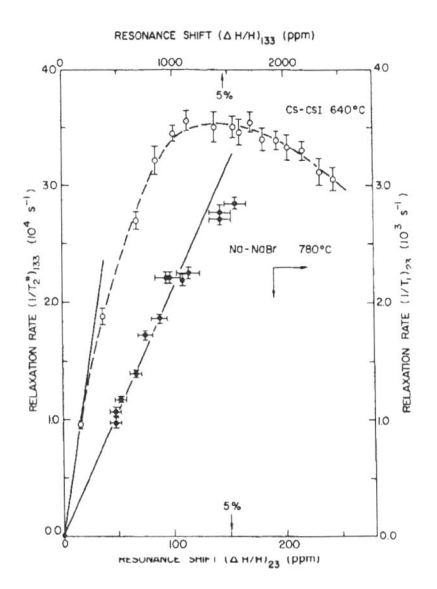

Figure 12. Plot of the spin-lattice relaxation rate versus NMR resonance shift (composition is the implicit parameter) for Na-NaBr and Cs-CsI[40].

For a diffusive electronic transport mechanism, the mobility is related to a jump time τ_j by an Einstein relation

$$\mu_e = ea^2 / 6k_B T \tau_j \tag{21}$$

where a is the mean jump distance. Identifying the jump time with the NMR correlation time leads to the conclusion that the mobility is independent of concentration in Na-NaBr. Since the conductivity curves downward with increasing metal concentration (Fig. 4), this analysis indicates that the increase in carrier concentration per unit of added metal is decreasing with metal concentration. Moreover, the shift was found[40] to be proportional to the conductivity. Thus, the missing electrons occupy states that are diamagnetic and do not contribute to the conductivity. Formation of these spin-paired, low mobility states is increasingly likely as the metal concentration increases. In this respect, Na-NaBr behaves much like K-KCl.

The foregoing has been confirmed nicely by the recent systematic study by von Blanckenhagen et al.[31] of optical absorption combined with *in situ* activity measurements of sodium, potassium and cesium metal/halide solutions. These authors concluded that F-centers are the predominant species formed by excess electrons in the cesium solutions, whereas stronger coulomb forces in the sodium solutions lead to formation of spin-paired species. The potassium solutions are intermediate, but clearly more like the sodium solutions than the cesium.

The apparent dominance of spin-pairing and constant mobilities of the smaller cation alkali metal/alkali halides solutions raises an obvious question. What about the continuous metal-nonmetal transition? The answer is that under the conditions of the experiments described above *there is no continuous metal-nonmetal transition*! These solutions exhibit liquid-liquid phase separation into poorly conducting salt-rich solutions and metallic metal-rich solutions (Fig. 3). Where a continuous metal-nonmetal transition does occur, i.e. cesium/cesium halide solutions, spin-paired species do not form and there is rapid delocalization of the F-centers with increasing metal concentration. This raises the question, unanswered so far, as to the relation between the thermal stability of the spin-paired species and the phase-separation. Are the spin-paired species unstable at temperatures above the consolute temperature where the metal-nonmetal transition is continuous?

4.3 ELECTRON-HOLE RECOMBINATION DYNAMICS

In § 4.1, I discussed the localized F-center analogs formed in dilute solutions of alkali metals in alkali halides. It is also possible to create F-centers by optical excitation of *pure* molten alkali halides. Intense illumination with light near the fundamental absorption edge creates electron-hole pairs in which the electron rapidly localizes in an F-center analog. The appearance of the F-centers and their subsequent disappearance by recombination can be monitored by means of transient absorption measurements at the wavelengths of the F-center absorption band.

The results of such an experiment[32] on KBr are shown in Figure 13. These data were obtained with 30 - 50 ps pulses at 266 nm generated by twice-doubling the 1064 nm fundamental output of an yttrium-aluminum-garnet laser. The absorption was monitored at the fundamental wavelength (1064 nm) which falls in the center of the F-center band in KBr. The data show the rapid initial rise of the F-band absorption and its subsequent, much slower decay. The decay does not follow a simple exponential time dependence and the rate of the decay is higher when the excitation intensity is higher. These features indicate that the recombination process is *non-geminate*, i.e. the localized electron does not combine with the hole (Br) from which it was excited. The electrons migrate away from their geminate ions.

A simple non-geminate recombination process would be described by the following expression for the time-dependent concentration of the absorbing species (F-centers):

$$n(t) = n_0 /(1 + k n_0 t) \qquad (22)$$

where n_0 is the initial ($t = 0$) concentration and k is a rate constant. However, this expression does not provide a good fit to the data of Figure 13. It was suggested[32] that the

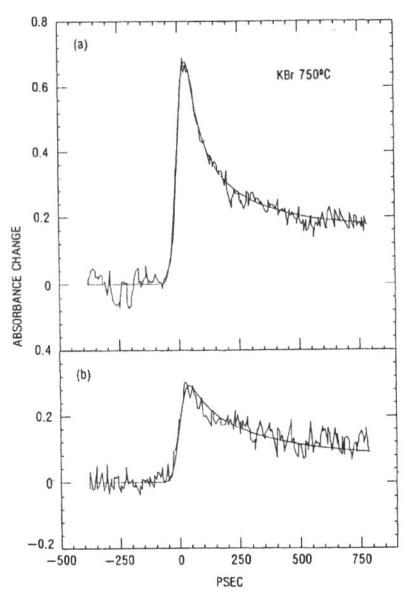

Figure 13. Transient absorption signals at 1064 nm from molten KBr at 750 °C at higher (a) and lower (b) 266 nm excitation pulse intensities. Solid lines are fits to the recombination reactions (25) and (26).

recombination process is more complex, involving the formation of intermediate halogen species according to the reactions

$$X^0 + X^- \rightarrow X_2^- \qquad (23)$$

$$2X_2^- \rightarrow X_3^- + X^- \qquad (24)$$

The dimer ion (X_2^-) is well known in solid alkali halide crystals as the V_k center, a hole polaron that complements the F-center electron state. Electron recombination with the dimer ions according to

$$e^- + X_2^- \rightarrow 2X^- \qquad (25)$$

is relatively fast, but once the dimers are exhausted by reactions (24) and (25), the third-order recombination with the trimer

$$2e^- + X_3^- \rightarrow 3X^- \qquad (26)$$

is relatively slow. This accounts for the long-time "tail" on the absorption decay curves, especially evident in Figure 13(a). Similar experiments on KCl yielded much slower decays, suggesting that the dimer recombination (25) may be inhibited by stronger coulomb repulsions with the smaller ions. Detailed analysis of the decay constants for KBr and KCl are consistent with the idea that the average mobility of the F-center electrons is 10 to 100 times the mobility of single ions in the molten salt. This explains why the electrons from a few atomic percent of added metal are sufficient to produce electronic conductivities comparable with the ionic conductivity (Figure 2).

5. Mixed Valence Metal/Molten Salt Solutions

Solutions of polyvalent metals with their molten salts present a variety of interesting behaviors. One example is the electrical conductivity of the Bi/Bi halides solutions illustrated in Figure 2. The bismuth systems have been extensively investigated and are discussed in various reviews[2, 41]. Because the consolute temperatures are relatively low, it is possible to study these solutions both above and below the onset of phase separation[41]. One of the common characteristics of the polyvalent metal/molten salt solutions is the presence of multiple valence states (sometimes described as subhalides) for the metal. Examples of multiple valences are the In^+ - In^{3+} and Bi^+ - Bi^{3+} couples.

5.1 "NEGATIVE U" EFFECTS

In the language of modern solid state physics, the mixed valence salts are examples of so-called "negative U" systems[42]. The Hubbard U is the electrostatic repulsion energy of two electrons occupying the same atom or ion. The normally positive (repulsive) energy U favors electronic configurations in which atoms are occupied by a single valence electron. It frequently happens, however, that a doubly-occupied configuration such as In^+ ($5s^2$ electronic configuration) is stabilized by a compensating structural relaxation or chemical bond formation. An important example of this in the solid state is the so-called "DX effect" in doped semiconductors. In this situation, the ground state of dopant impurities consists of pairs of singly- and doubly-occupied impurities rather than the conventional singly-occupied state found, for example, for electron donors in silicon. In CdTe, for example, Ga does not enter as a shallow donor that can introduce carriers in the semiconductor. Instead, the following reaction[43] is believed to occur:

$$Ga^{2+} + Ga^{2+} \rightarrow Ga^+ + Ga^{3+} + \text{local lattice relaxation} \qquad (27)$$

Similar effects occur in other II-VI compound semiconductors, in $Al_xGa_{1-x}As$, and CdF_2. The molten salt "InI_2" is also an example of the negative U effect. This compound is actually $In^+(InI_4)^-$ [44]. Here the doubly-occupied configuration (In^+) is stabilized by the binding energy of the complex anion $(InI_4)^-$.

5.2   CHARGE TRANSFER EXCITATIONS

A negative U system can be excited, i.e. reactions such as (27) can be reversed, either thermally or optically. The excited singly-occupied states are paramagnetic and can be detected in an NMR experiment. Figure 14 shows the ^{115}In spin-lattice and spin-spin relaxation rates[45] versus inverse temperature for molten "InI_2." There is a dramatic increase in relaxation rates at higher temperatures. Analysis of the data yields the thermal excitation energy for creation of pairs of In^{2+} ions. Similar results have been obtained[46] in "$InCl_2$."

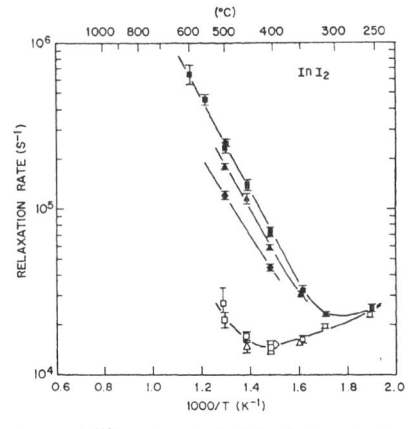

Figure 14. ^{115}In spin-spin (solid points) and spin-lattice (open points) relaxation rates in molten "InI_2" versus inverse temperature[45].

The optical excitation of In^{2+} pairs appears as a charge transfer band in the optical absorption of molten $InCl_2$. The band, corresponding to an optical excitation energy of 2.1 eV may be seen in Figure 15 just below the fundamental absorption edge[46].

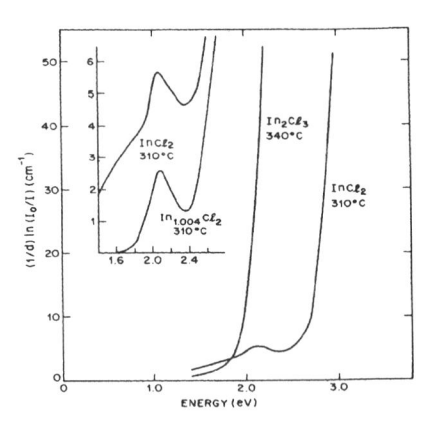

Figure 15. Charge-transfer bands in the optical extinction coefficient of $InCl_2$ and related salts[46].

6. Summary

The introduction of the various spectroscopic techniques described in these lectures has led to considerable progress in understanding the properties of metal/molten salt solutions. This is particularly true of the state of dilute electrons in very salt-rich alkali halide solutions. Despite the persistence of a number of unanswered fundamental questions about these interesting systems, however, there has been a decrease of activity in the very recent past. In concluding I would like to suggest a few extensions of the existing work that should be feasible with contemporary experimental techniques.

Suggestions for future work on the alkali metal/alkali halide solutions would include study of the thermal stability of bipolaronic species – extension of magnetic resonance and electrical measurements to the region about the phase separation in systems such as Na-NaCl. What is the structure of the bipolaronic species? Can one observe these with Raman spectroscopy operating in the region of low absorption between the F-center band and the fundamental absorption edge? In the pure molten salts under UV excitation, various intermediate species have been proposed to form and to react during the electronic recombination process. It would be interesting to try to observe these species directly using transient absorption spectroscopy.

In the mixed valence salts, it would be interesting to test the "negative U" model by detecting the polyatomic anion species, e.g. $(InCl_4)^-$, and then observing its dissociation under illumination at the wavelength of the charge transfer band. Further, can one observe the spin-paramagnetism of the thermally-generated singly-occupied species by high temperature magnetic susceptibility measurements, or do these dates relax before the magnetization becomes thermalized? In low temperature studies of negative U impurities in semiconductors, photo-generated magnetism has been observed. Can one observe the analogous effects in molten salts, say by observing photo-enhanced nuclear spin-lattice relaxation?

7. References

1. A. H. W. Aten, Z. phys. Chem. **66**, 641 (1909); **73**, 578 (1910).
2. M. A. Bredig, in *Molten Salt Chemistry*, edited by M. Blander (Interscience, New York, 1964), p. 367.
3. See, for example, N. F. Mott, *Metal-Insulator Transitions* (Taylor and Francis, London, 1974).
4. R. W. Schmutzler, et al., Phys. Lett. **55A**, 57 (1975); Ber. Bunsenges. Phys. Chem. **80**, 197 (1976).
5. H. P. Bronstein et al., J. Am. Chem. Soc. **80**, 2077 (1958); J. Chem. Phys. **37**, 677 (1962).
6. Cs-CsI: S. Sotier, H. Ehm, and F. Maidl, J. Non-Cryst. Solids **61-62**, 95 (1984).
7. L F. Grantham, J. Chem. Phys. **43**, 1415 (1965).
8. L F. Grantham and S. J. Yosim, J. Chem. Phys. **38**, 1671 (1963).
9. See, for example, F. Hensel and W. W. Warren, Jr., *Fluid Metals* (Princeton University Press, Princeton, 1999), p. 11 ff.
10. See, for example, C. Kittel, *Introduction to Solid State Physics* (Wiley, New York, 1986), 6th edition, p. 264 ff.
11. See, for example, J. M. Ziman, *Principles of the Theory of Solids* (Cambridge, London, 1964), p. 93.

12. N. F. Mott, Proc. Phys. Soc. A **62**, 416 (1949).
13. J. D. Corbett, in *Fused Salts*, edited by S. Sundheim (McGraw-Hill, New York, 1964), Chap. 6..
14. W. W. Warren, Jr., in *Advances in Molten Salt Chemistry*, edited by G. Mamantov and J. Braunstein (Plenum, New York, 1981), vol. 4, p. 1.
15. N. H. Nachtrieb, C. Hsu, M. Sosis, and P. A. Bertrand, in *Proc. Int Symposium on Molten Salts*, edited by J. P. Pemsler, J. Braunstein, and K. Nobe (Electrochemical Society, Pennington, 1976), p. 506.
16. A. S. Dworkin, H. R. Bronstein, and M. A. Bredig, Disc. Faraday Soc. **32**, 188 (1961); J. Phys. Chem. **70**, 2384 (1966).
17. H. H. Emons and D. Richter, Z. Anorg. Allgem. Chem. **353**, 148 (1967).
18. N. Nicoloso and W. Freyland, Z. phys. Chem. N.F. **135**, 39 (1983).
19. K. Grjotheim, H. A. Ikeuchi, and J. Krogh-Moe, Acta Chem. Scand. **24**, 985 (1970); K. Grjotheim, S. Dhabanandana, and J. Krogh-Moe, Acta Chem. Scand. **26**, 3427 (1972).
20. N. H. Nachtrieb, J. Phys. Chem. **66**, 1163 (1962).
21. B. Borresen, G. A. Voyiatzis and G. N. Papatheodorou, Phys. Chem. Phys. Chem. **1**, 3309 (1999).
22. See, for example, J. E. Enderby, J. Phys. C **15**, 4609 (1982).
23. A. B. Bhatia and D. E. Thornton, Phys. Rev. B **2**, 3004 (1970).
24. P. Chieux, P. Demay, J. Dupuy and J. F. Jal, J. Phys. Chem **84**, 1211 (1980);J. Jal, *Thesis*, Université Claude Bernard-Lyon I (1981); J. F. Jal, J. Dupuy, and P. Chieux, J. Phys. C **81**, 1347 (1985).
25. W. W. Warren, Jr., in Molten Salt Techniques, edited by R. J. Gale and D. G. Lovering (Plenum, New York, 1991), vol. 4, p. 111.
26. For a basic text on NMR, see for example, C. P. Slichter, Principles of Magnetic Resonance (Springer, Berlin, 1990), 3rd Edition.
27. E. Mollwo, Nachr. Ges. Wiss. Göttingen, Math—Phys. Kl, Fachgruppe II, **1**, 203 (1935).
28. D. M. Gruen, M. Krupelt, and I. Johnson, in *Molten Salts, Characterization and Analysis*, edited by G. Mamantov (Dekker, New York, 1969), p. 169.
29. W. Freyland K. Garbade, and E. Pfeiffer, Phys. Rev. Lett. **51**, 1304 (1983).
30. D. Nattland, T. Rauch, and W. Freyland, J. Chem. Phys. **98**, 4429 (1993).
31. B. von Blanckenhagen, D. Nattland, K. Bala, and W. Freyland, J. Chem. Phys. **110**, 2652 (1999)
32. W. W. Warren, Jr., B. F. Campbell, and G. F. Brennert, Phys. Rev. Lett. **58**, 941 (1987).
33. K. S. Pitzer, J. Am. Chem. Soc. **84**, 2025 (1962).
34. W. W. Warren, Jr., S. Sotier, and G. F. Brennert, Phys. Rev. B **30**, 65 (1984).
35. F. Hughes and J. G. Allard, Phys. Rev. 125, 173 (1962).
36. H. Seidel and H. C. Wolf, in *Physics of Color Centers*, edited by W. B. Fowler (Academic, New York, 1963), p. 538.
37. T. Schindelbeck and W. Freyland, J. Chem. Phys. **105**, 4448 (1996).
38. E. Fois, A. Selloni, R. Car, M. Parrinello, J. Phys. Chem. **92**, 3268 (1988).
39. Reference 26, p. 196.
40. W. W. Warren, Jr., S. Sotier, and G. F. Brennert, Phys. Rev. Lett. **50**, 1505 (1983).
41. W. W. Warren, Jr. in *Molten Salt Chemistry*,edited by G. Mamantov and R Marassi, NATO ASI, Series C, 202 (Reidel, Dordrecht, 1987), p. 237.
42. P. W. Anderson, Phys. Rev. Lett. **34**, 953 (1975).
43. C. H. Park and D. J. Chadi, Phys. Rev. B **52**, 11884 (1995).
44. P. L. Radloff and G. N. Papatheodorou, J. Chem. Phys. **82**, 992 (1980).
45. K. Ichikawa and W. W. Warren, Jr., Phys. Rev. B **20**, 900 (1979).
46. W. W. Warren, Jr., G. Schönherr, and F. Hensel, Chem. Phys. Lett. **96**, 505 (1983).

LIGHT SCATTERING FROM MOLTEN SALTS: STRUCTURE AND DYNAMICS

G. N. PAPATHEODOROU[1,2] and S. N. YANNOPOULOS[1]

[1] *Foundation for Research and Technology Hellas - Institute of Chemical Engineering and High Temperature Chemical Processes, P.O. Box 1414, GR-26500 Rio, Greece*
[2] *Department of Chemical Engineering, University of Patras, GR-26500, Patras, Greece*

1. Introductory Remarks

Much of our knowledge about structure and dynamics in solids, liquids or gasses comes from the interaction (scattering) between externally imposed stimuli and the system under study. Light has proved to be one of the best such stimuli that can furnish a great deal of important information over wide spatial and temporal ranges [1–4]. Undoubtedly, other scattering techniques like neutron and x–rays, applicable to other space and time regimes provide complementary data to light scattering [5].

In a scattering process (see Fig. 1) one usually interrogates the system that is under investigation by selecting suitable "particles", forwarding them onto the system and then examining the changes on "particles" properties such as intensity, polarization, energy etc. brought about after the interaction has occurred. Such suitable "particles"

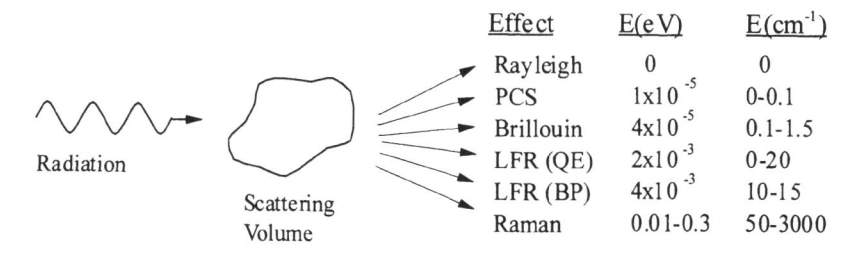

Effect	E(eV)	E(cm^{-1})
Rayleigh	0	0
PCS	1x10^{-5}	0-0.1
Brillouin	4x10^{-5}	0.1-1.5
LFR (QE)	2x10^{-3}	0-20
LFR (BP)	4x10^{-3}	10-15
Raman	0.01-0.3	50-3000

Figure 1. Schematic representation of the various light scattering effects; typical values for the energy involved in each of the processes is given in eV and cm^{-1} (1 meV = 8.07 cm^{-1} = 11.6 K). PCS: Photon correlation spectroscopy, LFR(QE): the quasi–elastic part of the low–frequency Raman spectrum and LFR(BP): the Boson peak contribution in the low–frequency Raman spectrum.

may be photons, neutrons, electrons, helium atoms or even phonons. Dispersion relations that link the energy of the particle with its wave vector characterize each kind of these particles. Therefore, the following expressions hold for various "particles":

Phonons: $\quad E_{phon} = \hbar\omega_j(\mathbf{q})$ (1a)

Photons: $\quad E_{phot} = \hbar c k$ (1b)

Neutrons, Electrons, Atoms: $\quad E_{part} = \hbar^2 k^2 / 2m$ (1c)

Here **q** is the wave vector and *j* the branch of the phonon; *k* is the magnitude of the wave vector for light *k*, and *m* the mass of the particle in question. In Fig. 2 the above dispersion relations are schematically illustrated in a double logarithmic plot (see figure caption for explanation). It is evident that light and neutron scattering probe quite different regions of wavenumber (*q*) and frequency. Broadly speaking, light scattering probes the region of low *q*, while neutron scattering probes the region of finite *q*. It is worth mentioned here that due to the reciprocity relation between the wave vector **q** and the real space dimension **r**, one can select to probe the desired spatial range ℓ of excitations in the studied medium by choosing the proper spectroscopy with wave vector $q \sim \ell^{-1}$.

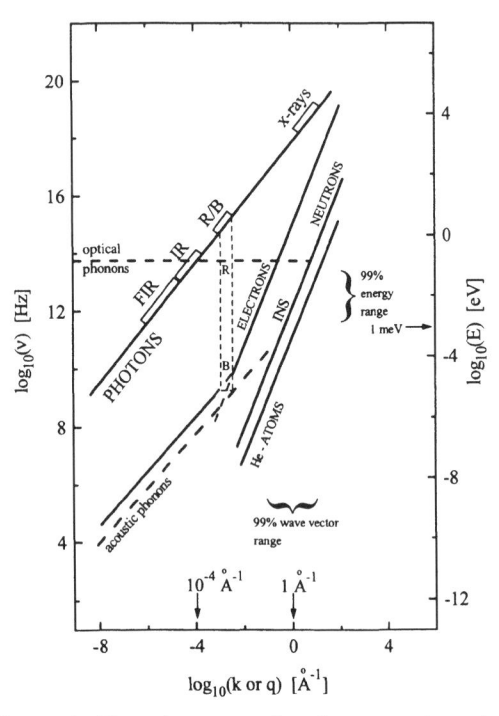

Figure 2. Dispersion curves for photons, neutron etc. Typical curves for acoustic and optical branches for particular solids are presented as dashed lines. The "99% ranges" indicate the energy and wave vector ranges required for probing the 99 % of phonons in a typical crystal, from Ref. [6]. R: Raman, B: Brillouin INS: Inelastic Neutron Scattering

A major advantage of light scattering is its ability to probe both *structure* and *dynamics*. While crystalline solids are adequately characterized after their structure has been determined, their amorphous counterparts or their melts are thoroughly understood if, in addition, their dynamics has been clarified. Consequently, Raman spectroscopy is mostly applicable to the study of the structure of solids (both crystalline and amorphous) and liquids while

Dynamic Light Scattering (DLS) is suitable for the elucidation of relaxation processes that occur in non–crystalline phases over a wide time scale.

The purpose of the present review is to present a general view of the various light scattering techniques that are able to furnish useful information concerning structural and dynamical aspects for simple and glass–forming molten salts. The structure of the review is as follows. Section 2 treats the subject of inelastic (Raman) scattering presenting some basic concepts, elementary experimental elements and a thorough account of experimental results concerning mono–, di–, tri–, and tetra–valent metal halides. In Section 3 we have attempted an introduction to the field of the low–energy modes in glass–forming salts. Numerous pertinent aspects of the low–frequency phenomenology are discussed in view of new experimental findings. Finally, section 4 is devoted to a brief survey on dynamic properties of glass–forming salts where Rayleigh–Brillouin scattering and Photon Correlation spectroscopy are introduced.

2. Inelastic Light Scattering or Raman scattering

2.1. BASIC CONCEPTS
2.1.1. Raman Scattering Theory

Due to the inelastic character of Raman Scattering (RS) the energy of the scattered photons is different from the energy of the exciting radiation. Attempting a microscopic study of the effect reveals that RS – or at least many of the most significant features of RS – can not revealed in a classical treatment, therefore extensive quantum mechanical concepts have to be employed. The fundamental physical principles underlying the classical theory of RS can be summarized as follows [1, 2]:

(i) The electromagnetic radiation induces in the system a time dependent dipole moment $\mathbf{P}(t)$. The oscillating dipoles re–emit secondary radiation or alternatively they "scatter" light.

(ii) The major source of light scattering in the visible and the near ultraviolet is the electronic cloud of the molecules and not the quasi–static nuclei.

(iii) RS is attributed to the coupling between the movement of nuclei and the movement of the electrons. Alternatively, the electrons "feel" an intramolecular field determined by the nuclei or the deformability of the electron cloud (*polarizability*) depends on the nuclear configuration at any time.

Considering that a sinusoidal electric field with instantaneous amplitude $\mathbf{E}(t) = \mathbf{E}_o \cos\omega_L t$ impinges on a molecule, the induced dipole moment is defined as $\mathbf{P}(t) = \overline{\alpha}\,\mathbf{E} + (1/2)\hat{\beta}\,\mathbf{E}^2 + ...$, where $\overline{\alpha}$ is the molecular polarizability and $\hat{\beta}$ the hyperpolarizability (responsible for the Hyper–Raman effect). $\overline{\alpha}$ is generally a second–rank tensor expressing the scattering anisotropy; namely, the components of $\mathbf{P}$ in some direction in general do not come from the corresponding ones of the electric field vector $\mathbf{E}$ in the same direction. $\overline{\alpha}$ can be written as a symmetric tensor:

$$\bar{\alpha} = \begin{bmatrix} \alpha_{xx} & \alpha_{xy} & \alpha_{xz} \\ \alpha_{yx} & \alpha_{yy} & \alpha_{yz} \\ \alpha_{zx} & \alpha_{zy} & \alpha_{zz} \end{bmatrix}$$

The polarizability tensor $\bar{\alpha}$ is often considered to be a combination of a symmetric and an asymmetric part. The symmetric part is described by the invariant, α

$$\alpha = \frac{1}{2}(\alpha_{xx} + \alpha_{yy} + \alpha_{zz}) = \frac{1}{3}\operatorname{Tr}\bar{\alpha} \qquad (2)$$

while the asymmetric part is described by the polarizability anisotropy, β

$$2\beta^2 = [(\alpha_{xx} - \alpha_{yy})^2 + (\alpha_{yy} - \alpha_{zz})^2 + (\alpha_{zz} - \alpha_{xx})^2 + 6(\alpha_{xy}^2 + \alpha_{yz}^2 + \alpha_{zx}^2)] \qquad (3)$$

A set of axes in the molecule can always be found such that, with these axes as the basis vectors, the polarizability tensor has only diagonal elements and the expression for $2\beta^2$ greatly simplifies. These axes define an ellipsoid and are called the principal axes of the polarizability. Along these axes **P** and **E** have the same direction. The polarizability ellipsoid will have the same symmetry as the charge distribution which will normally be the symmetry defined by the nuclear positions. It can be seen that a spherical top molecule (e.g., ZrCl$_4$(g)) has isotropic polarizability and β=0. The polarizability is anisotropic (i.e., β>0) when two or more axes are different.

Since $\bar{\alpha}$ depends on the distance between the nuclei it can be written as a function of a generalized vibrational coordinate that quantifies this dependence, then $\bar{\alpha} = \bar{\alpha}(q_i)$. An expansion of $\bar{\alpha}$ in a power series of the vibrational coordinate around its equilibrium value $\bar{\alpha}_0$ at $q_i = 0$ leads to the expression:

$$\bar{\alpha}(q_i) = \bar{\alpha}_0 + (\partial\bar{\alpha}/\partial q_i)_0 q_i + (1/2)(\partial^2\bar{\alpha}/\partial q_i^2)_0 q_i^2 + \ldots \qquad (4)$$

The linear term in q_i determines the *first-order* Raman effect while the quadratic one gives rise to the *second-order* Raman effect. Considering the first-order Raman scattering and assuming a sinusoidal change for q_i we arrive at:

$$\mathbf{P}(t) = \bar{\alpha}_0 \mathbf{E}_0 \cos\omega_L t + \frac{1}{2}\left(\frac{\partial\bar{\alpha}}{\partial q_i}\right)_0 \mathbf{E}_0 q_{i0} \cos[(\omega_L - \omega)t] + $$
$$+ \frac{1}{2}\left(\frac{\partial\bar{\alpha}}{\partial q_i}\right)_0 \mathbf{E}_0 q_{i0} \cos[(\omega_L + \omega)t] \qquad (5)$$

It is seen from the above equation that the induced dipole moment will fluctuate not only with the frequency ω_L of the incident light but also with the combination of the frequencies $\omega_L \pm \omega$ (Stokes Raman frequency $-\omega$, anti-Stokes Raman frequency $+\omega$). The latter frequencies are due to the modulation of the electronic polarizability $\overline{\alpha}$ by the vibrational motion of atoms, which is the prerequisite in order to observe the Raman effect. The classical treatment followed so far predicts correctly the occurrence of the inelastic processes but leads to incorrect ratio for their intensities, $I^{Stokes}/I^{anti-Stokes} = (\omega_L - \omega)^4/(\omega_L + \omega)^4$, which is less than unity contradicting the experimental value.

In a quantum–mechanical treatment Raman scattering is the result of an inelastic collision process between the photon and the elementary excitations of the medium. The photon either loses one or more quanta of vibrational energy (Stokes lines) or gains one or more such quanta (anti–Stokes lines). In *first–order* scattering only one phonon is involved; this corresponds to the term linear in the vibrational coordinate q_i in the expansion of $\overline{\alpha}(q_i)$. In *second–order* scattering two phonons are involved, corresponding to the term proportional to the term quadratic in q_i or to an anharmonic coupling of a phonon which is active in the first–order Raman scattering. The mechanism of the first order RS is schematically depicted in Fig. 3.

The energy and momentum conservation laws for the first and second order scattering processes are summarized as follows:

$$\omega_L = \omega_s, \quad q_L = q_s \quad \text{(Rayleigh scattering)} \tag{6a}$$

$$\left.\begin{array}{l} \omega_L = \omega_s \pm \omega \\ \mathbf{q}_L = \mathbf{q}_s \pm \mathbf{q} \end{array}\right\} \quad \begin{array}{l}\text{(Stokes and anti–Stokes)}\\ \text{Raman 1}^{st}\text{ order scattering)}\end{array} \tag{6b}$$

$$\left.\begin{array}{l} \omega_L = \omega_s \pm \omega \pm \omega' \\ \mathbf{q}_L = \mathbf{q}_s \pm \mathbf{q} \pm \mathbf{q}' \end{array}\right\} \quad \begin{array}{l}\text{(Stokes and anti–Stokes}\\ \text{Raman 2}^{nd}\text{ order scattering)}\end{array} \tag{6c}$$

The quantum mechanical treatment can be obtained by considering a model of the elementary perturbation process that involves a photon and a medium consisting of nuclei and electrons (i.e. a molecule, a crystal, an amorphous solid, a liquid, a melt...). If a photon of energy $\hbar\omega_L$ in the visible or UV region of the spectrum approaches the medium, it sets up a perturbation of the electronic wave functions in the medium, because only electrons are light enough to follow the fast–changing electric field of the photon. The wave functions of the perturbed system then acquire a mixed character and become linear combinations of all possible wave functions of the unperturbed medium with time–dependent coefficients. We may then regard the medium as having attained a non–stationary level of higher energy by considering the perturbation as belonging to the crystal. This description has nothing to do with the concept of energy levels that are used to describe the absorption processes. It has therefore been given the name *virtual state* in order to indicate that it is introduced only for a modellistic description of the perturbation process. In classical language a virtual level corresponds to a forced oscillation of the electrons with the frequency ω_L of the incident light. After the photon realizes that the system has no stationary sates of energy $\hbar\omega_L$, it leaves the unstable situation in which it was temporarily trapped. In analogy with the aforementioned we can consider that the photon is being emitted by the perturbed crystal, which jumps back to one of its stationary states, resulting thus in elastic or inelastic scattering.

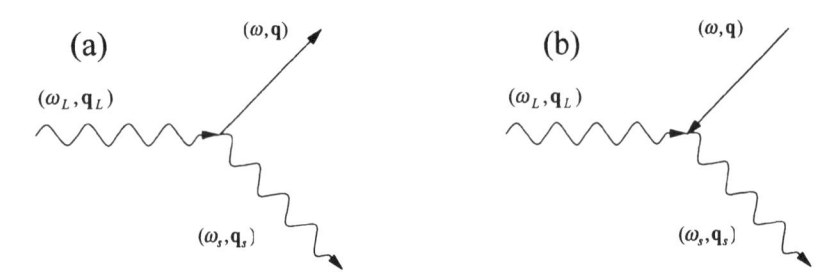

Figure 3. Scattering mechanisms in first–order RS: (a) anti–Stokes scattering, (b) Stokes scattering.

2.1.2. Isotropic and Anisotropic Scattering

The expression derived by Placzek [7] for the differential scattering cross section into a solid angle, Ω and frequency range [ω_s, ω_s+dω_s] about ω is related to the time dependent correlation function of the electronic polarizability:

$$\frac{d^2\sigma}{d\Omega d\omega_s} = \frac{\omega_s^{-4}}{2\pi} \int_{-\infty}^{\infty} dt\, e^{-i\omega_s t} <(\hat{\varepsilon}_{in} \cdot \overline{a}(0) \cdot \hat{\varepsilon}_{sc})(\hat{\varepsilon}_{in} \cdot \overline{a}(t) \cdot \hat{\varepsilon}_{sc})> = \omega_s^{-4} M(\omega) \qquad (7)$$

Where $\hat{\varepsilon}_{in}$, $\hat{\varepsilon}_{sc}$ are unit vectors defining the polarization directions of the incident and scattered light, $\overline{\alpha}$ is the polarizability tensor of the scattering medium and M(ω) is introduced for convenience and represents the integral function of the polarizability.

Experimentally and theoretically the scattering cross section may be divided into two parts, polarized and depolarized. Experimentally, the most common polarizations used in the study of molten salts and other isotropic liquids and glasses are those denoted as:

(i) VV: indicating that the polarization direction of both the incident and scattered radiation have the $\hat{\varepsilon}_{in}$ and $\hat{\varepsilon}_{sc}$ vertical to the scattering plane.

(ii) HV: indicating that relative to the scattering plane the unit vector $\hat{\varepsilon}_{in}$ is horizontal while the $\hat{\varepsilon}_{sc}$ is vertical.

The polarizability tensor $\bar{\alpha}$ can again be divided into a scalar part α and an anisotropic part $\bar{\beta}$

$$\bar{\alpha} = \alpha\,\hat{i} + \bar{\beta}$$

where: α has been defined in Eq. 2, $Tr\,\bar{\beta} = 0$ and $\hat{i}$ is the unit tensor. With these definitions the correlation function containing the $<\alpha(0)\,\alpha(t)>$ elements defines the isotropic scattering:

$$\omega_s^{-4}\,M_{ISO}(\omega) = \left[\frac{d^2\sigma}{d\Omega d\omega_s}\right]_{ISO} = \frac{\omega_s^{-4}}{2\pi}\int_{-\infty}^{\infty} dt\,e^{-i\omega_s t} <\alpha(0)\,\alpha(t)> \tag{8}$$

while the correlation function for the anisotropy of the polarizability is determined from

$$\omega_s^{-4}\,M_{ANISO}(\omega) = \left[\frac{d^2\sigma}{d\Omega d\omega_s}\right]_{ANISO} = \frac{\omega_s^{-4}}{2\pi}\int_{-\infty}^{\infty} dt\,e^{-i\omega_s t} <Tr\bar{\beta}(0)\,\bar{\beta}(t)> \tag{9}$$

The $M_{ISO}(\omega)$ and $M_{ANISO}(\omega)$ integrals depend respectively only on the diagonal and non-diagonal elements of the polarizability tensor. Both the VV and HV polarizations contribute to the ISO scattering cross section (Eq. 10) while the ANISO and HV scattering cross sections are proportional (Eq. 11).

$$\omega_s^{-4}M_{ISO}(\omega) = \left[\frac{d^2\sigma}{d\Omega d\omega_s}\right]_{ISO} = \left[\frac{d^2\sigma}{d\Omega d\omega_s}\right]_{(VV)} - \frac{4}{3}\left[\frac{d^2\sigma}{d\Omega d\omega_s}\right]_{(HV)} \tag{10}$$

$$\omega_s^{-4}\,M_{ANISO}(\omega) = \left[\frac{d^2\sigma}{d\Omega d\omega_s}\right]_{ANISO} = 1/10\left[\frac{d^2\sigma}{d\Omega d\omega_s}\right]_{(VV)} \tag{11}$$

Normally, one frequently sees Eqs. (10) and (11) written in terms of the intensities of the scattered light:

$$I_{ISO}(\omega) = I_{VV}(\omega) - \frac{4}{3}I_{HV}(\omega) \tag{12}$$

$$I_{ANISO}(\omega) = I_{HV}(\omega) \tag{13}$$

Where $I_\sigma(\omega)$ with (σ = ISO, ANISO, VV, HV) is the experimentally determined scattered light intensity which in general can be expressed as:

$$I_\sigma(\omega) = G^{expt}\, \omega\, (\omega_L \pm \omega)^{-4}\, M_\sigma\, B^{-1}(\omega) \qquad (14)$$

The first term G^{expt} is a constant characteristic of the experimental set-up used while the remaining terms are predicted theoretically. $B(\omega,T)$ is a temperature function related to the Boltzman population factor $n(\omega)$

$$B(\omega, T) = n(\omega) \qquad ; \qquad \text{anti-Stokes scattering} \qquad (15a)$$
$$B(\omega, T) = n(\omega) + 1 \qquad ; \qquad \text{Stokes scattering} \qquad (15b)$$

With

$$n(\omega) = [\exp(\hbar\omega/kT) - 1]^{-1} \qquad (16)$$

Finally, M_σ is defined above Eqs (7, 10, 11) and is the only function whose value depends on the polarizability characterizing the scattering medium.

If α and β are expanded in the vibrational normal coordinates of a molecule and only the linear terms are kept then:

$$\alpha \hat{i} = \alpha_0 \hat{i} + \sum_i \left[\frac{\partial \alpha_i}{\partial q_i}\right] q_i + ... \qquad (17a)$$

$$\beta = \beta_0 + \sum_i \left[\frac{\partial \beta}{\partial q_i}\right] q_i + ... \qquad (17b)$$

The α_0 and β_0 terms are the isotropic and anisotropic components of the molecular polarizability tensor evaluated at the equilibrium internuclear separation. They give rise to the polarized and depolarized Rayleigh scattering and rotational Raman scattering, which occur without change in the vibrational coordinates of the molecule. The terms dependent on the normal coordinates give rise to Raman scattering. Since $\partial\alpha/\partial q_i$ is independent of molecular orientation the $I_{ISO}(\omega)$ Raman spectrum is dependent only on vibrational motion; whereas, $\partial\beta/\partial q_i$ depends on molecular reorientation. Therefore I_{ANISO} depends on both vibrational and reorientational motions of the molecule. In addition to the familiar qualitative use of the *depolarization ratio*, $\rho = I_{HV}/I_{VV}$ to identify symmetric modes (i.e., $\rho < 0.75$) the quantitative studies of I_{ISO} and I_{ANISO} may provide detailed information about vibrational and reorientational relaxation.

2.1.3. Reduced Raman Spectra

Perhaps the most easily changing external thermodynamic parameter in a light scattering experiment from melts is temperature. Variations of temperature have proved quite useful in the elucidation of structure and often structural models are proposed based on the temperature dependence of the intensity, width, polarization, frequency shift etc. for particular vibrational lines. However, since Raman scattering is a purely quantum mechanical effect, it is expected that the elementary vibrational transitions taking place during the process depend upon the population of the involved energy levels. This contribution accounts for the B(ω,T) term in Eq. (14). It is therefore desirable to be able to disentangle the changes brought about to the vibrational lines by the temperature and by alterations of local species equilibria or modifications in structure. This is usually done by means of the so-called reduction procedures.

A reduced spectrum $R_\sigma(\omega)$ is calculated from the $I_\sigma(\omega)$:

$$R_\sigma(\omega) = (\omega_L \pm \omega)^{-4} \, \omega \cdot B(\omega,T) \, I_\sigma(\omega) \qquad (18)$$

where both a correction for the wavelength and the temperature dependence of the scattering are included. Due to its linear term in frequency, reduced spectra are particularly suppressed in the low frequency region being therefore not proper for analyzing low energy excitations. On the other hand, for spectra obtained at high temperature, as in the case of many molten salts, the reduced spectra are useful in comparing relative Raman intensities at frequencies below 1000 cm^{-1}. Having in mind the definition of σ, reduced intensities for different polarization configurations, namely R_{ISO}, R_{ANISO}, R_{VV} and R_{HV} can be calculated. Typical examples comprising the I_σ and R_σ spectral representation are given in Fig. 4. Finally, by comparing Eqs. 8, 9, 12 and 15, it should be noted that the reduced intensity, apart from an experimental constant, is only related to the polarizability function M_σ.

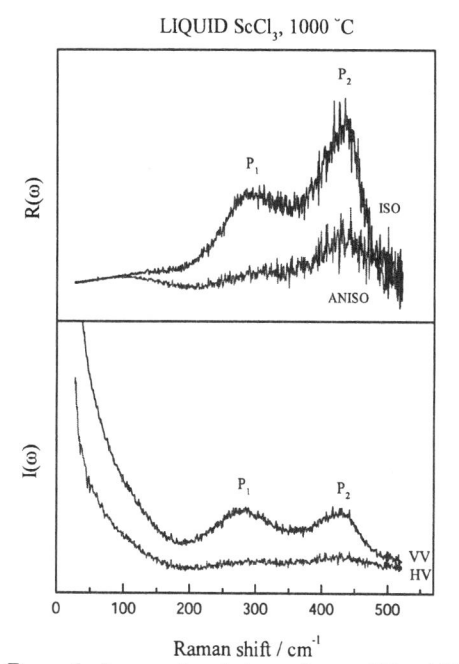

Figure 4. A comparison between the raw VV and HV spectra with the reduced ISO and ANISO spectra for molten ScCl₃ at 1000°C. Note the inversion of the relative intensities and the wave number shifts of the two main bands.

2.2. EXPERIMENTAL METHODS

A typical experimental set-up for measuring the Raman spectra from molten salts is shown schematically in Fig. 5. The scattering plane is that of the page. A right angle (θ=90°) scattering geometry is generally used but backscattering techniques (θ=180°) have been employed especially in cases of dark colored melts. Lines from visible Ar⁺

and Kr^+ ion lasers are usually the sources for the incident light. CW laser power ranges from a few mW (5-20 mW) up to a few W (1-3 W). The purpose of the focusing lens it to increase the laser power density at the scattering volume from where the scattered radiation is collected by a set of lenses. The aperture of the Collecting Lens determines the scattering collection angle Ω. Fiber optics as well as microscopies have been also used as CL systems from melts [8, 9]. The dispersive system are gratings in a single, double or triple monochromator. The detector system involves electronic amplification of signals obtained from a photomultiplier tube or a CCD detector. Quasi CW lasers, involving a chopper and lock-in amplifier as well as pulsed lasers and gated techniques have been also used [10]. Details for the currently used instrumentation can be found elsewhere [11].

It should be noted however that for a given melt with a specific scattering cross section and at fixed laser power there are two main factors improving the intensity of the Raman signal:

i. The spectrometer and optics transmission; this implies careful alignment and maching of all optical components, wide collection angle and stability of the overall system.
ii. The quantum efficiency of the detection systems; e.g. the use of high efficiency PMT/or intensified CCD detectors.

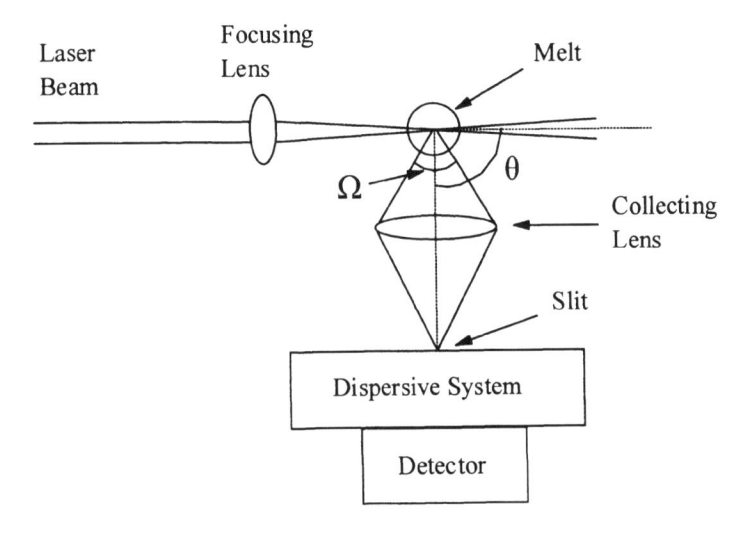

Figure 5. Schematic diagram of the experimental set-up used for measuring Raman spectra from molten salts.

High intensity Raman signals minimize the measurement time, which is a very important factor especial for the study of corrosive melts.

Sample preparation for Raman studies occasionally present serious difficulties. Even high purity inorganic salts available commercially may contain traces of organic impurities which upon melting the salt (e.g. NaCl) give a slight coloration (e.g. yellowish) and make the melt fluorescent. Oxide formation during drying process may also create problems. Filtration of fused salts through a sintered frit maybe desirable because the Tyndall scattering from solid particles usually increases the noise and background scattering will interferes with data at low Raman shifts. It is often used to treat the starting materials with activated charcoal in water to remove fluorescent and other organic impurities and then to recrystallize the salt from the aqueous solution. Zone refining by melt crystallization is also necessary for high melting corrosive melts like metal fluorides. For most salts however, the ideal means of purification is sublimation under vacuum. Fairly elaborate anhydrous preparation procedures are required for strongly acidic salts like the halides of aluminum, zirconium and zinc.

Two types of Raman optical cells have been used so far. Fused silica in the form of cylindrical tubes (2-10 mm ID) is the proper and simplest material for non-corrosive melts [12]. Windowless cells from graphite or noble metals have proved adequate for studying corrosive fluoride and/or oxide melts [13].

2.3. MOLTEN SALTS AS SCATTERING MEDIA

From a physicochemical point of view, molten salts are a class of liquids having many microscopic and macroscopic properties similar to the corresponding properties of other (molecular, atomic) liquids. In a simplified way, molten salts have been described as liquids composed of positive and negative ions that interact mainly via the strong (repulsive or attractive) long-range Coulomb potential. The experimental and theoretical evidence accumulated during the past 50 years, however, with the exception of a few cases, shows that molten salts exhibit individual or group peculiarities that complicate the understanding of their morphology.

The polarizability of the ions forming the melt gives rise to light scattering due to either electrostatic field fluctuation around the ion and/or to polarizability changes due to bond formation within the melt (i.e. formation of "complex" ions).

An isolated n-atom molecule has 3n degrees of freedom and 3n-6 vibrational degrees of freedom (or 3n-5 for a linear molecule). The collective motions of the atoms moving with the same frequency and in phase with all other atoms give rise to the normal modes of vibration. The determination of the form of the normal modes for any molecule requires, in principle, the solution of equation of motion appropriate to the n-atom system. Established methods for such analyses are available [14]. Molecular symmetry and methods of group theory are important in deriving the symmetry properties of the normal modes. With the aid of the character tables for point groups and the symmetry properties of the normal modes, the "selection rules" for Raman (R) and IR activity can be derived. For a molecule with a center of symmetry, e.g. LX_6, octahedral molecule none Raman active mode is also IR active, whereas for the MX_4 tetrahedral molecule some modes are simultaneously IR and Raman active. The vibrational properties of a group of isolated molecules change drastically when these molecules are "condensed" to form a crystalline solid. The influence of the neighboring "molecules' and the surrounding crystalline lattice will alter the vibrational modes of the molecules. The long-range order correlates the atoms in the crystal, and the vibrations are described in terms of lattice waves rather than by free molecular modes.

For a solid the lattice vibrations give rise to various "transversed" and "longitudinal" branches, where the IR and Raman activities are determined by the symmetry characteristics of the system and the space group of the crystal. Tables I gives two examples of the different vibrational modes and activities for the Cs$_2$NaLCl$_6$, L=Fe, Sc [12] and Cs$_2$FeCl$_4$ [8] solids. The number of normal modes is determined from the number of atoms in the Bravais lattice. Motions of ions or molecules related to each other within the unit cell give rise to the external (or lattice modes, which are subdivided into translational and librational (rotatory) modes. Translational vibrations arise from translational motions of the atoms in the unit cell with respect to each other (i.e., the cell remains stationary) and are different from the acoustical modes which are produced when the translational motions of the atoms in the unit cell all occur in the same direction (i.e., the whole unit cell moves in a specific direction).

When a molecule (or molecular ion) is introduced into the lattice, splitting of the degenerate internal (molecular) vibrations generally occurs; this is due to lower site symmetry and/or to interactions with internal vibrations of other molecules and ions within the same unit cell of the crystal. This is shown by the "correlation diagram" for the FeCl$_4^{2-}$ tetrahedral ion in the Cs$_2$FeCl$_4$ crystal. For the LCl$_6^{3-}$ octahedral ion in Cs$_2$NaLCl$_6$, however, no splitting occurs, since the molecular symmetry remains undistorted in the highly symmetrical space group of the crystal.

The Raman spectra [15] of Cs$_2$NaScCl$_6$ in Fig. 6 show, as predicted from Table I four bands, three of which are due to the internal modes of the ScCl$_6^{3-}$ octahedra

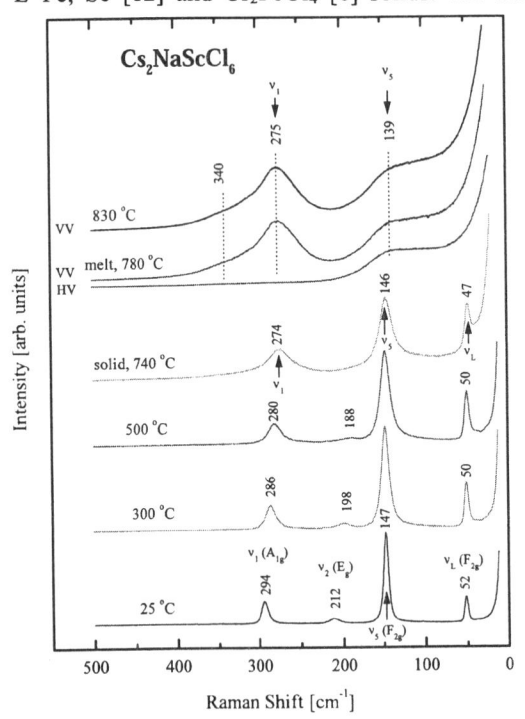

Figure 6. Temperature dependence of the Raman spectra Cs$_2$NaScCl$_6$ in the polycrystalline elpasolite (O_h^5) and the molten state.

Table I: Distribution of vibrational modes of solids with isolated octahedral and tetrahedral species

Octahedra in crystalline Cs_2NaLCl_6 (L= Sc, Fe)
Space group F_m3_m (O_h^5)
Four molecules per unit cell; one per Bravais lattice
3N = 30 normal modes

$\Gamma_{3N} = \Gamma_{acoustic} + \Gamma_{translatory} + \Gamma_{libration} + \Gamma_{internal}$
$\Gamma_{acoustic} = T_{1u}$ (IR)
$\Gamma_{translatory} = T_{2g}(R) + 2T_{1u}$ (IR)
$\Gamma_{libration} = T_{1g}$ (IA)
$\Gamma_{internal} = A_{1g}(R) + E_g(R) + T_{2g}(R) + 2T_{1u}$ (IR) $+ T_{2u}$(IA)

Correlation Diagram

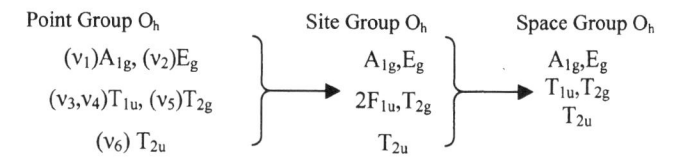

Point Group O_h	Site Group O_h	Space Group O_h
$(\nu_1)A_{1g}, (\nu_2)E_g$	A_{1g}, E_g	A_{1g}, E_g
$(\nu_3,\nu_4)T_{1u}, (\nu_5)T_{2g}$	$2F_{1u}, T_{2g}$	T_{1u}, T_{2g}
$(\nu_6) T_{2u}$	T_{2u}	T_{2u}

Tetrahedra in crystalline $CsFeCl_4$
Space Group P_{nma} (D_{2h}^{16})
Four molecules per unit cell; four per Bravais lattice
3N = 84 normal modes

$\Gamma_{3N} = \Gamma_{acoustic} + \Gamma_{translatory} + \Gamma_{libration} + \Gamma_{internal}$
$\Gamma_{acoustic} = B_{1u}(IR) + B_{2u}(IR) + B_{3u}(IR)$
$\Gamma_{translatory} = 6A_g(R) + 6B_{1g}(R) + 3B_{2g}(R) + 3B_{3g}(R) + 3A_u(IA) + 3B_{1u}(IR) + 6B_{2u}(IR) + 6B_{3u}(IR)$
$\Gamma_{libration} = A_g(R) + B_{1g}(R) + 2B_{2g}(R) + 2B_{3g}(R) + 2A_u(IA) + 2B_{1u}(IR) + B_{2u}(IR) + B_{3u}(IR)$
$\Gamma_{internal} = 6A_g(R) + 6B_{1g}(R) + 3B_{2g}(R) + 3B_{3g}(R) + 3A_u(IA) + 3B_{1u}(IR) + 6B_{2u}(IR) + 6B_{3u}(IR)$

Correlation Diagram

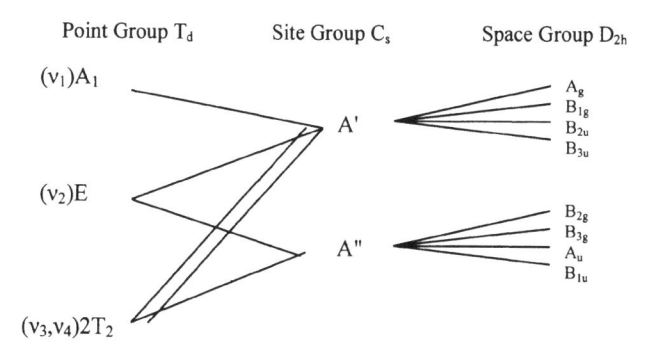

Point Group T_d Site Group C_s Space Group D_{2h}

$(\nu_1)A_1$ — A' — $A_g, B_{1g}, B_{2u}, B_{3u}$

$(\nu_2)E$

$(\nu_3,\nu_4)2T_2$ — A'' — $B_{2g}, B_{3g}, A_u, B_{1u}$

[$v_1(A_{1g})$ =294 cm^{-1}, $v_2(E_g)$=212cm^{-1} and $v_5(T_{2g})$=147 cm^{-1}] and one which is attributed to an external translatory mode [$v_{lattice}$ (T_{2g})=52 cm^{-1}]. For the dark colored iron compound Cs$_2$NaFeCl$_6$ only two octahedral modes [$v_1(A_{1g})$=290 cm^{-1} and $v_5(T_{2g})$ =165 cm^{-1}] can be seen in the spectra (Fig. 7). In the case of Cs$_2$FeCl$_{4(s)}$ (Fig. 7) not all the predicted Raman modes have been observed, but the four characteristic frequencies of the FeCl$_4^{2-}$ tetrahedral structure are the predominant modes at room temperature [$v_1(A)$=340 cm^{-1}, $v_2(E)$ ≈117 cm^{-1}, $v_3(F_2)$) ≈389 cm^{-1}, $v_4(F_2)$) ≈143 cm^{-1}]. Thus, the spectra for all compounds support the presence of discrete octahedral and tetrahedral units in Cs$_2$NaLCl$_6$ and Cs$_2$FeCl$_4$, respectively.

Upon melting, the long-range order and space symmetry of the solid are destroyed. In principle, however, the vibrational modes of the liquids can be considered as the long-wavelength limit of the solid vibrations, and thus certain internal and/or external modes may be present in the vibrational spectrum of the melt. The "internal" modes in melts have been investigated mainly by Raman spectroscopy in a variety of melt mixtures. The objective of these studies is the determination and characterization of possible discrete species (i.e., "complexes") in the melt.

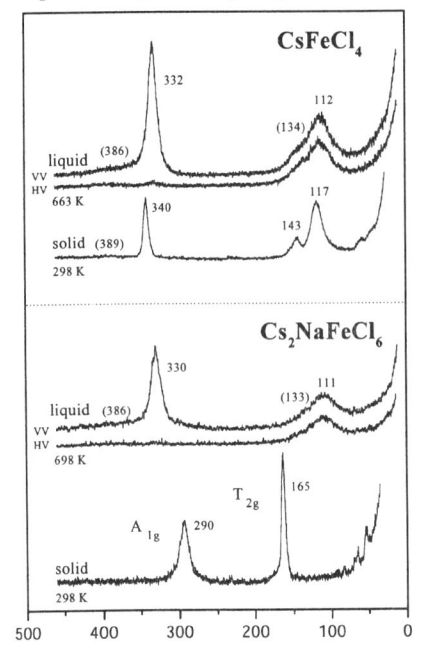

Figure 7. Raman spectra of polycrystalline CsFeCl$_4$ (D_{2h}^{16}) and Cs$_2$NaFeCl$_6$ (O_h^5) and of the corresponding melts.

Figure 6 shows that with increasing temperature the Raman bands of the solid Cs$_2$NaScCl$_6$ become broader and certain modes (e.g., the v_1 stretching mode) shift slowly to lower energies. Upon melting, the lattice modes disappear and the predominantly observed Raman bands are those attributed to octahedral (ScCl$_6^{3-}$) species.

Similar is the behavior of the CsFeCl$_4$ spectra (Fig. 7). The tetrahedral modes predominate the solid spectra and are preserved upon melting indicating the presence of the FeCl$_4^-$ "complex" ions. In contrast the melting of the Cs$_2$NaFeCl$_6$ is followed by drastic frequency shifts of the octahedral bands giving rise to a melt spectrum almost identical to that of molten CsFeCl$_4$. This suggest that an octahedral (FeCl$_6^{3-}$) to tetrahedral (FeCl$_4^-$) coordination change occurs upon melting.

It should be emphasized, however, that in a condensed phase such as the melt, the vibrational modes of the species cannot be treated as though they were isolated, that is, without accounting for the perturbations due to the environment. Thus, in certain common anion mixtures of the type NX_n-AX (N= polyvalent metal, A= alkali metal, X= halide) and at low concentrations of NX_n the forces within the species (e.g., $FeCl_4$, $ScCl_6^{3-}$) are stronger than the forces between the species and the neighboring ions, and it is then reasonable to consider the isolated species as complex ions for interpreting the spectra. On the other hand, when the interactions with the neighboring cations (e.g. A= Li or in mixtures rich in NX_n) are strong, the formation of discrete complex species will be drastically perturbed giving rise to other associated species and/or network-like structures.

Finally it is noteworthy that the vibrational modes measured by Raman spectroscopy may arise from short-lived local structures in the melt. If the lifetime of the structure is long enough (10^{-12} sec) so that there is time for vibration and interaction with the exciting light and if the structure has "bonds" (with a nonzero polarizability derivative), then Raman activity may arise. However, diffusion times in melts are of the order of the 10^{-11} sec, and thus the local structure may not maintain its identity for long before exchanging ions with its immediate environment. It appears from the above discussion that caution is required in interpreting Raman (and IR) spectra of melts; otherwise misleading information on the structural properties of the melts and mixtures may be obtained.

2.4. PROBING THE STRUCTURE OF MOLTEN HALIDE METAL HALIDES BY RAMAN SPECTROSCOPY

Among molten salts, metal halides have been extensively investigated by a variety of physicochemical methods. Thermodynamic and electrochemical properties of pure molten halides and their mixtures are easily available [16, 17]. Structural studies involving neutron electron and X-ray diffraction have been the subject of research papers and reviews [18, 19]. The vibrational spectroscopy of melts have been also reviewed in the past [20, 21]. The interest on these systems is twofold. First there are important industrial processes for metal production (e.g. Al, Mg) that utilize melt baths and melt electrolysis. An understanding of the physical properties and the chemistry taking place in these melts is required in order to improve the process. Second the variety of the metal halides available, from monovalent alkali halides to tetra and pentavalent havy metal halides, allows systematic investigation where factors like the charge, the polarizability and the size of both the anion and the cation can be changed and their effect on a specific property can be investigated. Thus, a better understanding of molten state can in general be obtained.

Most metal chlorides and bromides have melting points below 1000 °C and experimentally can be handled in fused silica containers. A large number of iodides behave similarly. However fluorides are very corrosive and can be handled only in graphite and noble metal containers. The degree of hygroscopicity of metal halides varies from salt to salt but for Raman spectroscopic measurements the secure way to handle halide salts is in vacuum sealed containers and/or glove boxes with low water and oxygen content (< 1 ppm).

The following subsections outline the Raman spectroscopic studies of four groups of metal halides involving cations with changes from one to four. The melt structures implemented by these studies will be presented and discussed.

2.4.1. Alkali Halides

Light scattering from molten alkali halides (AX: A=Li, Na, K, Rb, Cs; X= F, Cl, Br, I) shows an elastic (Rayleigh) component plus a relatively strong intensity inelastic (Raman) contribution which overlaps with the Rayleigh wing and decays almost exponentially, spreading to a wide spectral range covering several hundreds of wave number [22, 23]. A series of experimental measurements in single AX salts has been reported and their results have been discussed and interpreted in terms of dipole-induced-dipole and short-range order interactions between the ions in the melt. It is evident from these studies that the intensity of the scattered light increases with increasing polarizability of both the anion and the cation. For the LiX melts containing the high field and low polarizability Li^+ cation, the scattering mainly arises from the fluctuation anion polarizability while for the CsX melts both the high polarizable Cs^+ cation and the X^- anion contribute to the scattering intensity [22-26]. Theoretical and molecular dynamics simulation studies of the light scattering spectra of molten alkali halides have been reported [27]. The mechanism for light scattering from these melts is related to the fluctuations of the polarizability tensor Eqs. (10-13) expressed in terms of the ANISO and ISO contributions:

$$M_{ISO} = M_{ISO}^{SR} + M_{ISO}^{\gamma} \qquad (19)$$

$$M_{ANISO} = M_{ANISO}^{DID} + M_{ANISO}^{B} + M_{ANISO}^{SR} + M_{ANISO}^{\gamma} \qquad (20)$$

The different terms are: M^{DID} refers to the dipole-induced-dipole (DID) mechanism; M^{SR} a short-range term which describes the effect of near-neighbor interactions on the ionic charge clouds; M^B and M^γ give the changes in polarizability induced on a given ion by the Coulomb field gradient and the Coulomb field, respectively. A combination of the four processes contributes to the light scattering spectra but due to cross correlations there is a degree of mutual cancellations [27]. In certain cases these cancellations increase the uncertainties since the overall intensity is calculated as a difference between large intensities arising from the different scattering mechanisms.

As an example of light scattering from single alkali metal halides we present the case of molten fluorides. The $I_{ISO}(\omega)$ and $I_{ANISO}(\omega)$ spectra of CsF, KF and LiF melts are shown in Fig. 8. In the low frequency region ($\omega < 30$ cm^{-1}) the scattered Raman intensity is very sensitive to the thermal population factor Eq. (18) and involves uncertainties associated with the reproducibility of the light scattering geometry and the purity of the samples, thus its shape cannot be determined accurately. Furthermore, in the same frequency region the high intensity of the scattered light tends to saturate the CCD detector response. For these reasons our results and discussion will be concerned with the intermediate and high frequency region of the spectra (30 < ω/cm^{-1} < 800). The spectra in the reduced representation $R_{ISO}(\omega)$ and $R_{ANISO}(\omega)$ (see also Fig. 8) suppress the low frequency region, force R(ω) towards zero and assist in defining the intermediate frequency bands. In contrast, in the high frequency region where I(ω) reaches the baseline the R(ω) is characterized by a low signal to noise ratio.

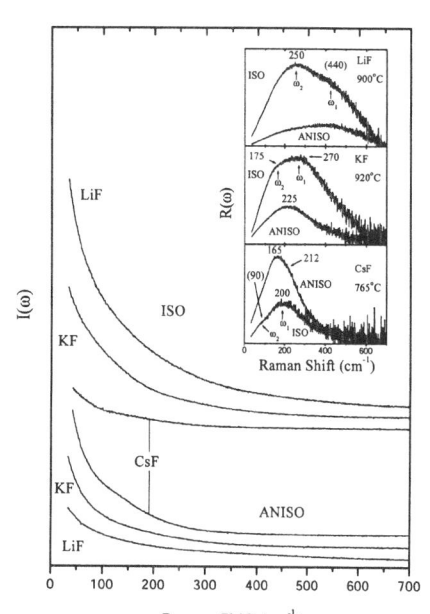

Figure 8. Raman scattering intensities for the ISO and ANISO (HV) configurations of molten LiF, KF and CsF at 900, 920 and 765 °C respectively. Inset: ISO and ANISO reduce Raman spectra for the same three melts.

The contributions of the different terms of Eq. (20) to the anisotropic scattering intensity have also been calculated [27]. Due to the small polarizabilities of both Li$^+$ (α_{Li^+} = 0.03 Å^3) and F$^-$ (α_{F^-} = 0.7 Å^3) the DID mechanism does not contribute much to the intensity. The Π^{SR} and Π^B terms are shown to give the largest contributions, but cancellation effects due to SR/B cross correlation drastically lower the overall intensity. A comparison between the theoretical and experimental anisotropic intensity for LiF is given in Fig. 9. It seems that the variation (slope) of log I with wavenumber is predicted relatively accurately but the theoretical overall relative to the anisotropic intensity is overestimated. The differences are emphasized in the insert of Fig. 9 where the experimental and theoretical depolarization ratios are compared. The experimental ρ(ω) is essentially constant and close to 0.15 which indicates that $I_{ANISO}(\omega)$ is approximately proportional to $I_{ISO}(\omega)$. Since the latter is dominated by short-range interactions, it appears that the anisotropic light scattering intensity also probes short-range interactions. For all three melts the maxima in the reduced spectra (Fig. 8) are different in the isotropic and anisotropic representations. The broad $R_{ISO}(\omega)$ spectra show two bands marked as ω_1 and ω_2 while only one band can be recognized in the $R_{ANISO}(\omega)$. As in cases of, LiCl and CsCl [24] the ω_1 band is associated with the short range cation-anion overlap and the Coulomb field interactions concerning the polarizability fluctuations of the fluoride anion, i.e. the M_{ISO}^{SR} and M_{ISO}^{γ} terms in Eq. (19). The rather uniform and symmetric field created around the F$^-$ by the neighboring cations gives rise to breathing-like fluctuation of the anion polarizability and enhances the isotropic scattering intensity. The A-F (A=Li, K, Cs) short range overlap is strengthened with increasing polarizing power ($1/r^+$) of the cation and thus, the contribution of the Π^{SR}_{ISO} term is expected to enhance the $R_{ISO}(\omega)$ intensity of the LiF melt relative to the other two salts. Such a tendency is evident in the reduced spectra where the $R_{ISO}(\omega)$ intensity diminishes relative to $R_{ANISO}(\omega)$ in the sequence Li>K>Cs. It is also noteworthy that the order of R is $R_{ISO}(\omega)$>$R_{ANISO}(\omega)$ for LiF and KF while for CsF the inequality is reversed.

It has been proposed that the second band (ω_2) is due to either polarizability fluctuation of the X⁻ arising from neighboring anions or/and to possible separation of the Π^{SR}_{ISO} and Π^{γ}_{ISO} fluctuation mechanisms contributing to light scattering with different energies [24, 25]. The trend of the relative intensities ω_2 to ω_1 as we go from LiF to CsF (Fig. 8) supports to a degree a fluctuation mechanism due to F-F neighbor anions. This becomes evident if we examine the local structure around the F⁻ anion in the three salts having the A⁺ cations as nearest neighbors and the F⁻ anions as second nearest neighbors. A close packing of the ionic "spheres" is assumed with a 4-6 coordination number for each ion. In the LiF the small cation radius allows the F⁻ anions to be almost in contact while for CsF the F⁻ anions are well separated by the large Cs⁺ cations. Thus, the anion polarizability fluctuation due to F⁻-F⁻ interactions should be more pronounced in the LiF than the CsF melts while the KF should stand in the middle. On going form LiF to CsF, the F⁻-F⁻ distance increases and the interaction between neighboring anions decrease affecting both the energy and the intensity of the polarizability fluctuations; i.e. the ω_2 band is expected to decrease its intensity and to shift to lower energies in the same sequence; this is in accordance with the experimental trends shown in Fig. 8.

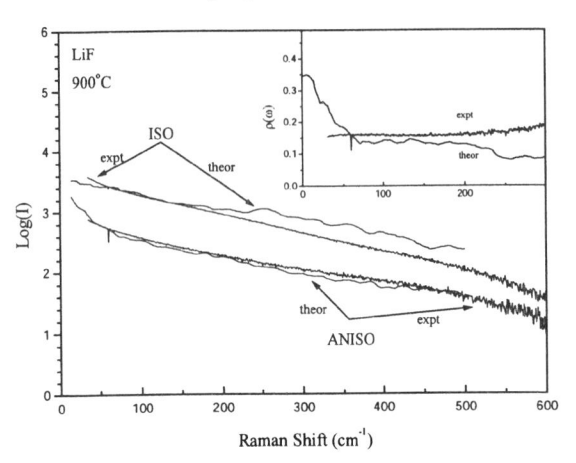

Figure 9. Comparison of the logarithmic experimental relative intensities with the simulated spectra of LiF. Inset: experimental and simulated depolarization ratios of LiF.

The anisotropic spectra [$R_{ANISO}(\omega)$] exhibit only one broad band with maxima close to the Debye frequency of the corresponding solid. All four mechanisms implied by Eq. (20) should contribute to the anisotropic scattering intensity. However, due to the small polarizability of the Li⁺ the anion-cation DID contribution (the M^{DID}_{ANISO} term) is expected to be weak for the LiF melt. In contrast the high polarizability of the Cs⁺

makes the anion-cation DID interactions rather important for this CsF melt and this is reflected in the spectra where the $R_{ANISO}(\omega)$ intensity is higher than that of $R_{ISO}(\omega)$. For all three salts the maxima of the $R_{ANISO}(\omega)$ are blue shifted relative to the ω_2 isotropic band indicating that the origin of these bands is different. Thus, we may exclude contributions to the anisotropic spectra from the M^{SR} and M^{γ} terms which appear at different energies in the isotropic spectra. This suggests that the M^{B}_{ANISO} term has a predominant contribution to the anisotropic scattering for LiF while for the KF and CsF salts both the M^{B}_{ANISO} and M^{DID}_{ANISO} may contribute to the scattering process. For the LiF melt having the smallest non-polarizable cation a high Coulomb field gradient is expected, i.e., the main contribution arises from the M^{B}_{ANISO} term, while for the CsF melt having the large higly polarizable cation the spectra are dominated by the M^{DID}_{ANISO} term. The size and polarizability of K^+ lie between those of Li^+ and Cs^+, thus both the M^{DID}_{ANISO} and M^{B}_{ANISO} terms are expected to contribute to the spectra of KF.

The Raman spectra of mixtures of alkali metal halides have been also reported and detail analysis in given elsewhere [24,25]. The studies concern the LiCl-CsCl,LiF-KF and LiF-CsF mixtures and the main findings can be summarized as follows:

(i) The scattering intensity from the LiX-AX (X=Cl, F;A=K,Cs) mixtures arises from contributions of the two cations which appear to occupy individual sites (cages) each possessing a characteristic frequency ω_{Li} and ω_A. The observed intense scattering in the ω_{Li} region and the frequency shifts on going from the pure components into the mixture, are mainly due to changes of short-range interactions and the loss of "symmetry" around the anion imposed by the local structure in the mixture. In contrast the observed intensity in the ω_{Cs} region and the associated with the formation of highly polarizable configurations (clusters) which have a relatively long lifetime at low temperatures.

(ii) The drastic increase of the isotropic scattering intensity with increasing temperature is associated mainly with the Li-X interactions and the "symmetry" of the local structures. Thus, increasing temperature strengthens the short-range overlap interactions and increases the local symmetry around the anion; both effects facilitate breathing like fluctuations of the polarizability and lead to increasing isotropic scattering intensity.

These findings yield a consistent picture of the systems investigated and help in separating the different interaction induced polarizability mechanisms. Furthermore, the analysis points out the important role of the local structure symmetry around the anion to the variation of the isotropic scattering intensity with composition and temperature.

2.4.2. Divalent Systems

Divalent metal halides and their mixtures with alkali metal halides have been among the first molten salt systems investigated by Raman spectroscopy [20, 21] Studies of molten mixtures of the type MX_2-AX (X= Cl (mainly), Br, I) have provided a means of identifying and characterizing the species that may exist in the mixture and have information regarding the liquid structure of the MX$_2$ component.

Binary mixtures involving MgX$_2$, CdX$_2$ and MCl$_2$ (M= first row transition metal) which are known to form octahedral layered structures in the solid state, are stabilized in the alkali halide rich melts by the formation of MX$_4^{2-}$ tetrahedra. For the same MX$_2$-AX systems the four-fold coordination appears to prevail the liquid structure in MX$_2$ rich mixtures including the structure of the pure divalent halide. Even for NiCl$_2$, which is known for its high-octahedral ligand field-stabilization energy, it has been found by both thermodynamic and neutron diffraction measurements that the four fold coordination of nickel predominates the structure of the NiCl$_2$-ACl melts at all compositions including pure NiCl$_2$ [28]. The situation is similar for the glass forming ZnX$_2$ (X= Cl, Br) melts where the four-fold coordinated ZnX$_4^{2-}$ species are stabilized in alkali halide rich mixtures, while a network like structure of ZnX$_4$ tetrahedra bridged mainly by edges characterize the pure ZnX$_2$ melts [29]. Finally, the melting of HgX$_2$ yields a molecular liquid involving X-Hg-X triatomic molecules, while in mixtures with AX both tetrahedral HgX$_4^{2-}$ and trigonal HgX$_3^-$ species have been identified [30]. One of the first fluoride molten system studies was the BeF$_2$-AF (A= Li, Na). The four-fold coordination predominates the melt structure with isolated BeF$_4^{2-}$ species formed in mixtures rich in AF. With increasing BeF$_2$ content bridging of the tetrahedral occurs and species like Be$_2$F$_7^{3-}$ and Be$_3$F$_{10}^{4-}$ were argued to exist.

In the following two recent examples of Raman studies for the MX$_2$-AX are presented (i.e. ZnF$_2$-AF (A= K, Cs) and BeCl$_2$-CsCl). It is shown that the systematic spectral changes with melt composition and temperature contribute to the understanding of the pure MX$_2$ structure.

2.4.2.1. Octahedral-tetrahedral species in ZnF$_2$ melts

Zinc fluoride is one of the main components in a series of fluoride multicomponent systems which are used for the production of new infrared transmitting glasses. Thus, the structural entities formed in the liquid are of particular interest to the understanding of the vibrational spectra and the structure of fluoride glasses.

Raman spectra of liquid ZnF$_2$-AF (A= K, Cs) mixtures have been measured at different ZnF$_2$ compositions and at temperatures up to 950°C. The spectral changes upon melting of polycrystalline K$_2$ZnF$_4$ were also measured. The K$_2$ZnF$_4$ crystal is tetragonal (D_{4h}^{17}/ I4/mmm) having the Zn cations in crystal sites of an octahedral symmetry. The "ZnF$_6$" octahedra form a two-dimensional corner sharing array in a layer structure. Factor group analysis of the vibrational modes of the K$_2$ZnF$_4$ crystal shows that only four Raman active modes are expected: Γ(Raman) = 2 A$_{1g}$ + 2 E$_g$

In Fig. 10 we present the Raman spectra of K$_2$ZnF$_4$ at 25 and 850 °C. In the room temperature spectrum three of the four expected Raman active modes were observed. In the liquid the Raman spectra are characterized by a strong polarized band at 471 cm^{-1}

and a broad depolarized band at about 170 cm^{-1}. If the coordination number of Zn^{2+} cation in the liquid was similar to that of the solid then a rather small "red" shift of the 356 cm^{-1} band is expected as we go into the melt. In contrast the relative large (~140 cm^{-1}) "blue" shift observed in the spectra indicates that the coordination number of Zn^{2+} in the melt is lower than six.

The composition dependence of the Raman spectra of ZnF$_2$-KF is shown in Fig. 11. The following general observations can be made.

(i) In the composition range $0 < X_{ZnF_2} \leq 0.33$ the spectra are characterized by a broad polarized band centered at 470 cm^{-1} (PT$_1$) and a broad depolarized at (DT$_1$) ~170 cm^{-1}. The relative intensities of these bands did not change with temperature variations.

(ii) In the region $0.33 < X_{ZnF_2} \leq 0.9$ the spectra are characterized by a polarized Rayleigh wing which probably overcomes one or more depolarized bands and by two polarized bands one at ~474 cm^{-1} (OT$_1$) and another at ~350 cm^{-1} (O$_1$). The O$_1$ band exhibits a small red

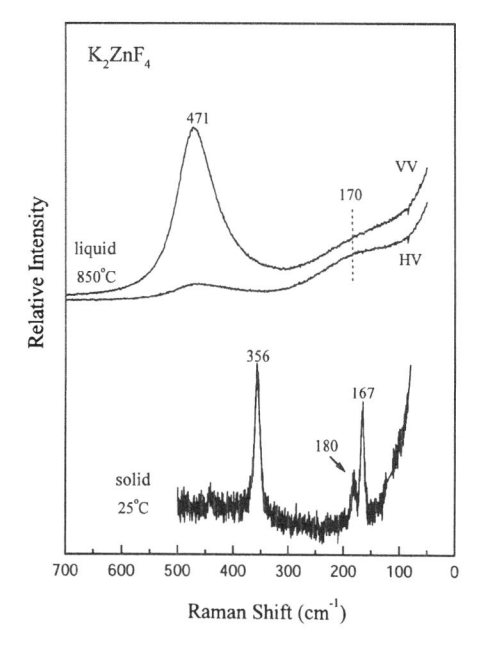

Figure 10. Raman spectra of K$_2$ZnF$_4$ solid (25°C) and melt (850°C).

shift with increasing ZnF$_2$ content while the OT$_1$ frequency position seems to be invariant to composition changes.

(iii) For 0.66 ZnF$_2$ - 0.33 CsF liquid mixture the relative intensities of the two polarized bands are affected by temperature variations [32]. Thus, with increasing temperature the intensity of OT$_1$ band increases relative to that of the O$_1$ band, while the frequencies at both bands remain constant. Such a variation of the relative intensities with temperature characterizes melt mixtures where two or more different species are in

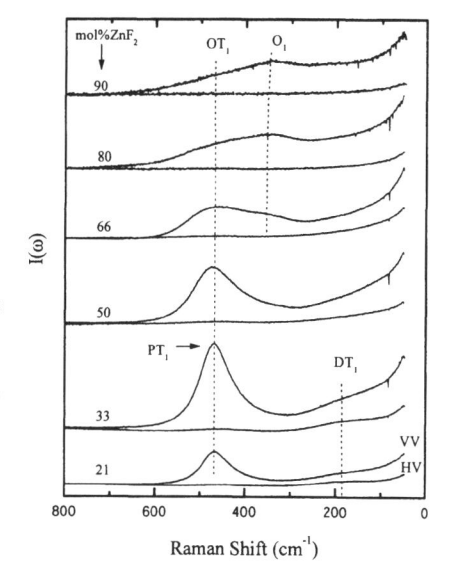

Figure 11. Composition dependence of the Raman spectra of molten ZnF$_2$-KF mixtures.

equilibrium.

Observation (i) implies that in melt mixtures rich in alkali fluoride, rather stable $ZnF_x^{(x-2)}$ species are present. The previous discussion regarding the spectra changes upon melting K_2ZnF_4 implies that x<6. Furthermore, semiquantitative spectra measurements indicate that the PT_1 band intensity increases with increasing ZnF_2 mole fraction. The increase is definitely observed up to 33 mol % ZnF_2, but within experimental error an intensity increase up to 50 mol % ZnF_2 cannot be excluded. Thus, the complex ions with stoichiometries either ZnF_4^{2-} or/and ZnF_3^{1-} are the predominant species in these melts. The appearance of stable tetrahedral ZnX_4^{2-} (X=Cl, Br) in liquid mixtures of ZnX_2 with alkali halides is well established [29] and support the presence of ZnF_4^{2-} tetrahedra in the fluoride systems. It is noteworthy that for the $HgCl_2$-ACl melt mixtures both the $HgCl_4^{2-}$ and $HgCl_3^-$ complexes are known to exist [30]. Since the ionic radii ratios are $Zn^{+2}/F^- \approx Hg^{+2}/Cl^- \approx 0.56$ then we may conclude that the structural behavior in the alkali halide rich region of the ZnF_2-KF and the $HgCl_2$-KCl melts may be similar. The tetrahedral ZnF_4^{2-} are the predominant species up to 33 mol % ZnF_2 while at higher mole fractions the ZnF_3^{1-} configurations may be also present.

From observations (ii) and (iii) and the above discussion it appears that in the mole fraction region $0.5 < X_{ZnF_2} < 1$ a new structural entity may be present in the melt mixtures. The O_1 band at ~350 cm^{-1} is assigned to this entity which presumably in the predominant "species" in compositions rich in ZnF_2. It has been argued [32] that the O_1 band can be associated with a fluoride configuration where the Zn^{2+} is six-fold coordinated. Due to the broadness of this main polarized band a wide range of distortions from octahedral symmetry is presumable present in melt mixtures rich in ZnF_2 and in pure molten ZnF_2.

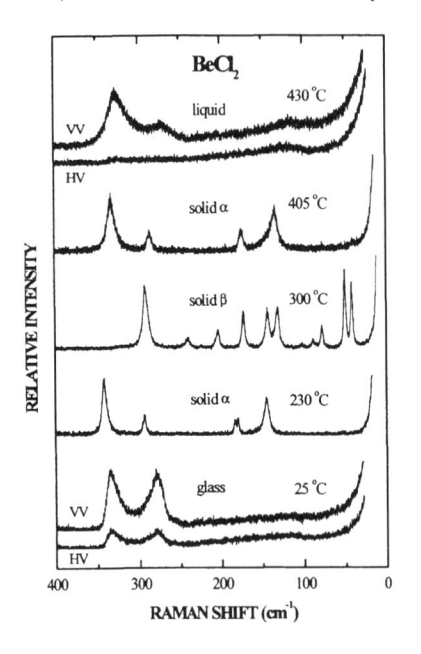

Figure 12. Raman spectra upon heating vitreous $BeCl_2$.

2.4.2.2. Tetrahedral polyions, chains and clusters in BeCl₂ melts

Phase transitions of BeCl₂. Pure beryllium chloride exists in three allotropic crystalline modifications and also forms a glass. Representative Raman spectra obtained by heating the glass are demonstrated in Fig. 12; only the frequency region below 400 cm^{-1} is presented, where all the intense bands are present. Increasing temperature devitrifies the glass at ~200 °C, resulting to the formation of the crystalline polymorph α. The α-form is transformed to the β-form at ~250 °C, which in turn is converted at ~405 °C again into the α modification. Finally, melting of this crystalline polymorph occurs at ~415 °C, giving a transparent, viscous liquid. Rapid or slow, cooling of the melt leads to the crystallization of BeCl$_2$ in the α modification. It has been also observed that the α-form is totally converted to the stable β-form after storage of the sample for a few months. Summarizing the above spectral data, we have concluded that the following phase transitions take place:

$$\text{glassy BeCl}_2 \xrightarrow[\text{cryst.}]{\sim 200\,°C} \alpha\text{-BeCl}_2 \xrightarrow{\sim 250\,°C} \beta\text{-BeCl}_2 \xrightarrow{\sim 405\,°C} \alpha\text{-BeCl}_2 \xrightarrow{\sim 415\,°C} \text{melt} \quad (21)$$

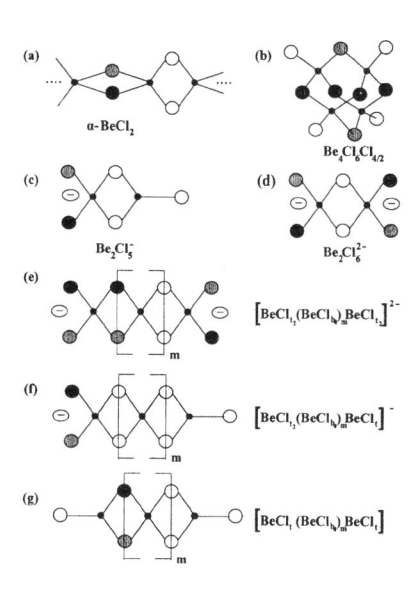

Figure 13. Structural models proposed for crystalline, vitreous and liquid BeCl$_2$ and for molten mixtures BeCl$_2$-CsCl (a) Linear chains of edge-sharing tetrahedral in solid form α-BeCl$_2$ (b) "supertetrahedra" cage like units Be$_4$Cl$_6$-Cl$_{4/2}$ in solid β-BeCl$_2$ (c) Isolated Be$_2$Cl$_5^-$ species in mixtures rich in CsCl (d) Isolated edge-bridged tetrahedra Be$_2$Cl$_6^{2-}$ species in mixtures rich in CsCl (e) Edge-bridged chain of BeCl$_4$ tetrahedra with both charged BeCl$_4$ ends in mixtures rich in BeCl$_2$ (f) Edge-bridged chain of BeCl$_4$ tetrahedra with one charged BeCl$_4$ end and the other terminated by neutral BeCl$_3$ unit in mixtures rich in BeCl$_2$ (g) Edge-bridged chain of BeCl$_4$ tetrahedra terminated by both neutral BeCl$_3$ units in pure liquid BeCl$_2$

The structure of α-BeCl$_2$ is orthorhombic [D_{2h}^{26} (Ibam)] and is composed of parallel linear chains of distorted edge-sharing tetrahedral units, as sketched in Fig. 13a. It is noteworthy, that such a "polymeric" structure is unique among metal halides MX$_2$, and isotypical to the chalcogenides SiS$_2$ and SiSe$_2$. The Raman spectra of polycrystalline α-BeCl$_2$ were measured in two temperature intervals. The first from LN$_2$ temperature to ~250 °C where the α → β phase transition takes place and the second from ~405 °C where the inverse β → α transition occurs up to the melting temperature (~415 °C). In both temperature ranges the spectra did not change; an expected band broadening was observed and all bands exhibited a red shift with increasing temperature.

A more recent x-ray diffraction study of the β-BeCl$_2$ [33] establishes a tetragonal structure (I4$_1$/acd). The major feature of the structure is the existence of "supper tetrahedra" cage like units [Be$_4$Cl$_6$Cl$_{4/2}$] as shown in Fig. 13b. Each Be atom is tetrahedrally coordinated to 4 Cl atoms and the BeCl$_4$ "tetrahedra" are linked by vertices while the arrangements of the Cl atoms around the Be is rather asymmetric.

Overall the differences between the two allotropic forms α- and β-BeCl$_2$ appear to arise from the different ways of linking between the BeCl₄ tetrahedra. Chains of edge bridged BeCl₄ form the low temperature metastable orthorhombic modification, which is transformed at intermediate temperatures to the tetragonal modification by connecting units of vertex sharing BeCl₄ tetrahedra. The inverse transformation of vertex-sharing to edge-bridged tetrahedra leads to the high temperature orthorhombic form.

The 340 cm⁻¹ band of the orthorhombic α-BeCl₂ corresponds to the breathing mode of the edge-bridged BeCl₄, while the corresponding mode of the vertex-bridged tetrahedra [34] in the tetragonal β-BeCl₂ form is at 300 cm⁻¹. These two bands denoted as v_1^α (=340 cm⁻¹) and v_1^β (=300 cm⁻¹) have the strongest Raman band intensity in the orthorhombic and tetragonal crystal respectively. Thus, the α → β transformation leads to a 40 cm⁻¹ downshift of the Be-Cl frequency which is associated with the edge- to vertex-bridging changes of the tetrahedra.

The Raman spectra of liquid and glassy BeCl₂ are rather similar (Fig. 12), characterized by two strong polarized bands between 250-350 cm⁻¹ and a broad depolarized band in the low frequency region. Two other weak bands were also measured above 400 cm⁻¹. The relative intensities of the two strong polarized bands change on going from the glass to the liquid; in the latter the observed change continuous in the same direction with increasing temperature. The close correspondence of all the bands observed in the glass and the liquid infers similarity in the structure of the two phases. However, there is no direct correlation of the glass/liquid spectra to those of the allotropic crystalline phases. The spectra of the α-form is closer to the spectra of the ~430 °C liquid, while a combination of the most intense v_1^α and v_1^β bands of the two solids may be considered to contribute to both the glass and the liquid spectra. This would imply a structure composed of a mixture of BeCl₄ tetrahedra linked both by edges (α-form) and by vertexes (β-form).

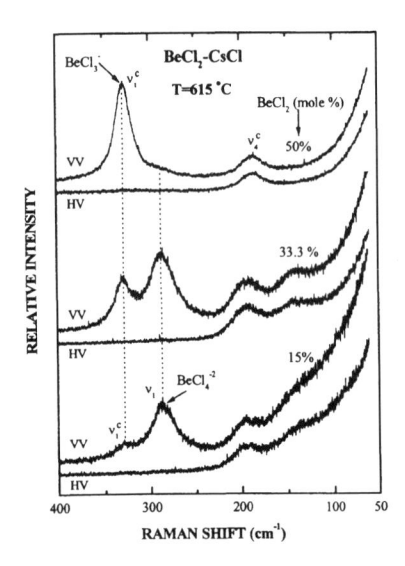

Figure 14. Raman spectra of molten BeCl₂-CsCl binary mixtures rich in CsCl

The CsCl-BeCl₂ melt mixtures. The systematics of the Raman spectra with BeCl₂ mole fractions up to 50 is shown in Fig. 14. At X_{BeCl_2} =0.15 the spectra are dominated by the vibrational modes of the $BeCl_4^{2-}$ tetrahedra plus the v_1^c band. At X_{BeCl_2} =0.5 the main characteristics of the spectra are a strong polarized and a depolarized band at ~330 and ~185 cm⁻¹ respectively. Based on the polarization, the relative positions and intensities, these bands are assigned to the v_1^c (A') and v_4^c (E') modes of a planar (D$_{3h}$) $BeCl_3^-$ species.

As seen from the spectra, the $BeCl_3^-$ is present along with the $BeCl_4^{2-}$ species in all melts with composition $X_{BeCl_2} \leq 0.5$ (Fig. 14). The temperature dependence of the spectra at compositions $X_{BeCl_2} = 0.5$ and $X_{BeCl_2} = 0.33$, as well as, the spectral changes with composition variation (Fig. 14) suggest [34] that these two predominant species participate in an equilibrium of the type:

$$BeCl_4^{2-} \rightleftarrows BeCl_3^- + Cl^- \qquad (22)$$

In melts dilute in BeCl$_2$ and at low temperatures the equilibrium shifts to the left, while in high temperature melts with composition near $X_{BeCl_2} = 0.5$ the $BeCl_3^-$ is the predominant species.

It should be pointed out that recent molecular dynamic simulations of the BeCl$_2$-KCl mixtures using the polarizable ionic potential [35] support the existence of the above equilibrium (22).

Fig. 15 shows the composition dependence of the spectra for mixtures with $X_{BeCl_2} = $ 0.6, 0.66, 0.75 and 1.0. The main features of the spectra are an extended polarized band in the frequency range 250-350 cm^{-1} - presumably covering more than one modes and a weak depolarized band centered below 200 cm^{-1}.

A detail analysis [34] gives an account of the observed spectral changes that may involve the following species and equilibria:
(i) for mixtures with X_{BeCl_2} between 0.5 and 0.66 the predominant species are the $BeCl_3^-$ in equilibrium with $Be_4Cl_{10}^{2-}$ in the "cluster" and "chain" forms [Fig. 13] and possibly with $Be_2Cl_5^-$. The concentration of $Be_2Cl_5^-$ and the "cluster" species decrease rapidly with increasing temperature and BeCl$_2$ mole fraction.

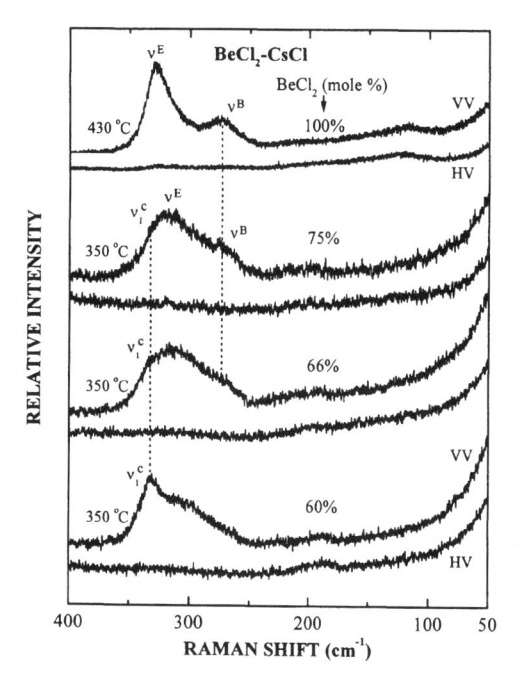

Figure 15. Raman spectra of molten BeCl$_2$-CsCl binary mixtures rich in BeCl$_2$ ($X_{BeCl_2} > 0.5$).

(ii) at $X_{BeCl_2} > 0.66$ equilibria are established between the edge-bridged chain (Fig. 13) and the vertex-bridged "cluster" like species. In both types of species the BeCl$_4$ tetrahedra are the major and principal units for building the structures, while the BeCl$_3$ terminating group of the "chains" or the "clusters" participates as a minor component.

The structure of liquid and glassy BeCl$_2$. The melting of α-BeCl$_2$ at 415°C is followed by a ~25% molar volume increase giving a rather viscous liquid with a low specific conductivity of ~0.01 ohm^{-1} cm^{-1}. These characteristics suggest that BeCl$_2$ has properties similar to those of glass forming inorganic liquids like ZnCl$_2$ and As$_2$O$_3$. Cooling of the melt yields the α-BeCl$_2$ crystals and not the glass. In contrast, the bulk glass is formed only through vapor transport, a situation which has been also observed for As$_2$O$_3$. It is noteworthy that both BeCl$_2$ and As$_2$O$_3$ form allotropic solid modifications; one at low temperatures consisting of superstructural units (β-BeCl$_2$ and arsenolite respectively), while at high temperatures the structure is consisted of chains (α-BeCl$_2$) or layers (claudetite).

Spectra of liquid BeCl$_2$ recorded up to 1100 cm^{-1} and the temperature variation of the relative intensities of the bands at ~328 and ~275 cm^{-1} have been measured for both the glass and the melt. The intensities of the glass from LN$_2$ up to the devitrification temperature (~200°C) do not change significantly in contrast to the liquid where rather fast and drastic changes take place.

It has been concluded [34] that for the glass/liquid the ~328 cm^{-1} band is due to the stretching vibration of the BeCl$_4$ tetrahedra participating in edge-bridged chains, while the ~275cm^{-1} band is associated with the same species involved in the construction of the cage like structure through vertex bridging. Thus, in the pure BeCl$_2$ liquid/glass the "chain" and the "cluster" structures exist which participate in a temperature dependent equilibrium. The high viscosity and low conductivity of the melt imply that the "chain" and "cluster" structures are neutral and of high molecular weight. To ensure the neutrality it is necessary to terminate these structures with BeCl$_3$ end units as shown in Fig. 13g. The ending of the cluster structure could be also attained with BeCl$_3$ units having one terminal chlorine and two others vertex linked to different BeCl$_4$ tetrahedra of the same or different clusters.

The "chain" and "cluster" structures participate with different concentrations in the glass and the liquid. The glass and the low temperature liquid favor the "cluster" configurations, while at high temperatures the edge-bridged "chains" dominate the liquid.

2.4.3. Trivalent Metal Halides

The structural properties of a large number of trivalent metal halide (LX$_3$)-alkali halide (AX) melt mixtures have been investigated by Raman spectroscopy at different compositions including the pure LX$_3$ component. Early studies of "model" systems like AlCl$_3$-ACl [20, 35], AlF$_3$-AF [13, 20] and YCl$_3$-ACl [12] have established certain common structural features as well as differences for these melts. For the AlCl$_3$-ACl melts composition and temperature dependence studies have shown that pure AlCl$_3$ is a molecular melt forming Al$_2$Cl$_6$ dimers which in the presence of alkali halides give AlCl$_4^-$ and Al$_2$Cl$_7^-$ species in equilibrium [20, 35]. For the AlF$_3$-AF system, changes of the relative Raman band intensities as a function of the melt composition suggested that at least two different coordination geometries for aluminium were present. Extensive studies have shown that for binary melts rich in alkali fluoride that AlF$_4^-$, AlF$_5^{2-}$ and AlF$_6^{3-}$ are the predominant species at equilibrium [13]. Due to the high corrosivity and volatility of AlF$_3$ at high temperatures, binary mixtures with compositions above 50% in AlF$_3$ cannot be practically investigated and thus the spectral aspects of pure molten AlF$_3$ have not been revealed yet.

The Raman spectroscopic studies for the YCl$_3$-ACl binary system have shown that the predominant species in melts rich in ACl (X_{YCl_3} <0.25, where X is the mole fraction) are the YCl$_6^{3-}$ octahedral [12]. The thermodynamics of mixing have suggested high stability and thus, long lifetimes for these species. With increasing YCl$_3$ content the structure of the melt mixture changes in a rather continuous way where the YCl$_6^{3-}$ octahedron starts sharing common chlorides (at X_{YCl_3} >0.25) forming polynuclear structures. Finally, comparison of the solid-to-liquid spectral changes of pure YCl$_3$ and the observation that there is practically zero change of molar volume upon melting YCl$_3$ suggested that the liquid structure is rather similar to that of the solid, i.e., it consists of distorted YCl$_6^{3-}$ octahedral sharing chlorides and forming a loose "network" structure. This structural model was further confirmed by neutron diffraction measurements in liquid YCl$_3$, where a direct determination of the local structure suggested a coordination number for Cl around Y of 5.9 [19].

2.4.3.1. Molecular melts and ionization; the FeCl$_3$-CsCl system

For the molten mixtures of FeCl$_3$-CsCl no changes in the Raman spectra have been observed on going from the rich in CsCl melts up to 50 mol% FeCl$_3$. Characteristic spectra of FeCl$_4^-$ (Fig. 7 and upper spectrum Fig. 16) have been obtained indicating that the "isolated" tetrahedral species are present at these compositions.

The addition of iron (III) chloride in the 50/50 FeCl$_3$-CsCl mixture changes drastically the FeCl$_4^-$ spectra. This is seen in Fig. 16 where we present the molten mixture spectra at different mole fractions X_{FeCl_3} up to pure iron (III) chloride. Certain changes observed in the spectra resemble those measured for the AlCl$_3$-ACl (A= alkali metal) system [20,35]. The bands marked with A,B,C,D gain their maximum intensity at $X_{FeCl_3} \approx 0.66$ and the measured frequencies scale well to the frequencies of Al$_2$Cl$_7^-$. Thus, these four bands are assigned to the Fe$_2$Cl$_7^-$ ion consisting of two tetrahedra bound by an apex.

Band T (Fig. 16) corresponds to the FeCl$_4^-$ species and coexists up to $X_{FeCl_3} \sim$ 0.66 with the Fe$_2$Cl$_7^-$ band. With increasing X_{FeCl_3}, a new band marked as K gains intensity, overlaps with the F band and eventually disappears in mixtures rich in FeCl$_3$. Temperature dependence measurements at X_{FeCl_3} = 0.66 have shown that bands T and K change their intensities in parallel relative to that of the A band, an observation indicating that an equilibrium is established where the Fe$_2$Cl$_7^-$ dissociates at high temperature giving FeCl$_4^-$ and the new species due to band K at ~390 cm^{-1}. By comparing known frequencies of iron (III) and iron (II) chloride systems the K band is assigned to the FeCl$_2^+$ cation. A bent triatomic structure is anticipated for the FeCl$_2^+$ species.

It therefore appears that the dissolution of FeCl$_3$ in mixtures with 0.5 < X_{FeCl_3} < 0.6-0.7 leads to an equilibrium involving prominently the Fe$_2$Cl$_7^-$, FeCl$_4^-$ and FeCl$_2^+$ ions. The existence of Fe$_2$Cl$_7^-$ and FeCl$_2^+$ persists even in the X_{FeCl_3} = 0.85 spectra where the intensity of the A band has diminished while the overlap of the K band at 395 cm^{-1} and the F band at 415 cm^{-1} of pure FeCl$_3$ gives rise to a broad band at ~405 cm^{-1} in the X_{FeCl_3} = 0.75 and 0.85 spectra.

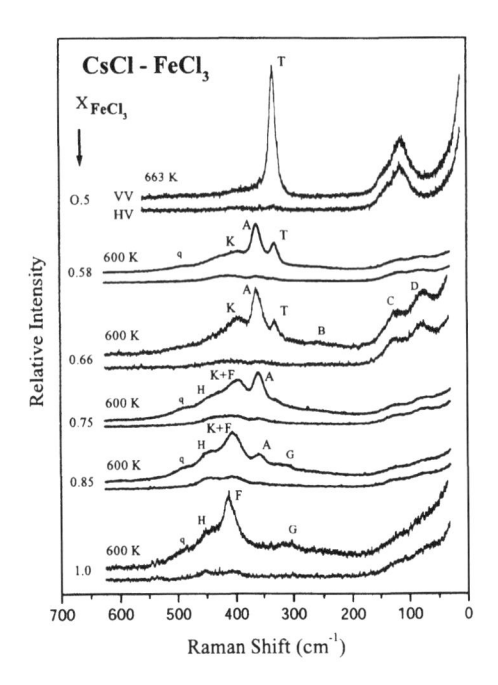

Figure 16. Polarized Raman spectra of CsCl-FeCl$_3$ molten mixtures in the 0.5 ≤ X_{FeCl_3} ≤ 1 concentration range. Frequencies assigned to FeCl$_4^-$: T; to Fe$_2$Cl$_7^-$: A,B,C,D ; to Fe$_2$Cl$_6$: K; to Fe$_2$Cl$_6$; F, G: and to Fe.Cl$_2^+$: H. Fused silica band: a.

The Raman spectra of dark brown-black iron (III) chloride obtained by micro-Raman techniques [8, 36] is shown in the lower part of Fig 16. The main polarized bands seen in the spectra are at 310 (G), 414(F) and 452(H) cm^{-1} and their relative intensities have been found to depend on the frequency of the laser excitation line (i.e. resonance-Raman behavior). The G and F bands have been assigned to the bridging and terminal frequencies of the Fe$_2$Cl$_6$ dimmer. From the trends of Fe-Cl frequencies observed for different coordination species it follows [5, 6] that the H band at ~450 cm^{-1} is probably due to Fe-Cl vibration in a trigonal geometry where the other two chlorides are bridged to another iron (III). An all iron (III) species compatible with the measured frequencies in the Fe$_2$Cl$_5^+$ consisting of an FeCl$_4$ tetrahedron bound by an edge with a trigonal FeCl$_3$

$$\left[\begin{matrix} Cl \\ Cl \end{matrix} \rangle Fe \langle \begin{matrix} Cl \\ Cl \end{matrix} \rangle Fe-Cl \right]^+$$

Such a species could be formed by self ionization of the molecular melt

$$2Fe_2Cl_6 \rightleftarrows Fe_2Cl_5^+ + Fe_2Cl_7^- \qquad (23)$$

In short, the Raman spectra of molten iron (III) chloride are best interpreted to indicate a structure where neutral Fe_2Cl_6 and charged $Fe_2Cl_5^+$ and $Fe_2Cl_7^-$ are the predominant species present. The dissociation of Fe_2Cl_6 and the presence of ionic species (reaction 23) accounts for the near ionic conductivity of the $FeCl_3$ melt. The strong Coulombic forces between the two ions involved and the spatial flexibility of the corner-connected tetrahedra of $Fe_2Cl_7^-$ will allow a more condense packing of the molecules and ions in molten iron (III) chloride, relative to the pure molecular aluminum chloride, and this also gives an account of the molar volume differences between the two melts. In addition, the coordination of iron (III) in $Fe_2Cl_7^-$ is, like in Fe_2Cl_6, four-fold while a mixture of three- and four-fold coordination exists in $Fe_2Cl_5^+$. Thus, in a total of eight iron (III) centers participating in reaction, seven are in a four-fold coordination and one in three-fold. In other words, the expected average coordination number for molten iron (III) chloride should be lower than four and this is in agreement with the finding of the neutron diffraction data [37] which give an average coordination number of 3.8.

2.4.3.2. Network structure of rare earth halides

Information regarding the structure of molten rare earth chlorides has been derived either directly from the scattering experiments (neutron, X-ray, Raman) or indirectly from the thermodynamic and transport properties [38]. Furthermore, during the past few years computer simulation have been used extensively to study the structure of pure LCl_3 [39-41]. It was first suggested by an extensive Raman study that the structure of molten YCl_3 may be a weak network of distorted chlorine-sharing octahedral [12]. This view was also supported by neutron diffraction studies [32, 42] on molten YCl_3. Raman studies on other rare earth fluorides, chlorides and bromides [43-47] indicate that the octahedra network-like structure is a general feature of all LX_3 (X= F, Cl, Br). The X-ray diffraction data also indicate that the six-fold coordination predominates the structure of all LCl_3 melts [48]. On the other hand, measurements of the total structure factor of molten LCl_3 show certain small but systematic differences depending on the rare earth cation size.

Detailed Raman spectroscopic measurements of a series of LX_3-AX (X= F, Cl, Br) [43-47] indicate that the spectral behavior and structure of these melts are very similar especially in the dilute in rare earth halide mixtures. Fig. 17 shows the spectra of

76

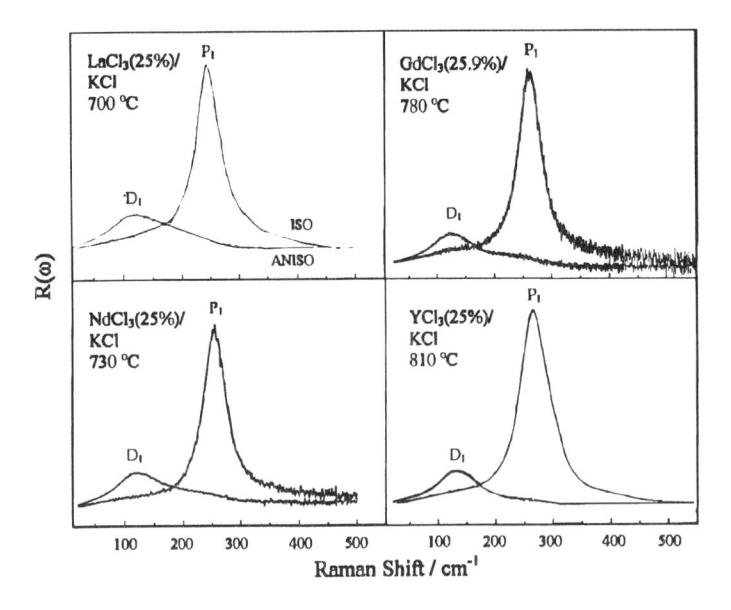

Figure 17. Reduced isotropic and anisotropic Raman spectra of LCl$_3$-KCl melts.

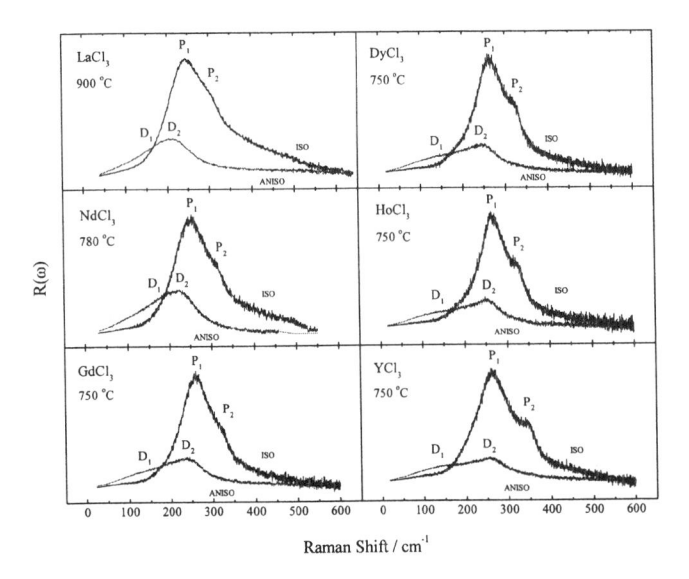

Figure 18. Reduced isotropic and anisotropic Raman spectra of pure molten rare earth chlorides.

chloride mixtures at LCl$_3$ mole fraction X$_{LCl_3}$ ≈ 0.25.

The two main bands, P$_1$ (polarized) and D$_1$ (depolarized) are assigned to the v_1 (A$_{1g}$) and v_5 (F$_{2g}$) modes of the LCl$_6^{3-}$ octahedra which are presumably the predominant species in these melts.

Studies of fluorides and bromides show the same behavior, with LF$_6^{3-}$ and LBr$_6^{3-}$ being the octahedral species present in melts rich in alkali halide [43, 45].

Due to their relative low melting point (below 1000 °C) the chloride and bromide systems have been investigated up to the pure LX$_3'$ (X' = Cl, Br) melt. With increasing LX$_3'$ mole fraction above 25 mol% a new polarized band (P$_2$) appears in the spectra which shifts continuously to higher frequencies while the P$_1$ (v_1) band remains practically unchanged. The systematic Raman spectral changes with composition and temperature observed for the systems involving LaCl$_3$, NdCl$_3$, GdCl$_3$, DyCl$_3$, HoCl$_3$, YCl$_3$, LaBr$_3$, GdBr$_3$, NdBr$_3$ and YBr$_3$ have been interpreted to indicate that the six-fold coordination around the rare earth cation is preserved at all mole fractions.

Spectra of pure molten LX$_3'$ (X' = Cl, Br) have been measured and are shown in Figs. 18 and 19. All spectra are characterized by the P$_1$ and D$_1$ bands observed in dilute LX$_3$-AX mixtures (Fig. 16) plus two new bands P$_2$ (polarized) and D$_2$ (depolarized). The "isomorphous" Raman patterns suggest that all these melts have similar network-like structures of distorted octahedra as suggested by the previous Raman studies [12] and the isotopic substitution neutron diffraction studies of DyCl$_3$ [39]. The resolution of the P$_2$ band (its relative intensity and separation from the P$_1$ band) decreases with increasing the L^{3+} cation size. Thus, for the YCl$_3$ and YBr$_3$ melts the P$_2$ band is well resolved while for the LaCl$_3$ and LaBr$_3$ melt the P$_2$ band appears as a strong shoulder on the P$_1$ band. The P$_1$-P$_2$ splitting has been recently correlated to the thermodynamics of mixing of the LCl$_3$-ACl binary systems and the cohesive (network-like) energies of the LCl$_3$ melts [49]. It appears that the network cohesive energy increases from LaCl$_3$ to YCl$_3$ and this is reflected to both the Raman spectra by an increase in the P$_1$-P$_2$ splitting and the enthalpy interaction parameter by more exothermic values in melts rich in YCl$_3$. Thus, the melts with the small L^{3+} cation size (i.e. Y, Dy, Ho) form a more rigid network with presumably long "LX$_6$" octahedral life times.

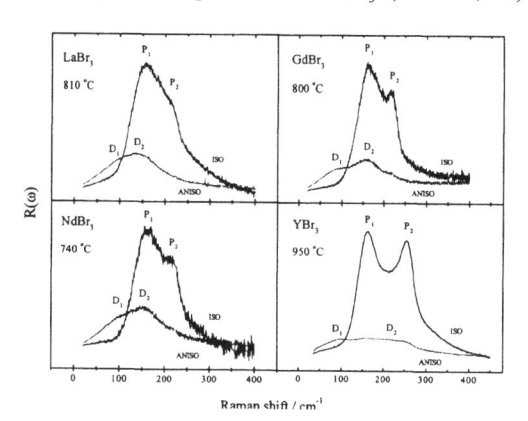

Figure 19. Reduced isotropic and anisotropic Raman spectra of pure molten rare earth bromides.

Raman spectra studies of fluoride mixtures rich in LF$_3$ are difficult to perform due both to their high melting points (above 1000 °C) and to the relative high vapor pressure. However, a comparative study for molten mixtures of YX$_3$-KX (X= F, Cl, Br) at composition near 50% of the fluoride mixture also shows four bands (Fig. 20), two depolarized at ~240 cm^{-1} (D$_1$), ~370 cm^{-1} (D$_2$) and two overlapping polarized bands ~440 cm^{-1} (P$_1$), ~460 cm^{-1} (P$_2$). The trends observed in the YX$_3$-KX spectra (Fig. 20) on going from the bromide to chloride to fluoride melts suggest that pure molten fluorides are likely to posses a similar "network" structure of edge-bridged distorted octahedral as in the case of molten chlorides and bromides [45,46].

2.4.4. Tetravalent Metal Halides

In contrast with the previous cases the Raman spectroscopic studies regarding the structure of molten tetravalent metal halides and their mixtures with alkali halides are limited. Solutions of ZrF$_4$ and ThF$_4$ in molten mixtures rich in LiF-KF eutectic have been investigated in the early seventies [50, 51]. These studies have argued the formation of ZrF$_6^{2-}$, ZrF$_8^{4-}$ and the ThF$_7^{3-}$ type of "complexes" in the mixtures. Due to the high melting points and volatilities of the tetravalent fluorides the investigation of these mixtures in a wide range of compositions was not possible and the structure of the pure component salts is unknown.

Recently systematic Raman spectroscopic investigations of the molten mixtures ZrCl$_4$-CsCl [52], ThCl$_4$-ACl [53] and ZrF$_4$-KF [54] have been reported providing structural information for both the melt mixtures and the pure component salts. In the following the findings of these studies are summarized.

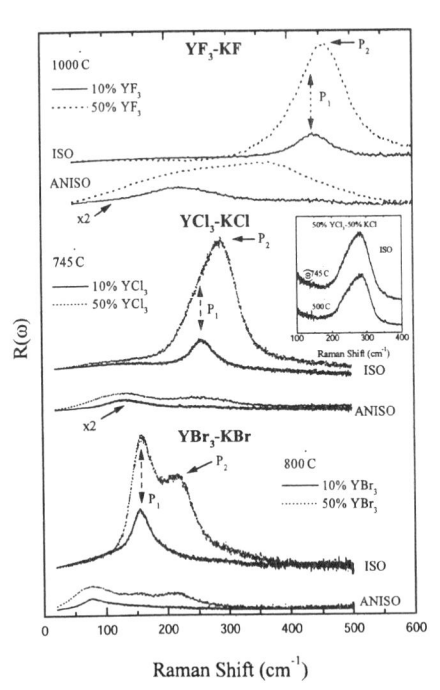

Figure 20. Reduced isotropic and anisotropic Raman spectra of KX-YX$_3$ (X= F, Cl, Br) (a) KF-YF$_3$, (b) KCl-YCl$_3$, (c) KBr-YBr$_3$. Solid line indicates 90% KX-10% YX$_3$; dot line indicates 50% KX-50% YX$_3$. Inset: isotropic Raman spectra of 50% KCl-50% YCl$_3$ molten mixture at 500 and 745°C.

2.4.4.1. Molecular/ionic clusters in molten ZrCl$_4$-CsCl mixtures

Crystalline zirconium tetrachloride is monoclinic with a P2/c (C$_{2h}^4$) space group having two molecules per unit cell. Twelve Raman active internal modes are expected for the solid spectra, $\Gamma_{int} = 6A_g + 6B_g$. The terminal Zr-Cl$_t$ frequency at ≈ 411 cm^{-1} is close to the analogous terminal frequency (≈ 412 cm^{-1}) of the vapor complex ZrAl$_2$Cl$_8$ in which Zr has a six-fold co-ordination. The bands at 283 and 220 cm^{-1} have been assigned to the two different types of bridging Zr-Cl$_b$ bonds. The main spectral characteristics of solid ZrCl$_4$ do not change with increasing temperature: all bands become broader and shift to lower energies (Fig. 21). At 430 °C, just below melting, seven bands are clearly seen. Upon melting, most of these modes appear to be transferred into the liquid. The deformation modes of the solid occurring in the 50-150 cm^{-1} region follow the "red shift" trends of the solid and appear in the liquid as broad and depolarized bands. The behavior of the terminal high-frequency polarized band v_t at 410 cm^{-1} is similar. In contrast, the v_{b_1} band is at ~295 cm^{-1}. Finally, a new high intensity polarized band appears in the spectra the frequency of which at 375 cm^{-1} is presumably that of the stretching frequency of the ZrCl$_4$ tetrahedra, indicating that monomers are also present in the liquid.

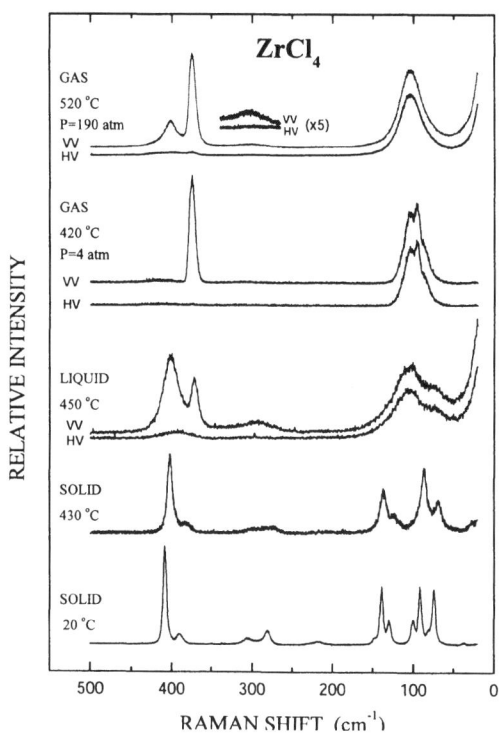

Figure 21. Raman spectra of solid, liquid and gaseous ZrCl$_4$.

A comparison of the solid and liquid spectra with the high density (supercritical) vapor spectra (upper spectrum Fig. 20) suggests that at least two different types of species are predominant in the liquid phase. Monomeric ZrCl$_4$ with a main polarized band at 375 cm^{-1} and "polymeric like" species with a main polarized band at 404 cm^{-1}. The proximity of the frequency of the latter band to the terminal frequency v_t of both the Zr$_2$Cl$_8$ gaseous dimer and of the $[(ZrCl_{4/2})Cl_2]_n$ chain in the solid cannot be used to characterize the size and the structure of the "polymeric" species. Measurement of the liquid-phase spectra at different temperatures (Fig. 22) confirms the presence of the monomers and suggest that an equilibrium may exist in the melt of the type:

$$(ZrCl_4)_n \text{ (l)} \rightleftarrows nZrCl_4 \text{ (l)} \tag{24}$$

The chemical equilibrium methods used in order to calculate the value of n in the vapor phase [53] are not applicable for the liquid and the extent of polymerization cannot be calculated. Upon melting ZrCl$_4$ a large volume expansion ($\approx$ 77%) occurs and a high-fluidity non-conducting liquid is formed. With increasing temperature the liquid continues to expand drastically and its fluidity increases. These observations are consistent with the above equilibrium and furthermore indicate that the polymerization (value of n) in the melt is rather small and more likely the molecular liquid is composed of either Zr$_2$Cl$_8$ dimers, as in the vapor phase, or Zr$_6$Cl$_{24}$ hexamers in equilibrium with monomers [53].

Measurements of Raman spectra of solid and molten Cs$_2$ZrCl$_6$ and CsZr$_2$Cl$_9$ at different temperatures show that the "isolated" molecular ions $ZrCl_6^{2-}$ and $Zr_2Cl_9^-$ are present in both phases. The $ZrCl_6^{2-}$ octahedra are the predominant species up to mole fraction of $X_{ZrCl_4} < 0.33$. In the composition range $0.33 < X_{ZrCl_4} < 0.66$ the spectral changes with composition and temperature suggest an equilibrium involving three ionic species : $ZrCl_6^{2-}$, $Zr_2Cl_9^-$ and $Zr_2Cl_{10}^{2-}$ (or $ZrCl_5^-$). At mole fractions rich in ZrCl$_4$ ($X_{ZrCl_4} > 0.66$) the spectra indicate an equilibrium between the ionic $Zr_2Cl_6^-$, the ZrCl$_4$ monomers and the (ZrCl$_4$)$_n$ polymer-like species. All data suggest that the value of n is small and most probably hexamers and or dimers are the predominant "polynuclear" species in melts rich in ZrCl$_4$ [53].

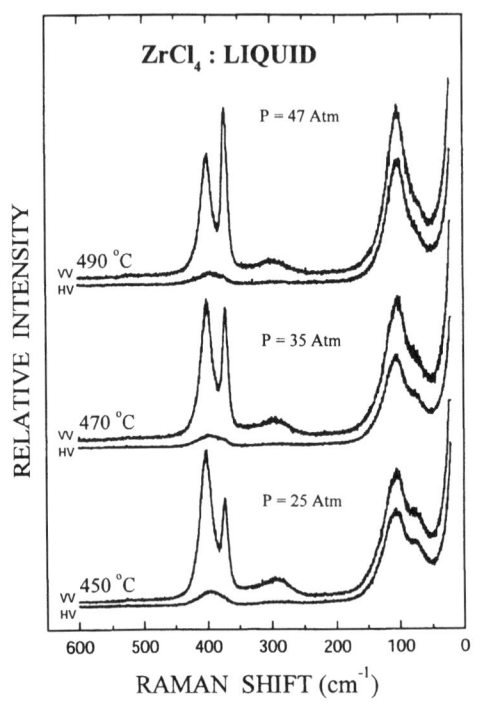

Figure 22. Temperature dependence of the Raman spectra of molten ZrCl$_4$.

2.4.4.2. Octahedral bridging in molten ZrF$_4$-KF and ThCl$_4$-ACl mixtures

Due to the fact that the ratio of ionic radii of Zr^{+4}/F^- almost equals to that of Th^{+4}/Cl^- the structural behavior of the corresponding binary melts are expected to be similar. Thus, it appears that an "isomorphism" exists in these melts.

Raman spectroscopic measurements for molten ThCl₄-ACl (A= Li, Na, K, Cs) were possible at all composition including the pure ThCl₄ melt. The data (Fig. 23) indicate that in molten mixtures rich in alkali metal chloride the predominant species are the $ThCl_6^{2-}$ octahedra [$v_1(A_{1g})$ 297 and $v_5(F_{2g})$ 125 cm^{-1}] in equilibrium with the $ThCl_7^{3-}$ [$v_1(A_1^{'})$ 280 cm^{-1}] pentagonal bipyramid (PB). The citation is similar for the alkali fluoride rich melts with ZrF₄. Both the ZrF_6^{2-} octahedra [$v_1(A_{1g})$ 570 and $v_5(A_1^{'})$ 248 cm^{-1}] and the PB ZrF_7^{3-} [$v_1(A_1^{'})$ 535 cm^{-1}] are formed. At mole fractions $X_{TX_4} < 0.33$ (TX₄ = ZrF₄, ThF₄) an equilibrium between the two species exists:

$$TX_6^{2-} + X^- \rightleftarrows TX_7^{3-} \tag{25}$$

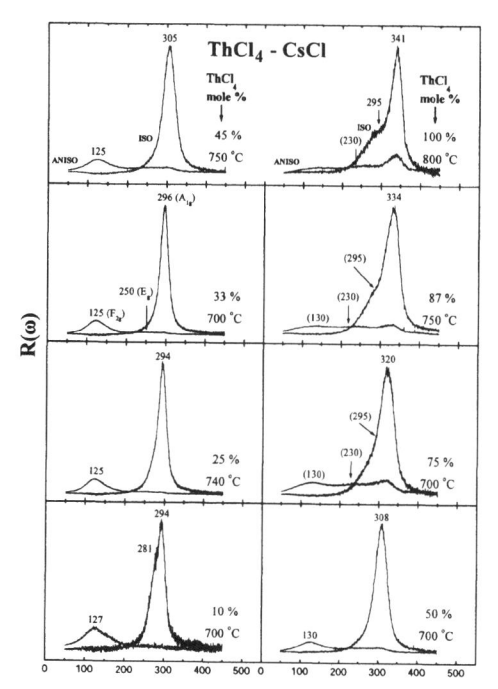

Figure 23. Reduced isotropic (ISO) and anisotropic (ANISO) spectra of ThCl₄-CsCl melt mixtures at different compositions.

The equilibrium shifts to the right with decreasing TX₄ mole fraction and increasing temperature (Fig. 24). Quantitative Raman measurements of the relative intensities of the octahedral $v_1(A_1)$ and PB $v_1(A_1^{'})$ bands as a function of temperature permits the evaluation of the enthalpy associated with the above reaction (see inset in Fig. 24). For both binary systems a approximately common enthalpy of reaction $\Delta H \sim 35$ KJ/mol has been found supporting the isostructural character of these melts.

For mole fractions up to 66 mol% TX₄ both binaries show similar behavior. Thus, with increasing TX₄ mole fraction above 0.33 the frequency of the $v_1(A_{1g})$ band shifts continuously to higher frequencies and new bands appears in the spectra. At about 66 mol% TX₄ the spectra are characterized by two polarized and two weak depolarized bands. The composition and temperature dependence of the Raman spectra for both systems were interpreted [52, 54] to support the formation in the melt mixture of bridged octahedral species $T_2X_{10}^{2-}$ and $T_3X_{14}^{2-}$ species in equilibrium with "free" TX_6^{2-} octahedra. In melts very rich in TX₄ ($X_{TX_4} > 0.7$) measurements were possible only for the thorium systems (see Fig. 23). The continuous band shifts observed support the view that the edge bridging of octahedra extends yielding chains of the type: $[T_nX_{4n+2}]^{2-}$ and $[T_nX_{4n-2}]^{2-}$ where the end T atoms of the chain are six- and four- fold co-ordinated for the chain anion and cation respectively.

The physicochemical properties (e.g. conductivity, molar volumes) of the ThCl$_4$-ACl melts suggest an ionic character which implies a rather small value of n for the chain species (e.g. n = 2,3,4).

Finally, the chain octahedral ionic structures appear to be the predominant species in pure molten ThCl$_4$. A mechanism creating opposite charged species having vibrational frequencies verified by the Raman spectra could involve a self ionization scheme of the type:

$$nThCl_4 \rightleftarrows \tfrac{1}{2}[Th_nCl_{4n-2}]^{2+} + \tfrac{1}{2}[Th_nCl_{4n+2}]^{2-} \qquad (26)$$

Temperature dependence measurements of the relative intensities of pure molten ThCl$_4$ support the above ionization. Increasing temperature induces breaking the bridges between the octahedra and either shifts reaction (26) to the left or/and lowers the n value. A similar self ionization scheme probably occurs for pure molten ZrF$_4$ [54].

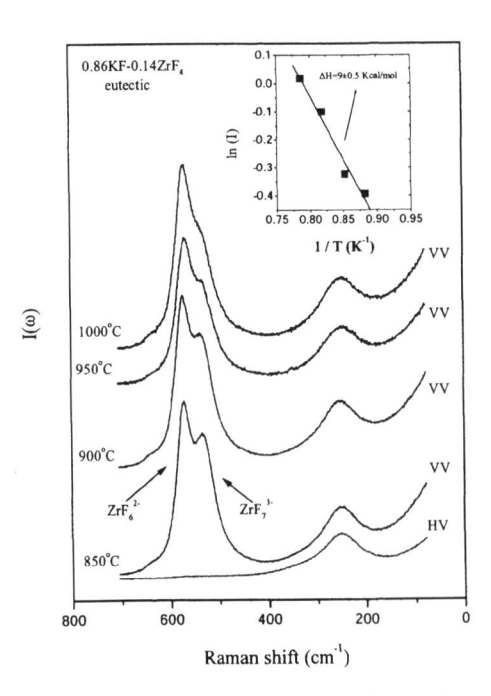

Figure 24. Temperature dependence Raman spectra of 0.14 ZrF$_4$-0.86 KF molten eutectic mixtures. Inset: ΔH calculations from an Arhenious like plot.

3. Low Frequency Raman Scattering from Supercooled Melts and Glasses

3.1. GENERAL REMARKS

Raman scattering and the spectroscopic techniques described in the next sections provide extensive information for two broad energy regions of the excitation spectrum in solids and liquids leaving however a gap between them. In particular, Raman spectroscopy probes high–energy excitations, i.e. *localized* molecular vibrations characterized by energies greater than ~100 cm^{-1}. On the other hand, Brillouin scattering and Photon Correlation spectroscopy probe very low-frequency *collective* fluctuations having energies below ~0.5 cm^{-1}. There exists, therefore, an energy gap not accessible by the aforementioned spectroscopic techniques. As has been mentioned in the introductory part, there is a reciprocity relation between the energy of an excitation and the spatial extent of the structural arrangement of the molecules that give rise to this excitation. Therefore, the high–energy excitations probed by Raman spectroscopy are signatures of the sort range order while the collective modes carry information for the extended range order. The gap associated with structure and dynamics in the intermediate range order is bridged partly by the low-frequency part of the Raman spectrum, i.e. below 100 cm^{-1}.

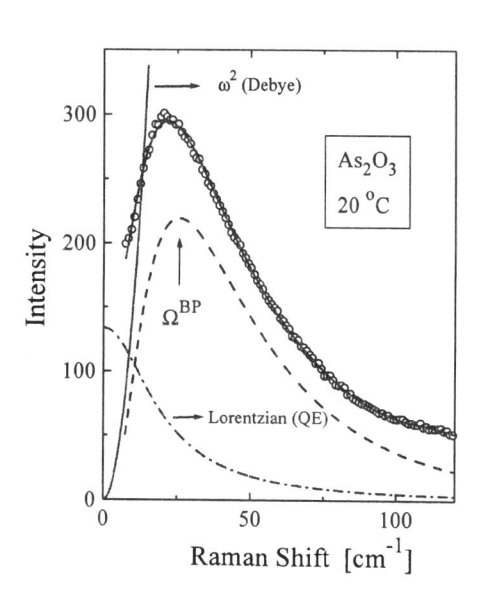

Figure 25. A typical low–frequency Stokes side Raman spectrum of As$_2$O$_3$ at room temperature. The solid line through the experimental points is a fit to the spectrum. The dashed line represents the Boson peak contribution and the dashed–dotted line stands for the QE line. The ω^2 dependence that the vibrational density of states should follow at such very low frequencies (Debye behavior) is also shown.

Low energy excitations in amorphous solids have proved a very fertile field of research since the pioneering works of theoreticians and experimentalists in the early seventies [55]. In their effort to comprehend the excess scattering at low frequencies of amorphous materials, they employed ideas and models akin to the traditional solid state physics at that time. Their primary goal, which still remains as the motive for new investigations, was the formulation of an analytic expression for the frequency dependence of the scattered intensity. Since then, a plethora of experimental data and particularly empirical attempts have appeared [56].

It is now a well–established fact that the low–frequency Raman pattern of an amorphous solid consists of two main contributions: the quasi–elastic (QE) scattering or relaxational part, and the Boson peak (BP) or vibrational component, see Fig. 25. The former appears usually as a single Lorentzian line centered at zero wavenumber, with a half–width hardly exceeding 10 cm^{-1}. The physical origin of the quasi-elastic scattering has been mainly pursued in two directions. In the first, double well potentials – either symmetric or asymmetric – have been employed to play the role of localized defects that couple to light. In the second, two phonon scattering processes are envisaged to take place. Second order scattering predicts much stronger temperature dependence of the total quasi-elastic line intensity than that measured experimentally. The Boson peak is observed as a broad, asymmetric line starting from zero wavenumber and extending up to 100 – 150 cm^{-1}. This feature was initially assigned to the acoustic modes, which become active in Raman due to a breakdown of selection rules in amorphous materials compared to the respective periodic solids, and scales according to the Bose factor [57]. Further, the microscopic origin of this feature has been also related to localized vibrations [58], to propagating transverse acoustic vibrations [59], and to strongly hybridized with acoustic waves short wavelength acoustic- or optic-like vibrations [60]. Both relaxational dynamics (QE scattering) and the excess density of vibrational states relative to the Debye case (BP) are also seen in inelastic neutron scattering spectra. A typical temperature dependent low–frequency Raman study for a molten salt is shown in Fig. 26.

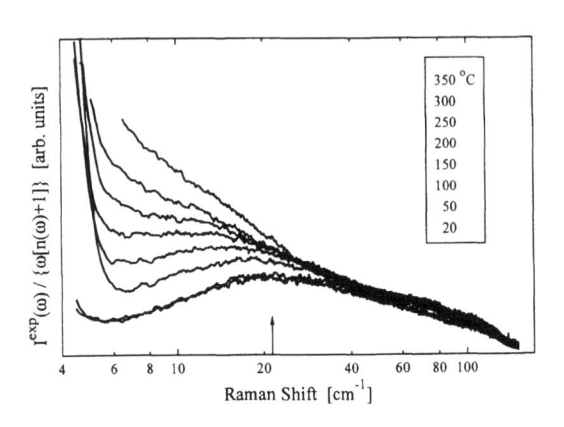

Figure 26. Bose factor scaled, low–frequency, Stokes side Raman spectra of $0.8ZnCl_2 - 0.2AlCl_3$ as a function of temperature. The arrow indicates the position of the BP while the QE component located between 5 and 10 cm^{-1} presents much stronger temperature dependence.

Due to the inherent difficulties that are present in the non–crystalline state, theoretical attempts have to rely on experimental data. Therefore a well–established phenomenology must be available to facilitate theoretical attempts. The three most widely discussed aspects of the low–frequency Raman phenomenology comprise: (i) a putative correlation between the relative weight of the relaxational to vibrational contribution at the glass transition temperature T_g and the dynamic character of the material, (ii) the frequency dependence of the photon–phonon coupling coefficient, and

(iii) the attempt to find a functional form that will be able to describe the frequency dependence of the scattered intensity. Before proceeding to the discussion of these issues we will briefly review some important theoretical contributions in this field.

3.2. THEORETICAL APPROACHES TO THE LFR SCATTERING

Electrical as well as *mechanical* disorder have long ago been recognized as important aspects that must be taken into account in the construction of models being able to account for the LFR spectra. Electrical disorder is manifested as spatially fluctuating polarizability [61]. Mechanical disorder is related to the departure of the vibrational modes from the plane wave form, as usually are described in the crystalline state. The consequences of mechanical disorder have been examined by Shuker and Gammon (SG) in Ref. [57].

The seminal work of SG has been the incentive for intense studies in the field of low–frequency Raman scattering in amorphous solids and supercooled liquids. Working in the harmonic approximation they showed that the **first order** Stokes scattered intensity in amorphous solids can be cast in the form:

$$I_{\exp}^{\alpha\beta}(\omega,T) = C^{\alpha\beta}(\omega)g(\omega)\omega^{-1}[n(\omega,T)+1] \qquad (27)$$

where $C^{\alpha\beta}(\omega)$ is the photon-phonon (Raman) coupling coefficient that reflects the activity of the vibrational excitations in scattering light, and $g(\omega)$ is the vibrational density of states. The different components labeled $\alpha\beta$ denote particular polarization geometries (VV and HV). According to its definition $C^{\alpha\beta}(\omega)$ is proportional to the spatial Fourier Transform (FT) of the space dependent "optical dielectric modulation" correlation function, i.e.

$$C^{\alpha\beta}(\omega) \propto FT\left\langle \frac{\partial \varepsilon_\alpha(\mathbf{r}')}{\partial Q_j} \frac{\partial \varepsilon_\beta(\mathbf{r}'+\mathbf{r})}{\partial Q_j} \right\rangle \qquad (28)$$

where $\partial\varepsilon/\partial Q_j$ represents the strength of the dielectric modulation of mode j. It should be emphasized here that $C^{\alpha\beta}(\omega)$ is the only one quantity involved in the scattered intensity expression that carries polarization information and hence the depolarization ratio $\rho(\omega)$ between depolarized (HV) and polarized (VV) geometries can be identified with the ratio $C^{HV}(\omega)/C^{VV}(\omega)$. The frequency dependence of $C^{\alpha\beta}(\omega)$ is a major challenge for the researchers in this field. Although considerable efforts have been undertaken to address this problem, there seems so far no consensus and hence the question for the frequency dependence of $C^{\alpha\beta}(\omega)$ still remains not satisfactorily explained. These approaches are briefly discussed in the following section.

It is important to open a parenthesis here in order to define the two most widely used reduced forms under which the low–frequency Raman spectra appear. These are based on the onset of Eq. (27) and having as a goal to isolate the function g(ω) it is frequently made use of the reduced Raman intensity and the susceptibility form $\chi''(\omega)$, which are defined as:

$$I_{red}^{\alpha\beta}(\omega) = \frac{I^{\alpha\beta}(\omega)}{\omega[n(\omega,T)+1]} = C^{\alpha\beta}(\omega)g(\omega) \tag{29}$$

$$\chi''(\omega) = \frac{I^{\alpha\beta}(\omega)}{[n(\omega,T)+1]} \tag{30}$$

Both contributions from electrical and mechanical disorder were considered in a model by Martin and Brenig (MB) where a continuum picture for the disordered–induced scattering from plane waves in a distorted elastic medium is put forward [62]. The MB theory is based on the fact that the lack of periodicity in the glass and the subsequently generated electrical and mechanical disorder give rise to spatial fluctuations in the elastic and the photo–elastic constants, which after being properly weighted by Gaussian distributions result to the following expression:

$$C_{HV}(\omega) = x^2 [g_{TA}(x) E^{TA} + g_{LA}(x) E^{LA}] \tag{31}$$

where $x = R_c \omega / c_{LA}$, $g_{LA}(\omega) = \exp(-x^2)$, $g_{TA}(x) = (c_{LA}/c_{TA})^5 \exp[-x^2(c_{LA}/c_{TA})^2]$, R_c stands for a structural correlation length, c_{LA} and c_{TA} are the longitudinal and transversal sound velocities, and the E values with respective indices are specific combinations of the photo–elastic coefficients.

Equation (31) predicts that the Boson peak has an asymmetric quasi–Gaussian profile with an ω^2 decay to zero intensity for $\omega \to 0$ and a short high-frequency tail. However, this expression has proved inadequate to fit a long body of experimental data for various substances. On the other hand, a proper modification of the $g_i(\omega)$ (i=LA, TA) functions in Eq. (31) changes the profile of the Boson peak from Gaussian–like to Lorentzian–like and seems to conform better with the measured spectra [63].

On the other hand, as far as the QE contribution is concerned, a model has been proposed that attributes the quasi–elastically scattered light to the population interplay between two equal energetically defect states described by different polarizability and represented by symmetric double well potentials [64]. They expression for the *polarized* Raman intensity in this case reads as:

$$I_{VV}^{QE}(\omega) \propto \omega \left[n(\omega) + 1\right]_0 \int \frac{\tau(V)}{1+\omega^2 \tau^2(V)} P(V) dV \approx \omega \left[n(\omega) + 1\right]_0 \frac{\tau}{1+\omega^2 \tau^2} \tag{32a}$$

where the distribution of activation energies or barriers V have a Gaussian distribution,

$$P(V) = \frac{1}{\sqrt{2\pi}V_0} \exp\left[\frac{-(E-E_m)^2}{2E_0^2}\right] \tag{32b}$$

and the relaxation time (residence time of a particle in the well) follows an Arrhenius temperature dependence,

$$\tau(V) = \tau_0 \exp(V/k_B T) \tag{32c}$$

It is customary to use the second equality in equation (32a), which in fact presumes that all V values are the same and the $1/\tau$ designates the half width at half height of the QE line.

From a different point of view, an attempt has been made to explain the low–energy vibrational features of amorphous materials as manifestation of medium–range order maintenance [65]. In this model, intrinsic density fluctuation domains are responsible for phonon scattering giving rise to the excess low frequency intensity. By utilizing the well–known Ioffe – Regel localization condition $q\ell \approx 2\pi$ the limiting value of the mean-free path ℓ can be estimated and the size of the medium-range order domains, where the density fluctuation-phonon interaction takes place, can be determined. By embedding a "mean" sound velocity, c, in the phonon wavevector q, a relation of the frequency maximum for the Boson peak Ω^{BP} and the diameter of the domain size $2R$, can be obtained: $\Omega^{BP} \approx c/2R$. This expression provided with experimental values for Ω^{BP} and c estimates R in the range 10–15 Å for a variety of amorphous solids [65]. Finally, it should be kept in mind that according to this model the Boson peak is the result of an enhancement of the vibrational density of states and is not due to an increase in the Raman coupling coefficient as was the main idea in the MB model.

3.3. LFR PHENOMENOLOGY BASED ON EXPERIMENTAL DATA.

In Sec. 3.1. we mentioned that there are a few puzzling topics where their unraveling if possible would lead to a proper understanding the microscopic nature of the low–energy modes in non–crystalline media. In what follows we will attempt to briefly survey these issues and to provide a critical evaluation of them in light of new experimental data recently obtained in our laboratory.

3.3.1. Boson peak, quasi–elastic scattering and the fragility of the supercooled liquid.

Some years ago a phenomenological correlation was proposed in order to classify glass–forming liquids regarding their *fragile* or *strong* character [66]. Fragility is a concept characterizing supercooled liquids introduced by Angell in order to describe the difference that various liquids exhibit in the temperature dependence of their viscosity (or structural relaxation time). Alternatively, fragility is defined as the easiness with which a supercooled liquid readjusts its structure to a new one (energetically favored) under the changes of temperature just above the glass transition temperature, T_g. Thus, fragility can be identified with the slope,

$$m = d\log\langle\tau,\eta\rangle/d(T_g/T),$$

of the logarithm of the dynamic variable (relaxation time or viscosity) near T_g, in a modified Arrhenius plot. A new classification has appeared according to which the ratio of the relaxational (intensity around the minimum located at the low frequency side of the Boson peak) to the vibrational (intensity around the maximum of the Boson peak) contribution, determined at T_g, $R(T_g) = I^r_{\min}/I^r_{\max}$, relates to the fragility of the liquid [67]. Specifically, the less fragile the glass is, the higher the vibrational contribution appears. This correlation has been considered as universal and has been amply used, alongside with the putative constancy of the depolarization ratio in the BP and QE region, as the experimental basis on which theoretical models have been used to describe the frequency dependence of the low–frequency Raman spectrum [68].

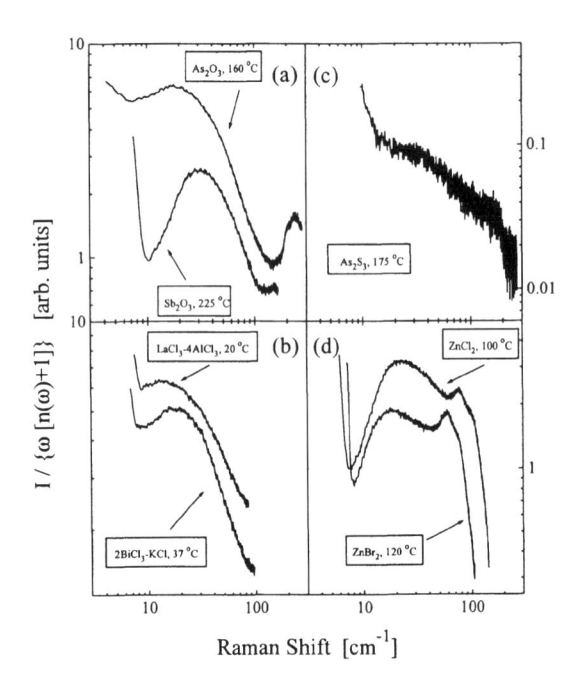

Figure 27. Log–log plots for the reduced low frequency Raman spectra for various glasses described in the figures, measured at their T_g dependence.

It has been recently shown that if new experimental data are taken into account the mentioned correlation between the strength of the relaxational to vibrational dynamics with fragility does not hold [69, 70]. This becomes evident in Fig. 27. Indeed, it can be seen, Fig. 27(a), that As$_2$O$_3$ which is one of the strongest glasses ($m \approx 19$) [71, 72], assumes near T_g the value $R(T_g) \approx 0.85$. This value is almost equal to the corresponding one for the fragile liquid 2BiCl$_3$-KCl ($m\approx 90$), Fig. 27(b). Further, the data for Sb$_2$O$_3$, Fig. 27(a), yield $R(T_g) \approx 0.37$ despite its close similarity in dynamical character with As$_2$O$_3$ [73]. Halide glasses, i.e. ZnCl$_2$ and ZnBr$_2$, Fig. 1(d), with $m\approx 30$ and $m\approx 45$ correspondingly [74], seem extremely dissimilar in the low frequency region when compared to As$_2$S$_3$, Fig. 27(c) ($m\approx 30$), despite their similar fragility. ZnCl$_2$ and

ZnBr$_2$, present well-defined Boson peaks even above their melting points, while in As$_2$S$_3$ the vibrational contribution is completely masked already 30 degrees below its T_g. Additionally, the $R(T_g)$ for LaCl$_3$–4AlCl$_3$ glass [75], Fig. 27(b), is twice the corresponding of ZnBr$_2$ while their fragility is almost the same [76]. All the aforementioned facts are summarized in Fig. 28 where we provide an illustration for the "correlation" between m and $R(T_g)$. It is unambiguously seen that the new results entirely confuse the picture and indicate that no systematic similarity between dynamic properties and low frequency Raman and the correlation between specific low-frequency features and supercooled liquid dynamics is not universal. Therefore, it would be probably more instructive somebody to test this idea of grouping glasses into families with alike structural features, where each subclass should follow its own slope in a $R(T_g)$ vs m plot, should be an alternative attempt. Possible criteria to categorize viscous liquids in the mentioned subclasses could be topological features of the short-range order, dimensionality of materials' structure (3D networks, layered materials, etc.) or some other substance specific property.

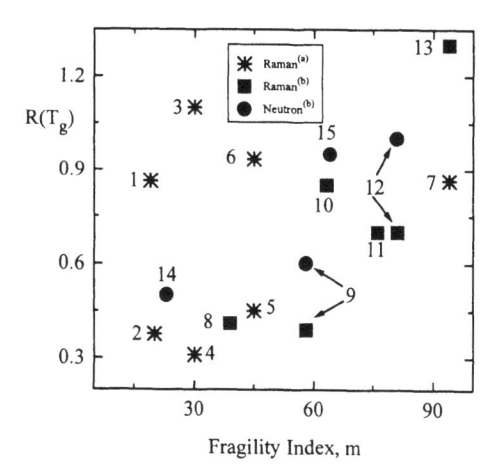

Figure 28. Plot of m versus $R(T_g)$ for various glasses: (a) this work, (b) data taken from Ref. [67]. The numbers denote: As$_2$O$_3$ (1), Sb$_2$O$_3$ (2), As$_2$S$_3$ (3), ZnCl$_2$ (4), ZnBr$_2$ (5), LaCl$_3$–4AlCl$_3$ (6), 2BiCl$_3$-KCl (7), B$_2$O$_3$ (8), Glycerol (9), Salol (10), m–TCP (11), OTP (12) CKN (13), SiO$_2$ (14), and poly-butadiene (15). We have omitted data from Ref. [67] with values of $R(T_g)$ larger than ~1.3 due to the high uncertainty in their estimation.

3.3.2. *The frequency and polarization dependence of the Raman coupling coefficient.*

The Raman coupling coefficient has been introduced in Eqs. (27, 28). From the experimental point of view many attempts have exploited the fact that incoherent neutron scattering probes exactly the vibrational density of states $g(\omega)$ and hence the coupling coefficient can be determined by means of a simple comparison of neutron data and the reduced Raman intensity, defined in Eq. (29). A recent review on the experimental and theoretical determination of $C^{\alpha\beta}(\omega)$ alongside with the relevant references can be found in Ref. [77]. It will be presented here a brief survey of these works without giving all details.

Direct comparisons between reduced Raman and neutron spectra have revealed linear frequency dependence for $C^{\alpha\beta}(\omega)$ in the case of harmonic acoustic vibrations, and a $C^{\alpha\beta}(\omega) \propto \omega^{0.5}$ dependence in the quasi-elastic scattering regime. Other workers have found a frequency independent coupling coefficient at frequencies below the Boson peak maximum and an almost linear ω-dependence for higher frequencies. In the most recent experimental data it has been shown that: (i) $C^{\alpha\beta}(\omega)$ is frequency independent for very low energies (below ~1meV) but assumes the form $C^{\alpha\beta}(\omega) \propto \omega^{1.3}$ for frequencies above the Boson peak maximum and (ii) the coupling coefficient has a non-vanishing value for ω→0, e.g. below ~30 cm^{-1} that departs from the linear frequency dependence observed above this interval.

Theoretically, the first attempts to clarify the frequency dependence for $C^{\alpha\beta}(\omega)$ have been initiated by the study of orientationally disordered crystals (electrical disorder) where an ω^2 behavior for the acoustic branch was found for the coupling coefficient [61]. The consequences of structural (mechanical) disorder were treated by SG where they considered that $C^{\alpha\beta}(\omega)$ was constant for all the modes in a given band (e.g. stretching bands, bending bands, etc.) and differs from band to band without specifying its spectral form [57]. Then Martin and Brenig (MB) considering both forms of disorder arrived at the expression given in Eq. 31. In other phenomenological approaches like the soft potential model where the main contribution to the spectra come from localized excitations in soft potentials it is expected that $C^{\alpha\beta}(\omega)$=const.

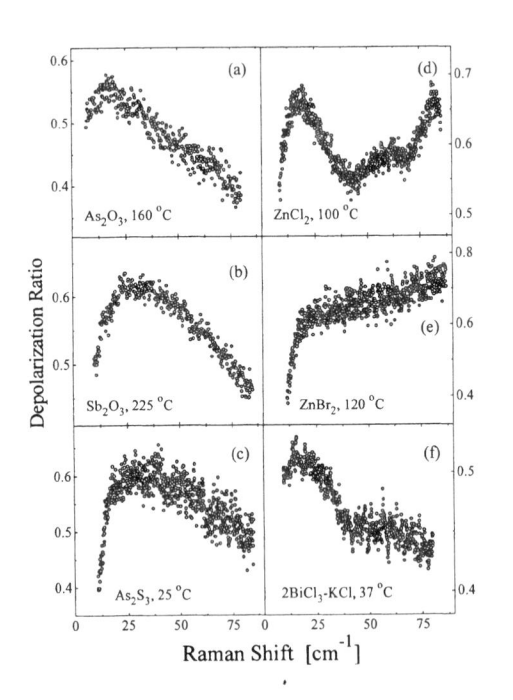

Figure 29. Frequency dependent depolarization ratios in the low frequency Raman region for various glasses. From Ref. [70].

From a completely different point of view fractal models have proved a valuable apparatus providing a concise framework for the elucidation of the low frequency Raman spectra [78]. The impetus has been given by the theoretical result that the vibrational density of states in a fractal medium can be written as, $g(\omega) \propto \omega^{\tilde{d}-1}$, where $\tilde{d}$ is the fracton or spectral dimension with a value very close to $\tilde{d} = 4/3$ [79]. All theoretical fractal approaches yield the power law form $C^{\alpha\beta}(\omega) \propto \omega^{\nu}$, where ν is a particular combination of fractal exponents and dimensionalities, that its form depends on the model used. In one of the simplest treatments it has been shown that $v = 2\tilde{d}/D_f$, with D_f being the fractal dimensionality; after using the relation $\tilde{d} = 2D_f/d_w$, where d_w is the random walk fractal dimension it yields: $v = 4/d_w$. It is therefore natural to assume that no polarization properties of the coupling coefficient are expected, because other wise one has to adopt polarization dependent d_w exponents.

So far, the Raman coupling coefficient was determined by considering only the depolarized reduced Raman intensity. This was based on another incorrect universality (mentioned in Sec. 3.3.1.) that the depolarization ratio is constant, i.e. independent of the scattering frequency. However, this was proven again not to be true as can be seen in Fig. 29. In all cases, $\rho(\omega)$ is clearly different between the quasi-elastic region, $\omega<15$ cm^{-1} and the Boson peak frequency, $\omega>15$ cm^{-1}, exhibiting further a non–monotonous frequency dependence. Another important observation is the continuously changing, i.e. decreasing $\rho(\omega)$ for the whole frequency region of the Boson peak.

In Fig. 30 we present the depolarization ratio spectra for As$_2$O$_3$ at various temperatures. It is clearly evidenced that the depolarization ratio is *strongly* frequency dependent implying that the two components – polarized and depolarized – of the Raman coupling coefficient have quite different spectral features. As far as the temperature dependence is concerned it appears that the depolarization ratio remains

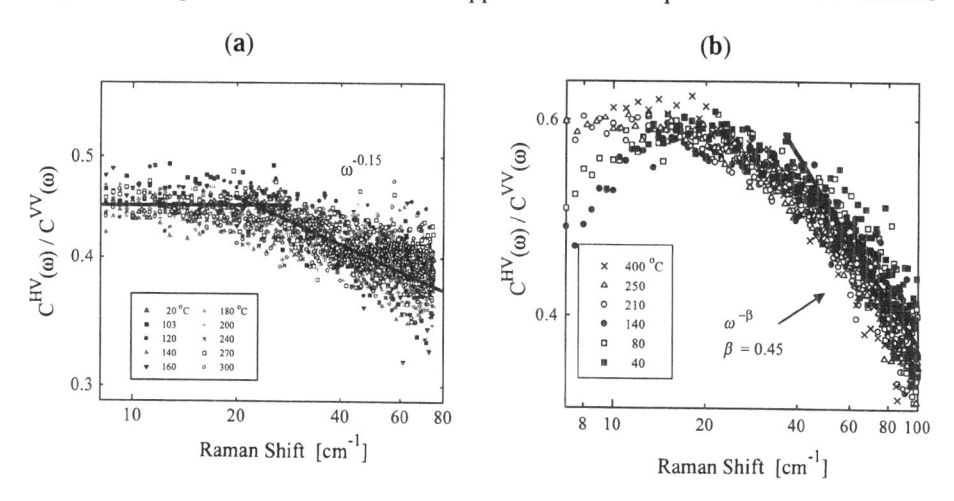

Figure 30. Double logarithmic plot of the depolarization ratio spectra for (a) 2BiCl$_3$–KCl and (b) arsenic trioxide at various temperatures showing their not constant dependence on frequency. The high frequency regime follows the power law $\omega^{-0.15}$ for (a) and $\omega^{-0.45}$ for (b).

almost unchanged in a wide temperature range at least in the high frequency part. Modest changes are observed in the low frequency range where most probably reflect the modification of the quasi–elastic component as the temperature evolves.

The experimental data presented above provide clear evidence for a frequency dependence of the depolarization ratio that is a manifestation of the polarization dependence of the Raman coupling coefficient $C^{\alpha\beta}(\omega)$. These findings pose serious doubts on the experimental approaches that determine the frequency dependence of the coupling coefficient through a direct comparison of Raman and neutron scattering data without taking account for the polarization dependence of $C^{\alpha\beta}(\omega)$. On the theoretical side, particular models that are attempting to describe the low energy excitations in supercooled liquids are based on incorrectly settled universalities e.g. constant (ω–independent) depolarization ratio [68]. Our data imply that a reconsideration of these approaches is needed. Finally, it is not at all obvious that all materials should exhibit the same frequency dependence for the coupling coefficient as universalities usually state. This is especially true since $C^{\alpha\beta}(\omega)$ is a very specific property of the studied system related to the way that vibrational motion modulates the polarizability of the medium and as such differs among various liquids.

3.3.3. The spectral features of the quasi–elastic line.

It has been recently shown that particular experimental data support the idea that the BP and the QE line are not convoluted but simply superposed [80]. This means that the total low–frequency Raman spectrum can be written as the sum of these two contributions. Further, in every approach that has appeared so far aiming at disentangling BP and QE scattering mechanisms the latter has been always subjected to the same Bose factor scaling usually employed to scale vibrational spectra. Although the reduction schemes presented in Eqs. (29, 30) are predominately valid *only for the first order* for the vibrational part for the spectrum, however they are applied also to the relaxational contribution for which an analogous "relaxational density of states" and a "photon–relaxator" coupling coefficient can not apparently conceived and further not obvious to define.

It seems at a first glance not plausible why such a central (ω=0) spectral line, assigned mainly to very fast relaxational processes that originate from local density fluctuations, should obey Bose statistics. It should be seen instead as a "classic" type of light scattering contrasted to the quantum nature accompanying the inelastic scattered light from vibrational modes. This idea is further supported by an analogous case, namely from the spectral form of the so–called "Mountain mode". It is well–known [3, 4] that viscous liquids, which consist of molecules with internal degrees of freedom, present in the low–frequency, low–wavenumber range in addition to the elastic or Rayleigh peak (due to entropy fluctuations) and the Brillouin doublet (due to propagating pressure fluctuations) an extra central peak the Mountain mode (due to structural relaxation). Although near the glass transition temperature the Mountain peak is much narrower than the QE peak observed in the low–frequency Raman spectrum, at higher temperatures the Mountain peak broadens considerably becoming thus comparable even with the QE contribution. However, the spectral dependence of the Mountain mode is described as a symmetric one,[13] i.e. no asymmetry in Stokes and anti–Stokes sides is invoked although it attains the same degree of "quasi–elastic" character (~ 1 THz) with the QE line. Concluding, it is not obvious why all quasi–inelastically scattered photons are true Raman signal.

Let us now proceed to a quantitative test of the aforementioned argumentation. We consider that the total low–frequency scattered intensity can be written as a sum of a symmetric part $R(\omega,T)$ representing all zero centered modes and the asymmetrical vibrational contribution $V(\omega)$. Then one can express the total scattered intensity in the Stokes and anti–Stokes region,

$$I_{\exp}^{S}(\omega,T) = R(\omega,T) + [n(\omega,T)+1]V(\omega) \tag{33a}$$

$$I_{\exp}^{aS}(\omega,T) = R(\omega,T) + n(\omega,T)V(\omega). \tag{33b}$$

As is easily discerned from the above equations the "pure" vibrational spectrum, free of other contribution and the temperature as well, can be obtained as

$$V(\omega) = I^{S}(\omega,T) - I^{aS}(\omega,T) \tag{34}$$

which after a multiplication with the $[n(\omega,T)+1]$ factor can yield back the Stokes component of the scattered intensity originating solely from the first order vibrational contribution, i.e.

$$I_{BP}^{S}(\omega,T) = V(\omega)[n(\omega,T)+1]. \tag{35}$$

Then the symmetric part is easily obtained as the difference

$$R(\omega,T) = I_{\exp}^{S}(\omega,T) - I_{BP}^{S}(\omega,T) \tag{36}$$

which is simply another form of Eq. (33a).

The above arguments imply that if the assumption of a symmetric central part – which we have put forward in Eqs. (33) – is correct then the excess scattering will disappear when the intensity described by Eq. (35) is plotted, and the following approximation should hold,

$$\lim_{\omega \to 0} I_{BP}^{S}(\omega,T) \to 0. \tag{37}$$

The last relation simply states that the vibrational part of the light scattering spectrum has to vanish at zero wavenumber.

To apply the procedure mentioned above one has to employ spectra with very high signal to noise ratio. Such spectra have been obtained by repetitive accumulation of data. The depolarized spectra for two molten salts are shown in Fig. 31. The

94

Figure 31. Stokes and anti–Stokes Raman spectra for the depolarized (HV) scattering geometry for As₂O₃, ZnCl₂ and 2BiCl₃–KCl.

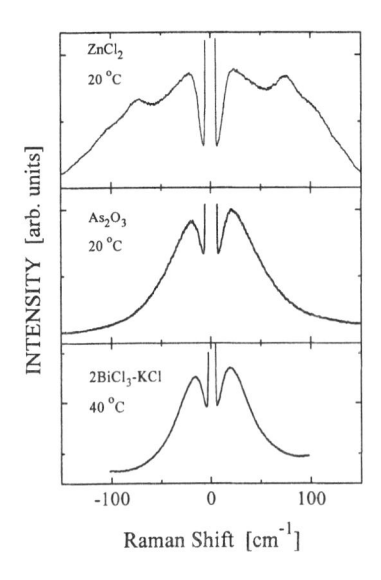

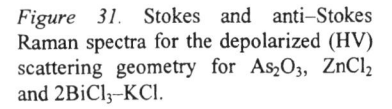

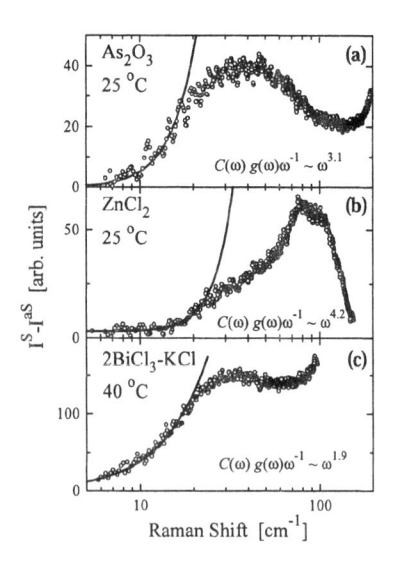

Figure 32. Semi–logarithmic plot of the reduced depolarized Raman spectra (open circles) for (a) As₂O₃, (b) ZnCl₂, and (c) 2BiCl₃–KCl. The solid lines represent power law fits to the data as described in the text.

$I^S(\omega,T) - I^{aS}(\omega,T)$ spectra have been obtained and are illustrated in Fig. 32 for the lowest temperatures. This figure reveals the following facts. First, the limiting low–frequency spectra do not show remnants of the QE component suggesting that it is rather a symmetric line. Second, the frequency dependence of the Raman coupling coefficient can be elucidated. Indeed, by combining Eqs. (27) and (34) after considering the QE as symmetric we obtain $I^S(\omega,T) - I^{aS}(\omega,T) = C^{\alpha\beta}(\omega)g(\omega)\omega^{-1} \propto \omega^x$.

We have made use of a power law proportionality in the above equation and have tried to obtain the overall exponent x. The high–frequency cut–offs of the fitting procedure that we have used were at about 20 cm⁻¹ for all glasses studied. The resultant values for the power law exponent do not show common or universal behavior, namely $x(As_2O_3)=3.1\pm0.1$, $x(ZnCl_2)=4.2\pm0.2$, and $x(2BiCl_3-KCl)=1.9\pm0.1$. Accordingly, the exponent for the net $C^{\alpha\beta}(\omega)g(\omega)$ product assumes mean values close to 4, 5, and 3 for As₂O₃, ZnCl₂ and 2BiCl₃–KCl respectively. The frequency dependence of the Raman coupling coefficient has been thoroughly discussed in the previous Section. In our case, based on the rigid conception that at such very low frequencies the vibrational density of states follows fairly well the Debye law, i.e. $g(\omega) \propto \omega^2$, we are led to the outcome that the low frequency behavior of the Raman coupling coefficient follows a power law with exponents close to 2, 3 and 1 for As₂O₃, ZnCl₂ and 2BiCl₃–KCl respectively.

Proceeding one step beyond, we have attempted to reconstruct from the pure vibrational parts of the spectra (or reduced spectra), i.e. $I^S(\omega,T) - I^{aS}(\omega,T)$, the corresponding Stokes side BP spectra by using Eq. (35). The obtained results for the 2BiCl$_3$–KCl salt are depicted as open circles in Fig. 33. It is obvious that the noise of the reconstructed spectra is higher than that of the raw data that are also shown in this figure as a solid line for comparison. However, all the reconstructed spectra share a common feature: they seem to follow rather well the predictions of Eq. (37). Alternatively, the light scattering vibrational spectrum tends to almost zero intensity at $\omega \to 0$. This observation is a clear evidence of the validity of our initial assumption concerning the symmetrical nature of the QE component.

As a further merit of the approach that we have been employing one can calculate the QE component – or any symmetric zero centered contribution – simply by using Eq. (36). Therefore, after subtracting the pure vibrational spectra, $I_{BP}^S(\omega,T)$ from the Stokes side experimental data we have obtained the QE spectra (open circles) for the materials under study; Fig. 34 shows the results for the salt glass 2BiCl$_3$–KCl. Lorentzian lineshapes have been used to model the spectral form of the QE contribution

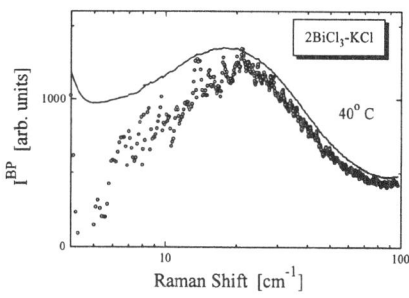

Figure 33. Semi–logarithmic plot of the depolarized "pure" Boson peak spectrum (open circles) for 2BiCl$_3$–KCl. The Stokes–side raw spectrum is also plotted for comparison.

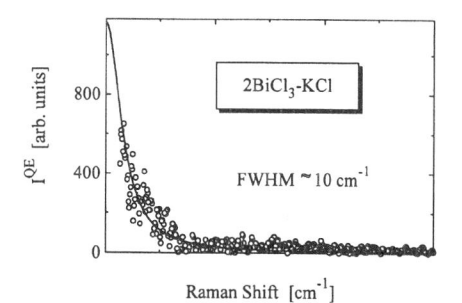

Figure 34. Pure quasi–elastic intensities (open circles) for 2BiCl$_3$–KCl obtained after subtracting the BP from the raw spectra.

(solid line in Fig. 34). It is evident that the Lorentzian line fits nicely to the experimental data resulting to physically accepted values for the fullwidths at half height, namely ~24 cm^{-1} for As$_2$O$_3$, ~30 cm^{-1} for ZnCl$_2$, and ~10 cm^{-1} for 2BiCl$_3$–KCl. More details for the kind of low–frequency data analysis can be found elsewhere [81].

The results from all studied systems suggest that the quasi–elastic component seems indeed to be a symmetric line, appearing rather as a "classic" type of scattering that does need to be treated under the same temperature basis as the Boson peak. The advantage of this finding is that one can get rid of the QE part or any other symmetric line (including constant unwanted background) of the spectrum obtaining thus the pure vibrational contribution.

4. Dynamic Light Scattering from Relaxational Modes

4.1. INTRODUCTORY REMARKS

The term dynamic light scattering is being used here mainly to emphasize the distinction between light scattering arising from vibrational degrees of freedoms (Raman) to that coming from local fluctuations of some particular quantities or relaxation processes. *Relaxation* – a central concept in the description of liquid dynamics – in its general sense is the time dependent response of a materials' dynamic property following an external perturbation. Historically this kind of light scattering was given the name "molecular light scattering" [82]. The spectra are understood to be changes in the frequency of the exciting light, which are induced by particular time variation of fluctuations representing different physical quantities acting as *optical heterogeneities*.

The field became active when at the beginning of the century the idea of *density fluctuations* was put forward to explain the phenomenon of critical opalescence. Then it was Einstein who employed the idea of density fluctuations and suggested their expansion in a three dimensional Fourier series. The fact that density fluctuations may exist in a variety of frequencies and wave vectors simply means that such fluctuations exist in a medium in a broad distribution of spatial extent and time scales. The amplitude in these expansions made possible the calculation of the scattering intensity. It is worth mentioning here that the Einstein Fourier components are another view of the Debye's thermal elastic waves (normal modes) employed to account for the specific heat in solids.

Let us now consider a continuous medium in which optical heterogeneities arise as a consequence of the statistical nature of the motion that the constituent particles execute. Only physical quantities that their fluctuations lead to optical inhomogeneity e.g. the dielectric constant $\varepsilon(\mathbf{r}, t)$ would significantly couple to light. Fluctuations in $\varepsilon(\mathbf{r}, t)$ include *density* and *temperature* fluctuations where the former depend upon pressure and entropy fluctuations. Pressure fluctuations arise when particles with momenta somewhat smaller or greater than the volume–average momentum accumulate in a particular point at a definite time while temperature or entropy fluctuations involve the situation where particles with a kinetic energy different from that of the volume–average have been congregated in a small region at a particular time. Pressure fluctuations are independent of temperature or entropy fluctuations. Formally, one can write,

$$\Delta\varepsilon(\mathbf{r},t) = \left(\frac{\partial\varepsilon}{\partial\rho}\right)_T \delta\rho(\mathbf{r},t) + \left(\frac{\partial\varepsilon}{\partial T}\right)_\rho \delta T(\mathbf{r},t) \tag{38}$$

It has been found experimentally that in many liquids the second term is negligible compared to the first one and therefore can be neglected.

The calculation of the scattered intensity is based on the assumption that the scattering due to pressure or entropy fluctuations at two different space–time points are not related. Such a calculation has resulted in the famous Einstein formula for the total (including both adiabatic and isobaric fluctuations) scattered intensity [3, 82]:

$$I \propto \left(\frac{\partial \varepsilon}{\partial \rho}\right)_T^2 V \rho^2 \beta_T k_B T, \qquad (39)$$

where V depends on the scattering angle, and ρ, β_T, are the density and the isothermal compressibility of the medium, respectively.

Considering that the medium density at any space–time point can be written as a sum of a mean value and a fluctuating part

$$\rho(\mathbf{r},t) = \delta\rho(\mathbf{r},t) + \rho_0 \qquad (40)$$

one can subsequently calculate a quantity of crucial importance in the study of dynamics of many–body systems, the so-called *dynamic structure factor*,

$$S(q,\omega) = \frac{1}{2\pi} \int_{-\infty}^{+\infty} dt \, e^{-i\omega t} \left\langle \delta\rho(\mathbf{q},t) \delta\rho^*(\mathbf{q},0) \right\rangle. \qquad (41)$$

The dynamic structure factor has proved quite useful in studies of non–crystalline materials since it carries all the salient information concerning both structure and dynamics [3–5]. It is the experimentally obtained quantity since it is just proportional to the scattered light intensity. In the next paragraphs we will briefly describe how this quantity can be obtained experimentally and what kind of information it can furnish about the dynamics and/or the structure of the medium.

4.2. RAYLEIGH–BRILLOUIN SPECTROSCOPY

As has been mentioned before, the fluctuations with which one deals in molecular light scattering are of the order of $q^{-1} \sim 1000$ Å, this corresponds to a wavelength appreciably larger than the intermolecular separations. These fluctuations involve the collective motions of large numbers of molecules, and consequently can be described by the laws of macroscopic physics, i.e. thermodynamics and hydrodynamics. Therefore, for a non–relativistic liquid the basic hydrodynamic equations (i) mass conservation, (ii) momentum conservation, and (iii) energy conservation provide the starting point for a rigorous derivation of the light scattering spectrum. In 1914 L. Brillouin predicted that light can be scattered from these long wavelength fluctuations and a doublet should appear in the spectrum, see Fig. 35.

The spectrum of the isotropic component consists of two central lines and the Brillouin doublet shifted symmetrically around the frequency of the incident light. One of the zero centered lines (Rayleigh peak) arising from the entropy fluctuations is always present and usually accounts for a small fraction of the central peak. The second central line (Mountain peak) caused by structural relaxation (for viscoelastic fluids) can be resolved under certain conditions. The Brillouin doublet due to propagating pressure fluctuations is shifted to $\omega_B = \pm c_B q$ with the wavevector defined as $q = (4\pi n/\lambda_0)\sin(\theta/2)$; c_B is the sound velocity, θ is the scattering angle, n is the refractive index of the medium and λ_0 is the excitation light wavelength. We present below the final result for the isotropic light scattering intensity; all the relevant details can be found elsewhere [3–5],

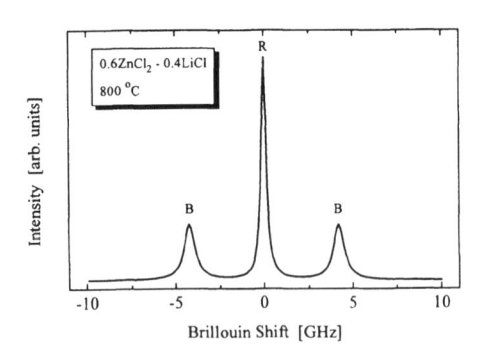

Figure 35. A typical Rayleigh–Brillouin spectrum for the molten salt 0.6ZnCl$_2$–0.4LiCl. R denotes the Rayleigh line and B the Brillouin doublet.

$$S(q,\omega) = \frac{1}{\pi} V\rho^2 \beta_T k_B T \left\{ \frac{\gamma-1}{\gamma} \frac{2\chi q^2}{\omega^2 + (\chi q^2)^2} + \frac{1}{\gamma}\left[\frac{\Gamma q^2}{(\omega+c_B q)^2 + (\Gamma q^2)^2} + \frac{\Gamma q^2}{(\omega-c_B q)^2 + (\Gamma q^2)^2} \right] \right.$$

$$\left. + \frac{1}{\gamma}[\Gamma + (\gamma-1)\chi]\frac{q}{c}\left[\frac{\omega+c_B q}{(\omega+c_B q)^2 + (\Gamma q^2)^2} - \frac{\omega-c_B q}{(\omega-c_B q)^2 + (\Gamma q^2)^2} \right] \right\} \quad (42)$$

where χ is the thermal diffusivity of the medium, and Γ is the width of the Brillouin peaks. The central peak (Rayleigh) is represented by the first term, while the Brillouin doublet by the second and third Lorentzian lines. The amplitude of the last two terms is by several orders of magnitude smaller than the amplitude of the Brillouin peaks. These terms yield s-shaped curves centered at the positions of the Brillouin doublet and cause a weak asymmetry in the Brillouin peaks, which induces a slight asymmetry of their positions towards the central peak. They are usually ignored in the data analysis since they are hardly observed in the experiment.

The above treatment is valid for simple or monatomic liquids. In molecular liquids rotational and vibrational relaxation can affect density fluctuations. The Brillouin spectrum is more complicated than that predicted by Equation (42). Molecular degrees of freedom couple to translational motion of the molecules leading to additional relaxation mechanisms. Such a treatment has been worked out by Mountain [83] where after taking into account a weak coupling between density fluctuations and a frequency dependent bulk viscosity he arrived at the prediction of another extra mode, frequently called Mountain or *relaxation* mode. The relaxation mode is strongly dependent on temperature and therefore its observability in the Brillouin spectra depends on the relation between ω and τ^{-1} (i.e. the on conditions $\omega\tau \gg 1$, $\omega\tau \approx 1$, and $\omega\tau \ll 1$), where ω corresponds roughly to the probing frequency, $\omega \sim 10^9$ Hz, and τ denotes the relaxation time associated with this new mode.

The propagation of density fluctuations is accompanied by dissipation in their energy content. Entropy fluctuations are diminished due to heat diffusion processes, while pressure fluctuations or sound waves are mainly subjected to viscous dissipation. The inverse of the Brillouin peak linewidth (full width at half maximum), $\Gamma_B = a_B c_B / \pi$, is a measure of the phonon lifetime; a_B being the sound attenuation coefficient. Temperature, as expected, plays a dominant role on the sound wave characteristics; thus the increase of temperature results to a decrease in the hypersound velocity, due to a softening of the elastic constants of the medium and due to the dispersion, see Fig. 36. The sound attenuation passes through a maximum at T_{max}, as the temperature evolves above T_g, and finally decreases to low values in the normal liquid state. The maximum in the attenuation coefficient at a given frequency ω_B, and the strong decrease in the speed of the sound signalize the onset of a *relaxation process*. Such a process occurs when energy is exchanged with a high rate, τ^{-1}, between structural units and propagative sound waves. Energy is transferred from the internal degrees of freedom to translational motion modifying the phonon characteristics. Therefore, a substantial amount of information concerning details even on a molecular scale can be gained by measuring hypersonic properties of viscous liquids. The inverse of the Brillouin shift frequency at T_{max} is a good estimate for the structural relaxation time characterizing the process, $\tau \sim \omega_B^{-1}$.

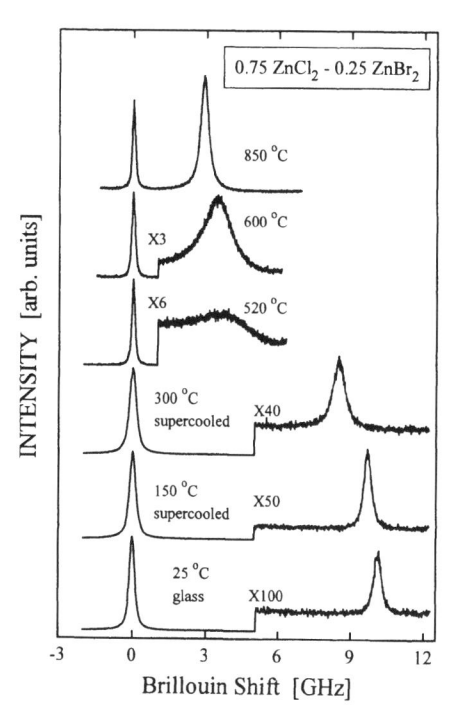

Figure 36. Stokes side Rayleigh–Brillouin spectra for the glass forming salt mixture 0.75ZnCl_2–0.25ZnBl_2 as a function of temperature. Magnification factors for the Brillouin peaks are included. From Ref. [84].

In the more sophisticated treatment mentioned above $S(q,\omega)$ (for long wavelength density fluctuations) after neglecting the thermal diffusivity mode (Rayleigh line) has the form [3, 4]:

$$S(q,\omega) = \frac{2v_0 q^2}{\omega^2} \, \text{Im} \left[\omega^2 - q^2 c_0^2 - i\omega q^2 \, \Phi_L(q,\omega) \right]^{-1} \tag{43}$$

where v_0 is the thermal velocity, c_0 is the adiabatic sound velocity and $\Phi_L(q,\omega)$ is the Fourier transform of the $\Phi_L(q, t)$, the memory function for the longitudinal kinematic viscosity. Until $\Phi_L(q, \omega)$ is specified explicitly no approximation is being involved. Thus, the specific features of the problem under study should be incorporated in the modelization of the memory function. Various models have been so far utilized for $\Phi_L(q,\omega)$, including single exponential relaxation, streching effects, empirical forms, slow and fast processes.

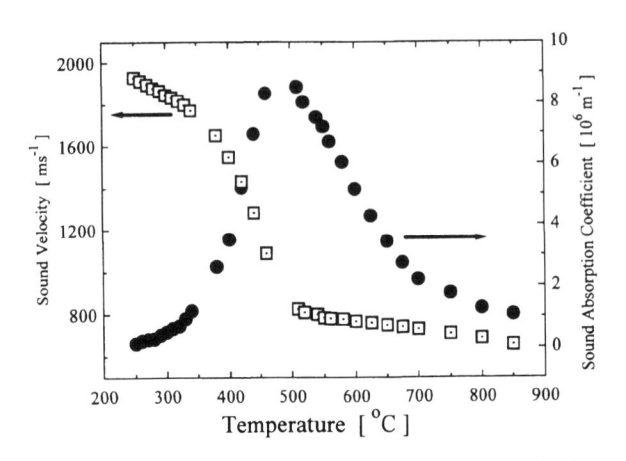

Figure 37. Temperature dependence of the sound velocity (open squares) and the sound absorption coefficient (filled circles) for the glass forming salt 0.75 ZnCl$_2$ – 0.25 ZnBr$_2$. From Ref. [84].

The most general way to model $\Phi_L(q,\omega)$ is to account for both the fast and slow process. Such a phenomenological ansatz can be bulit by using a weighted sum of two terms representing α– and β– relaxations [72]. Therefore, the memory function for the longitudinal kinematic viscosity can be written as:

$$\Phi_L(q\,\omega) = \left(c_\infty^2 - c_0^2\right) q^2 \left[\tilde{f} \, \Phi_\alpha(\omega) + \left(1 - \tilde{f}\right) \Phi_\beta(\omega) \right] \tag{44}$$

with Φ_α and Φ_β being respectively the parts of the memory function describing the slow (α) and the fast (β) processes, which are considered as Debye–like ones,

$$\Phi_i = \frac{1}{\omega} \left[\frac{i\omega\tau_i}{1 - i\omega\tau_i} \right] \qquad i = \alpha, \beta \tag{45}$$

where $\tilde{f}$ is a parameter accounting for the relative weight of the two processes. More details concerning such kind of analysis can be found elsewhere [72].

4.3. PHOTON CORRELATION SPECTROSCOPY

As has been mentioned above, the magnitude of the product $\omega\tau$ compared to unity determines the observability of the structural relaxation in Brillouin spectra. Specifically, at high temperatures the condition $\omega \ll \tau^{-1}$ holds since the relaxation rate τ^{-1} becomes extremely large compared to the frequency of the Brillouin peaks. Then the Mountain mode appears as a very broad structureless background in the Brillouin spectra. As temperature is lowered the relaxation rate becomes slower and at some particular temperature the condition $\omega\tau \sim 1$ is fulfilled. In this temperature region Brillouin spectra are seriously distorted due to the appearance of the easily observed Mountain mode with a width comparable to the shift of the Brillouin peak. At still lower temperatures, where structural relaxation follows macroscopic time scales the width of the Mountain mode becomes extremely narrow to be observed in Brillouin scattering and the best tool for its study is a time domain technique called Photon Correlation Spectroscopy (PCS), known also as Dynamic Light Scattering (DLS).

PCS is effective in the range 10^{-7}–10^{3} s. The strong temperature dependence of the relaxation time in this interval renders the dynamic behavior of the systems significant and necessitate the use of PCS that is capable to monitor the drastic temperature evolution of dynamic properties. Besides, the data obtained by PCS can be directly treated to give the width of the distribution for the relaxation times without being subjected to assumptions that could lead to misinterpretation of the data. This emerges from the rather simple three–parameter fitting used to the correlation function, as will be explained later. PCS probes both density and orientation fluctuations depending on the selected scattering geometry. The isotropic part of the correlation function is associated with entropy fluctuations that obey diffusive law and with the structural relaxation. On the other hand, the anisotropic part of the correlation function contains information concerning the orientational dynamics that arise mainly from fluctuations in the molecular optical anisotropy, as has been mentioned previously. Both reorientation and shear translational motion contribute to the depolarized dynamic light scattering spectrum. The coupling between these two contributions renders the interpretation of the VH intensity quite complex.

Polarized and depolarized (density and orientation) autocorrelation functions can be measured in a broad time scale (almost ten decades, 10^{-7}–10^{3}s) with a full multiple taut digital correlator. Under the assumption of homodyne detection and Gaussian scattered lightconditions, the desired normalized electric field time correlation function $g(\mathbf{q},t) = <E(\mathbf{q},t)E^*(\mathbf{q},0)> / <|E^*(\mathbf{q},0)|^2>$ is related to the measured function $G(t)$ through the relation:

$$G(\mathbf{q},t) = A\left[1 + f^*\left|g(\mathbf{q},t)\right|^2\right] \tag{46}$$

where A describes the long delay time behavior of $G(t)$, usually $A=1$. If all relaxation processes contributing to $g(\mathbf{q}, t)$ relax within the time window of the correlator, then the instrumental factor f^* represents the short time intercept of the measured net function $[G(\mathbf{q}, t)/A - 1]$, see Fig. 38.

The information obtained by DLS from an isotropic medium depends on the polarization of the incident in respect the scattered light. Fluctuations in dielectric constant, $\delta\varepsilon_{zz}(q,t)$, determine the electric field correlation function $g_{VV}(q, t)$ of the polarized scattered light. In a single component system $\delta\varepsilon_{zz}(q,t)$ is proportional to the q th component of the number density fluctuations $\delta\rho(q,t) = \sum_j \exp(iqr_j(t))$, with $r_j(t)$ being the position of the j th molecule:

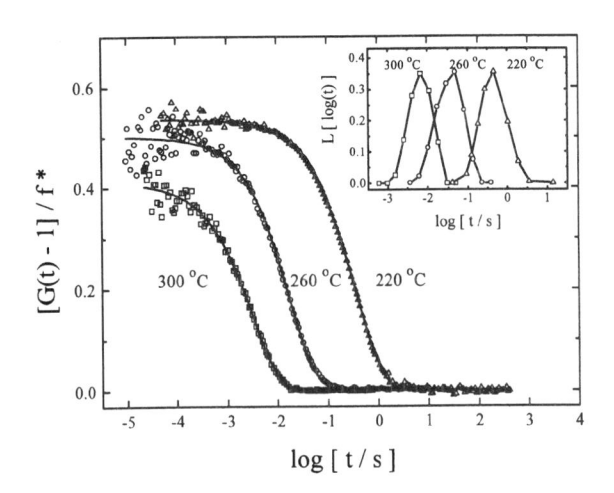

Figure 38. Depolarized intensity correlation functions, for supercooled As$_2$O$_3$, fitted with the *KWW* equation for three different temperatures. *Inset:* Distribution of the relaxation times obtained by inverse Laplace transform analysis, Eq. 50. From Ref. [72].

$$g_{VV}(q,t) = \left(\frac{\partial\varepsilon}{\partial\rho}\right)^2 \langle\delta\rho(q,t)\delta\rho(q,0)\rangle / \langle|\delta\rho(q,0)|^2\rangle \quad (47)$$

In the depolarized geometry, scattered light is determined by the fluctuations in $\delta\varepsilon_{yz}(q,t)$ which arises from the fluctuations $\delta\alpha_{yz}(q,t) = \sum_j \alpha_{yz}^j(t)\exp(iqr_j(t))$ in the molecular optical anisotropy $\alpha_{yz}(q,t)$, where α_{yz}^j is the yz component of the j th molecule. Then the corresponding normalized correlation function reads:

$$g_{VH}(q,t) = \langle\delta\alpha_{yz}(q,t)\delta\alpha_{yz}(q,0)\rangle / \langle|\delta\alpha_{yz}(q,0)|^2\rangle \quad (48)$$

The stretched exponential or Kohlrausch–Williams–Watts (KWW) equation is a frequently adopted empirical scheme employed to describe structural relaxation. KWW is used as a three parameter, τ^*, β_{KWW} and A, fitting function.

$$g(t) = A \exp[-(t/\tau^*)^{\beta_{KWW}}] \qquad (49)$$

The fractional exponent β_{KWW} ($0 < \beta_{KWW} < 1$) is a measure of the departure from the single exponential or Debye relaxation and has been correlated to the non-Arrhenius character of the α-process; τ^* is the structural relaxation time characterizing the process under study and A the amplitude of $g(t)$. The microscopic origin of the above equation remains still a matter of intense discussion. Two possible explanations have been put forward: (i) the heterogeneous scenario where different regions relax exponentially with a distribution in the time scales, and (ii) the homogeneous scenario where intrinsically non–exponential decay takes place in every region of the medium. Although there is still no consensus on the subject, the heterogeneous scheme seems as the most probable.

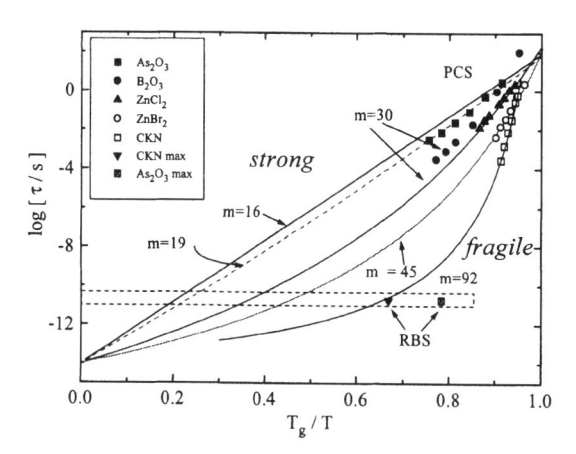

Figure 39. Modified Arrhenius (fragility) plot for typical inorganic glass–forming liquids. Relaxation times close to T_g have been obtained from PCS, whereas the dotted box denotes the time scale of the RBS.

Formulating $g(t)$ in the spirit of the heterogeneous scenario mentioned above the following expression is obtained:

$$g(t) = \int L(\log \tau) f(t,\tau) d\log \tau \qquad (50)$$

where $f(t,\tau) = \exp(-t/\tau)$. Therefore, the field correlation function is the weighted sum, $L(\log \tau)$, of independent contributions, each one caused by an individual microdomain. Solving equation (5) for $L(\log \tau)$ is a difficult task due to the ill–posed nature of the inverse Laplace transform (ILT) issue. Equation (50) is uniquely solved only in the case where no noise is present in the data. The most widely accepted method followed is the regularization procedure of the CONTIN code. The advantage following this approach is that no assumptions are made on any particular form of $L(\log \tau)$, except that $f(t,\tau)$ is simply exponential. Figure 38 shows typical experimental time correlation functions analyzed with the two methods described above. More details about application of dynamic light scattering to glass forming molten salts can be found elsewhere [72, 74].

Finally, it is worth mentioning that PCS is one of the few experimental techniques that can provide dynamic information over a wide time range close to the glass transition temperature of supercooled liquids. When such data are plotted in a modified Arrhenius plot, see Fig. 39, as proposed by Angell [66], one can distinguish between the so-called *strong* and *fragile* liquids. This classification has proved quite useful in the glass transition community where phenomenologies rather than theories seem frequently more valuable in assessing the dynamic properties of glass forming liquids.

5. Summary

In these lectures we have discussed a variety of experimental (light scattering) results from simple and glass–forming molten salts. The elementary theoretical principles and the data analysis procedures for Raman scattering, Rayleigh–Brillouin spectroscopy and Dynamic Light Scattering (Photon Correlation Spectroscopy) have also been presented. It has become evident that structural and dynamical features are interrelated and the interpretation of many changes that occur in the non–crystalline state can be better understood if a combined study for both structure and dynamics is undertaken.

References

1. Long, D.A. (1977) *Raman spectroscopy* Mc Graw-Hill, New York
2. Ferraro J.R., and Nakamoto K., (1994) *Introductory Raman spectroscopy* Academic Press, New York.
3. Berne, B.J. and Pecora, R. (1976) *Dynamic Light Scattering*, Wiley, New York.
4. Boon, J.P. and Yip S. (1991) *Molecular Hydrodynamics*, Dover, New York.
5. Hansen, J.P. and Mc Donald I.R., (1986) *Theory of Simple Liquids*, Academic Press, London.
6. Brüesch, P. (1986) *Phonons: Theory and Experiments II*, Springer–Verlag, Berlin.
7. Placzek, G. (1934) *Rayleigh-Streuung und Raman-Effect*, Leipzig.
8. Voyiatzis, G.A., Kalampounias, A.G. and Papatheodorou, G.N. (1999) *Phys. Chem. Chem. Phys.* **1**, 4797 and refs therein
9. Dai, S., Young, J.P., Begun, G.M., Coffield, G.M. and Mamantov, G. (1992) *Microchemica Acta* **108**, 261.
10. Iida, Y., Furukawa, M. and Morikawa, H. (1997) *Appl. Spectrsc.* **51**, 1426 and refs therein.
11. Laserna, J.J. (1966) *Modern Techniques in Raman spectroscopy* John Wiley and Sons, New York.
12. Papatheodorou, G.N. (1977) *J. Chem. Phys.* **66**, 2893.
13. Gilbert, B. and Materne, T. (1990) *Appl. Spectrosc.*, **44**, 299 and refs therein.
14. Cotton, F.A. (1990) *Chemical Applications of Group Theory*, Wiley-International, New York.
15. Zissi, G.D. and Papatheodorou, G.N. (1999) *Chem. Phys, Lett.*, **308**, 51.
16. See the biannual *Proceedings on Molten Salts* published by the Electrochemical Society Incorp., (1998) New York (e.g. Proc. Vol. 98-11
17. e.g. see "Progress in Molten Salt Chemistry" Vol. 1, Elsevier, New York (2000)

18. Enderby, J.E., and Biggin, S., in Advances in Molten Salt Chemistry ed. G. Mamantov, Elsevier, New York 1983 vol. 5 and refs therein
19. Neilson, G.W., and Adya, A.K. (1977) *Annual Reports C: Royal Soc. Chem.*, **93**, 101.
20. Brooker, H.M. and Papatheodorou, G.N. in Advances in Molten Salt Chemistry, ed. G. Mamantov, Elsevier, New York, 1983 vol. 5 and refs therein.
21. Papatheodorou, G.N. in Comprehensive Treatise of Electrochemistry, ed. B.E. Conway, J. O'M. Bockris and E.Yeager, Plenum Press, New York, 1983, vol. 5 p. 399.
22. Raptis, C., Mitchell, E.W.J. and Bunten, R.A.J. (1983) *J. Phys. C* **16**, 5351.
23. Raptis, C. and McGreevy, R.L. (1992) *J. Phys. Condens. Matter.* **4**, 5471 and references therein.
24. Papatheodorou, G.N., Kalogrianitis, T.G. Mihopoulos, S.G. and Pavlatou, E.A. (1996) *J. Chem. Phys.*, **105**, 2660.
25. Papatheodorou, G.N. and Dracopoulos, V. (1995) *Chem. Phys. Lett.*, **24**, 345.
26. Papatheodorou, G.N. and Dracopoulos, V. (2000) *Phys. Chem. Chem. Phys.*, **2**, 2021.
27. Madden, P.A., O'Sullivan, K.F., Board, J.A.B. and Fowler, P.W. (1991) *J. Chem. Phys.* 94, 918 (1991)
28. Badyal, Y.S. and Howe, R.A. (1993) *J. Phys. Condens. Matter.* **5** 7189.
29. Pavlatou, E.A. and Papatheodorou, G.N. Proc. VIII Int. Symp. Molten Salts, Electrochemical Society, 1992 vol. 92-16 p.72
30. Voyiatzis, G.A. and Papatheodorou, G.N. (1994) *Ber. Bunsenges Phys. Chem.* **98** 683.
31. Quist, A.S., Bates, J.B. and Boyd, G.E. (1972) *J. Phys. Chem.* **76** 78.
32. Dracopoulos, V. and Papatheodorou, G.N. Proc. XI Int. Symp. Molten Salts, Electrochem. Society 1998 vol. 98-11 p. 554
33. Spundflasche, E., Fink, H. and Seifert, H.J. (1995) *Z. Anorg. Allg. Chem.* **621**, 1723.
34. Pavlatou, E.A. and Papatheodorou, G.N. (2000) *Phys. Chem. Chem. Phys.* **2**, 1035.
35. Wilson, M. and Ribeiro, M.C.C. (1999) *Mol. Phys.* **96**, 867 and refs therein
36. Papatheodorou, G.N. and Voyiatzis, G.A. (1999) *Chem.Phys. Lett.* **303**, 151.
37. Price, D.L., Saboungi, M.L., Wang, J., Moss, S.C., Leheny, R.L. (1998) *Phys. Rev.* **B57**, 10496.
38. Tosi, M.P., Price, D.K. and Saboungi, M.L. (1993) *Ann. Rev. Phys. Chem.* **44**, 173.
39. Tagaki, R., Hutchinson, F., Madden, P.A., Adya, A.K. and Gaune-Escard, M. (1999) *J. Phys. Condens. Matter.* **11**, 645.
40. Pavlatou, E.A., Madden, P.A. and Wilson, M. (1997) *J. Chem. Phys.* **107**, 10446.
41. Hutchinson, F., Rowley, A., Walters, M.K., Wilson, M., Madden, P.A., Wasse, J.C. and Salmon, P.A. (1999) *J. Chem. Phys.* **111**, 2028.
42. Wasse, J.C. and Salmon, P.S. (1999) *J. Phys. Condens. Matter* **11**, 1381 ; (1999) **11**, 2171 and (1999) **11**, 9293 and refs therein.
43. Dracopoulos, V., Gilbert, B. and Papatheodorou, G.N. (1998) *J. Chem. Soc., Faraday Trans.* **94**, 1601.
44. Dracopoulos, V., Gilbert, B., Borresen, B., Photiadis, G.M. and Papatheodorou, G.N. (1997) *J. Chem. Soc., Faraday Trans.*, **93**, 3081.
45. Photiadis, G.M., Borresen, B. and Papatheodorou, G.N. (1998) *J. Chem. Soc., Faraday Trans.* **94**, 1605.
46. Chrissanthopoulos, A. and Papatheodorou, G.N. (2000) *Phys. Chem. Chem. Phys.* **2**, 3709.
47. Metallinou, M.M., Nalbandian, L., Papatheodorou, G.N., Voigt, W. and Emons, H.H. (1991) *Inorg. Chem.* **30**, 4260.
48. Iwadate, Y., Okado, N., Koyama, Y., Kubo, H. and Fukushima, K. (1995) *J. Mol. Liq.* **65-66** 369 and refs therein.
49. Chrissanthopoulos, A., Zissi, G.D. and Papatheodorou, G.N. (2000) *Proc. High Temp. Materials Chem.* (IUPAC).
50. Toth, L.M., Quist, A.S. and Boyd, G.E. (1973) *J. Phys. Chem.* **77**, 1384.
51. Toth, L.M. and Boyd, G.E. (1973) *J. Phys. Chem.* **77** 2654.
52. Photiadis, G.M. and Papatheodorou, G.N. (1999) *J. Chem. Soc. Dalton Trans.* 3541.
53. Photiadis, G.M. and Papatheodorou, G.N. (1998) *J. Chem. Soc., Dalton Trans.*, 981.

54. Dracopoulos, V., Vagelatos, J. and Papatheodorou, G.N. (2001) *J. Chem. Soc., Dalton Trans.* 1117.
55. Jäckle, J. in *Amorphous Solids: Low–Temperature Properties*, edited by W. A. Phillips (Springer, Berlin, 1981) p. 151.
56. See for example the relevant works in the proceedings of the Third International Discussion Meeting on *"Relaxations in Complex Systems"*, ed. by Ngai, K.L., Riande, E. and Ingram, M.D. (1998) *J. Non–Cryst. Solids* **235-237**.
57. Shuker, R. and Gammon, R. (1970) *Phys. Rev. Lett.* **25**, 222.
58. Foret, M., Courtens, E., Vacher, R. and Suck, J.-B. (1996) *Phys. Rev. Lett.* **77**, 3831.
59. Horbach, J., Kob, W. and Binder, K. (1998) *J. Non–Cryst. Solids* **235-237**, 320.
60. Taraskin, S.N. and Elliott, S.R. (1997) *Europhys. Lett.* **39**, 37 ; (1999) *Phys. Rev.* B **59**, 8572.
61. Whalley, E. and Bertie, J.E. (1967) *J. Chem. Phys.* **46**, 1264.
62. Martin, A.J. and Brenig, W. (1974) *Phys. Stat. Solid* (b) **64**, 163.
63. Malinovski, V.K. and Sokolov, A.P. (1986) *Solid State Commun.* **57**, 757.
64. Theodorakopoulos, N. and Jäckle, J. (1976) *Phys. Rev.* B **14**, 2637.
65. Elliott, S.R. (1992) *Europh. Lett.* **19**, 201.
66. For a review see, Angell, C.A. (1991) *J. Non–Cryst. Solids* **131-133**, 15.
67. Sokolov, A.P., Rössler, E., Kisliuk, A. and Quitmann, D. (1993) *Phys. Rev. Lett.* **71**, 2062.
68. See for example, Novikov, V.N. (1998) *Phys. Rev.* B **58**, 8367 and references therein.
69. Yannopoulos, S.N., Papatheodorou, G.N. and Fytas, G. (1997) *J. Chem. Phys.* **107**, 1341.
70. Yannopoulos, S.N. and Papatheodorou, G.N. (2000) *Phys. Rev.* B **62**, 3728.
71. Yannopoulos, S.N., Papatheodorou, G.N. and Fytas, G. (1996) *Phys. Rev.* E **53**, R1328.
72. Yannopoulos, S.N., Papatheodorou, G.N. and Fytas, G. (1999) *Phys. Rev.* B **60**, 15131.
73. Yannopoulos, S.N. and Kastrissios, D.T. unpublished data.
74. Pavlatou, E.A., Yannopoulos, S.N., Papatheodorou, G.N. and Fytas, G. (1997) *J. Phys. Chem.* **101**, 8748.
75. Zissi, G.D. private communication.
76. Zissi, G.D. and Yannopoulos, S.N. submitted to Phys. Rev. E.
77. Yannopoulos, S.N. (2000) *J. Chem. Phys.* **113**, 5868.
78. For a review on dynamical properties of fractals see: T. Nakayama, K. Yakubo and Orbach, R. (1994) Rev. Mod. Phys. **66**, 381.
79. Alexander, S. and Orbach, R. (1982) *J. Phys. (Paris) Lett.* **43**, L625.
80. Kirillov, S.A. and Yannopoulos, S.N. (2000) *Phys. Rev.* B **61**, 11391.
81. Kastrissios, D.T. and Yannopoulos, S.N. submitted to *J. Chem. Phys.*
82. Fabelinskii, I.L. Molecular Scattering of Light, (Plenum, New York, 1968)
83. Mountain, R.D. (1966) *Rev. Mod. Phys.* **38**, 205.
84. Yannopoulos, S.N. (1996) PhD Thesis, University of Patras.

NEUTRON SCATTERING:
TECHNIQUE AND APPLICATIONS TO MOLTEN SALTS

ASHOK K. ADYA

School of Science & Engineering
University of Abertay Dundee
Bell Street, Dundee, Scotland
United Kingdom DD1 1HG

www.abertay.ac.uk

E-mail: a.k.adya@abertay.ac.uk

Neutron scattering is a powerful, versatile and well-established technique capable of revealing the structural and dynamic properties of materials of ever increasing complexity at the atomic level. This article, resulting from a series of lectures given by the author at the NATO-ASI (advanced studies institute) on molten salts, consists of two parts. The first part gives an overview of the techniques of neutron scattering, its underlying theory, and methods of data collection and analyses. The major contribution of these techniques is the ability to determine, by using neutron diffraction isotopic substitution (NDIS) experiments, the individual partial structure factors (PSFs), $S_{\alpha\beta}(Q)$, and pair distribution functions (PDFs), $g_{\alpha\beta}(r)$, which is crucial in obtaining structural details of high spatial resolution. Since these distribution functions are the first in a hierarchy of inter-atomic correlations, they are the only ones directly accessible from experiments, computer simulations, and theory. The information obtained from such experiments can thus provide a critical test of the model potentials and liquid state theories. The NDIS methods can also assist in the analysis of spectroscopic (*e.g.*, Raman) data, which are targeted at the identification of chemical species in a liquid. However, if these species are short-lived and not dominant, they will not be detected by the NDIS because the $g_{\alpha\beta}(r)$ are average functions over all possible configurations and do not contain any information on individual species other than those that are long-lived and exhibit high degree of correlation.

The second part of this article deals with the application of the above techniques of neutron diffraction (ND) and neutron diffraction with isotopic substitution (NDIS) to a variety of molten salts aimed at revealing the structural details of these technologically important systems. It is worth emphasising that the dynamics of these systems form an integral and important part of a full structural description, and although neutrons provide a unique tool for their determination, due to space restrictions these aspects are not discussed here.

1. Introduction

Neutron scattering is a versatile and well-established technique, which allows us to explore the structure and dynamics of increasingly complex materials at the atomic level. It has several advantages over other diffraction techniques. These advantages are a direct consequence of the properties of the neutron, which are ideally suited to perform these measurements.

1.1. NEUTRON AND ITS PROPERTIES

Since neutrons are uncharged particles they can penetrate deeply into the target material without the need to overcome any Coulomb barrier. The neutrons, thus, interact mainly with the nuclei of the system *via* the strong nuclear force. The mass of the neutron results in neutron wavelengths of the order of inter-atomic separations, and the resulting interference effects yield information on the structure of the scattering material. The information contained in a diffraction pattern can be directly related to the inter-nuclear or inter-atomic structure of the system. Neutrons are only weakly absorbed by most materials and hence large samples can be examined. Moreover, by mixing different elements in the correct proportion it is possible to fabricate the so-called 'null alloy' containers, which are transparent to neutrons. Containers for work at elevated pressures and temperatures can thus be easily constructed and they do not contribute to the structure contained in the diffraction pattern. The most common of these is an alloy, $Ti_{0.68}Zr_{0.32}$, of titanium and zirconium, which is corrosion resistant and has the strength of intermediate steel.

In contrast to X-ray diffraction where X-rays are scattered more strongly by atoms with higher atomic number, neutrons are coherently scattered equally strongly by light or heavy elements. Thus, neutrons can probe the structure of materials containing light elements, such as lithium and hydrogen, in the presence of heavier atoms. Additionally, for neutrons there is a lack of form factor, which is characteristic of X-ray diffraction. The most significant scattering property of neutrons from structural viewpoint is, however, the isotope dependence of the scattering amplitude or scattering cross-section, which may differ considerably for different isotopes of the same element. This permits application of difference techniques of neutron diffraction with isotopic substitution (NDIS) for a unique structure determination. Also, since the interaction of neutrons with matter is weak, the neutron scattering can be conveniently treated theoretically on the basis of the first-order Born approximation [1].

The energy of thermal neutrons is similar to that of many excitations in condensed matter, which makes it possible to probe simultaneously the spatial and temporal correlations by determining the energy transfers in inelastic neutron scattering experiments. Neutrons have a magnetic moment. Thus, they can also be magnetically scattered *via* interaction between their magnetic moment and those of unpaired electrons in the sample. While inelastic and quasi-elastic neutron scattering techniques are powerful tools to study the dynamic behaviour of condensed matter, magnetic scattering gives useful information on magnetism – magnetic structures, magnetic moment distributions, and magnetic excitations. However, due to space restrictions none of these topics will be covered in this article, which will be mainly focused on the structural aspects only.

1.1.1. *The Scattering Vector*

The velocity of a neutron with mass, m (= 1.675×10^{-24} g) is directly related to its wavelength (or energy) through the de Broglie's relation, $\lambda = h/mv$, where h is the Planck's constant. The wavelength, λ, is a scalar quantity. However, the neutron beam, having a definite direction given by its velocity, requires a more appropriate quantity called the wave vector $\boldsymbol{k}$ to represent it. It has a magnitude $k = |\boldsymbol{k}| = 2\pi/\lambda = mv/\hbar$, so that the associated energy, $E = \hbar^2 k^2 / 2m$.

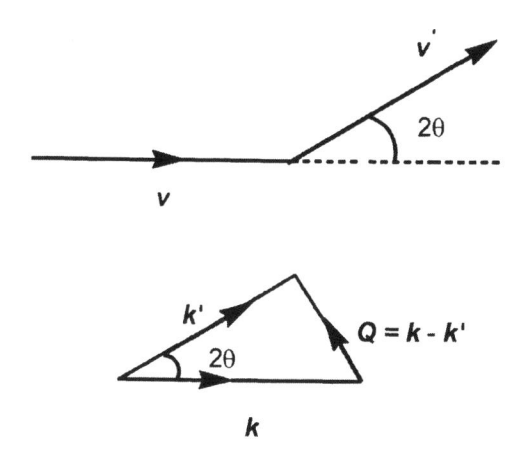

Figure 1. Definition of the neutron scattering vector, $\boldsymbol{Q}$.

In an actual scattering experiment, incident neutron beam travelling with velocity v (wave vector $\boldsymbol{k}$) strikes a target sample and is scattered with velocity v' (wave vector $\boldsymbol{k'}$) at an angle 2θ with the incident beam (see Figure 1). The transfer of energy ($\Delta E = \hbar \omega$) to the target sample and the transferred wave vector ($\boldsymbol{Q}$) in the scattering process are,

$$\Delta E = E - E' = \hbar \omega = \frac{1}{2}(mv^2 - mv'^2) = \frac{\hbar^2}{2m}(k^2 - k'^2), \qquad (1)$$

and, $$Q = k - k'. \tag{2}$$

The scattering vector, Q, has the dimensions of reciprocal length (Å^{-1}), the magnitude given by,

$$Q^2 = k'^2 + k^2 - 2k'k\cos(2\theta), \tag{3}$$

and can be considered as a momentum transfer vector since transfer of momentum to the target sample is given by $\hbar Q = mv - mv'$.

It is to be noted that the motion of atoms and their spatial correlations are revealed in a neutron scattering experiment when ΔE (or $\hbar\omega$) and $Q = |Q|$ match the corresponding energies and wave vectors involved in the scattering process. It is also clear from the above relations (Equations (1) an (2)) that the energy, E (or ω) and Q are related, and this relation imposes "kinematic" constraints on the scattering experiments. For instance, to gain access to a domain of large energy change (or large ω) but modest Q one has to use incident neutrons with very high energies and to perform measurements at small scattering angles. In other words, one can never access the whole of ω (or E) – Q space in a single neutron scattering experiment. It, thus, follows that the characteristics of a neutron source as well as that of the instrument used for neutron scattering measurements limit the range of experiments that can be performed.

1.1.2. Types of Scattering

Elastic Scattering. The energy change (ΔE) in elastic scattering is zero, so that $E' = E$ and $k' = k$. Since the wave-vector changes not in magnitude but only in direction, the magnitude of the scattering vector, Q becomes (from Equation (3)),

$$Q = |Q| = 2k\sin\theta = \frac{4\pi}{\lambda}\sin\theta. \tag{4}$$

Inelastic Scattering. When neutrons are scattered by the nuclei in a condensed system some of the scattering, resulting from the inelastic collisions (interactions) of the neutrons with vibrational and rotational modes of motion in atoms and molecules, is inelastic. The energy change (ΔE) in such a scattering process is finite and can be accurately measured. By examining the energy distribution of inelastically scattered neutrons, it is possible to gain substantial information about the various kinds of motions in condensed matter. However, at the same time, since such neutrons have wavelengths comparable to inter-atomic separations, the scattering is characteristic of both the spatial and dynamic correlations in the material under investigation.

Quasielastic Scattering. Quasielastic scattering occurs as a result of the interactions of neutrons with nuclei of atoms (in the sample) undergoing stochastic or random motions in a relaxation process. The energy change (ΔE) in this process is relatively small, and such scattering gives information about translational and rotational diffusion in condensed matter.

1.2. NEUTRON SOURCES AND INSTRUMENTATION

Matter can be investigated on a microscopic scale by different types of radiation such as X-rays (synchrotrons), light (lasers), electrons, and neutrons. With the exception of neutrons for which large scale facilities are required, all other types are also available on the small-laboratory scale. Two types of neutron sources are currently in use for the neutron scattering work, and these are the *nuclear reactors* and *accelerator-based spallation sources*.

Nuclear Reactors. The neutrons are produced in the core of the reactor by fission. An example of such a reactor is the Institut Laue Langevin (ILL) high-flux (58 MW) reactor (see, www.ill.fr) in Grenoble, France. It is, at present, the most powerful source of neutrons in the world dedicated for scientific use. It has the average source strength of 1.2×10^{15} neutrons cm^{-2} s^{-1}. Most *reactor* sources in the world are *continuous* sources, which deliver strong steady fluxes of neutrons. For the neutron scattering work, a reactor should have a high thermal flux, which is generally achieved by having a compact core.

Spallation Sources. Here, neutrons are produced by bombarding a heavy metal target with highly energetic particles from a powerful accelerator. Most *spallation* sources in the world are *pulsed* sources, which deliver short bursts of neutrons to a suite of instruments installed at a given facility. ISIS facility (see, www.nd.rl.ac.uk) at the Rutherford Appleton Laboratory, Didcot, UK, is the strongest pulsed neutron source in the world at present. The H$^-$ beam emerging from an ion source is accelerated in a linac to ~70 MeV. The beam, on entry to a synchrotron, is stripped to protons by the removal of its electrons in a 0.3 µm thick aluminium oxide foil. Its injection allows accumulation of ~2.5×10^{13} protons into two bunches, which are accelerated to ~800 MeV in approximately 10000 revolutions of the synchrotron. The accelerated proton beam is then deflected in a single turn into the extraction line where it is transported to a heavy metal non-fissioning target such as tantalum. This process is repeated 50 times a second giving a pulse frequency of ~50 Hz and a pulse width of ~0.4 µs.

The protons have sufficient energy to excite the nuclei of the target material into highly excited states, which release energy by "boiling off" neutrons. Each incident proton produces typically 15-20 neutrons.

Moderators. The neutrons produced in both types of sources are too energetic to be of use for thermal neutron scattering work in condensed matter. They must be slowed down or moderated to lower energies, and this is achieved by passing them through hydrogenous materials (moderators) placed around the reactor core or the spallation target. The large scattering cross-section of hydrogen and the multiple collisions of the passing energetic neutrons with the hydrogen nuclei result in their thermalisation to energies close to that of the moderator material. The beam characteristics (*e.g.*, spectral energy distribution, pulse shape, *etc.*) of the neutrons produced depend on the type and temperature of the moderator, which can be tailored for different types of experiments.

The moderators in use at ISIS are water at ambient temperature (H_2O, 316 K), liquid hydrogen (H_2, 20 K) and liquid methane (CH_4, 100 K). The neutron spectrum produced comprises of an epithermal region where the intensity varies as $1/E$ and a thermal region where a Maxwellian "hump" occurs whose maximum corresponds to the average temperature of the moderator. At the high-flux reactor at the ILL, 25 litres of liquid deuterium used as a cold moderating source provides a Maxwellian energy distribution with $T \sim 25$ K, while a block of graphite acts as a hot source providing energy distribution with $T \sim 2000$ K.

Collimators. Neutrons emerge from the moderator in all the directions. They are collimated to define the beam, and the simplest way to achieve this is with a series of apertures. On the D4C (upgraded recently from D4B) diffractometer (optimised for structural investigations of liquids and amorphous materials) at ILL, for instance, the vertical and horizontal slits produce a rectangular-shaped beam on the sample kept inside an evacuated sample chamber (bell jar). On SANDALS (Small Angle Neutron Diffractometer for Amorphous and Liquid Samples) diffractometer at ISIS, a circular aperture of diameter 3.2 cm defines the full sized beam. However, if fine collimation is required, Soller collimators comprising of a number of neutron absorbing blades arranged appropriately are used.

Monochromators. Bragg reflection by a suitable single crystal is the most commonly used technique for monochromatising the neutron beam on a steady state reactor source. By varying the monochromator angle, $2\theta_M$, monochromatic neutrons of varying wavelengths can be selected in a practically continuous way. For instance, on D4C at ILL, a cubic shaped holder capable of rotating around a vertical axis holds three different monocrystals on its lateral faces: Cu (200), Cu (220) and Cu (331) with d-spacing between their planes being 1.907, 1.278 and 0.829 Å, respectively. The rotating monochromator support enables selection of the suitable monocrystal in the reflecting mode with the possibility to focus the neutrons in the sample region, allowing selection of three different neutron wavelengths: 0.7 Å and 0.5 Å by using $\lambda/2$ filters (Ir, Rh), and 0.35 Å without the need of a $\lambda/2$ filter.

Suitable wavelength neutrons can also be produced by passing the beam through a helical velocity selector, which is a rotating cylinder of neutron absorbing material having many slots cut in its periphery along the major axis on a helical path. By choosing the speed of rotation of the cylinder about its major axis, neutrons of a particular wavelength can be continuously transmitted. However, the shortest wavelength that can be selected with this technique is ~ 4 Å, and the resolution, $\Delta\lambda/\lambda$ is generally greater than 3%.

The use of a single chopper on a pulsed source has a similar effect in producing single wavelength neutrons, but the beam is not continuous and the phase of the chopper relative to the beam pulse has to be accurately determined and maintained. By adjusting the position of the chopper with respect to the pulsed source one can choose a suitable wavelength whose resolution, related to the opening period, can be reduced by increasing the rotational velocity of the chopper, but the mechanical design becomes cumbersome and expensive.

1.2.1. *Instrumentation*

Two types of instruments are currently in use in neutron research: time-of-flight and time-average instruments. In principle, both types of sources can serve both types of instruments, but time-average instruments are found mainly at continuous sources because the continuous beams such as those at ILL are much stronger and deliver, on average, ~30 times more neutrons per second than do the pulsed beams such as those at the ISIS.

The necessary pulsing for the time-of-flight instruments at ILL is achieved however, mechanically by periodic chopping of the continuous beam. For the time-of-flight instruments, therefore, the spallation pulsed sources such as the ISIS have an edge over the continuous sources since neutron pulses from the spallation source are stronger than those produced by mechanical beam-chopping on a continuous source. In addition, while the advantages of the time-of-flight diffractometers at ISIS are the stationary and highly symmetric instrument configuration, and high resolution, those of the time-average diffractometers at the ILL are the better source stability, calibration, background and signal shape.

1.3. SAMPLE ENVIRONMENT

The stability of the sample, sample positioning and sample environment are crucial, and they play a vital role in determining the final accuracy of the diffraction data. The choice of the container material is dictated by the requirement that it must not react with the sample, and it should mostly scatter incoherently so that no additional structure is added to the scattering pattern. Moreover, it should have a low absorption cross-section to avoid excessive attenuation of the beam, and be sufficiently thin to minimise multiple scattering effects. At the same time, it is desirable to keep the sample in a sealed environment and attain a good statistics in the data acquisition.

Neutrons being uncharged particles, their penetration through most materials is much larger than other charged particles such as X-rays, electrons or protons. This makes it relatively easy to design container cells, which enable studies of liquid samples particularly in environments other than ambient. It is possible to fabricate titanium-zirconium null alloy (Ti$_{0.68}$Zr$_{0.32}$) containers, which being transparent to neutrons do not contribute to the structure contained in the diffraction pattern of a sample. Such containers are corrosion resistant, have the strength of intermediate steel, can be made with either cylindrical or flat-plate (parallelepiped) geometry, and used for both ambient and high-pressure studies. Vanadium, having a largely incoherent scattering cross-section, is also a suitable candidate to fabricate thin-walled (0.1 mm) containers (generally cylindrical) used for ambient and low-temperature experiments.

In addition, containers made from silica (quartz) or sapphire are used for furnace experiments where the samples are doubly contained in the quartz or sapphire tube inside a Ti/Zr or V-heater (furnace). On D4C at ILL, for instance, a normal V-furnace and a high-temperature furnace are available to work in the temperature ranges 100–1300 °C and 100–1800 °C, respectively. The other sample environments such as cryostat (2–350 K) and cryofurnace (-25–175 °C) are also catered for. Since silica has a very strong scattering of its own, which is highly Q-dependent, and as this is a major source of uncertainty when extracting the contribution of the sample (*e.g.*, a molten salt) from the observed diffraction pattern, care is necessary to properly subtract the silica (container) scattering.

1.4. NEUTRON DETECTORS

For good performance, an ideal neutron detector should have high neutron detection efficiency, a low intrinsic detector background, a low sensitivity to non-neutron events (in particular γ-rays), and a good stability. Since neutrons are neutral particles, their detection is based on nuclear interactions, such as, $^{3}\text{He} + {}^{1}\text{n} \rightarrow {}^{3}\text{H} + {}^{1}\text{p} + 0.77\,\text{Mev}$, or $^{10}\text{B} + {}^{1}\text{n} \rightarrow {}^{7}\text{Li} + {}^{4}\text{He} + 2.3\,\text{Mev}$, induced by their absorption by the detector material, producing energetic charged particles. Two types of neutron detectors have mainly been in use: the proportional gas detectors and the scintillation detectors. The former consist of a metallic (steel or copper) cylinder filled with ^{3}He (at ~16 atmospheric pressure) or BF$_3$ gas enriched with ^{10}B. A thin wire running axially along the cylinder acts as anode while the cylinder's outer case acts as a cathode, and a voltage of a few kV is applied between the two electrodes. The energetic charged particles produced by the nuclear reactions cause ionisation along their track, producing a detectable pulse of charge. The electrons produced in the ionisation are accelerated by the applied voltage and produce further ionisation. This results in a larger pulse proportional to the initial ionisation, and can be detected.

In scintillation detectors the absorber material (more commonly ^{6}Li embedded in glass or plastic) being in solid form, is much denser and results in higher detector efficiency as compared to the gas detectors. While the minimum thickness of a gas detector for good neutron efficiency is around 1 cm, the scintillation detectors can be made, within a few mm thickness, at least twice as efficient. This allows the installation of a large number of detectors on the instrument (such as those on SANDALS and GEM at the ISIS), thereby improving the statistics during the data acquisition. The nuclear reaction, $^6Li + {}^1n \rightarrow {}^4He + {}^3He + 4.79\,Mev$, leads to a large amount of energy to be deposited in the scintillator material (ZnS(Ag)) or the active component (Ce^{3+}) in glass, which in turn, emit a flash of light detected by a photomuliplier tube. Higher count rates are possible with the scintillation detectors than with the gas detectors because of shorter dead time. But, no matter how good a detector is, it is always "dead" for a short while (typically, 2–10 µs for a ZnS detector) after a neutron event has occurred, a dead-time correction [2] has to be made.

The position sensitive detectors (PSDs) in "banana" form in one-dimension or area detectors in 2-dimensions, containing hundreds or thousands of individual scintillator elements have also been installed on many instruments. Here, flexible optical fibres are used to link the scintillator elements to the photomultipliers in a coded sequence.

The fixed detectors on the SANDALS instrument at ISIS are situated on a constant resolution surface, and give the angular coverage from 3.8° to 37° in 2θ. Their concentration at low scattering angles optimises the use of epithermal neutrons and minimises the requirement to make inelasticity corrections, which become crucial while studying samples containing light elements. There are currently 1180 zinc sulphide ZnS(Ag)/^{6}Li scintillation detectors combined into 18 detector groups, each associated to a constant angle 2θ. With the incident flight path (distance between the moderator and the sample) of 11 m and the final flight path (distance from the sample to the different detectors) in the range 0.75 – 4.0 m, the available Q-range extends from 0.1 to 50 Å^{-1}.

On D4C, a two-axis instrument at ILL, nine microstrip detectors each consisting of 64 cells detect the diffracted neutrons after they fly through the evacuated collimation tubes of each detector. The high gas-pressure, 15.4 bar ^{3}He + 0.6 bar CF_4, gives good detection efficiency even at small wavelengths (91% at 0.7 Å, 82% at 0.5 Å). During the standard scans, the whole ensemble of detection array can be rotated around the sample from 1.25° to 140° covering overlapping angular ranges. The detector collimation and two $^{10}B_4C$ beam-stops placed after the sample avoid direct beam and parasitic peaks originating from evacuated aluminium bell jar containing the sample and furnace, *etc.*, on all the detectors. The detector collimation, however, restricts the maximum dimensions of a typical cylindrical sample as seen by the detectors to be 20 mm diameter and 50 mm height. Using Equation (4), and taking into account the angular range (1.25° < 2θ < 140°), the accessible Q-ranges for the three fixed wavelengths available on this instrument are: 0.7 Å (0.2 < Q/Å^{-1} < 17), 0.5 Å (0.3 < Q/Å^{-1} < 23), and 0.35 Å (0.4 < Q/Å^{-1} < 33).

Both D4C and SANDALS are also equipped with an incident beam monitor placed just before the sample, and this allows measuring the total number of incoming neutrons or recording the incident spectrum. On D4C, for instance, it is practically a transparent ^{3}He gas detector. In addition, on some instruments transmission beam monitors are also installed, and SANDALS has one. Together with the incident monitor, it may be used for measuring the total (coherent + incoherent + absorption) cross-section of the sample as a function of λ.

2. Overview of Theory

Variety of measurements can be made on the scattered neutrons after their interaction with the scattering system (a liquid, an amorphous solid, a crystal or a gas), but the results are usually expressed in terms of cross-sections – the quantities actually measured in a scattering experiment. Experimental cross-sections are generally quoted per atom (as for example in molten salts or ionic solutions) or per molecule (as for example in molecular liquids), and the theory deals with deriving theoretical expressions for them. Several textbooks [3, 4] and original research articles give excellent account of the theory of neutron scattering and the reader is encouraged to refer to them. In this section, however, we shall simply give an overview of some relevant theoretical results after introducing the basic quantities – the scattering cross-sections.

2.1. THE SCATTERING CROSS-SECTIONS

If a detector subtending a solid angle $d\Omega$ on the sample is set up (see Figure 2) to count the number of neutrons scattered in a given direction (θ, φ) as a function of their energy, E', the *partial differential cross-section*, $d^2\sigma/(d\Omega dE')$, having dimension of area/energy, gives the fraction of neutrons of incident energy E, scattered into $d\Omega$ in the direction (θ, φ), with final energies between E' and $E' + dE'$. However, if instead of analysing the energy of the scattered neutrons, we simply count all the neutrons scattered into the solid angle $d\Omega$ in the direction, θ, φ, the cross section corresponding to these measurements, called the *differential cross-section*, $d\sigma/d\Omega$, gives the fraction of neutrons scattered into $d\Omega$ in the direction, θ, φ. The *total scattering cross-section*, σ_{tot}, then gives the fraction of incident neutrons scattered in all the directions. The three cross-sections are related since,

$$\frac{d\sigma}{d\Omega} = \int_0^\infty \left(\frac{d^2\sigma}{d\Omega dE'} \right) dE', \tag{5}$$

and,
$$\sigma_{tot} = \int_{\theta,\varphi} \left(\frac{d\sigma}{d\Omega}\right) d\Omega. \qquad (6)$$

2.1.1. *The Scattering Lengths*

Because the wavelength (~10^{-10} m) of thermal neutrons is much larger than the range (~$10^{-14} - 10^{-15}$ m) of neutron-nucleus interactions, the scattering analysed in terms of partial waves [1] can only contain s-waves so as to be isotropic, and is characterised by a single complex parameter b, called the scattering length whose imaginary part corresponds to absorption. For some nuclei, the scattering length, b varies rapidly with the energy of the neutron, and the imaginary part of their scattering lengths is large. Such nuclei strongly absorb neutrons and the scattering from them leads to a resonance phenomenon associated with the formation of a compound nucleus. For majority of nuclei under discussion in this article, however, the imaginary part of the scattering length and hence absorption is small, and the scattering length is independent of the

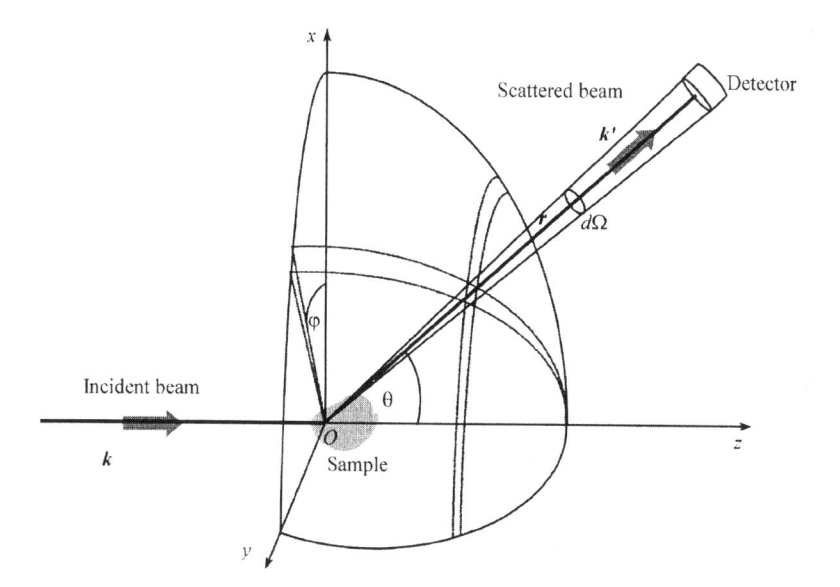

Figure 2. The scattering geometry in which the incident beam is scattered by a sample placed at the origin. The detector, defined by the polar angles, θ and φ, placed far from the zone of interaction of the neutron-target potential measures the number of neutrons, dn_i, scattered per unit time into the solid angle $d\Omega$.

energy of the neutron. The scattering lengths, in general, vary not only for different atomic types but also for different isotopes of the same element, and these depend also on the combination of the neutron spin with the nuclear spin (if it exists).

The form of the scattering potential, $\hat{V}$, that, under the Born approximation, gives isotropic scattering is a delta function in the form,

$$\hat{V}(r) = \frac{2\pi\hbar^2}{m} b\delta(r-R), \qquad (7)$$

and is known as the Fermi pseudopotential. R defines the position vector of the nucleus, and r that of the neutron. It is easy to show then that for scattering from a single fixed nucleus, the cross-section $d\sigma/d\Omega = |b|^2$, and hence the total cross-section, $\sigma_{tot} = 4\pi |b|^2$.

Coherent and Incoherent Scattering Cross-sections. If we now consider the scattering from a sample consisting of an assembly of N nuclei of a *single element*, defined by the position vectors, R_j ($j = 1,...N$), the scattered waves from different nuclei in different spatial positions will interfere, and the scattering is said to be *coherent* when all the nuclei in the sample have the same spin and isotope. Since the interference effects are given only by the average scattering potential, which is proportional to $\bar{b}$, the coherent scattering cross-section is proportional to $|\bar{b}|^2$, and is given by,

$$\sigma_{coh} = 4\pi |\bar{b}|^2. \qquad (8)$$

The deviations from the average potential are randomly distributed due to random distribution of scattering lengths or spin combinations. Such random distributions cannot give interference effects; they give rise to *incoherent* scattering with an isotropic incoherent cross-section, σ_{incoh} proportional to the mean-square deviation, $\overline{|b-\bar{b}|^2}$ of the scattering length, and given by,

$$\sigma_{incoh} = 4\pi \left\{ \overline{|b-\bar{b}|^2} \right\} = 4\pi \left\{ \overline{|b|^2} - |\bar{b}|^2 \right\}. \qquad (9)$$

$\bar{b}$ and $\overline{b^2}$ are respectively, the mean and mean-squared neutron scattering lengths for the target nuclei, averaged over all the isotopes and spin states. The cross-sections are usually given in units of barn, where 1 barn = 10^{-24} cm^2. The total scattering cross-section can, thus, be subdivided into coherent and incoherent parts,

$$\sigma_{tot} = \sigma_{coh} + \sigma_{incoh}. \qquad (10)$$

Sears [5] has listed the neutron scattering lengths and cross-sections of the various elements and their isotopes.

2.2. THE SCATTERING FUNCTIONS

In a scattering process, let the system (comprising of N nuclei) change from an initial state λ with energy, E_λ to a state λ' with energy $E_{\lambda'}$. By using Fermi's golden rule in quantum mechanics to predict the number of transitions from the initial state (k, λ) to the final state (k', λ'), and by applying energy conservation, which dictates the energy distribution of the scattered neutrons to be a δ-function of the form $\delta(E_\lambda - E_{\lambda'} + E - E')$ in which $E - E' = \hbar\omega$, such that $\int \delta(E_\lambda - E_{\lambda'} + \hbar\omega)dE' = 1$, it can be shown that the partial differential scattering cross-section (defined before, see Equation (5), Section 2.1) is given by,

$$\frac{d^2\sigma}{d\Omega\, dE'} = \frac{k'}{k} N \left[\overline{b}^2 S_{coh}(Q,\omega) + \left(\overline{b^2} - \overline{b}^2 \right) S_{incoh}(Q,\omega) \right]$$

(11)

$$= \frac{k'}{k} N \left[\frac{\sigma_{coh}}{4\pi} S_{coh}(Q,\omega) + \frac{\sigma_{incoh}}{4\pi} S_{incoh}(Q,\omega) \right].$$

$S_{coh}(Q,\omega)$ and $S_{incoh}(Q,\omega)$ are the coherent and incoherent *scattering functions* of the system, and these are the Fourier transforms *in time* of intermediate scattering functions, $I(Q,t)$ and $I_s(Q,t)$, respectively.

$$S_{coh}(Q,\omega) = \frac{1}{2\pi\hbar} \int I(Q,t) \exp(-i\omega t) dt. \quad (12)$$

$$S_{incoh}(Q,\omega) = \frac{1}{2\pi\hbar} \int I_s(Q,t) \exp(-i\omega t) dt. \quad (13)$$

The *intermediate scattering functions* are defined as,

$$I(Q,t) = \frac{1}{N} \sum_{j,j'} \left\langle \exp[-iQ.R_{j'}(0)] \exp[iQ.R_j(t)] \right\rangle, \text{ and} \quad (14)$$

$$I_s(Q,t) = \frac{1}{N} \sum_j \left\langle \exp[-iQ.R_j(0)] \exp[iQ.R_j(t)] \right\rangle. \quad (15)$$

The summation in $I(Q,t)$, (see Equation (14)) is carried out over all the atom pairs, j, j', and this function describes the time evolution of the relative separations of the atom pairs. The "self" intermediate scattering function, $I_s(Q,t)$ (see Equation (15)) however, describes the self-movement of the individual atoms with time.

The first term on the right hand side of Equation (11) shows that the cross-section is essentially a product of two factors: σ_{coh}, which depends on the interaction of the neutron with the individual particles in the scattering system, and $S_{coh}(Q,\omega)$, which depends not at all on the properties (intrinsic or otherwise) of the neutron. The coherent scattering function, $S_{coh}(Q,\omega)$, is thus a property only of the scattering system, and it contains information about the structure (relative positions) and motion of atoms relative to each other. The incoherent scattering function, $S_{incoh}(Q,\omega)$, reveals information on the motion of individual atoms in space and time. If the sample contains more than one element, generalisation of Equation (11) can be readily made, but then $S_{coh}(Q,\omega)$ becomes a weighted average over all the element pairs, and $S_{incoh}(Q,\omega)$ an average over all the elements in the scattering system. It should be noted that since crystalline materials have preferred orientations, the vector character of Q must be maintained. However, for isotropic substances such as liquid and amorphous materials, which are under discussion in this article, Q can be considered a scalar quantity since orientational averages are taken.

2.3. THE PAIR CORRELATION FUNCTIONS

We can define the time dependent pair-correlation function, $G(r, t)$ of the scattering system as,

$$G(r,t) = \frac{1}{(2\pi)^3} \int I(Q,t) \exp(-iQ \cdot r) dQ, \qquad (16)$$

such that the scattering function, $S_{coh}(Q,\omega)$ is the Fourier transform of $G(r, t)$ in space and time,

$$S_{coh}(Q,\omega) = \frac{1}{2\pi\hbar} \int G(r,t) \exp(-iQ \cdot r - \omega t) dr\, dt. \qquad (17)$$

It is easy to show then, that elastic scattering is directly related to the functions $I(Q, \infty)$ and $G(r, \infty)$ at $t = \infty$. Thus, if pure elastic scattering can be determined, it will give a time averaged distribution function, $G(r, \infty)$, describing the structure. $G_s(r, t)$ can be defined similarly, and it can be shown that for the incoherent elastic cross-section, the Fourier transform of $G_s(r, \infty)$ is the Debye-Waller factor.

2.4. STATIC APPROXIMATION AND NEUTRON DIFFRACTION

It should be noted, however, that pure elastic scattering does not occur for liquids and gases since the atoms in them are not localised in space. An important problem in treating these systems is the inelastic contribution to the scattering. When total (elastic + inelastic) scattering (called *neutron diffraction*) is measured by a quantity, called the differential scattering cross-section (DCS), $d\sigma/d\Omega$ (already defined in Equation (5)), it represents the scattered intensity summed over all possible energy transfers.

It is appropriate, to discuss at this stage, a limiting case known as the *static approximation* (SA) in which the term $E_\lambda - E_{\lambda'}$ in the energy conserving delta function, $\delta(E_\lambda - E_{\lambda'} + E - E')$ introduced in Section 2.2 is ignored. The cross-section then becomes a δ-function in $E - E'$ so that the scattering is zero unless $k = k'$. The condition for the validity of static approximation is $\Delta k \ll k$, which means that the energy transferred to or from the system is much smaller as compared to the energy of the incident neutron. Under these conditions, the only contribution to the integral involved in defining $S_{coh}(Q,\omega)$ in Equations (12) and (14), or (16) and (17), comes from $Q = Q_e$ (the value of Q when $k = k'$). We may, thus, use Equations (5) and (11) to write,

$$\left(\frac{d\sigma}{d\Omega}\right)^{SA} = \int_0^\infty \left(\frac{d^2\sigma}{d\Omega dE'}\right)^{SA} dE'$$
$$= N\left[\frac{\sigma_{coh}}{4\pi}S(Q_e) + \frac{\sigma_{incoh}}{4\pi}\right] = N\left[\overline{b}^2 S(Q) + \left(\overline{b^2} - \overline{b}^2\right)\right]$$
(18)

The subscript e is usually ignored, and the *static structure factor*, $S(Q)$, is obtained by integrating the coherent scattering function, $S_{coh}(Q,\omega)$ of Equation (11) over frequency,

$$S(Q) = \int_{-\infty}^{\infty} S_{coh}(Q,\omega) d(\hbar\omega)$$
$$= 1 + \rho \int (g(r) - 1)\exp(i\mathbf{Q}\cdot\mathbf{r}) d\mathbf{r},$$
(19)

where ρ, the average number density for a liquid is uniform, and given by $\rho = \langle \rho(r) \rangle = N/V$ (N = total number of particles and V = total volume). Since we confine are discussion to liquids, which are isotropic, $g(r)$ depends only on $|r|$, so that,

$$S(Q) = 1 + 4\pi\rho \int_0^\infty r^2 dr (g(r) - 1) \frac{\sin(Qr)}{Qr},$$
(20)

in which arguments of the functions are scalar quantities, r and Q (= $(4\pi/\lambda)\sin\theta$, see Equation (4)). The static pair distribution function, $g(r)$ can be obtained by back Fourier transformation of Equation (19),

$$g(r) = 1 + \frac{1}{2\pi^2 \rho} \int_0^\infty Q^2 dQ (S(Q)-1) \frac{\sin(Qr)}{Qr}. \tag{21}$$

The limit of $S(Q)$ for $Q \to 0$ (from Equation (19)) is,

$$\lim_{Q \to 0} [S(Q)-1] = \rho \int (g(r)-1) dr = \rho k_B T \chi_T - 1 \tag{22}$$

where the integral on the right hand side represents *density fluctuations* in the liquid, and it can be shown that this limit is related to the isothermal compressibility, χ_T.

2.4.1. *Placzek Corrections*

Unfortunately, the static limit (discussed above) cannot be realised in practice, and the static approximation fails when the scattering centres, due to their weak confining potential, can be easily knocked out of their positions by the momentum of the scattering collision. Corrections, usually called Placzek corrections [6], must be introduced in the data analysis to overcome these inelastic effects, and in its most simple form we may use Equation (18) to write for the differential cross-section,

$$\left(\frac{d\sigma}{d\Omega}\right)_{measured} = \left(\frac{d\sigma}{d\Omega}\right)^{SA} + \text{inelastic correction}$$

$$= N\bar{b}^2 S(Q) + N\left(\overline{b^2} - \bar{b}^2\right) + \text{inelastic correction} \tag{23}$$

Such a recoil effect will be small for a heavy nucleus, but for lighter nuclei such as H and D, the recoil is large. For samples containing H and/or D, the inelastic corrections are normally taken care of empirically by treating the second and third terms on the right hand side of Equation (23) as a slowly varying background, where polynomials in Q are fitted to remove the Placzek droop in the measured DCS. However, for heavier atoms, the Placzek [6] mass expansion may be used.

For the reactor based experiments, the procedure used to tackle inelasticity corrections may be checked by running the experiment at a slightly different neutron incident energy (or wavelength), and confirming that they lead to similar results. For the time-of-flight experiments on pulsed sources, however, a similar approach would require changing the neutron moderator, but this cannot be done routinely. Nevertheless, since the data recorded at different detector banks (scattering angles) require different recoil corrections, the correctness of the procedure can be established if at the end of the analysis there is a high degree of overlap between the data from different detector groups.

2.5. SCATTERING FROM A POLYATOMIC (MULTICOMPONENT) SYSTEM

The above treatment, restricted to systems with a single component, can be easily generalised to a multi-component system, and it is possible to write for the total differential scattering cross-section,

$$\frac{d\sigma}{d\Omega} = N \sum_\alpha c_\alpha \overline{b_\alpha^2} + N \sum_\alpha \sum_\beta c_\alpha c_\beta \overline{b_\alpha} \overline{b_\beta} (S_{\alpha\beta}(Q) - 1), \quad \text{or} \quad (24)$$

$$\frac{d\sigma}{d\Omega} = N \sum_\alpha c_\alpha \left(\overline{b_\alpha^2} - (\overline{b_\alpha})^2 \right) + N \sum_\alpha c_\alpha (\overline{b_\alpha})^2 + N \sum_\alpha \sum_\beta c_\alpha c_\beta \overline{b_\alpha} \overline{b_\beta} (S_{\alpha\beta}(Q) - 1). \quad (25)$$

The total DCS is, thus, a sum of incoherent, self and distinct contributions,

$$\left(\frac{d\sigma}{d\Omega}\right)_{tot} = \left(\frac{d\sigma}{d\Omega}\right)_{incoh} + \left(\frac{d\sigma}{d\Omega}\right)_{coh}^{self} + \left(\frac{d\sigma}{d\Omega}\right)_{coh}^{dist}, \quad \text{or} \quad (26)$$

$$\left(\frac{d\sigma}{d\Omega}\right)_{tot} = N \sum_\alpha c_\alpha \frac{\sigma_{\alpha,incoh}}{4\pi} + N \sum_\alpha c_\alpha \frac{\sigma_{\alpha,coh}}{4\pi} + N \sum_\alpha \sum_\beta c_\alpha c_\beta \overline{b_\alpha} \overline{b_\beta} (S_{\alpha\beta}(Q) - 1). \quad (27)$$

Since $\lim_{Q \to \infty} S_{\alpha\beta}(Q) = 1$, the total DCS should oscillate about the sum of self plus incoherent terms. The last term on the right hand side of Equation (27) defines the total structure factor, $F(Q)$,

$$F(Q) = \sum_\alpha \sum_\beta c_\alpha c_\beta \overline{b_\alpha} \overline{b_\beta} (S_{\alpha\beta}(Q) - 1), \quad (28)$$

which is a linear combination of contributions from pairs of different atom types (α, β) weighted by their scattering lengths ($\overline{b_i}$) and concentrations (atomic fractions), c_i. Fourier transforms of the partial structure factors, $S_{\alpha\beta}(Q)$ yield the partial pair distribution functions, $g_{\alpha\beta}(r)$,

$$g_{\alpha\beta}(r) = 1 + \frac{1}{2\pi^2 \rho} \int_0^\infty Q^2 dQ \left(S_{\alpha\beta}(Q) - 1\right) \frac{\sin(Qr)}{Qr}, \qquad (29)$$

where ρ is the average atomic number density. The function $g_{\alpha\beta}(r)$ represents the probability that there is an atom of type β at a distance r if there is one of type α at the origin. The peak positions in $g_{\alpha\beta}(r)$ correspond to preferred inter-atomic correlations between α and β atoms. The coordination number, $\bar{n}_\alpha^\beta$ for species of type β around a species of type α in the range $r_1 < r < r_2$ is related to the area under the peak, and can be determined from the pair distribution function (PDF), $g_{\alpha\beta}(r)$,

$$\bar{n}_\alpha^\beta = 4\pi \rho c_\beta \int_{r_1}^{r_2} r^2 g_{\alpha\beta}(r) dr. \qquad (30)$$

The Fourier transform of $F(Q)$, the total structure factor, is $G(r)$, the total radial distribution function (RDF), which can be written as a linear combination (or weighted sum) of all the partial PDFs, $g_{\alpha\beta}(r)$,

$$G(r) = \sum_\alpha \sum_\beta c_\alpha c_\beta \overline{b_\alpha b_\beta} (g_{\alpha\beta}(r) - 1). \qquad (31)$$

2.6. METHOD OF ISOTOPIC SUBSTITUTION

While it is still often possible to derive useful structural information from the total RDF, $G(r)$, the situation gets worse, particularly when there are too many partials, $g_{\alpha\beta}(r)$, contributing to it, making it extremely difficult to interpret the total $G(r)$ in terms of the individual pair contributions. Nevertheless, the unique sensitivity of the neutron scattering to isotopic composition of the sample means that appropriate experiments, in which the $\bar{b}$ values are changed by isotopic substitution, leaving everything else unchanged, permits the determination of the individual partial structure factors (PSFs), $S_{\alpha\beta}(Q)$ and PDFs, $g_{\alpha\beta}(r)$.

It may be recalled that these PDFs are the first in a hierarchy of inter-atomic correlations, and are the only ones directly accessible from experiment, computer simulations and theory. Knowledge of the $g_{\alpha\beta}(r)$ functions, thus, enables us not only to characterise the local structure of a liquid but also provides a useful and critical test of the model potentials used in simulation studies and liquid state theories.

For a system containing two distinct species, e.g., a simple molten salt, MX_n, a binary mixture or a diatomic molecular liquid, there are three PDFs, $g_{MX}(r)$, $g_{XX}(r)$ and $g_{MM}(r)$, and this number becomes proportionately greater in more complex systems.

For example, for a system comprising of m independent species the number of $g_{\alpha\beta}(r)$ are $m(m+1)/2$. The difference methods of neutron diffraction and isotopic substitution (NDIS) are ideally suited to determine these functions either individually or as linear combination of the form $G_\alpha(r)$ which is specific to the substituted species, α.

Under the assumption that the isotopic substitution does not affect the liquid structure, the intensities obtained from neutron diffraction experiments performed on two samples, identical except for the isotopic composition of one of the species, may be combined to give the first-order difference function,

$$\Delta_\alpha(Q) = F(Q) - F'(Q)$$
$$= \sum_\beta A_\beta \left[S_{\alpha\beta}(Q) - 1 \right]$$
(32)

with
$$A_\beta \begin{cases} = 2 c_\alpha c_\beta \overline{b_\beta} \left(\overline{b_\alpha} - \overline{b_{\alpha'}} \right), \beta \neq \alpha \\ = c_\alpha^2 \left[(\overline{b_\alpha})^2 - (\overline{b_\beta})^2 \right], \beta = \alpha, \end{cases}$$
(33)

where $\overline{b_\alpha}$ and $\overline{b_{\alpha'}}$ are the mean coherent scattering lengths of the substituted species in the two samples. The Fourier transform of $\Delta_\alpha(Q)$ gives $\overline{G}_\alpha(r)$,

$$\overline{G}_\alpha(r) = \frac{1}{2\pi^2 \rho} \int_0^\infty \Delta_\alpha(Q) Q^2 dQ \frac{\sin(Qr)}{Qr}$$
$$= \sum_\beta A_\beta \left[g_{\alpha\beta}(r) - 1 \right],$$
(34)

which is specific to the isotopically substituted species, α, and is a weighted combination of the corresponding partial distribution functions, $g_{\alpha\beta}(r)$. An important property of the difference function, $\Delta_\alpha(Q)$, is that the inelasticity corrections to a first-order approximation, being identical for both the samples, should cancel to give a negligible contribution from the inelastic term in Equation (32). The validity of this approximation can be checked as follows. Since $g_{\alpha\beta}(r) = 0$ at r-values below the distance of closest approach of the α and β species (i.e., r_{min}), $\overline{G}_\alpha(r)$ should be equal to the calculated limit,

$$\overline{G}_\alpha(0) = -\sum_\beta A_\beta,$$
(35)

at low r. In practice, $\overline{G}_\alpha(r)$ will oscillate about this value owing to statistical noise on the Q-space data, truncation of the data at finite Q-value, and systematic errors. However, if these effects are small, setting $\overline{G}_\alpha(r)$ equal to the $\overline{G}_\alpha(0)$ limit for $r \leq r_{min}$ and back Fourier transformation will yield the Q-space functions in close agreement with the original first-order difference functions from which they are derived. If a peak in $\overline{G}_\alpha(r)$ can be identified with some specific interaction and is completely separable then suitable integration limits can be chosen to obtain the coordination number,

$$^{eff}\overline{n}_\alpha^\beta = \frac{4\pi \rho c_\beta}{A_\beta} \int_{r_1}^{r_2} \left[\overline{G}_\alpha(r) - \overline{G}_\alpha(0)\right] r^2 \, dr. \tag{36}$$

The superscript '*eff*' indicates the possibility of multiple contributions to the peak. It may be stressed that this method of evaluating the coordination number gives the correct value if and only if the following conditions are satisfied:
(i) Left- and right-hand wings of the peak should return to their $\overline{G}_\alpha(0)$ values,
(ii) there are no spurious shoulders on either side of the concerned peak,
(iii) the peak is not affected by the spurious oscillations caused by the Fourier transformation of the truncated Q-space data,
(iv) low- and high-limits, r_1 and r_2, of integration should be exactly known, and
(v) especially, if no other atomic-pair contributes to the same peak.

Second- or higher-order differences between two $\Delta_\alpha(Q)$ in a manner similar to the first-order differences can, in theory, be performed to extract the individual $g_{\alpha\beta}(r)$. However, extreme care is required while working with higher order differences because the contribution of an individual partial to the difference function becomes, in least favourable cases, comparable to the errors associated with it. Thus, small errors in calculating the corrections to the observed intensity, in data normalisation, or presence of some systematic errors may prevent accurate determination of an individual $g_{\alpha\beta}(r)$.

The success of neutron diffraction isotopic substitution (NDIS) experiments depends crucially on several other factors. These include, (*i*) high quality and well characterised samples whose composition is accurately known in terms of atomic concentration and isotope content, (*ii*) a high flux neutron source which can provide sufficient statistical accuracy in the data, and (*iii*) stable instrumentation so that data are highly reproducible during the course of an experiment, which takes typically *ca.* 10-12 h per sample.

3. Data Collection and Analyses

For liquids and amorphous materials diffraction, irrespective of whether it is a constant wavelength (as on D4C at ILL) or time-of-flight (TOF) (as on SANDALS or GEM at

ISIS) diffraction, the scattering intensity, which is proportional to the structure factor $F(Q)$, is measured as a function of the momentum transfer, $\hbar Q$, where for elastic (no exchange of energy) scattering, $Q = (4\pi/\lambda)\sin\theta$ (see Equation (4)). In fixed wavelength experiments, λ, the neutron wavelength is fixed and the scattering angle, 2θ is varied. However, in the case of TOF experiments, λ is measured (see later) by the time-of-flight with θ held constant. An essential difference between the constant wavelength diffraction on a reactor source and the TOF diffraction on a pulsed source is that the TOF technique obviates the need to monochromatise the neutron beam which means that the full spectrum of neutron energies (or wavelengths) is incident on the sample.

The time-of-flight, t_{obs}, taken by a neutron to travel the total flight path from moderator to a detector (from the moderator to the sample, L_0, + sample to a detector, L_1) via the sample is given by $t_{obs} = (m/\hbar)[(L_0/k) + (L_1/k')]$. Assuming the static approximation to be valid (i.e., initial and final neutron energies to be the same, $E = E'$), $|k| = |k'| = k_e$, and the above expression becomes: $t_{obs} = (m/\hbar)[(L_0 + L_1)/k_e]$, where m is the neutron mass and k_e is the elastic wave vector for a particular time channel. It is then straightforward to determine the neutron wavelength, λ, for a detector forming an angle 2θ between a transmitted and scattered neutron wave vector from the relation: $\lambda = (h t_{obs})/[m (L_0+L_1)]$ (see Figure 3). The modulus of the transferred wave vector $\mathbf{Q} = \mathbf{k} - \mathbf{k'}$ becomes within the static approximation, Q_e, which is given by $Q_e = (4\pi/\lambda_e)\sin\theta$ (Equation (4)).

3.1. DATA COLLECTION

Usually, a set of data acquisition electronics is employed to store the data. In fixed wavelength experiments on reactor sources, scattered intensity is simply stored as a function of the scattering angle. However, the data acquisition using a TOF neutron diffraction experiment is more complex. It consists in recording each neutron event and giving it a label corresponding to the number of the detector in which it occurred and to the time of its arrival at the detector. The clock measuring this time of arrival is set to 0 every time a pulse of protons hits the target (e.g., tantalum), and at the ISIS this occurs 50 times a second. The counting of this event is incremented for each detector, pulses after pulses, so as to build a histogram of events. The counts are sorted into discrete time channels. Apart from the detection and collection of the scattered neutrons, the brightness of successive pulses may vary, for instance due to a small change in the moderator temperature. Also, from time to time, protons beam steering can modify the energy dependence of the spectrum. However, an incident monitor placed just before the sample (as discussed before) records the incident spectrum to counteract these effects.

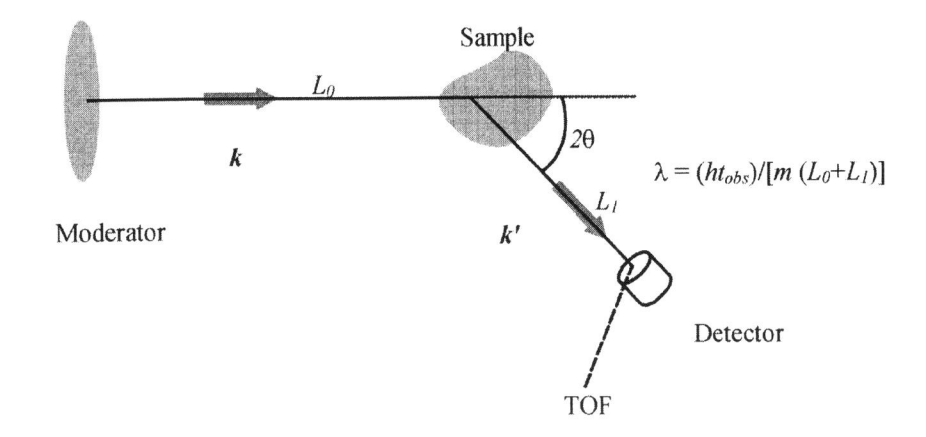

Figure 3. The schematics of a neutron diffractometer on a pulsed source.

3.2. DATA ANALYSES

Various types of corrections need to be applied to obtain the normalised differential scattering cross-section of a molten salt sample from the experimentally measured intensity data. Since the melt is usually doubly contained within the container kept inside a heater (furnace), the incident neutron beam could be attenuated by the heater, the container and the sample. The scattered beam originating within the sample or the container, or the heater, could be attenuated in the heater and either the container or the sample, or both. The data must be corrected for such attenuation effects. Additionally, since we wish to extract the total DCS corresponding to single diffraction events, the data have to be corrected for multiple scattering events. Practically, the absorption corrections are treated independently from the multiple scattering corrections. Also, since the data are to be normalised to the scattering from a standard vanadium calibration sample, and the background (when there is nothing in the beam) originating from neutrons and other forms of radiation must be subtracted, five measurements are needed:

    (i)   the sample(s) + container + heater ($S+C+H$)
    (ii)   the container + heater ($C+H$)
    (iii) the heater (furnace) alone (H)
    (iv) the vanadium (V), and
    (v)   the background (B)

Following the method of Paalman and Pings [7], the measured scattering intensities for a fixed wavelength experiment on a reactor source, for instance, can be written as:

$$I_H^E(\theta) = I_H(\theta) A_{H,H}(\theta), \tag{37}$$

$$I_{C+H}^E(\theta) = I_H(\theta) A_{H,CH}(\theta) + I_C(\theta) A_{C,CH}(\theta), \tag{38}$$

$$I_{S+C+H}^E(\theta) = I_H(\theta) A_{H,SCH}(\theta) + I_C(\theta) A_{C,SCH}(\theta) + I_S(\theta) A_{S,SCH}(\theta). \tag{39}$$

$I_H(\theta)$, $I_C(\theta)$ and $I_S(\theta)$ are the theoretical neutron intensities that would originate from the heater, container, and the sample, respectively, if no attenuation effects were present. The terms, $A_{ij}(\theta)$ are the attenuation factors, which when linearly combined with $I_H(\theta)$, $I_C(\theta)$ and $I_S(\theta)$, yield the experimentally measured intensities with superscript, E. In the attenuation factors, $A_{ij}(\theta)$, subscript i indicates the region in which the scattering occurs and the string j of characters in the subscript indicates the region(s) where attenuation through absorption or scattering, or both, occurs. Equations (37)-(39) yield the required attenuation corrected intensity, $I_S(\theta)$, which would be scattered by the melt in isolation.

$$I_S(\theta) = T_1(\theta) \left[I_{S+C+H}^E(\theta) - T_2(\theta) I_{C+H}^E(\theta) - T_3(\theta) I_H^E(\theta) \right], \tag{40}$$

where,

$$T_1(\theta) = 1/A_{S,SCH}(\theta),$$
$$T_2(\theta) = A_{C,SCH}(\theta)/A_{C,CH}(\theta), \tag{41}$$
$$T_3(\theta) = 1/A_{H,H}(\theta) \left[A_{H,SCH}(\theta) - \frac{A_{H,CH}(\theta) A_{C,SCH}(\theta)}{A_{C,CH}(\theta)} \right].$$

The Paalman and Pings attenuation coefficients, $A_{ij}(\theta)$, are determined by the Gauss method of integration described by Poncet [8] taking into account the scattering geometry. The intensities, $I_{S+C+H}^E(\theta), I_{C+H}^E(\theta)$ and $I_H^E(\theta)$ appearing in Equation (40) are free of background counts, $I_B(\theta)$, which means that background corrections are assumed to have already been made to the measured spectra before $I_S(\theta)$ is computed. The attenuation corrected spectra are corrected for multiple scattering [9] and then put on an absolute scale of barns steradian^{-1} (b sr^{-1}) by normalising it to the corrected spectrum of the vanadium standard, $I_V(\theta)$, given by,

$$I_V^E(\theta) = I_V(\theta) A_{V,V}(\theta), \tag{42}$$

and corrected for multiple scattering events. If, for instance, the vanadium calibration sample is placed inside the heater (at room temperature), its spectrum will have to be corrected for the various attenuation factors. The attenuation-corrected spectrum in this case is given by,

$$I_V(\theta) = \frac{1}{A_{V,VH}(\theta)} \left[I_{H+V}^E(\theta) - \frac{A_{H,VH}(\theta)}{A_{H,H}(\theta)} I_H^E(\theta) \right], \tag{43}$$

which, in turn, is corrected for the multiple scattering events. $I_{H+V}^E(\theta)$ is the experimentally measured intensity for the vanadium calibration sample inside the heater. Again, a background correction of the measured spectra is performed before computing $I_V(\theta)$.

It is likely that during the long runs some detectors will become noisy, unstable or even dead. It is necessary, therefore, to remove such bad detectors from the raw data files. Once that is done, the procedure of obtaining accurate structure factor entails making corrections to the data for counter dead time, and detector efficiencies in addition to the background scattering, multiple scattering and attenuation, discussed above. Moreover, the intensity data for all the measurements must be normalised to the incident beam monitor.

Let us now illustrate the above procedure for a time-of-flight (TOF) experiment on a pulsed source. The scattering intensity, $I^E(\theta, Q_e)$ for each detector group defined by the scattering angle, θ, and given by neutron counts per Q_e-bin, is proportional to the incident neutron flux, $\Phi(k_e)$, the solid angle subtended by the detector opening surface, $d\Omega$, the detector efficiency, $e(k_e)$, and the total TOF DCS (uncorrected for absorption and multiple scattering) of the sample, $\Sigma^{\text{uncorr.}}(\theta, Q_e)$, and can be written as,

$$I^E(\theta, Q_e) = \Phi(k_e)\, e(k_e)\, d\Omega\, N\, \Sigma^{\text{uncorr.}}(\theta, Q_e), \tag{44}$$

where N is the number of scattering units in the neutron beam. The incident flux and detector efficiency represented here as a function of k_e, emphasise that they are not a function of the scattering angle, θ. Using Equation (44) and taking into account the attenuation and multiple scattering corrections, we may write,

$$I_{S+C+H}^E(\theta,Q_e) = \Phi(k_e)e(k_e)d\Omega \left[\begin{array}{l} N_S \Sigma_S(\theta,Q_e) A_{S,SCH} + N_C \Sigma_C(\theta,Q_e) A_{C,SCH} \\ + N_H \Sigma_H(\theta,Q_e) A_{H,SCH} + M_{SCH}(k_e) \end{array} \right],$$

$$I_{C+H}^E(\theta,Q_e) = \Phi(k_e)e(k_e)d\Omega \begin{bmatrix} N_C \Sigma_C(\theta,Q_e)A_{C,CH} + N_H \Sigma_H(\theta,Q_e)A_{H,CH} \\ + M_{CH}(k_e) \end{bmatrix}, \text{and}$$

$$I_H^E(\theta,Q_e) = \Phi(k_e)e(k_e)d\Omega[N_H \Sigma_H(\theta,Q_e)A_{H,H} + M_H(k_e)],$$
(45)

where $\Sigma_S(\theta,Q_e), \Sigma_C(\theta,Q_e)$ and $\Sigma_H(\theta,Q_e)$ are the unattenuated single scattering total TOF DCS of the sample, container and heater, respectively. N_S, N_C and N_H are, respectively, the number of scattering units of the sample, the cell (container) and the heater. The Paalman and Pings coefficients have the same meaning as described before, and M_{SCH}, M_{CH} and M_H are the total multiple scattering DCS for the sample + can + heater, can + heater and heater, respectively.

A characteristic of neutron scattering is the ability to perform an independent estimate of the instrument calibration, and this is done by removing the unknown quantity $e(k_e) d\Omega/e_m(k_e)$ appearing when a detector spectrum after subtraction of the background is normalised to the incident monitor. The intensity measured by the incident monitor has the form,

$$I_m(k_e) = \Phi(k_e) e_m(k_e),$$
(46)

where $e_m(k_e)$ is the efficiency of the incident monitor detectors. For normalising to the incident monitor counts the spectrum for each measurement is divided by that of the incident monitor, so that,

$$^{Norm}I(\theta, k_e) = I^E(\theta, k_e) / I_m(k_e).$$
(47)

The estimate of the instrument's calibration is achieved by measuring the TOF spectrum of the vanadium standard because it gives a signal, which is almost elastic and scatters largely incoherently,

$$I_V^E(\theta,Q_e) = \Phi(k_e)e(k_e)d\Omega[N_V \Sigma_V(\theta,Q_e)A_{V,V} + M_V(k_e)].$$
(48)

Again, it is assumed, as before, that the background intensity, $I_B(\theta, Q_e)$ has already been subtracted from each of the measurements, including that of the vanadium standard. After removing the background from the vanadium spectrum, Chebyshev polynomials are fitted to the vanadium data, in order to remove the statistical noise and small Bragg's peaks, thus leaving the underlying structure. Once the attenuation and multiple scattering coefficients are evaluated, the vanadium calibration $C_V(\theta, Q_e)$ is obtained by using the smoothed vanadium spectrum, $^{Smooth}I_V(\theta, Q_e)$,

$$C_V(\theta, Q_e) = {}^{Smooth}I_V(\theta, Q_e)/I_m(k_e)$$
$$= [N_V \Sigma_V(\theta, Q_e)A_{V,V} + M_V(k_e)] \qquad (49)$$
$$= e(k_e)d\Omega/e_m(k_e).$$

Thus, the sequence of steps to determine the normalised TOF DCS of a sample is:

Step I. subtract background,

$$^{Total}I_J(\theta, Q_e) = I_J^E(\theta, Q_e) - I_B^E(\theta, Q_e), \qquad (50)$$

Step II. normalise to the incident beam monitor,

$$^{Norm}I_J(\theta, Q_e) = {}^{Total}I_J(\theta, Q_e)/C_V(\theta, Q_e), \qquad (51)$$

Step III. calculate single scattering events by subtracting the multiple scattering,

$$^{Single}I_J(\theta, Q_e) = {}^{Norm}I_J(\theta, Q_e) - M_J(Q_e), \qquad (52)$$

for the subscript $J = S + C + H$, $C + H$ and H in each of the Steps I, II and III above.

Step IV. Subtract heater from S+C+H and C+H,

$$^{Single}I_{S+C}(\theta, Q_e) = {}^{Single}I_{S+C+H}(\theta, Q_e) - {}^{Single}I_H(\theta, Q_e)\frac{A_{H,SCH}}{A_{H,H}},$$
$$\qquad (53)$$
$$^{Single}I_C(\theta, Q_e) = {}^{Single}I_{C+H}(\theta, Q_e) - {}^{Single}I_H(\theta, Q_e)\frac{A_{H,CH}}{A_{H,H}},$$

Step V. apply absorption corrections,

$$^{Single}I_S(\theta, Q_e) = \frac{{}^{Single}I_{S+C}(\theta, Q_e) - {}^{Single}I_C(\theta, Q_e)\frac{A_{C,SCH}}{A_{C,CH}}}{A_{S,SCH}}. \qquad (54)$$

It can be easily seen that Steps IV and V, together, lead to similar expression as Equation (40).

Step VI. Normalise to the amount of sample in the beam to calculate the total TOF DCS (b Sr^{-1}) within the static approximation,

$$\left(\frac{d\sigma}{d\Omega}\right)(\theta, Q_e) = \Sigma_S(\theta, Q_e) = \frac{1}{N_S}\left(^{Single}I_S(\theta, Q_e)\right), \quad (55)$$

which is given by,

$$\frac{d\sigma}{d\Omega} = \sum_\alpha c_\alpha \overline{b_\alpha^2} + \sum_\alpha \sum_\beta c_\alpha c_\beta \overline{b_\alpha} \overline{b_\beta}(S_{\alpha\beta}(Q) - 1). \quad (56)$$

The inelasticity corrections are then applied as discussed before (see Section 2.4.1) to each detector group (bank) separately to yield the distinct DCS for each bank. These are then merged to form a single composite structure factor $F(Q)$ (see Equation (28)). On SANDALS diffractometer at ISIS, this leads to Q-space data in the range ~ 0.1 Å^{-1} for the lowest angle detector group to ~ 50 Å^{-1} for the highest.

3.2.1. *Fourier Transformation*

The structure factors (total or partial) can be converted to r-space distribution functions, $G(r)$, $\overline{G}_\alpha(r)$ or $g_{\alpha\beta}(r)$ (see Sections 2.5 and 2.6) by direct Fourier transformation as discussed before (see, *e.g.*, Equations (29) and (34)). However, the real space distribution functions, so obtained, normally contain spurious features. One of the main causes of these errors is the statistical noise in the Q-space data, which, when two or more signals are combined together (as in the case of difference techniques of neutron diffraction by isotopic substitution), becomes especially significant. It is worth noting that the statistical noise at high-Q contributes significantly to the spurious low-r ripples because of the Q-dependence of the Fourier transform. Thus, by varying Q_{max} (the maximum Q-value for Fourier transformation), it is possible to distinguish between spurious features and the real data in the resulting transforms.

Additionally, even if inelasticity corrections are carried out with great care, some residual inelasticity may still remain in the data. Since these effects are expected to be slowly varying with Q (usually a low-order polynomial), they manifest themselves in the real-space data as a high frequency ripple at low r. For an experiment performed on a pulsed source the available Q-range is normally large enough to ensure that any scattering function has gone to zero before the Q_{max} is reached. Thus, the oscillations resulting from the termination step of the experimental data have very little influence on the r-space functions as compared to the data from a reactor source, where the data are smoothly forced to zero by the last experimental point by fitting some sort of window functions. Other methods of inverting the experimental scattering functions such as the minimum noise reconstruction technique [10-12] and the maximum entropy methods [13] have been proposed and widely used. These are particularly well suited for transforming noisy signals such as the higher-order difference functions.

The structural parameters, such as inter-atomic distances and coordination numbers (see, e.g., Equations (30) and (36), and Sections 2.5 and 2.6) can be determined directly from the pair distribution functions, $g_{\alpha\beta}(r)$, if these can be satisfactorily extracted by the isotopic substitution experiments. When this is not feasible, especially when the structural parameters are derived from the total structure factor, $F(Q)$, it is usual to fit this function with various models of the liquid structure. One need to be careful here because without the benefits of isotopic substitution, the total structure factor comprising of several partial structure factors may produce significant cancellation effects with much of the structure remaining hidden. The total diffraction patterns from ionic melts with strong charge ordering can thus be easily misinterpreted due to charge-induced mutual cancellation of important structural features. When the structural parameters are derived from the total radial distribution functions, $G(r)$, or $\overline{G}_\alpha(r)$ (see Sections 2.5 and 2.6) where several atom-pairs may contribute to a given peak, the functions $G(r) - G(0)$ or $\overline{G}_\alpha(r) - \overline{G}_\alpha(0)$ are usually fitted [14, 15] by a sum of Gaussians of the form,

$$f_i(r) = \frac{A}{\sigma\sqrt{2\pi}} \exp\left[-\frac{(r-r_{max})^2}{2\sigma^2}\right]. \tag{57}$$

The fitted parameters A, σ and r_{max} of each Gaussian then yield the area, half-width at half-maximum (HWHM) and mean position, respectively, of each peak.

4. Applications to Molten Salts

Molten salts are predominantly Coulombic type of liquids, with physical properties closely related to their microscopic structure. Although, detailed and unique structural information about them at an atomic level can be obtained from the NDIS methods [16-18], the acquisition of individual partial structure factors, $S_{\alpha\beta}(Q)$, and pair distribution functions, $g_{\alpha\beta}(r)$ between all the pairs of elements of a molten salt still represents a major problem in the structural elucidation of these disordered systems. In favourable cases, full information can be obtained if suitable isotopes with relatively large contrast variations and low absorption coefficients are available and affordable. The extensive application of this technique has led to a much better understanding of the 1:1 and 2:1 molten halide systems at a microscopic level with high precision. Since the two stable isotopes of chlorine (^{35}Cl and ^{37}Cl) have a large difference in neutron scattering lengths, it is not surprising that the most extensive data in the literature exist for the chlorides. Page and Mika [19] were the first to apply the NDIS technique [16] to the structural study of molten CuCl. Subsequently, Edwards et al. [20] applied this technique to some other 1:1 and then also to 2:1 halide melts followed by other workers in the field. With the exception of molten $DyCl_3$ for which partial PDFs were also obtained [21, 22] by the NDIS technique, a large number of other 3:1 halide melts have been investigated recently by neutron diffraction (ND) at the total structure factor or RDF level. Howe et al. [23] have extended these structural investigations to the molten halide mixtures. Among others, Adya and Neilson [24–26] applied this technique to pure and mixed molten nitrates. All these results are well documented and serve to illustrate the extent to which our knowledge of these systems has improved over the past three decades.

Since the aim of this article is not to give a review of these studies, in this section we shall choose some specific examples to illustrate how the ND and NDIS techniques have helped in revealing the microscopic structural details, structural trends and structure-property relations in these systems, with a view to encourage and stimulate new applications of these techniques either on their own or in conjunction with other experimental, theoretical or modelling techniques.

4.1. MOLTEN METAL HALIDES

For a binary molten salt with two components, α and β, the matrix elements describing the weighting of the three PSFs to the total $F(Q)$ (see Equation (28)) can be written as,

$$F(Q) = c_\alpha^2 \bar{b}_\alpha^2 [S_{\alpha\alpha}(Q)-1] + c_\beta^2 \bar{b}_\beta^2 [S_{\beta\beta}(Q)-1] + 2 c_\alpha c_\beta \bar{b}_\alpha \bar{b}_\beta (S_{\alpha\beta}(Q)-1), \quad (58)$$

If scattering lengths of one or both the components can be varied in such a way as to obtain three experimental total structure factors, $F_1(Q)$, $F_2(Q)$ and $F_3(Q)$, which are sufficiently different to enable three linear equations to be simultaneously solved for $S_{\alpha\alpha}(Q)$, $S_{\beta\beta}(Q)$ and $S_{\alpha\beta}(Q)$, all the PSFs may be extracted. In matrix notation, we may write,

$$|C||X(Q)| = |F(Q)|, \qquad \text{where,} \qquad (59)$$

$$|C| = \begin{vmatrix} c_\alpha^2 \bar{b}_{\alpha,1}^2 & c_\beta^2 \bar{b}_\beta^2 & 2c_\alpha c_\beta \bar{b}_{\alpha,1} \bar{b}_\beta \\ c_\alpha^2 \bar{b}_{\alpha,2}^2 & c_\beta^2 \bar{b}_\beta^2 & 2c_\alpha c_\beta \bar{b}_{\alpha,2} \bar{b}_\beta \\ c_\alpha^2 \bar{b}_{\alpha,3}^2 & c_\beta^2 \bar{b}_\beta^2 & 2c_\alpha c_\beta \bar{b}_{\alpha,3} \bar{b}_\beta \end{vmatrix} = \begin{vmatrix} c_{11} & c_{12} & c_{13} \\ c_{21} & c_{22} & c_{23} \\ c_{31} & c_{32} & c_{33} \end{vmatrix}, \quad (60)$$

$$|X(Q)| = \begin{vmatrix} S_{\alpha\alpha}(Q)-1 \\ S_{\beta\beta}(Q)-1 \\ S_{\alpha\beta}(Q)-1 \end{vmatrix} = \begin{vmatrix} X_1(Q) \\ X_2(Q) \\ X_3(Q) \end{vmatrix}, \qquad \text{and} \qquad (61)$$

$$|F(Q)| = \begin{vmatrix} F_1(Q) \\ F_2(Q) \\ F_3(Q) \end{vmatrix}. \qquad (62)$$

The inverse matrix for the solution of these equations,

$$|X(Q)| = |C|^{-1} |F(Q)|, \qquad (63)$$

can be worked out to extract the individual PSFs, $S_{\alpha\beta}(Q)$.

4.1.1. 1:1 Molten Metal Halides

Let us begin by choosing the simplest example of a common salt, NaCl, which is typical of a molten alkali metal chloride. Edwards et al. [20] and later Biggin and Enderby [27] determined the total $F(Q)$ for three isotopic samples, Na^{35}Cl, Na^{37}Cl and Na^{N}Cl at 875° C. They employed the NDIS techniques, as described in earlier sections, to extract the three PSFs, which are shown in Figure. 4. As can be seen, a deep valley – "Coulombic dip" in $S_{\text{NaCl}}(Q)$ at ~2 Å^{-1} lies in phase with the main "Coulombic" peaks in $S_{\text{ClCl}}(Q)$ and $S_{\text{NaNa}}(Q)$. This feature is representative of simple Coulombic ordering in 1:1 fused salts. The PDFs, $g_{\text{NaCl}}(r)$, $g_{\text{ClCl}}(r)$ and $g_{\text{NaNa}}(r)$ obtained by Fourier transformation of the corresponding PSFs are compared in Figure 5 with those obtained from molecular dynamics (MD) computer simulations [28] and liquid-state structural theory [29]. The agreement seems to be very good. An important feature of these functions is that the cation-cation, $g_{++}(r)$, and anion-anion, $g_{--}(r)$, PDFs are phased to give complete charge cancellation for $r \geq 4$ Å. This feature is characteristic of complete ionisation in the melt. Structural properties of all the molten alkali metal chlorides, LiCl [30,31], NaCl [20,27], KCl [32], RbCl [33], CsCl [32,34], typical of loosely coordinated structure, are well documented, and it is now known that their structural features can be reasonably reproduced (see e.g., Figure 5) by theoretical models based on Fumi-Tosi potentials.

The results of Eisenberg et al. [35], who used three isotopic samples, Cu^{35}Cl, Cu^{37}Cl and Cu^{N}Cl, to determine the PSFs and PDFs in molten CuCl by the NDIS technique, are generally similar though of higher accuracy than those of Page and Mika [19], and these indicate that the system is only weakly ionic with a definite tendency for charge ordering at long distances. These characteristics tend to favour low coordination number (3 ± 0.7) of nearest-neighbour unlike atoms and shorter Cu-Cu distances. In fact, the PDFs show (Figure 6) a deep penetration of Cu^+ ions into the first coordination shell of a Cu^+ ion, a considerable asymmetry between Cu-Cu and Cl-Cl PDFs, and almost a featureless (disordered) $g_{++}(r)$ and the corresponding PSF (not shown). By the same methods molten CuBr [36] and AgCl [37] are also shown to possess characteristics typical of weakly ionic systems similar to those of molten CuCl, but unlike those of alkali metal halides discussed above. Although, theoretical models based on simple ionic pair-potentials can still reproduce the gross structural features of these melts, one begins to feel the need to include polarisation effects in the model potentials.

4.1.2. *2:1 Molten Metal Halides*

A number of 2:1 halide melts [38-45] such as $MgCl_2$, $CaCl_2$, $BaCl_2$, $SrCl_2$, $ZnCl_2$, NiX_2 (X = Cl, Br and I) have been studied by the NDIS experiments while some others such as $MnCl_2$, $ZnBr_2$, and ZnI_2 have been investigated by the total neutron diffraction (ND) measurements. McGreevy and Mitchell [40] used the three isotopic samples of $SrCl_2$ enriched with ^{35}Cl, ^{37}Cl and ^{N}Cl, to determine the PSFs and PDFs in

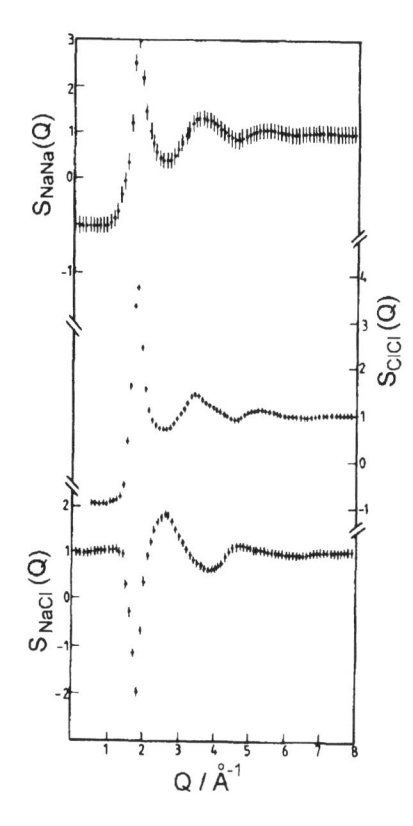

Figure 4. The partial structure factors, $S_{\alpha\beta}(Q)$ of molten NaCl at 875° C (after Biggin *et al.* [27]).

its molten state by the NDIS technique. These results are compared in Figures 7 and 8 with those obtained from MD simulations [46] and liquid-state structural theory [47]. The contrast between the structural features in the PSFs of molten NaCl *versus* those of $SrCl_2$ can be clearly seen (compare Figures 4 and 7). For instance, unlike that in NaCl, the principal peak in $S_{SrSr}(Q)$ lies neither in phase with the peak in $S_{ClCl}(Q)$ nor with the valley in the cross-term structure factor, $S_{SrCl}(Q)$, which in turn lies at Q-values intermediate between the positions of the main peaks in the like-ion structure factors. Also, the peak in $S_{SrSr}(Q)$ is significantly sharper and higher than that in $S_{ClCl}(Q)$. This behaviour, which is similar to that observed for other molten alkaline-earth metal chlorides such as $BaCl_2$ for which Edwards *et al.* [41] applied the NDIS technique to determine the Ba-Ba, Cl-Cl and Ba-Cl PSFs, suggests that the Coulombic ordering of the two ionic species clearly observed (see section 4.1.1) in the case of the alkali metal halides is disturbed to some extent by the relatively strong Coulombic repulsions between the divalent cations. Such repulsions should lead to

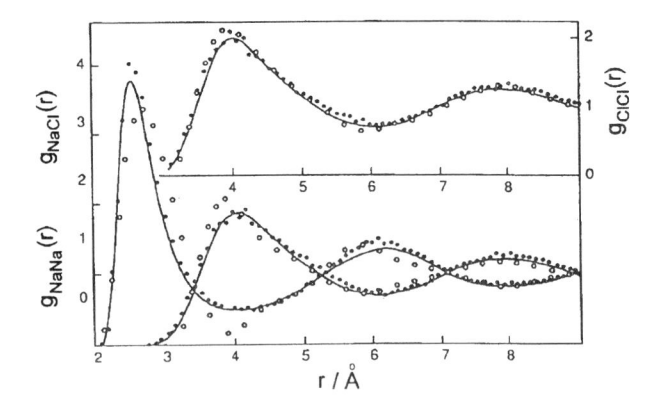

Figure 5. The PDFs of molten NaCl at 875° C from NDIS [27] (circles), liquid-state theory [29] (lines), and computer simulations [28] (dots) (after Ballone *et al.* [29]).

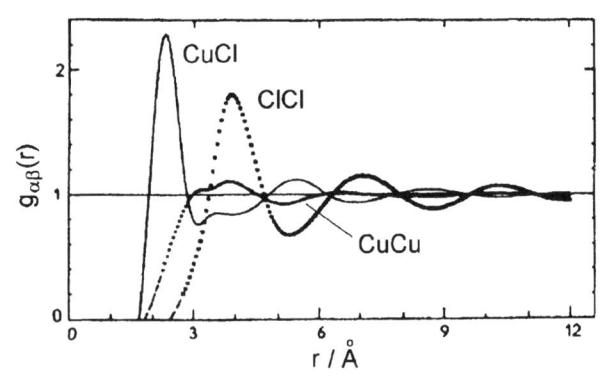

Figure 6. The PDFs of molten CuCl at 500° C from NDIS (after Eisenberg *et al.* [35]).

relatively large cation-cation distances as has actually been observed (~5 Å in BaCl$_2$ and SrCl$_2$ compared to ~ 4 Å in NaCl) in the $g_{++}(r)$ (see Figures 5 and 8 (*a*)). It may be noted (see Figures 7 and 8) that although a simple ionic pair-potential is still able to reproduce the broad features of structural ordering in SrCl$_2$ melt, significant discrepancies at a quantitative level have now appeared between the computed and experimental data.

We now consider the structural behaviour of molten ZnCl$_2$, which is quite different from the other 2:1 melts such as BaCl$_2$ and SrCl$_2$ discussed above. Biggin and Enderby

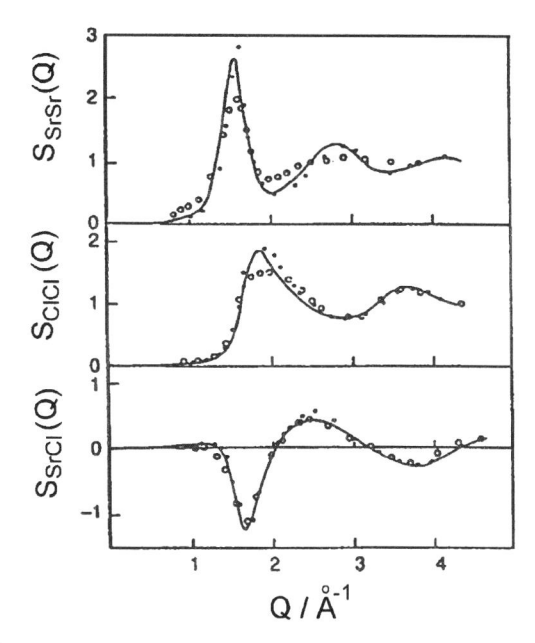

Figure 7. The PSFs of molten SrCl$_2$ at 925° C from NDIS (circles) [40], liquid structural theory (curves) [47] and MD simulations (dots) [46] (after Pastore *et al.* [47]).

[42] extracted the PSFs and PDFs of molten ZnCl$_2$ by using three samples with different isotopic enrichments on the chlorine atom. The PSFs reveal (see Figure 9) that the main peak in $S_{ClCl}(Q)$ is now sharper and higher than that in $S_{ZnZn}(Q)$ (which is in contrast to the pattern observed in molten BaCl$_2$ and SrCl$_2$, see *e.g.*, Figure 7). The principal peaks in $S_{ClCl}(Q)$ and $S_{ZnZn}(Q)$ are again, as in the molten NaCl (see Figure 4), in phase with each other and also with the valley in the cross-term structure factor, $S_{ZnCl}(Q)$. Interestingly, a new peak called the First Sharp Diffraction Peak (FSDP) or pre-peak has now appeared at ~ 1 Å^{-1} in nearly all the PSFs, and this is more prominent in the metal-metal PSF, $S_{ZnZn}(Q)$. The prepeak is felt to be a signature of intermediate range order (IRO) corresponding to inter-atomic correlations on a scale of $\sim 2\pi/Q \approx 6$ Å. While in the case of BaCl$_2$, SrCl$_2$ and CaCl$_2$, penetration effects (see *e.g.*, Figure 8 (*a*) and (*b*)) play a major role in determining their static structure and dynamic behaviour, the PDFs of molten ZnCl$_2$ show (Figure 8 (*c*)) virtually no such penetration of like ions into the nearest-neighbour shells. Also, the PDFs of molten ZnCl$_2$ reveal that ~ 4 chloride ions provide a tetrahedral site for each zinc ion, that the nearest neighbour distances, $r_{Zn-Zn} = r_{Cl-Cl}$, and that there is no exchange of ions between the first-shell and the surrounding liquid. These results are a clear testimony for the existence of highly stable tetrahedral, $ZnCl_4^{2-}$ structural units, which tend to form a network *via*

chlorine-sharing linkages. Such a network should result in a state of pronounced IRO giving rise to the appearance of the FSDP, actually observed in the structure factor data.

The structural results at the PDF level obtained from NDIS studies on 2:1 chloride melts also reveal that with decrease in the cation size from Ba (1.35 Å) → Sr (1.12 Å) → Ca (0.99 Å) → Zn (0.74 Å), the coordination number, n_{+-} decreases from 7.7, 6.9, 5.4 to ~ 4, the first peak in $g_{++}(r)$ shifts markedly inwards relative to that in $g_{--}(r)$ to an extent that in CaCl$_2$ and ZnCl$_2$ these peaks coincide, which is contrary to the requirements of maximising the separation between doubly charged cations. Also with decrease in the cation size, the nearest-neighbour shell in $g_{+-}(r)$ becomes progressively much well defined with lesser penetration of like ions into the first coordination shell of unlike pairs, and the FSDP indicative of the presence of IRO appears in the cation-cation PSFs in smaller cation (Ca, Mg, and Zn) systems. Computer simulations based on simple ionic pair potentials [48,49] have failed to reproduce such an intriguing evolution

Figure 8. The PDFs, of molten (*a*) SrCl$_2$ from NDIS (circles) [40], liquid-state theory [47] (curves) and MD simulations [46] (dots): on left, $g_{+-}(r)$, on right, $g_{--}(r)$ and inset, $g_{++}(r)$ (after Pastore *et al.* [47]). The PDFs of molten (*b*) CaCl$_2$ [39] and (*c*) ZnCl$_2$ [42] $g_{+-}(r)$, $g_{--}(r)$, $g_{++}(r)$: (full, broken and dotted curves, respectively).

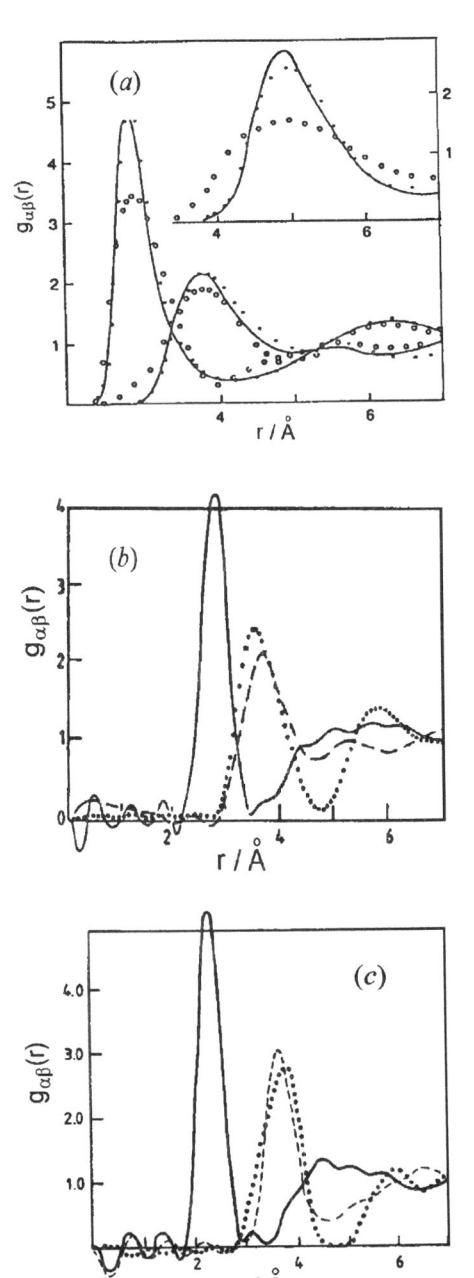

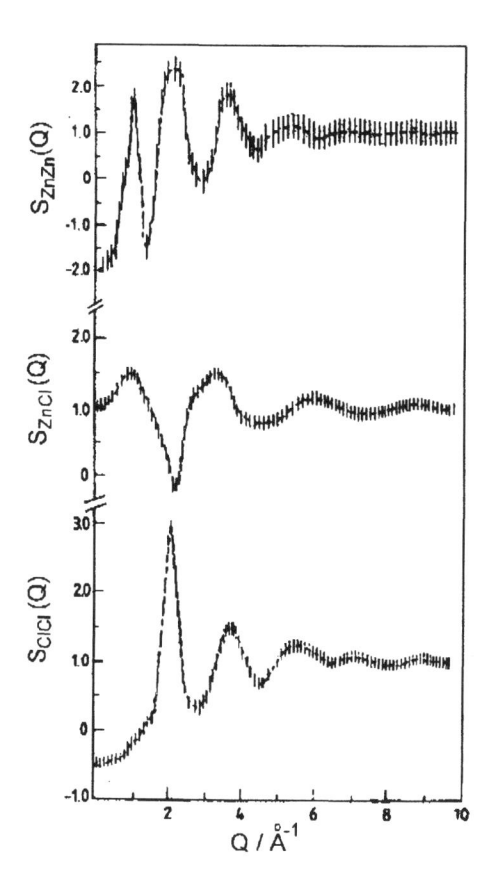

Figure 9. The PSFs, $S_{\alpha\beta}(Q)$, of molten $ZnCl_2$ (after Biggin and Enderby [42]).

of short- and intermediate-range order in the structure of these 2:1 melts with decreasing cation size. The results of total ND measurements on molten ZnX_2 (X = Cl, Br, and I) by Allen *et al.* [44] confirm that the tetrahedral coordination of Zn^{2+} ions is stable to both changes in the anionic species and temperature.

The three PDFs, $g_{\alpha\beta}(r)$ relating to Ni–Ni, Ni–X and X–X (X = Cl, Br and I) in NiX_2 melts have also been determined by using ^{N}Ni, ^{62}Ni and ^{0}Ni enriched isotopic samples. The results on $NiCl_2$ suggest that although the local structure around Ni^{2+} ion resembles that around the Zn^{2+} ion, there are differences. For instance, unlike Zn^{2+}, there is absence of intermediate range correlations of the Ni^{2+} ions, and the shallow first minimum in Ni-Ni PDF suggests a weaker first-coordination shell and an enhanced mobility of the cations in molten $NiCl_2$. The Ni^{2+} ions in the three melts [43,50,51] are 4-fold coordinated with the ratios r_{-}/r_{++} close to (1.63) being truly tetrahedral. The

$g_{++}(r)$ and $g_{--}(r)$ bear a close resemblance to each other without any significant first shell penetration for both $NiCl_2$ and $NiBr_2$. In contrast, the cation-cation PDF of molten NiI_2 shows a broad peak similar to the one observed previously in molten CuCl, and the degree of overlap with the principal peak of the cation-anion distribution is so large that it leads to an extensive penetration of the nickel species into the first coordination shell.

An interesting aspect of ND structural data on 2:1 and also 3:1 (see later) molten halides is the appearance of a FSDP at ~ 1 $Å^{-1}$ which is absent in the diffraction patterns of 1:1 halide melts. Since the FSDP occurring at small Q-values is given a relatively small weighting in the Fourier transform procedure (see Equation (29)), the real space effect of this prepeak is spread out and not well understood. As mentioned earlier, its presence has been linked to the existence of intermediate range order (IRO). For example, all the three Zinc halide melts show a well resolved prepeak at ~ 1 $Å^{-1}$ in the total structure factors, $F(Q)$. A significant shift in its position to a lower Q-value and an increase in its height occur with increase in the anion size or with decreasing ionicity from chloride to bromide to iodide. This trend reflects highest degree of IRO in the iodide where the short range ordering appears to be the weakest. A similar trend is also found in the prepeaks observed at Q = 0.99, 0.92 and 0.88 $Å^{-1}$ in molten $NiCl_2$, $NiBr_2$ and NiI_2, respectively. The most successful attempts to explain the structural evolution in 2:1 halide melts have been by Wilson and Madden [52] who used a simple polarisable ion model (PIM) without involving the concept of charge transfer. Computer simulations based on this model unlike those based on the simple ionic pair potentials – rigid ion model (RIM) are found to reproduce qualitatively the observed experimental features such as the overlap of the principal peaks in $g_{++}(r)$ and $g_{--}(r)$ as well as the FSDP in $S_{++}(Q)$.

4.1.3. *3:1 Molten Metal Halides*

Although a large number of 3:1 halide melts of type MX_3, spanning a wide range of ionic radii, have now been investigated, the NDIS technique has been applied only to molten dysprosium chloride ($DyCl_3$) in which isotopic substitution on Dy was employed to satisfactorily extract [18,21,53] the three PSFs, $S_{Dy-Dy}(Q)$, $S_{Dy-Cl}(Q)$ and $S_{Cl-Cl}(Q)$, and the corresponding PDFs, $g_{\alpha\beta}(r)$. The experimental structural results have been interpreted with the help of RIM and PIM computer simulations [22] in a truly complementary way.

The experimental total $F(Q)$ for the three isotopic samples of $DyCl_3$ show that the main or predominant contribution to the FSDP at $Q \approx 1$ $Å^{-1}$ is the metal-metal term, and these Dy-Dy correlations exist on the scale of IRO through chlorine sharing linkages. While the RIM and PIM simulated PDFs, $g_{Dy-Cl}(r)$ and $g_{Cl-Cl}(r)$ reproduce equally well the experimental PDFs, and have similar shapes from the two models, the results for Dy-Dy partial PDF obtained from the two models are quite different (see Figure 10). The NDIS results show that the first peak in $g_{Dy-Dy}(r)$ is split into two component distances and the PIM results seem to reproduce the experimental function very nicely,

Figure 10. The partial PDFs for pure molten $DyCl_3$: experimental data (open circles), RIM results (solid line) and PIM results (dashed line) for (*a*) Cl-Cl, (*b*) Dy-Cl and (*c*) Dy-Dy correlations [22].

highlighting the importance of including polarisation effects in the model potential used for simulations. These results show that (*i*) the distribution of anions is hardly affected by the polarisation effects and, (*ii*) although repulsion between triply charged Dy^{3+} ions should maximise the separation between the cations (similar to that observed in the RIM results), this effect is offset by the polarisation of anions by triply charged cations, thus leading to shortened cation-cation distances due to Dy-Cl-Dy 'bond bending'. The first peak in $g_{Dy-Cl}(r)$ yields a coordination number of 6, which shows that octahedral coordination in molten $DyCl_3$ is preserved on melting.

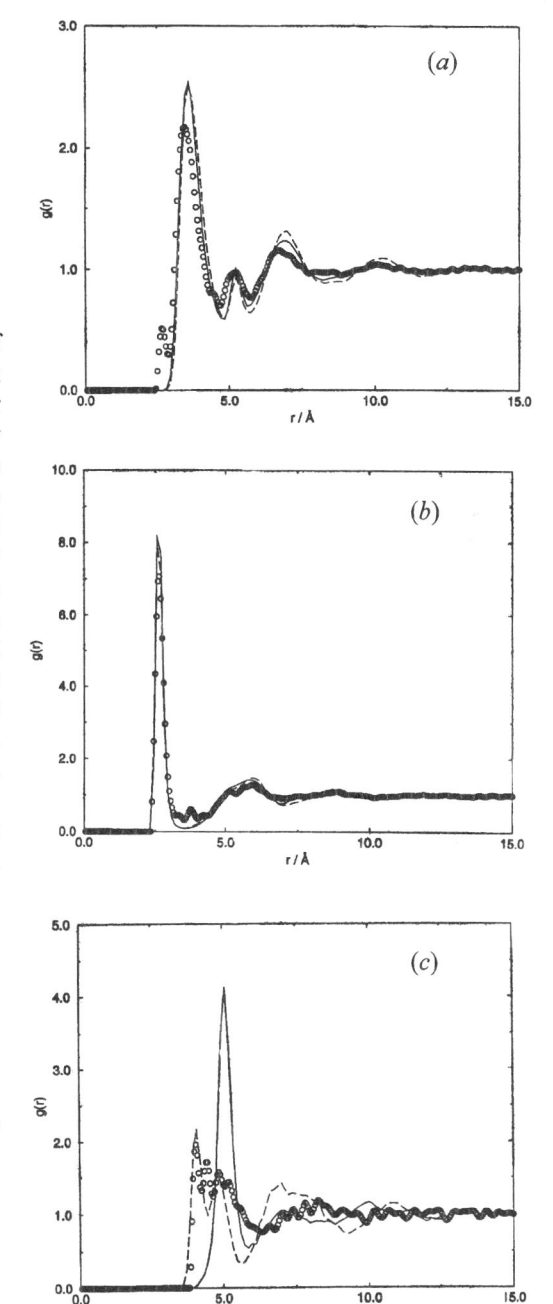

Another trend worth noticing (see Figure 10) is that the main peak in $g_{Dy-Dy}(r)$ is a doublet. The low-r feature is associated with a pair of cations connected to each other *via* a pair of chloride ions while the large-r feature arises from a pair of cations sharing a single anion. It is clear that the shift of the first feature to substantially smaller separations is caused due to the two cations being drawn together by the edge-sharing arrangement between the connected polyhedra.

All the other trivalent halide melts have been analysed at the total $F(Q)$ and $G(r)$ level [54-71], and due to space restrictions these will not be discussed here.

4.2. MOLTEN NITRATES

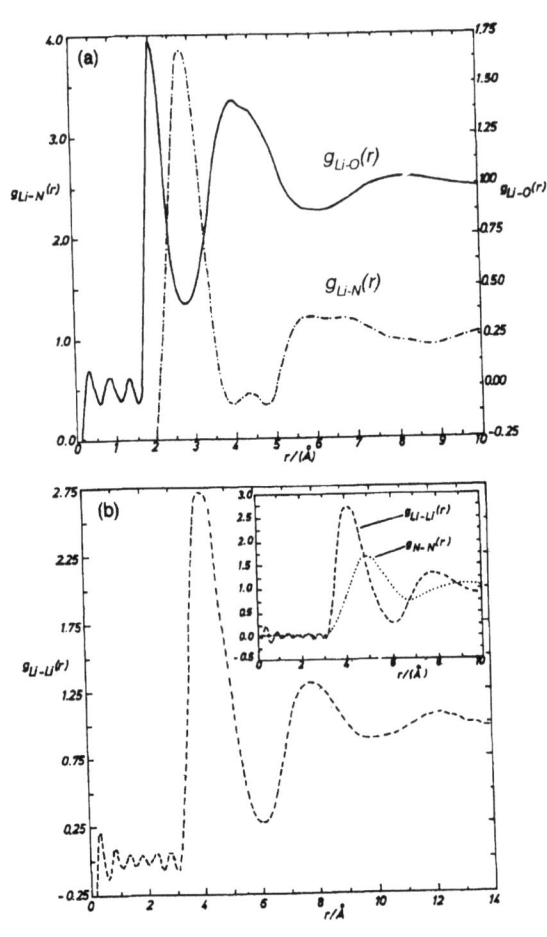

Figure 11. The PDFs [25] (*a*) $g_{Li-N}(r)$, $g_{Li-O}(r)$, and (*b*) $g_{Li-Li}(r)$ in molten LiNO$_3$ compared with $g_{N-N}(r)$ [72] in inset (after Adya *et al.* [25]).

Molten nitrates have relatively low melting points; provide industrially important low temperature baths; show superior glass forming capabilities owing to the tendency of the nitrate ion to resist reorientation in the disordered phase; exhibit rich phase behaviour on mixing with other nitrates and are essential ingredients in many industrial fertilisers and explosives. Although, a number of ND studies on the MNO$_3$ (*e.g.*, with M = Li, Na, K, Rb, Cs, Ag, Tl, and NH$_4$) melts have been reported, the results from most of them have been analysed at the total $F(Q)$ or $G(r)$ level only. Molten LiNO$_3$ and NH$_4$NO$_3$ are the only systems on which Adya *et al.* [24-26, 73] applied the NDIS techniques to determine the local coordination around the component ions. Here we discuss the results of molten LiNO$_3$ in which isotopic substitution on Li and N-nuclei have been used to determine the ion-ion and ion-counterion partial structure factors relating to the $g_{Li-Li}(r)$, $g_{Li-N}(r)$, and $g_{Li-O}(r)$ functions (shown in Figure 11) using the nuclear reactor source [25], and those relating to the $g_{Li-N}(r)$, $g_{Li-O}(r)$, $g_{N-N}(r)$, $g_{N-O}(r)$, and $g_{O-O}(r)$, using the TOF ND measurements on a spallation source [72]. The results reveal that a central Li^+ is surrounded on an average by about four nearest-neighbouring NO_3^- ions, one oxygen in each of these nitrate ions facing towards Li^+, with the nearest-neighbour distances, r_{Li-O} = 2.1 Å, r_{Li-N} = 2.8 Å, r_{Li-Li} = 4.1 Å, r_{N-N} = 4.8 Å. The space and time averaged local structure in the molten state is found to be appreciably different from that in the crystalline state. The discrepancies between these results and those reported by MD computer simulations by Yamaguchi et al. [74], where the reported nearest-neighbour Li-O and Li-Li distances of 1.86 Å and 3.86 Å and the corresponding coordination numbers are far less, highlight the uncertainties in the effective pair potentials used in the MD study.

5. Summary and Future Work

Our knowledge of the structural aspects of molten salts has greatly improved due to significant advancements already made in the neutron scattering techniques. These techniques in conjunction with computer simulations and liquid-state structural theory will continue to play a significant role in deepening our understanding of the molten salt systems. The direct determination of the pair radial distribution functions by neutron diffraction isotopic substitution experiments not only offers a means of characterising the different structures but also provides theorists with information to construct more realistic potential models, which can be used to explore both microscopic and macroscopic properties in regimes not currently accessible to experiments.

It is anticipated that the neutron scattering techniques will continue to be further developed and applied to a wide variety of more complex systems, such as molten oxides, fluorides, cryolites and molten salt mixtures of importance in nuclear industry, electrochemical devices and energy storage systems. Container-less scattering will become a routine with the easy availability of levitation furnaces, and high-pressure studies of molten salts will also be undertaken. The construction and commissioning of new diffractometers with higher count rates and more stable detectors of improved design, such as D4C at ILL and GEM at ISIS with an optimised sample environment to work under non-ambient conditions will enable more extensive research and new applications of the neutron techniques.

With the recent launch of the "New Millennium" programme at the ILL and the planning of the next generation European Spallation Source (ESS) already in an advanced stage, new sources of much higher flux and stable instrumentation will open further new opportunities to explore increasingly complex systems. For instance, as the technology develops it will become feasible in the longer term to investigate changes of structure as chemical reactions proceed in the molten state. Finally, with the development and availability of new powerful X-ray sources, such as the ESRF at Grenoble, greater use will be made in the next decade of combining neutrons and X-ray data, and this is already beginning to happen.

Acknowledgements

The author would like to thank his many colleagues to whose work he has referred in the preceding pages. He is particularly grateful to Professor John E. Enderby for his enthusiastic support and insightful comments on many aspects of the neutron work, Dr. Marcelle Gaune-Escard for her invitation to give lectures at the NATO-ASI 2001 in Turkey, which have resulted in this publication. He would like to acknowledge his co-workers and collaborators, Professors Marcelle Gaune-Escard, Paul A. Madden, (late) Ryuzo Takagi and Drs. Haruaki Matsuura, George W. Neilson and Adrian C. Barnes, for innumerable discussions and access to unpublished and about-to-be published results. He would also like to thank his colleagues at the neutron institutions where many of the experiments were conducted. The services of Drs. Pierre Chieux, Henry E. Fischer (ILL) and Drs. Spencer Howells, Alan Soper (ISIS) have been essential to the success of most of the experiments described above. Finally, he thanks the EPSRC for the continuing support of his neutron research programme.

References

[1] Sears, V. (1989) Neutron Optics, An Introduction to the Theory of Neutron Optical Phenomena and their Applications, Oxford University Press.
[2] Soper, A.K., Howells, W.S. and Hannon, A.C. (1989) ATLAS –Analysis of Time-of-Flight Diffraction Data from Amorphous and Liquid Samples, R.A.L. Report No. 89-046.
[3] Squires, G.L. (1978) Introduction to the Theory of Thermal Neutron Scattering, Cambridge University Press.
[4] Lovesey, S.W. (1986) Theory of Neutron Scattering from Condensed Matter Vol. 1, International Series of Monographs on Physics 72, Oxford Science Publications.
[5] Sears, V.F. (1992) Neutron Scattering Lengths and Cross-sections, *Neutron News* **3**, 26.
[6] Placzeck, G. (1952) *Phys. Rev.* **86**, 377.
[7] Paalman, H.H. and Pings, C.J. (1962) *J. Appl. Phys.* **33**, 2635.
[8] Poncet, P.F.J. (1976) Doctoral Thesis, University of Reading; (1978) ILL Report 78P0875, ILL.
[9] Blech, I.A. and Averbach, B.L. (1965) *Phys. Rev. A* **137**, 1113.
[10] Soper, A.K., Andreani, C. and Nardone, M. (1993) *Phys. Rev. E* **47**, 2598.
[11] Soper, A.K. (1986) *Chem. Phys.* **107**, 61.
[12] Soper, A.K. (1990) Neutron Scattering Data Analyses, Instr. Phys. Conf. Ser. 107, Ed. Johnson, M.W., Bristol IOP, p. 57.
[13] Allen, M.P. and Tildesley, D.J. (1987) Computer Simulation of Liquids, Clarendon Press, Oxford.
[14] Kalugin, O.N. and Adya, A.K. (2000) *Phys. Chem. Chem. Phys.* **2**, 11-22.

[15] Adya, A.K. and Kalugin, O.N. (2000) *J. Chem. Phys.* **113**, 4740-4750.
[16] Enderby, J.E., North, D.M. and Egelstaff, P.A. (1966) *Philos. Mag.* **14**, 961.
[17] Neilson, G.W., and Adya, A.K. (1997) Neutron Diffraction Studies on Liquids, *Annual Reports C: Royal Soc. Chem.* **93**, 101-145.
[18] Adya, A.K., Takagi, R. (1998) Unravelling the Internal Complexities of Molten Salts, *Z. Naturforsch* **53a**, 1037-1048.
[19] Page, D.I., Mika, K. (1971) *J. Phys. C.* **4**, 3034.
[20] Edwards, F.G., Enderby, J.E., Howe, R.A. and Page, D.I. (1975) *J. Phys. C.* **8**, 3483.
[21] Adya, A.K., Takagi, R., Sato, Y., Gaune-Escard, M., Barnes, A.C. and Fischer, H.E., to be submitted.
[22] Takagi, R., Hutchinson, F., Madden, P. A., Adya, A. K. and Gaune-Escard, M. (1999) *J. Phys.: Condens. Matter* **11**, 645-658.
[23] Allen, D.A., Howe, R.A., Wood, N.D. and Howells, W.S. (1992) *J. Phys.: Condens. Matter* **4**, 1407.
[24] Adya, A.K. and Neilson, G.W. (1990) *Molec. Phys.* **69**, 747.
[25] Adya, A.K., Neilson, G.W., Okada, I. and Okazaki, S. (1993) *Molec. Phys.* **79**, 1327.
[26] Adya, A.K. and Neilson, G.W. (1990) *Molec. Phys.* **71**, 1091.
[27] Biggin S. and Enderby, J.E. (1982) *J. Phys. C: Solid State Phys.* **15**, L305.
[28] Allen, D.A., Howe, R.A., Wood, N.D. and Howells, W.S. (1991) *J. Chem. Phys.* **94**, 5071.
[29] Badyal, Y.S. and Howe, R.A. (1993) *J. Phys.: Condens. Matter* **5**, 7189.
[30] McGreevy, R.L. and Howe, M.A. (1989) *J. Phys.: Condens. Matter* **1**, 9957.
[31] Howe, M.A. and McGreevy, R.L. (1988) *Phil. Mag. B* **58**, 485.
[32] Derrien, J.Y. and J. Dupuy, (1975) *J. Phys. Paris* **36**, 191.
[33] Mitchell, E.W.J., Poncet, P.F.J. and Stewart, R.J. (1976) *Phil. Mag. B* **34**, 721.
[34] Locke, J., Messoloras, S., Stewart, R.J., McGreevy, R.L. and Mitchell, E.W.J. (1985) *Phil. Mag. B* **51**, 301.
[35] Eisenberg, S., Jal, J.-F., Dupuy, J., Chieux, P. and Knoll, W. (1982) *Phil. Mag. A* **46**, 195.
[36] Allen, D.A. and Howe, R.A. (1992) *J. Phys.: Condens. Matter* **4**, 6029.
[37] Derrien, J.Y. and J. Dupuy, (1976) *Phys. Chem. Liq.* **5**, 71.
[38] Biggin, S., Gay, M. and Enderby, J.E. (1984) *J. Phys. C: Solid State Phys.* **17**, 977.
[39] Biggin, S. and Enderby J.E. (1981) *J. Phys. C: Solid State Phys.* **14**, 3577.
[40] McGreevy, R.L. and Mitchell, E.W.J. (1982) *J. Phys. C* **15**, 5537.
[41] Edwards, F.G., Howe, R.A., Enderby J.E. and Page, D.I. (1978) *J. Phys. C: Solid State Phys.* **11**, 1053.
[42] Biggin, S. and Enderby J.E. (1981) *J. Phys. C: Solid State Phys.* **14**, 3129.
[43] Newport, R.J., Howe, R.A. and Wood, N.D. (1985) *J. Phys. C: Solid State Phys.* **18**, 5249.
[44] Allen, D.A., Howe, R.A., Wood, N.D. and Howells, W.S. (1991) *J. Chem. Phys.* **94**, 5071.
[45] Badyal, Y.S. and Howe, R.A. (1993) *J. Phys.: Condens. Matter* **5**, 7189.
[46] de Leeuw, S. (1978) *Molec. Phys.* **36**, 103 and 765.
[47] Pastore, G., Ballone, P. and Tosi, M.P. (1986) *J. Phys. C: Solid State Phys.* **19**, 487.
[48] Woodcock, L.V.C., Angell, A. and Cheeseman, P. (1976) *J. Chem. Phys.* **65**, 1565.
[49] Gardner, P.J. and Heyes, D.M. (1985) *Physica B* **113**, 227.
[50] Wood, N.D. and Howe, R.A. (1988) *J. Phys. C: Solid State Phys.* **21**, 3177.
[51] Wood, N.D., Howe, R.A., Newport, R. J. and Faber Jr., J. (1988) *J. Phys. C: Solid State Phys.* **21**, 669.

[52] Wilson, M. and Madden, P.A. (1993) *J. Phys.: Condens. Matter* **5**, 6833; (1994) **6**, 159.
[53] Adya, A.K., Takagi, R., Sakurai, M. and Gaune-Escard, M., (1998) Proc. 11th Int. Symp. Molten Salts, Ed. P.C. Trulove, H. De Long and S. Deki, Electrochem. Soc. Inc., Pennington, **98-11**, 499-512.
[54] Triolo, R. and Narten, A.H. (1978) *J. Chem. Phys.* **69**, 3159.
[55] Johnson, E., Narten, A.H., Thiessen W.E. and Triolo, R. (1978) *Farad. Discuss. Chem. Soc.* **66**, 287.
[56] Badyal, Y.S., Allen, D.A. and Howe, R.A. (1994) *J. Phys.: Condens. Matter*, **6**, 10193.
[57] Price, D.L., Saboungi, M.L., Hashimoto, S. and Moss, S.C. (1992) Proc. 8th Int. Symp. Molten Salts, Ed. R.J. Gale, G. Blomgren and H. Kojima, Electrochem. Soc. Inc., Pennington, **92-16**, 14.
[58] Price, D.L., Saboungi, M.L., Badyal, Y.S., Wang, J., Moss, S.C. and Leheny, R.L. (1998) *Phys. Rev. B* **57**, 10496.
[59] Fukushima, Y., Misawa, M. and Suzuki, K. (1975) *Res. Rep. Lab. Nucl. Sci.* (Tohoku University) **8**, 113.
[60] Price, D.L., Saboungi, M.L., Howells, W.S. and Tosi, M.P. (1993) Proc. Int. Symp. Molten Salts, Ed. M.L. Saboungi and H. Kojima, Electrochem. Soc. Inc., Pennington, **93-9**, 1.
[61] Saboungi, M.L., Howe, M.A. and Price, D.L. (1990) Proc. 7th Int. Symp. Molten Salts, Ed. C.L. Hussey, S.N. Flengas, J.S. Wilkes and Y. Ito, Electrochem. Soc. Inc., Pennington, **90-17**, 8.
[62] Tosi, M.P., Pastore, G., Saboungi, M.L. and Price, D.L. (1991) *Physica Scripta* **T39**, 367.
[63] Saboungi, M.L., Price, D.L., Scamehorn, C. and Tosi, M.P. (1991) *Europhys. Lett.* **15**, 283.
[64] Wasse, J.C. and Salmon, P.S. (1999) *J. Phys.: Condens. Matter* **11**, 1381-1396.
[65] Wasse, J.C. and Salmon, P.S. (1998) *Physica B* **241-243**, 967-969.
[66] Wasse, J.C. and Salmon, P.S. (1999) *J. Phys.: Condens. Matter* **11**, 9293-9302.
[67] Wasse, J.C. and Salmon, P.S. (1999) *J. Phys.: Condens. Matter* **11**, 2171-2177.
[68] Wasse, J.C., Salmon, P.S. and Dalaplane, R.G. (2000) *J. Phys.: Condens. Matter* **12**, 9539-9550.
[69] Wasse, J.C., Salmon, P.S. and Dalaplane, R.G. (2000) *Physica B* **276-278**, 433-434.
[70] Adya, A.K., Matsuura, H., Takagi, R., Rycerz, L. and Gaune-Escard, M. (2000) Proc. XII Int. Symp. Molten Salts, Electrochem. Soc. Inc. Pennington, **99-41**, 341-355; Adya, A.K., Matsuura, H., Hutchinson, F., Madden, P.A. and Gaune-Escard, M., to be submitted.
[71] Adya, A.K., Matsuura, H., Hutchinson, F., Gaune-Escard, M., Madden, P.A., Barnes, A.C. and Fischer, H.E. (2000) Progress in Molten Salt Chemistry 1, Ed. R.W. Berg and H.A. Hjuler, Elsevier, p 37-44.
[72] Kameda, Y., Kotani, S., and Ichikawa, K. (1992) *Molec. Phys.* **75**, 1.
[73] Adya, A.K. and Neilson, G.W. (1996) *J. Non-crystalline Solids* **205-207**, 168-171.
[74] Yamaguchi, T., Okada, I.,Ohtaki, H., Mikami, M. and Kawamura, K. (1986) *Molec. Phys.* **58**, 349.

METAL-MOLTEN SALT INTERFACES:
WETTING TRANSITIONS AND ELECTROCRYSTALLIZATION

W. FREYLAND
Institute of Physical Chemistry, University of Karlsruhe
D-76128 Karlsruhe, Kaiserstr. 12, Germany

1. Introduction

Interfaces play an important role in various areas of physical chemistry, both in fundamental and applied research. This includes phenomena in catalysis, colloid chemistry, tribology and, more generally, in interfacial thermodynamics and electrochemistry, to name only a few aspects. During the last two decades two novel interfacial topics have attracted considerable attention: phase transitions at interfaces or wetting phenomena and microscopy of interfaces, in particular, *in situ* investigations of electrocrystallization on the nanometer scale. Both topics will be introduced in this lecture and first results and open problems will be discussed focusing on ionic liquids.

Since the stimulating work of Cahn [1] and Widom [2] in the seventies of the last century a vast number of publications on wetting phenomena has appeared and a variety of discoveries has been made – for reviews see e.g. [3] - [6]. In general wetting transitions are characterized by structural changes of the intrinsic interface between two phases α and β when as a function of the bulk state a third phase γ intrudes into the α/β interface. As an example we consider a gas (β) in contact with the walls (α) of a closed container. By increasing the pressure so that bulk gas (β) and bulk liquid (γ) coexistence is approached, a liquid film will nucleate at the walls. At coexistence the walls are wet by a liquid film of macroscopic thickness. Similar wetting phenomena can occur with varying temperature, pressure or chemical potential in binary mixtures of immiscible liquids, at the fluid/wall or fluid/vapour interface – see below. Specific examples of such wetting transitions are surface melting and surface freezing. Surface-induced melting can be regarded as a wetting transition whereby the solid is wet by its own melt. A thin liquid-like film forms at the solid/vapour interface well below the triple point temperature and it increases and eventually diverges approaching the bulk melting point. These phenomena have been studied for a variety of metal surfaces – see e.g. [7], [8] - for solid rare gas films [9] and, partly controversial, for crystalline ice – see reviews [10], [11]. Although it is believed to be a commonly occurring phenomenon, experimental or theoretical studies on surface melting of ionic solids are still missing. In surface freezing a solid-like film forms on top of the bulk liquid at temperatures above the bulk triple point. It has been observed in liquid crystalline systems [12] or in liquid alkanes [13] and, more recently, in liquid alloys like Ga-Pb [14] and Ga-Bi [15]. In the latter system a wetting transition precedes the surface freezing of a Bi-rich solid-like film. Again, surface-induced freezing in ionic liquids has not been studied yet.

So it can be stated that very little is known about wetting phenomena in classical molten salts or ionic liquids like the so called room temperature molten salts. This concerns the nature of wetting transitions in these systems, the specific influence of Coulomb interactions on the wetting film thickness and structure, and, last not least, the correlation between the bulk and surface phase behaviour. Related with this problem is the question: how does the microscopic structure of an ionic liquid change across the liquid/vapour interface? In van der Waals type systems the forces in the solid, liquid and vapour phases have the same character and so it is not surprising that the densities along the normal to the respective interfaces vary continuously on a microscopic scale. At a liquid metal/vapour interface an electronic phase transition occurs resulting in a stratified layering structure normal to the interface [16]. In the case of an ionic liquid/vapour interface again a drastic change of the interatomic potentials has to be considered and so a peculiar variation of the microscopic structure can be expected. A similar problem arises if one considers the ionic liquid/wall or electrode interface. This is of great fundamental and technological interest for applications like electroplating or electrowinning of metals or semiconductors from molten salt electrolytes. In contrast to aqueous electrolytes the molten salt/electrode interface is not well characterized and understood. The equivalent of the double layer structure is not known.

The discovery of scanning tunneling microscopy (STM) by Binnig and Rohrer in the eighties of the last century [17] enabled a completely new insight into interfacial electrochemistry and *in situ* studies of electrodeposition and electrocrystallization on the nanometer scale. First applied to surface science under UHV-conditions it was soon employed in experiments of the aqueous electrolyte/electrode interface [18], [19]. Details of nucleation, growth morphology and 2D and 3D phase formation can be imaged in real space by these electrochemical scanning probe techniques – for reviews see [20], [21]. Besides this progress in electrochemical research, new developments in electrochemical nanostructuring and fabrication have been made [22], [23]. More recently the extension of these methods to study ionic liquid/electrode interfaces has been demonstrated [24], [25]. Doubts on the feasibility of such experiments existed since in comparison to aqueous electrolytes the conductivity of ionic liquids or molten salts is much higher, 10^{-2} to 10^{0} Ω^{-1} cm^{-1}, so that it was unclear if the low tunneling currents of the order of 10^{-9} A can be detected. A great advantage of these experiments with ionic electrolytes is, that the potential range for electrodeposition of elements and alloys, metals or semiconductors, is significantly enhanced. This is not the case with the reduced electrochemical window of aqueous solutions. As for the mechanism of electron tunneling in these liquids, this still remains unknown.

In the following we deal with two topics of ionic liquid interfaces. In section 2 a brief introduction of some basic concepts of interfacial thermodynamics is given, spectroscopic methods to study fluid interfaces at high temperatures are described, and then the wetting characteristics of an ionic fluid/wall interface are presented and discussed for the metal-molten salt system K-KCl. In section 3 we focus on some recent results of electrocrystallization from ionic electrolytes studied with scanning probe techniques. After a brief description of some basic aspects of electrodeposition and of the electrochemical scanning probe technique, several examples of underpotential (UPD) and overpotential (OPD) metal, alloy, and semiconductor electrodeposition are presented and the informations with respect to phase formation and growth are discussed. The concluding remarks summarize the present state of research activities with ionic liquids and sketches some perspectives.

2. Interfacial Phase Transitions in Metal-Molten Salt Solutions

2.1. THERMODYNAMICS OF INTERFACES – SOME BASIC CONCEPTS

Interfaces are fascinating.[*] This impression is given by the illustration in Fig. 1, where a child plays with soap bubbles. This early experience with surface or interfacial tension or energy tells us that interfaces are real objects, in general have a finite thickness, are part of a heterogeneous thermodynamic system, and that the reversible work necessary to increase their size is proportional to the area A. i.e.

Figure 1: Early experience with surface tension or energy σ

$$dW_{rev} \equiv dG = \sigma\, dA, \quad \text{at const } p, T \qquad (1)$$

where G is the free enthalpy or Gibbs function and σ is the surface or interfacial energy. From this definition it is clear that σ has the SI-units J m^{-2}. From the simple experiment sketched in Fig. 1 we also learn that for a curved interface like spherical bubbles a pressure difference exists across the interface: the smaller the bubble the greater the pressure of the air inside relative to the outside. We observe that a smaller bubble in contact with a larger one will be absorbed by the latter. This result is explained by the Young-Laplace Equation for a spherical interface, i.e. $\Delta p = 2\sigma R^{-1}$, where R is the radius of the bubble.

---

[*] The notation "surface" typically is used for the contact area of a condensed phase (solid, liquid) with a gas or vapour phase or vacuum; the "interface" or "interphase" describes the phase of in general microscopic thickness between two condensed phases in contact; sometimes these terms are used synonymously.

In the following we first restrict to heterogeneous systems with planar interfaces. Whereas in homogeneous systems the thermodynamic variables like density ρ or mole fraction x_i are independent of the position coordinates x, y, z, this is not the case in heterogeneous systems. It is illustrated in Fig. 2 for two examples:

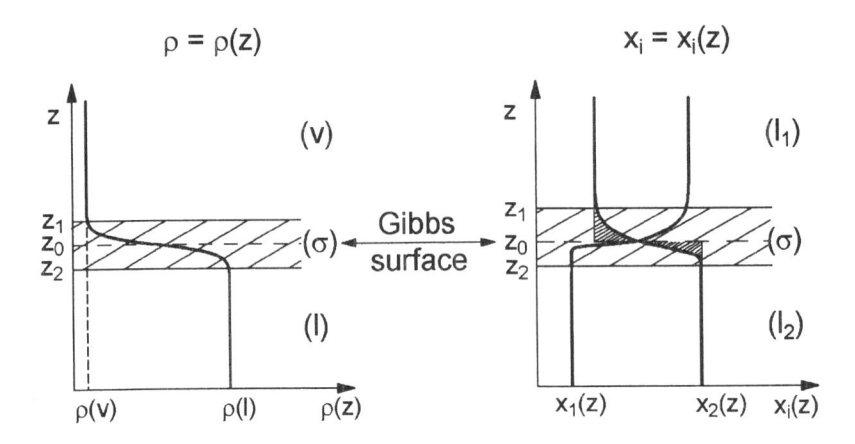

Figure 2: Heterogeneous systems and definition of the Gibbs dividing surface. *Left part*: variation of density ρ(z) across a liquid (l)/vapour (v) interface; the interfacial region (σ) is given by the interval $z_1 \leq z \leq z_2$ where ρ(z) deviates from the constant bulk values. *Right part*: variation of the mole fractions $x_i(z)$ across the liquid (l_1) / liquid (l_2) interface of a binary mixture with a miscibility gap; if the location z_0 of the Gibbs dividing surface is fixed with respect to $x_2(z)$ so that the surface excess $\Gamma_2 \equiv 0$, i.e. $\int_{z_2}^{z_0}(x_2(l_2) - x_2(z))dz = \int_{z_0}^{z_1}(x_2(z) - x_2(l_1))dz$, see dashed areas of the $x_2(z)$-profile, then $\Gamma_1 \neq 0$. The Gibbs dividing surface for the liquid / vapour interface is constructed accordingly.

a liquid/vapour interface of a one component system and a liquid l_1/liquid l_2 interface of a binary mixture exhibiting a miscibility gap at given p, T. For simplicity we consider simple liquids here, e.g. van der Waals type liquids with a smooth change of the thermophysical properties across the interfacial region (σ) separating the respective bulk equilibrium phases (α) and (β). In general the thickness of the interfacial phase (σ) can range from one to several interatomic distances at conditions remote from the respective critical points. In defining the contribution of the interphase to the thermodynamic quantities of the total heterogeneous system, a problem arises which is due to the continuous variation of these quantities across the interface. For this reason Gibbs first introduced the concept of a geometrical dividing surface of zero thickness which is parallel to the surface of tension. In this way, both bulk phases (α) and (β) are assumed to be homogeneous up to the hypothetic Gibbs dividing surface and the interfacial contributions are determined from the difference of the quantities of the total system and those of phase (α) and (β). For instance, with the symmetric density profile ρ(z) plotted in Fig. 2 the Gibbs dividing surface is pinned to the inflection point of ρ(z) so that both the liquid and gas density are extrapolated up to the dividing surface. Similarly the Gibbs surface is constructed for the binary system (Fig. 2), whereby the surface adsorption or surface excess Γ for component 2 is $\Gamma_2 \equiv 0$, whereas $\Gamma_1 \neq 0$.

With these introductory considerations the thermodynamic contributions or excess quantities of the interface can be defined. Selecting the Gibbs dividing surface such that the total volume V is given by

$$V = V^{(\alpha)} + V^{(\beta)}, \tag{2}$$

the interfacial free energy, $F^{(\sigma)}$, or entropy, $S^{(\sigma)}$, are defined as:

$$\begin{aligned} F^{(\sigma)} &= F - F^{(\alpha)} - F^{(\beta)} \\ S^{(\sigma)} &= S - S^{(\alpha)} - S^{(\beta)} \end{aligned} \tag{3}$$

where the quantities without index refer to the total system. At thermal equilibrium of a planar interface, the intensive variables are equal, in particular, the following equality holds for the chemical potentials:

$$\mu_i^{(\sigma)} = \mu_i^{(\alpha)} = \mu_i^{(\beta)} = \mu_i. \tag{4}$$

In analogy with homogeneous systems we can write for the total differential of the interfacial free energy:

$$dF^{(\sigma)} = -S^{(\sigma)} dT + \sigma dA + \sum \mu_i \, dn_i^{(\sigma)} \tag{5}$$

where the $n_i^{(\sigma)}$ are given by:

$$n_i^{(\sigma)} = n_i - n_i^{(\alpha)} - n_i^{(\beta)}. \tag{6}$$

The homogeneity relation for $F^{(\sigma)}$ together with Equ. (5) yields the following Gibbs-Duhem relation:

$$S^{(\sigma)} dT + A d\sigma + \sum n_i^{(\sigma)} d\mu_i = 0, \tag{7}$$

and, by simple rearrangement, the Gibbs-adsorption equation is obtained:

$$d\sigma = -s^{(\sigma)} dT - \sum \Gamma_i^{(\sigma)} d\mu_i. \tag{8}$$

Here $s^{(\sigma)} \equiv S^{(\sigma)}/A$ is the relative interfacial entropy, which can be obtained from measurements of the temperature dependence of σ, $s^{(\sigma)} = -(\partial\sigma/\partial T)_{x_i}$, and $\Gamma_i^{(\sigma)} \equiv n_i^{(\sigma)}/A$ is the relative adsorption or interfacial excess in moles per m^2 which is determined from isothermal measurements of $\sigma(x_i)$ with knowledge of $\mu_i = \mu_i(x_i; T)$. Undoubtedly, the Gibbs adsorption equation and its relevant quantities are of central interest in elucidating interfacial phenomena and their thermodynamic characteristics. As a simple example we consider an interfacial phase transition whereby the surface excess Γ_i changes discontinuously from low to high adsorption at constant T with increasing chemical potential. This transition is signaled directly by a change of slope in the Gibbs adsorption isotherm as depicted schematically in Fig. 3.

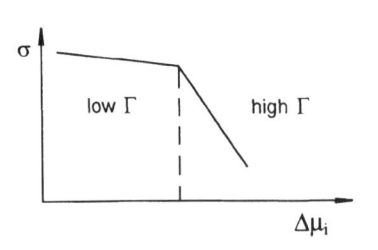

Figure 3: Surface energy σ vs chemical potential $\Delta\mu_i$ at constant temperature; schematic variation of the Gibbs adsorption isotherm for a system undergoing an interfacial phase transition from low to high adsorption indicated by a jump of the surface excess $\Gamma_i = -(\partial\sigma/\partial\Delta\mu_i)_T$ of component i.

The brief description given so far has focused on a few basic aspects of surface thermodynamics of planar interfaces. There are numerous review articles and books which treat this subject in detail, however we can give here only a few selected references which are recommended for further reading [26] - [30], starting with the classics of Prigogine and Defay [26].

We continue with some basic aspects of wetting phenomena, starting with the wetting of a planar solid substrate by a liquid drop, see Fig. 4.

Partial wetting, $0° < \Theta \leq 90°$ Complete wetting, $\Theta = 0°$

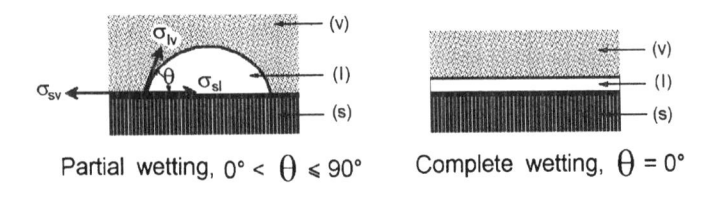

Figure 4: Wetting of a solid substrate (s) by a liquid drop (l) in equilibrium with its vapour (v); *left*: partial wetting configuration; *right*: complete wetting

The equilibrium configuration of the partially wetting drop is given by the balance of the interfacial tensions which leads to the Young equation, i.e.:

$$\sigma_{sv} = \sigma_{s\ell} + \sigma_{\ell v} \cdot \cos\Theta, \qquad (9)$$

where the contact angle Θ for the partial wetting configuration has values of $0 < \Theta \leq 90°$. Partial drying occurs for $90° < \Theta \leq 180°$. Equ. (9) can also be derived from the minimum of $F^{(\sigma)}$ or $G^{(\sigma)}$, see [27]. Changing the thermodynamic state of the liquid drop with varying p, T, a transition from partial to complete wetting ($\Theta = 0$) occurs. (Fig. 4). Very similar wetting scenarios exist in fluid mixtures, at the fluid/wall or fluid/vapour interfaces, see also [28]. A familiar example of the latter is an oil droplet on top of a liquid water/air interface. In these cases of deformable interfaces the Young equation has to be replaced by the Neumann equation, see e.g. [28]. Discussing wetting phenomena questions of particular interest are: when and where in the bulk phase diagram does a wetting transition occur and of what type is it, first or second order? In the case of critical point wetting as first considered by Cahn [1] a simple illustration can be given based on Equ. (9). Approaching the critical point of a pure fluid σ_{lv} follows a scaling law, $\sigma_{\ell v} \propto ((T-T_c)/T_c)^\gamma$ where $\gamma > 1$. At the same time $(\sigma_{sv} - \sigma_{s\ell}) \propto (\rho_v - \rho_\ell) \propto ((T-T_c)/T_c)^\beta$ with $\beta < 1$, where ρ_v, ρ_ℓ are the respective vapour and liquid densities and T_c is the critical temperature. It follows that there must be a wetting temperature $T_w < T_c$ at liquid-vapour coexistence, where $\cos\Theta = 1$ and thus a transition from partial to complete wetting occurs.

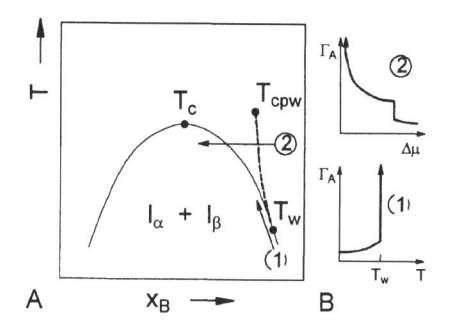

Figure 5: Schematic bulk phase diagram of a binary liquid mixture exhibiting a miscibility gap with an upper critical point at T_c (full line in the temperature-mole fraction plane). Included in this figure is the surface phase diagram with a wetting transition at coexistence at T_w and prewetting transitions from low to high adsorption along the prewetting line (dashed line). Changes of wetting behaviour and surface excess Γ_A along path (1) and (2) are explained in the text

With respect to the second question above, we restrict to the example of a binary liquid mixture exhibiting a miscibility gap with an upper critical point, see Fig. 5. We focus on two changes of the thermodynamic states indicated by path (1) and (2) in the bulk phase diagram and concentrate on the behaviour of the surface excess of component A, Γ_A. If at coexistence along path (1) Γ_A increases continuously for $T < T_w$ and diverges at T_w this is called a critical wetting transition. If, on the other hand, Γ_A is finite (low) for $T < T_w$ and for an infinitely large system jumps to an infinite value at T_w this is the signature of a first order wetting transition. It is accompanied by so called prewetting transitions, whereby a jump from microscopically thin to thick adsorption films occurs in the neighbouring homogeneous phase. The corresponding prewetting line merges the coexistence curve tangentially at T_w and ends at a critical prewetting point at T_{cpw} in the homogeneous phase, see Fig. 5. Considering the minimum of the surface free energy of the system it can be shown that prewetting transitions can exist only on one side of the miscibility gap, see also [28]. Along path (2) in Fig. 5, Γ_A shows a discontinuous but finite (microscopic) change when crossing the prewetting line and then diverges continuously at coexistence. This marks a complete wetting transition. Divergence of the surface excess Γ at a first order or a complete wetting transition implies an infinitely large heterogeneous system and in reality leads to macroscopically thick wetting films. However, prewetting films have a microscopic thickness ($T_w < T < T_{cpw}$) with typical values of the order of a nanometer for van der Waals liquids, see e.g. He on Cs [31].

For a quantitative discussion of different wetting scenarios various theoretical approaches of statistical thermodynamics have been developed, see e.g. [3] - [5]. An essential role plays the effective interfacial potential $\Omega^{(\sigma)}$, which in mean field approximation of the grand canonical potential can be written as [3]:

$$\Omega^{(\sigma)}(d; T, \mu) = \sigma_{\alpha\beta} + \sigma_{\alpha\gamma} + \Delta\mu d + \omega(d), \qquad (10)$$

where gravitational influences have been neglected. Here the σ_{ik} are the interfacial energies, (β) being the non-wetting phase, d is the wetting film thickness, $\Delta\mu$ denotes the deviation of μ relative to that at coexistence, and $\omega(d)$ describes the correction of the surface energies due to a finite thickness of the wetting films. From Equ. (10) the thickness of the wetting films is obtained from $\partial\Omega^{(\sigma)}/\partial d = 0$. Thus, for a van der Waals type liquid, where $\omega(d) = H/d^2$ (H = Hamaker constant, see also [29]) the prewetting film thickness is:

$$d = (2H/\Delta\mu)^{1/3}. \qquad (11)$$

For a Yukawa liquid with $\omega(d) = \sigma_0 \exp(-\lambda d)$, the corresponding prewetting films have a logarithmic divergence at coexistence $(\Delta\mu = 0)$ according to:

$$d = \lambda^{-1} \ln(\lambda\sigma_0/\Delta\mu). \qquad (12)$$

An overview of the qualitative changes of $\Omega^{(\sigma)}$ passing a prewetting transition at constant T is given in Fig. 6. For $\Delta\mu < \Delta\mu^{pw}$ the stable prewetting film has a thickness d_2, microscopically thick film,

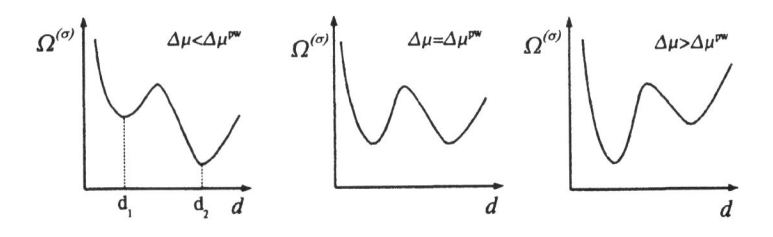

Figure 6: Effective interfacial potential $\Omega^{(\sigma)}$ at conditions off of coexistence. Shown is the variation of $\Omega^{(\sigma)}$ for a prewetting transition with $\Delta\mu^{pw}$ measuring the distance of the prewetting line from the coexistence curve at constant T (see text).

whereas for $\Delta\mu > \Delta\mu^{pw}$ only a thin adsorbed film is stable with $d = d_1$. For a detailed description of the different wetting scenarios reference is given to the review by Dietrich [3].

2.2. SURFACE SENSITIVE PROBES FOR THE STUDY OF FLUID INTERFACES AT HIGH TEMPERATURES

A variety of surface sensitive tools exists to study the microscopic and electronic structure of interfaces – see e.g. [30], [32]. Most of these methods use electrons as probing particles which requires ultra high vacuum (UHV) conditions and restricts their application to solid and liquid samples with low vapour pressures. In general, high temperature measurements of ionic liquids or metal-molten salt solutions are made difficult due to their elevated vapour pressures, high reactivity or corrosivity so that a vacuum tight containment is necessary. So the interfacial properties at the fluid/vapour or fluid/wall interface have to be probed through a corresponding window. This reduces interfacial studies to a few techniques like X-ray and neutron reflection and grazing incidence diffraction, Raman scattering, transmission and reflection spectroscopy, and ellipsometry and second harmonic generation measurements. The last two methods have been employed here to study the liquid alkali metal-alkali halide/sapphire interface. In the following we briefly describe the principle of these methods, their potential and limitations for interfacial studies at high temperatures.

Second harmonic generation (SHG) is a method of nonlinear optics, see [33], [34]. If a laser pulse of high intensity impinges on a medium the macroscopic polarization P_i is no longer proportional to the electric field strength $E_i(\omega)$ of the incident light at frequency ω, but contains a contribution with a quadratic dependence on $E_i(\omega)$:

$$P_i^{(2)}(2\omega) = \chi_{ijk}^{(2)} \cdot E_j(\omega) \cdot E_k(\omega). \qquad (13)$$

Here $P_i^{(2)}(2\omega)$ is the second order polarization which determines the intensity of photons reflected at the second harmonic, $I(2\omega)$, and $\chi_{ijk}^{(2)}$ is the second order susceptibility, a tensor of third rank. For symmetry reasons SHG is forbidden inside media with inversion symmetry like liquids or gases [33]. However, this symmetry is necessarily broken across an interface with the consequence that SHG occurs at the interface and is highly surface specific. The SH intensity is proportional to the square of the nonlinear surface susceptibility χ^s [33] which depends on the electronic structure of the surface states, i.e. the SH signal of a liquid metal/sapphire interface is clearly stronger than that of the corresponding molten salt interface. Thus a salt rich wetting film intruding at the bulk liquid metal/sapphire interface is clearly distinguished by the corresponding change of the SH intensity. In this way we have used SHG to probe the interface continuously as a function of temperature and bulk composition. For details of the experimental setup and intensity measurements see [35].

Ellipsometry measures the change of polarization when polarized light is reflected from or transmitted by an interface. For strongly absorbing media reflection ellipsometry is used. Varying the wavelength of the incident light a spectroscopic characterization of the sample is possible; this mode is called spectroscopic ellipsometry. For the simple case of reflection at a planar interface between two isotropic media the change of polarization is described by the ratio ρ of the complex Fresnel reflection coefficients for the parallel, r_p, and vertical, r_s, polarization, see [36]:

$$\rho \equiv r_p / r_s = |\rho| \exp(j\Delta), \qquad (14)$$

where Δ is the phase change on reflection.

If the optical constants, refractive index n_0 and absorption coefficient κ_0, of one medium, e.g. sapphire window (0), are known, those of the second medium defining the interface, e.g. bulk liquid (1), can be determined from the measured ellipsometric reflectivity ρ with the aid of the Fresnel equations, see e.g. [36]:

$$n_1 - jk_1 = (n_0 - jk_0)\sin\Phi_0\left(1 + \frac{(1-\rho)^2}{(1+\rho)^2}\tan^2\Phi_0\right)^{1/2}, \quad (15)$$

where Φ_0 is the angle of incidence.

For the ellipsometric characterization of a mesoscopic wetting film – its optical constants and thickness – the configuration depicted in Fig. 7a may be considered:

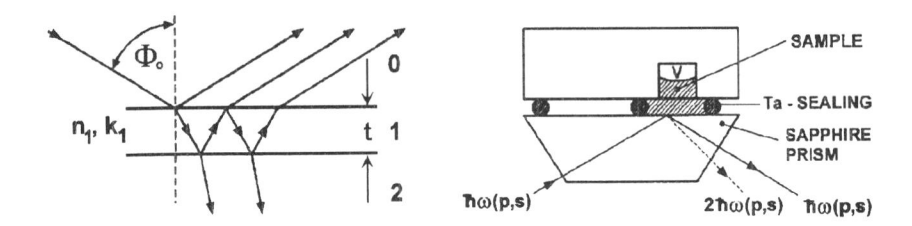

Figure 7: a) Reflection and transmission of light at the sapphire (0)/wetting film (1)/ bulk fluid (2) configuration; Φ_0 is the angle of incidence, n_1 and k_1 are the optical constants of the film, t its thickness b) High temperature sapphire cell with Ta-wire sealing

Sapphire (0), saltrich wetting film (1), metalrich bulk liquid (2). Within this three phase or Drude slab model it is assumed that the optical constants vary discontinuously across the interfaces which surely is a problematic approximation for microscopically thin films. Addition of the partial waves leads to an infinite geometrical series for the reflection coefficient ρ with the following general dependence [36]:

$$\rho = \rho(N_0, N_1, N_2, t, \Phi_0, \lambda), \quad (16)$$

with $N_i = n_i - jk_i$. Knowing the optical constants of the sapphire window, N_0, and the bulk liquid, N_2, ρ depends on three parameters of the film: its optical constant n_1, k_1 and the film thickness t. These are accessible by measurements of ρ at different Φ_0. However, the resulting system of equations cannot be solved analytically as is the case with the two phase model above. For details of this data analysis and the high temperature ellipsometric measurements see [37], [38]. A final remark concerns the experimental difficulties with these spectroscopic studies of metal-molten salts at high temperatures. First, a vacuum tight containment has to be achieved. If sapphire is corrosion resistant,, the cell construction sketched in Fig. 7b for optical measurements can be used. The two sapphires are sealed by squeezed Ta or Nb wires. This technique we have employed several times in different constructions of ellipsometry and SHG experiments.. The disadvantage of sapphire, especially in ellipsometric measurements, is its birefringence. This necessitates an *in situ* calibration [37] of the profile of birefringence which is a nontrivial experiment.

2.3. WETTING AND PREWETTING TRANSITION AT THE FLUID/WALL INTERFACE OF K-KCL MIXTURES

Wetting phenomena in systems with van der Waals type interaction have been investigated intensively in recent years. A few experimental studies also exist of metallic liquid alloys [39] - [41]. However, information on wetting transitions in ionic systems is restricted so far to spectroscopic studies on alkali metal – alkali halide (M-MX) solutions, which is reviewed here. The main motivation for studying these systems has been:
- They exhibit a metal-nonmetal transition with varying composition which is continuous at elevated temperatures and occurs in the saltrich fluid phase [42]. In different experiments on the electronic and thermodynamic properties we have made the necessary experience to work with these systems at unusual conditions.
- The optical constants of the components, molten salt and liquid metal, are extremely different and thus facilitate the distinction of a saltrich film on top of a metalrich bulk phase. Furthermore, the non-metallic saltrich phase is uniquely characterized by the liquid F-centre absorption band which for e.g. K-KCl lies at 1.3 eV with a half width of ~ 1 eV [42].
- The phase diagrams of the M-MX systems exhibit a large miscibility gap for the lighter alkali metals which is reduced for K-KCl and gives way to complete miscibility in the Cs-CsCl system [43]. So within the same class of systems the influence of strong variations in the bulk phase behaviour on the wetting characteristics can be investigated.

The following discussion of the wetting characteristics in M-MX systems will focus on the K_xKCl_{1-x} melts. Its bulk phase diagram is given in Fig. 8 whereby the metal mole fraction x is stretched logarithmically since the wetting transition occurs on the metal-rich end. For comparison the inset shows the phase diagram on a linear scale. Included in this figure are the wetting characteristics: the apparent wetting temperature T_w, as first detected by optical reflectivity measurements [44], and the prewetting transition line, dashed curve [35]. The prewetting line has been derived from the distinct changes of the SH intensities measured in the x – T plane of the homogeneous metalrich phase. The changes of the SH intensities are consistent with a salt-rich microscopic wetting film intruding between the metalrich liquid and the sapphire window [35].

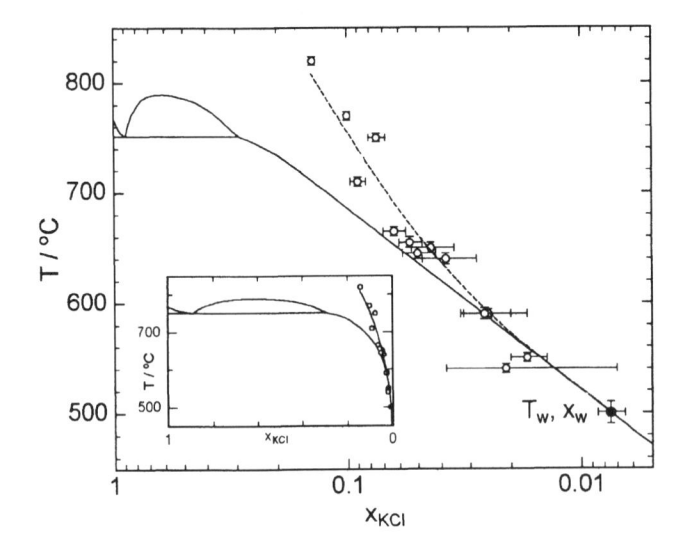

Figure 8: Bulk phase diagram of K-KCl in a semilogarithmic plot; the dashed line represents the prewetting line as determined from SHG-measurements, open circles [35]; the apparent wetting temperature $T_w \sim 500°C$ refers to older studies along the coexistence curve [44] not taking into account tetra point wetting.

In order to get spectroscopic evidence of the prewetting transition in K_xKCl_{1-x} and to get further insight into the wetting characteristics including the wetting film thicknesses, we have performed spectroscopic ellipsometry measurements of K_xKCl_{1-x} for $0.80 \leq x \leq 0.98$ at temperatures up to 730 °C [38], [48]. Typical spectra are shown in Fig. 9, where that in Fig. 9a is representative of measurements above the prewetting line, whereas that in Fig. 9b has been obtained for conditions near the prewetting line. Spectra similar to that in Fig. 9b have been observed along the coexistence curve. The clearly metallic characteristics of the spectra in Fig. 9a prove the expected behaviour above the prewetting line. On the other hand, the F-centre fingerprint of Fig. 9b gives clear evidence, that a saltrich prewetting film has formed. The main results of these investigations can be summarized as follows [38], [48]:

- Analysis of the absorption coefficients of the F-centre band yields a composition of the wetting films of ~ 90 mole% KCl.
- Observation of the liquid state F-centre at conditions at and off of coexistence at temperatures as low as 530 °C indicates a strong undercooling of the wetting films below the monotectic temperature of 751 °C, see Fig. 8.
- Along coexistence the thickness of the wetting films increases with temperature from ~ 30 nm at 560 °C to ~ 300 nm at 720 °C; an approximate value of the prewetting film thickness at 730 °C is 60 nm.

- The half width of the F-centre band observed for the wetting films is clearly reduced in comparison to that of the bulk liquid F-centre; this can be explained by a reduction of the Madelung potential fluctuations due to charge ordering inside the wetting films which is consistent with predictions of Heyes and Clarke [45] based on MD calculations on an electrified liquid KCl film of 6 nm thickness.

- In comparison with van der Waals systems the thickness of the ionic prewetting films in K-KCl are bigger by more than an order of magnitude. This magnitude is estimated according to Equ. (12) if one assumes a complete surface charging at the sapphire interphase and takes a screening length of $\lambda \sim 3$ nm consistent with the MD calculations [38]. Here σ_0 is approximated by $\sim 10^2$ m J/m^{-2}, a value typical of the surface energy of a molten salt.

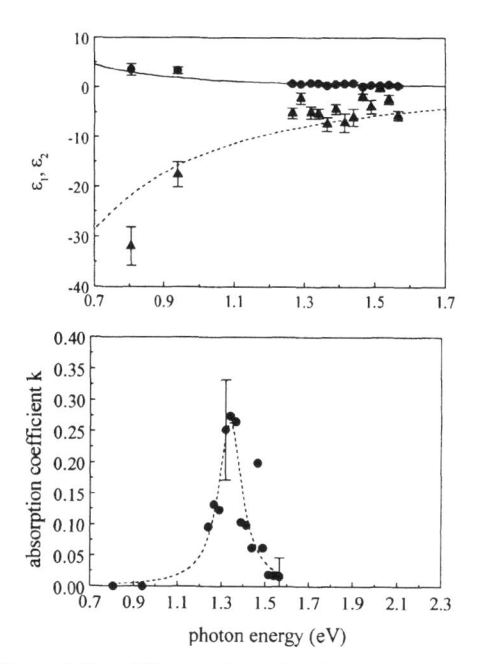

Figure 9: Top: Ellipsometric results of the real, ϵ_1, and imaginary, ϵ_2, part of the complex dielectric function in liquid K$_x$KCl$_{1-x}$ (x = 0.98, T = 600°C); symbols: experimental data; curves: Drude model calculations. Bottom: F-band absorption determined by spectroscopic ellipsometry of a prewetting film in liquid K$_x$KCl$_{1-x}$ at x = 0.95 and T = 660°C [48]

In conclusion, these observations, in particular, the strong undercooling and the variation of wetting film thicknesses with temperature indicate that tetra point wetting [46] occurs in the K-KCl system, i.e. complete wetting accompanied by macroscopic thick wetting films is reached along the metastable extension of the liquid-liquid coexistence curve. So the true wetting temperature should be lower than the $T_w \sim 500$ °C which has been assigned with reference to the coexistence curve of the metalrich solutions. Finally, it should be pointed out that earlier ellipsometric measurements of Cs-CsCl solutions give no indication of a wetting transition in this completely miscible system [47].

3. Electrocrystallization at the Ionic Liquid/Substrate Interface – Phase Formation and Growth on the Nanometer Scale

3.1. ELECTRODEPOSITION AND ELECTROCRYSTALLIZATION – SOME BASIC CONCEPTS

What the significance of the chemical potential μ_i is in thermodynamics, the electrochemical potential η_i is in electrochemistry. The electrochemical potential of species i in phase (α) is defined by:

$$\eta_i^{(\alpha)} = \mu_i^{(\alpha)} + z_i F \varphi^{(\alpha)}, \tag{17}$$

where z_i = charge number on ion I, F = Faraday constant, and $\varphi^{(\alpha)}$ = Galvani potential in phase (α). Thus, the Gibbs adsorption equation for electrochemical conditions is simply rewritten as:

$$d\sigma = -s^{(\sigma)} dT - \sum \Gamma_i^{(\sigma)} d\eta_i, \tag{18}$$

where now the interfacial adsorption or excess $\Gamma_i^{(\sigma)}$ includes ionic species. Equ. (18) plays an important role for electrocapillarity and electrowetting phenomena, see e.g. [49].

For electrochemical deposition of a metal Me on its own substrate – in general the substrate material can be different from the deposited metal – the following reaction at the electrolyte/substrate interface is of interest:

$$Me^{Z+} + ze^- \rightleftharpoons Me. \tag{19}$$

At the electrode-electrolyte equilibrium, $\eta_i^{(\alpha)} = \eta_i^{(\beta)}$; with $\eta_i = \mu_i^\theta + RT \ln a_i + zF\varphi$ the potential difference across the electrode/electrolyte interface is given by the Nernst equation following from the electrochemical equilibrium:

$$\Delta\varphi \equiv \varphi^{(Me)} - \varphi^{(Me^{z+})} = \Delta\varphi^\theta + \frac{RT}{zF} \ln(a_{Me^{z+}}), \tag{20}$$

where $\Delta\varphi^\theta$ = standard Galvani potential and a_{Me} = 1 for the pure metal. The other symbols have their usual meaning. For measurements of $\Delta\varphi$, two electrodes (I and II) are needed leading to an electrochemical cell and the electromotoric force, E:

$$E \equiv \Delta\varphi(I) - \Delta\varphi(II) = E^\theta + \frac{RT}{nF} \ln K, \tag{21}$$

Here K is the equilibrium constant of the cell reaction, n is the number of exchanged electrons, and E^θ is the Nernst standard equilibrium potential.

The experimental deposition potentials often differ from the Nernst potential according to Equ. (21), $\eta = E_{dep} - E_{Nernst} \neq 0$. When $\eta > 0$ deposition from an undersaturated solution occurs at anodic potentials, this is called underpotential deposition (UPD). It is favoured if the metal-substrate bonding is stronger than the metal-metal bonding. In the case of overpotential deposition (OPD), $\eta < 0$, the deposition potential is reduced due to kinetic effects like charge transfer overpotential, concentration or diffusion overpotential. For further details see the introductory and advanced textbooks [50] - [52].

Electrocrystallization involves several distinct steps which are summarized as follows [52], [53]:
- Diffusion of ions in the bulk electrolyte towards the electrode interphase;
- Electron transfer reaction at the electrolyte/electrode interface; this reaction is complex, e.g. it is not generally clear if electron transfer sets in first with subsequent release of solvation energy or if this process occurs vice versa; differences between an aqueous and ionic electrolyte have to be considered, whereby the microscopic structure of an electrified ionic interface is not clear – see chapter 2.3
- Formation of ad-atoms;
- Diffusion of ad-atoms on the substrate surface; the role of adsorbed ions is a matter of ongoing research;
- Clustering of ad-atoms and heterogeneous nucleation:
- Formation of critical nuclei and growth of two (2D) or three dimensional (3D) deposits.

Space does not allow to discuss these initial stages in more detail here and so we refer to the comprehensive and recent book of Budevski, Staikov and Lorenz [54]. Only the concept of critical nucleation shall be touched briefly. For this we confine to a simple model of homogeneous nucleation and assume that the contact area of a spherical cluster of radius R with the electrode is small. The free energy change associated with the formation of the cluster contains two contributions, a surface term, $4\pi R^2 \cdot \sigma$, and a volume term, $4\pi/3 \cdot R^3 \cdot nF\eta/V_m$, where η is the overpotential and V_m is the molar volume of the cluster. Thus the overall free energy change is:

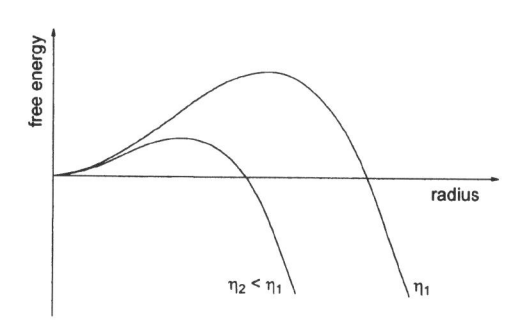

Figure 10: Homogeneous nucleation of a spherical cluster: schematic dependence of the free energy of nucleation on the cluster radius at different overpotentials η

$$\Delta F_{nucl} = 4\pi R^2 \sigma + (4/3)\pi R^3 (nF\eta/V_m). \qquad (22)$$

This is plotted in Fig. 10 as a function of R for two different overpotentials η. The main conclusion from this graph is that a thermodynamically stable cluster or nucleus has to overcome the energy barrier $\Delta F_{nucl}(R_c)$ with a radius $R > R_c$ where R_c is obtained from $\partial \Delta F/\partial R = 0$. This yields the critical radius $R_c = (-2V_m\sigma/nF\eta)$ resulting in a dependence on the overvoltage η of $R_c \sim \eta^{-1}$. Since η is determined by kinetic effects (see above) and these should depend on electrolyte properties it must be expected that the nucleation kinetics in for example an ionic and aqueous electrolyte can be different.

3.2. ELECTROCHEMICAL SCANNING PROBE TECHNIQUES WITH IONIC ELECTROLYTES

Several of the initial steps of electrocrystallization nowadays can be imaged directly by scanning tunneling microscopy (STM) or scanning probe microscopy (SPM) techniques. Electrochemical changes at the interface are probed *in situ* (EC-STM) whereby the morphology is scanned in real space with nanometer and partly atomic resolution and with time resolution down to approximately milliseconds. In principle, STM is based on the quantum mechanical effect of electron tunneling through a potential barrier. Accordingly the transmission probability of a free electron with kinetic energy E impinging on a potential barrier of height V_B and thickness d is given by

$$T(E, V_B, d) \propto \exp(-\kappa d), \tag{23}$$

with

$$\kappa^2 = 2m(V_B - E)/\hbar^2. \tag{24}$$

Here m = electron mass and $\hbar$ = Planck constant $/2\pi$. For example, if $\Phi = (V_B - E) \sim 1$ eV and d ~ 2 Å then T ~ 0.37, i.e. an appreciable tunneling current results. In a tunneling microscope electrons tunnel between a metal tip – typically Pt/Ir or W – and the substrate which are separated by a potential barrier – vacuum or electrolyte – at a distance d. The corresponding energy diagram is shown schematically in Fig. 11.

If tip and substrate are in thermal equilibrium the respective Fermi energies $\varepsilon_F(A)$ and $\varepsilon_F(B)$ are equal and are separated from the vacuum levels by their respective work functions Φ_i. Applying a bias voltage V between tip and substrate – which must be controlled independently from the deposition potential η by a bipotentiostat – the Fermi levels are shifted relative to each other effecting the tunneling transmission probability. Furthermore, the influence of the respective density of states – dashed curves in Fig. 11 – has to be taken into account. In summary, the tunneling current j can be written as, see e.g. [55]:

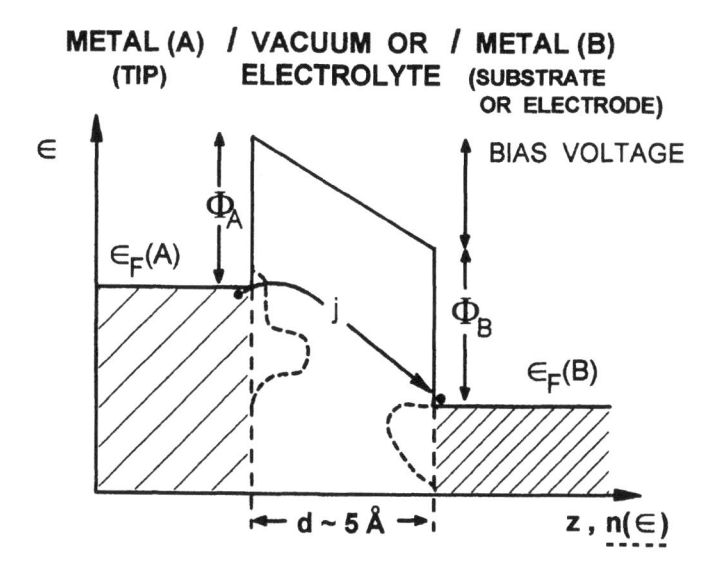

METAL (A) / VACUUM OR / METAL (B)
(TIP)   ELECTROLYTE (SUBSTRATE
                         OR ELECTRODE)

Figure 11: Tunneling microscopy and spectroscopy: Schematic energy diagram of tip and sample in thermal equilibrium and with applied bias voltage; the respective Fermi energies are $\varepsilon_F(A)$ and $\varepsilon_F(B)$, the work functions Φ_A and Φ_B; the density of states, $n(\varepsilon)$, is sketched by dashed lines. (see text)

$$j \propto \int_0^{eV} n_A(E) n_B(E-eV) T(E, eV) dE. \qquad (25)$$

With further simplification for the $n_i = \tilde{n}_i =$ const and the transition probability T, this yields [55]

$$j \propto V \tilde{n}_A \tilde{n}_B \exp\left(-\text{const} \cdot d \cdot (\Phi_{eff} - V)^{1/2}\right), \qquad (26)$$

so that the tunneling current is mainly determined by the distance d and the effective barrier height, Φ_{eff}. In the microscopy mode V is kept constant. During scanning in the x, y-directions, the z-movement of the STM tip is controlled by a feedback circuit in such a way that j remains constant, too. Thus information on the surface topography d(x, y) in real space is provided. In the spectroscopy mode the distance d(z) is fixed at a selected point at the surface and j is measured at different V from which the effective work function or barrier height is determined according to Equ. (26). Thus an *in situ* analytical information on the nanometer scale is possible.

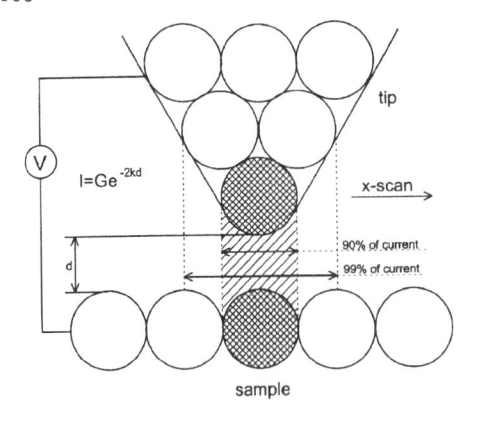

Figure 12: Principle of scanning tunneling microscopy with atomic resolution (see text)

In order to achieve atomic resolution with STM the tip should have an ideal shape as sketched in Fig. 12. Since the current decays exponentially with d, in this case ~ 90% of the electrons would tunnel between the outermost tip atom and the adjacent atom of the substrate. In reality, this configuration is seldom reached. By electrochemical etching tips with a radius of curvature r of the order of 1 nm can be obtained. If the shape of this tip is approximated by a parabolic curve and the tip minimum is located at x = 0 at a distance d, the current flowing at a point x ≠ 0 will be:

$$j(d + x^2/2r) \propto \exp(-\kappa x^2/r). \quad (27)$$

The half width of this Gaussian is ~ $(r/\kappa)^{1/2}$ and with $\kappa \sim 1$ Å^{-1} and r ~ 10 Å this yields a resolution in the x-y-plane of ~ 3 Å. Values of this magnitude –sometimes better – can be achieved in EC-STM experiments. The resolution in the z-direction typically is clearly better. Besides the tip problem, as sketched here, the resolution is further influenced by the quality of vibration damping, piezo crystals and tip holder and thermal stabilization [55].

A crucial problem in an EC-STM experiment is the electrical insulation of the tip in contact with the electrolyte. Since the tunneling currents are of the order of 10^{-9} A and the Faraday currents are orders of magnitude larger, that part of the tip inside the electrolyte must be insulated with exception of the very end of several nm length. This problem is more severe in ionic than in aqueous electrolytes due to the higher conductivities of ionic liquids. Coating with electropaints is generally employed for temperatures near room temperature. At higher temperatures (≤ 500 K) glass coating has been successfully tested in contact with NaCl-AlCl$_3$ melts [56]. A drawing of the principle arrangement of the STM tip plus piezo scanner and the electrochemical cell is shown in Fig. 13.

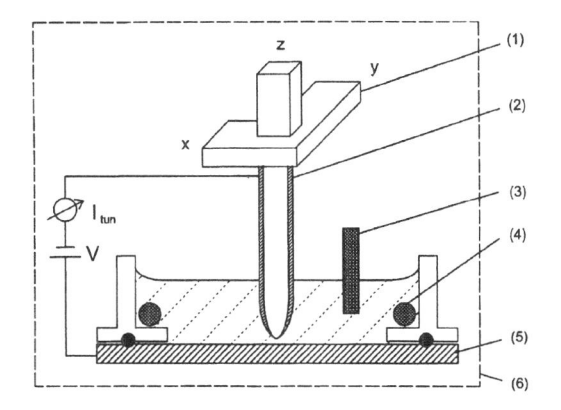

Figure 13: Schematic drawing of electrochemical STM set up; x,y,z-piezo scanner (1), STM tip with electrical insulation (2), reference electrode (3), counter electrode (4), working electrode (5), vacuum tight housing (6)

This set up has been employed for measurements on ionic liquids sensitive to air and humidity [24], [25]. The electrochemical cell and the scanner are contained in a vacuum tight chamber (dashed lines). The construction of Fig. 14 has been developed for high temperature studies. The electrochemical cell is tightly sealed with the scanner via a flexible stainless steel bellow [56].

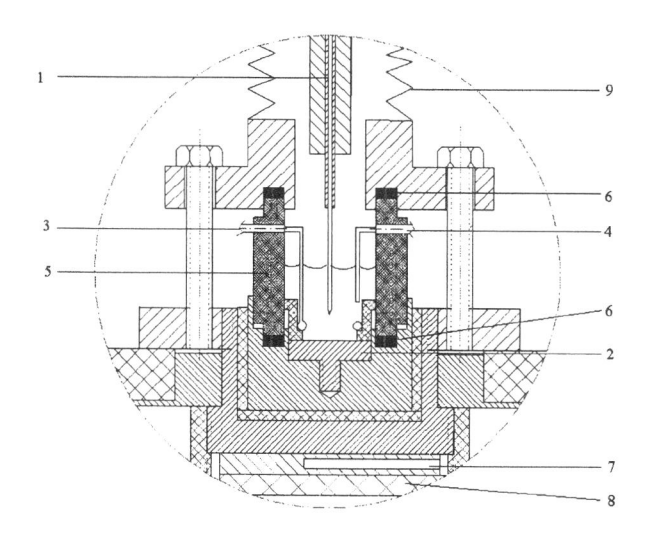

Figure 14: High temperature electrochemical STM cell; tip holder (1), working electrode (2), counter electrode (3), reference electrode (4), glass cylinder (5), graphite sealing (6), thermocouple boring (7), heating element (8), stainless steel bellow connection with scanner (9)

3.3. UNDERPOTENTIAL DEPOSITION FROM ROOM TEMPERATURE MOLTEN SALTS ON AU(111) SUBSTRATES

In this section 2D phase formation during electrodeposition from ionic liquids at room temperature is considered. Fig. 15 shows a sequence of three STM images of Ag deposition on a Au(111) surface from an acidic aluminium chloride/1-methyl-3-butylimidazolium chloride (MBIC)-5 mmol/l AgCl melt at three different potentials of $\eta = 0.2$ V (b), $\eta = 0.05$ V (c) and $\eta = 0.03$ V (d). The corresponding cyclic voltammogram (cv) measured at a scan rate of 100 mV/s at 298 K is presented in Fig. 15a. It exhibits essentially five redox processes marked A – E which have been assigned on the basis of CV, current transients, and *in situ* EC-STM measurements to the following reactions:

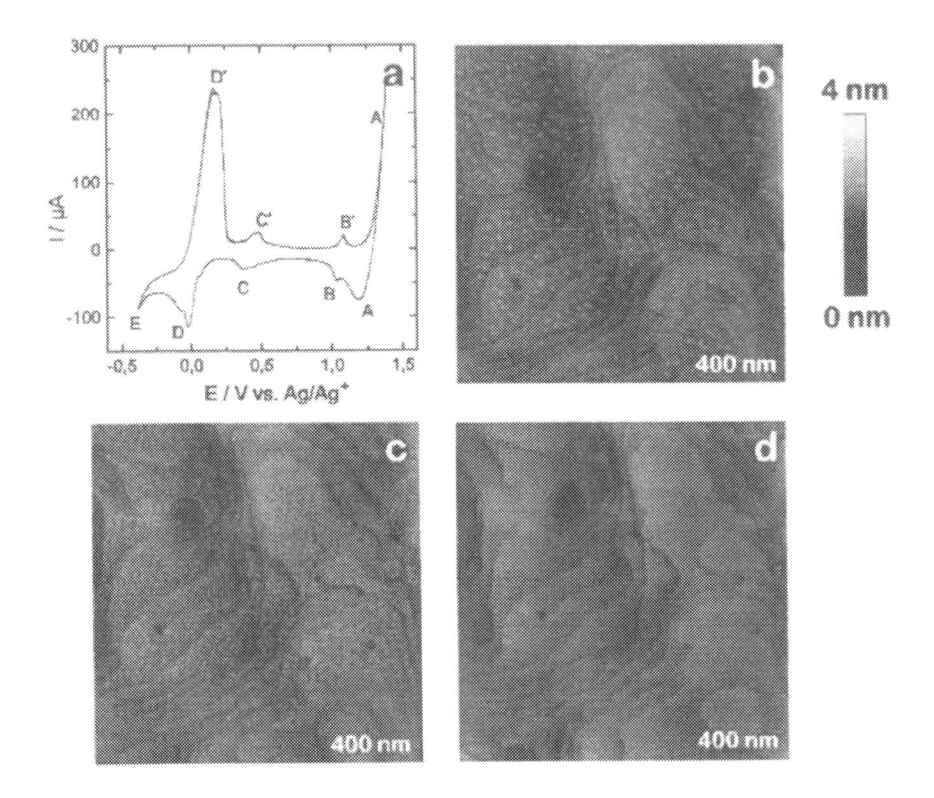

Figure 15: Monolayer formation of Ag at the solid Au(111)/liquid AlCl$_3$-MBC(2:1) – 5 mmol l^{-1} AgCl interface. a) cyclic voltammogram vs Ag/Ag$^+$ reference electrode, v = 100 mV/s; *b)* STM image of Ag islands forming on the monoatomic terraces of the Au(111) surface; $\eta = 0.2$ V vs Ag/Ag$^+$, V bias = 0.15 V, I tunneling = 5 nA; scan rate = 3 Hz; *c)* same as b), but $\eta = 0.05$ V vs Ag/Ag$^+$, V bias = 0.0 V; *d)* completed Ag monolayer; $\eta = 0.03$ V vs Ag/Ag$^+$, V bias = 0.07 V. [24]

Au bulk oxidation (A), step edge oxidation of Au terraces (B), Ag adsorption and monolayer formation (C), 3D Ag cluster formation and layer by layer growth (D), and bulk Ag deposition including Ag-Al alloying (E); see Ref. [24] for further details. We focus here on process (C) and have a closer look at the information from STM pictures for $0.2\,V \geq \eta \geq 0.03\,V$. In Fig. 15b Ag islands of monoatomic height and with diameters of 5 – 7 nm are seen on top of the Au terraces with monoatomically high steps. The average distance between the Ag islands is 16 ± 3 nm with an average coverage of the Au(111) surface of ~ 30%. At constant potential no change of the islands occurs over a period of about one hour. Reducing η to a potential of 0.05 V the islands grow and the coverage clearly increases (Fig. 15c). At a potential of 0.03 V a complete Ag monolayer covers the Au(111) surface exhibiting the original topography of the uncovered Au(111) surface. In the interval $0.03\,V \geq \eta \geq 0\,V$ the formation of a second Ag monolayer can be observed in a similar manner. Comparing these results with measurements of Ag deposition from aqueous electrolytes, a similar picture appears for the monolayer formation, whereas distinct differences seem to occur at higher potentials, for example, a superstructure near 0.5 V has been reported [57].

The process of Ag monolayer formation is found to be completely reversible. This has been documented by STM in several cycles. However, after extended polarization at potentials close to the Nernst potential, dissolution of the Ag deposit reveals small holes of ~ 250 pm depth in the Au(111) surface which heal after about half an hour – see Fig. 16.

This indicates that surface alloying between the Ag deposit and the top layer of the Au substrate takes place. After dissolution of Ag these holes are filled via surface diffusion of Au atoms which have been expelled before by surface alloying. Similar investigations of Al electro-deposition in the UPD range give a completely different picture [24]. At a potential of 0.9 V vs Al/Al^{3+} two dimensional Al islands are formed exhibiting a characteristic Ostwald ripening. They are not dissolved at anodic potentials up to 1.3 V and do not change their size and shape

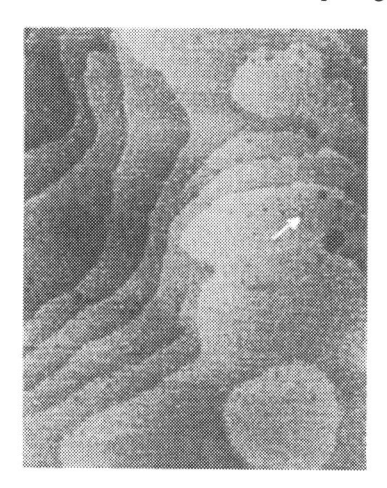

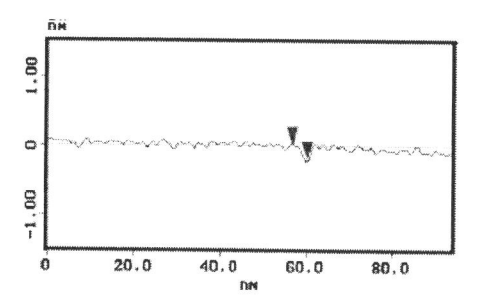

Figure 16: Ag-Au surface alloying. Holes of nanometer diameter and atomic diameter depth are seen after Ag deposition at $\eta = 0.03$ V and subsequent dissolution at $\eta = 0.5$ V vs Ag/Ag$^+$. In the lower part a typical depth profile is shown corresponding to the hole marked by the arrow in the upper STM image [24]

decreasing the potential down to 0.2 V. This indicates that a stable Al-Au compound is formed at the Au surface.

The nucleation and growth of monolayers as shown for the example of Ag underpotential deposition is typical of metal deposition on substrates with commensurable structures and is referred to as 2D electrocrystallization. The kinetics of 2D electrocrystallization of Ag on Au(111) were found to be controlled by an adsorption reaction with a Langmuir type adsorption mechanism consistent with the exponentially decaying current transients [24].

Totally different surface structures and growth behaviour can be observed at the interface of a semiconductor/ionic liquid. This is shown for the Ge deposition from MBIPF$_6$ + 1 mmol GeI$_4$ on Au(111), see Fig. 17. At a potential of –2V vs. Pt quasi referefence the surface of the deposit shows a terrace-like structure with an average height of the terraces of 330 ± 30 pm [58]. Oxidizing the surface for about 10 s and reversing the electrode potential back to –2V a wormlike vacancy structure is evident. The vacancies cluster with time and pinch-off events are clearly visible, Fig. 17b. This behaviour is characteristic of a transport mechanism via periphery diffusion and strongly depends on the line tension along the step edges [59]. This mechanism is clearly different from what is typically considered in 2D nucleation and growth of metals.

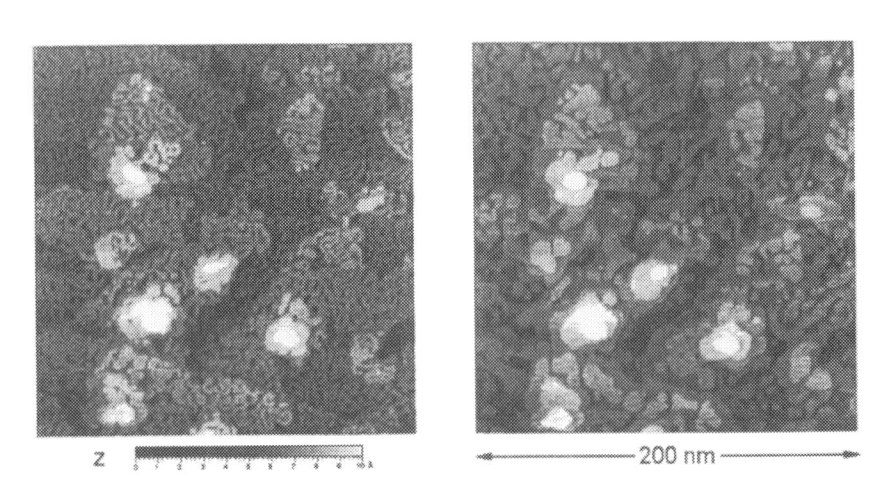

Figure 17: Evolution of wormlike structures at the Ge deposit/ionic liquid interface after a short period of oxidation; a) oxidation at –1.5 V for 10 s, STM image taken 2 min later at –2.0 V; b) same as a) but STM image taken 45 min later [58]

3.4. INITIAL STAGES OF PHASE FORMATION DURING ELECTRODEPOSITION OF NI AND NI-AL ALLOYS: STM AND STS STUDIES

Nickel electrodeposition from aqueous electrolytes on different single crystal substrates like Au(111) and Ag(111) has been investigated at length by EC-STM, see e.g. [60]. Only a few points of these studies shall be pointed out which are relevant to the following. In the underpotential range no Ni deposition is found on reconstructed Au(111) electrodes. The first Ni monolayer is formed at an overpotential of ~ 0.1 V. Due to the mismatch of the Ni monolayer adlattice (d_{Ni-Ni} = 2.5 Å) and the hexagonal lattice of the Au(111) surface (d_{Au-Au} = 2.9 Å) the first Ni monolayer is manifested by a hexagonally ordered Moiré pattern with a nearest neighbour spacing of ~ 22 Å [60]. With cathodic increase of the deposition potential a layer-by-layer growth is observed leading to smooth Ni films of a well defined thickness.

Our interest in studying the UPD- and OPD of Ni and Ni-Al electrodeposition from a molten salt electrolyte was diverse. First, the influence of the ionic electrolyte on electrocrystallization should be elucidated. Secondly, the alloying with Al has not been studied by STM before which cannot be performed in aqueous solutions due to their low decomposition potential. Finally, we aimed at characterizing the alloy formation *in situ* by measuring the voltage tunneling spectra which give at least a qualitative analytical information.

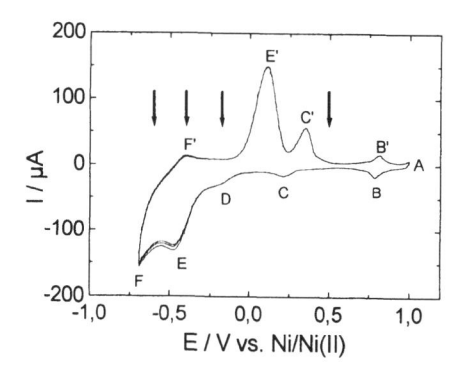

Figure 18: Cyclic voltammogram of the system Ni^{2+}/Au(111) in the $AlCl_3$/MBIC (58-42 room temperature molten salt at 298 K, v = 100 mV/s. [25]

The cyclic voltammogram of $AlCl_3$/MBIC + 5 mmol Ni(II) is shown in Fig. 18. The relevant redox processes and potentials vs Ni/Ni^{2+} are: bulk Au oxidation at the anodic limit of 1.0 V (A), Au step edge oxidation at 0.8 V (B), Ni adsorption in the UPD range (C), 3D Ni cluster formation near –0.2 V in the OPD-range (D), Ni_xAl_{1-x} alloy formation (E), and bulk Al deposition starting at –0.7 V (F). All these processes have been visualized by STM and STS which shall be demonstrated for a few examples.

Fig. 19 presents a selection of STM images corresponding to UPD- and OPD conditions. At a potential of 0.11 V vs Ni/Ni(II) the first Ni monolayer is completed which is evidenced by the appearance of a Moiré pattern; it has a clear hexagonal superstructure with a nearest neighbour distance of 23 ± 1 Å [61] - see Fig. 19a. Reducing the potential towards the Nernst potential Ni islands start to grow from the Au step edges, Fig. 19b. 3D step decoration of these islands occurs at slightly cathodic potentials. At a potential of –0.2 V (see process D) columnar clusters grow exclusively along the Au step edges with a height of ~ 30 Å, Fig. 19c. Stepping the potential to –0.2 V the growth behaviour of Ni clusters in the OPD range has been determined; this is shown in Fig. 19 d for the example of the aspect ratio (width/height) as a function of time. The latter observation is comparable to Ni nucleation in aqueous electrolytes on a Ag(111) substrate [60]. In summary, these images give clear evidence from STM for the assignment of processes C and D given above.

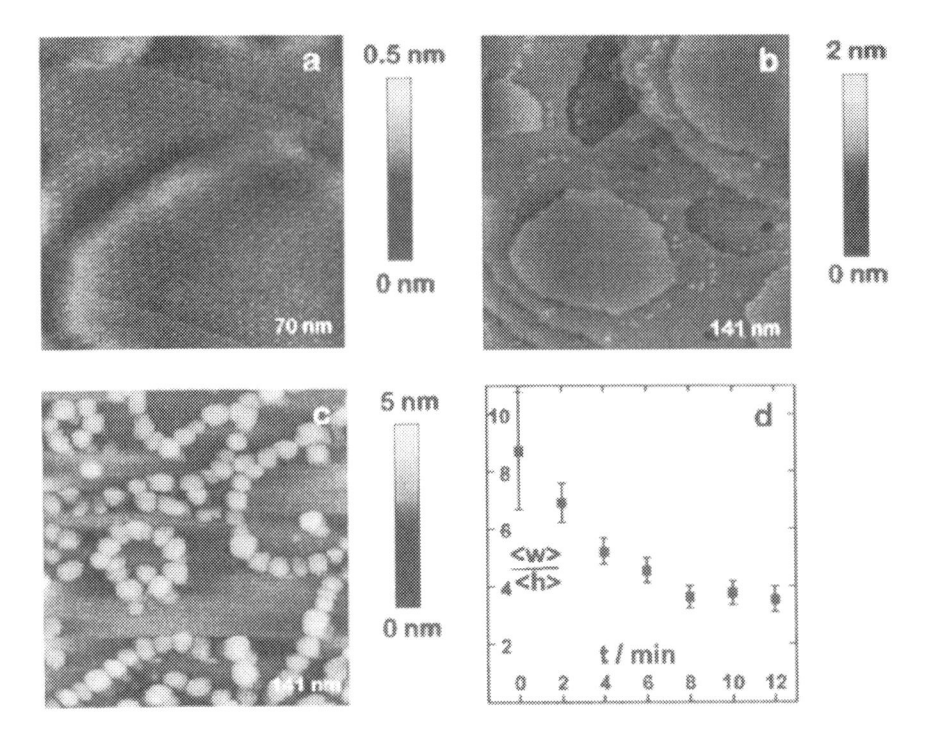

Figure 19: STM images of 2D and 3D Ni deposition and growth. *a)* Moiré pattern of the first Ni monolayer in the UPD range, η = 0.11 V, V bias = 0.2 V, I tunneling = 3 nA, scan rate = 3 Hz. *b)* Growth along Ni islands and steps, η = 0.0 V, else see a) *c)* Columnar growth of Ni clusters in the OPD range, η = -0.2 V, else see a) *d)* Time dependence of Ni cluster growth, aspect ratio vs time at η = -0.2 V (see text). [61]

A I – V curve corresponding to columnar Ni clusters at a potential of –0.2 V is shown in Fig. 20. The tunneling current is measured as a function of the bias voltage at a constant tip-substrate distance of ~ 6 Å at a fixed cluster marked by a (+) in the upper part of the corresponding STM image. From a fit of the I – V curve by Equ. (28):

$$I = const \int_0^V \exp\left(-2(2m)^{0.5}\hbar^{-1}(\Phi_{eff} - Ve/2)^{0.5} d\right) dV ,\tag{28}$$

the effective work function Φ_{eff} or tunneling barrier of these Ni clusters has been determined to be ~ 1 eV. Similar measurements made at deposition potentials of –0.4 V but else identical experimental conditions (process E in Fig. 18) yield an Φ_{eff} ~ 0.4 eV, whereas at –0.7 V –Al deposition – a value of 0.2 eV has been found [25]. This trend in the reduction of the effective work function strongly supports the analysis of the redox process E by bulk Ni-Al formation. It is also apparent in the STM images of Fig. 20, where in the case of Ni_xAl_{1-x} deposition clusters are homogenously spread, whereas the pure Ni clusters at –0.2 V are arranged in the form of chains along the Au steps.

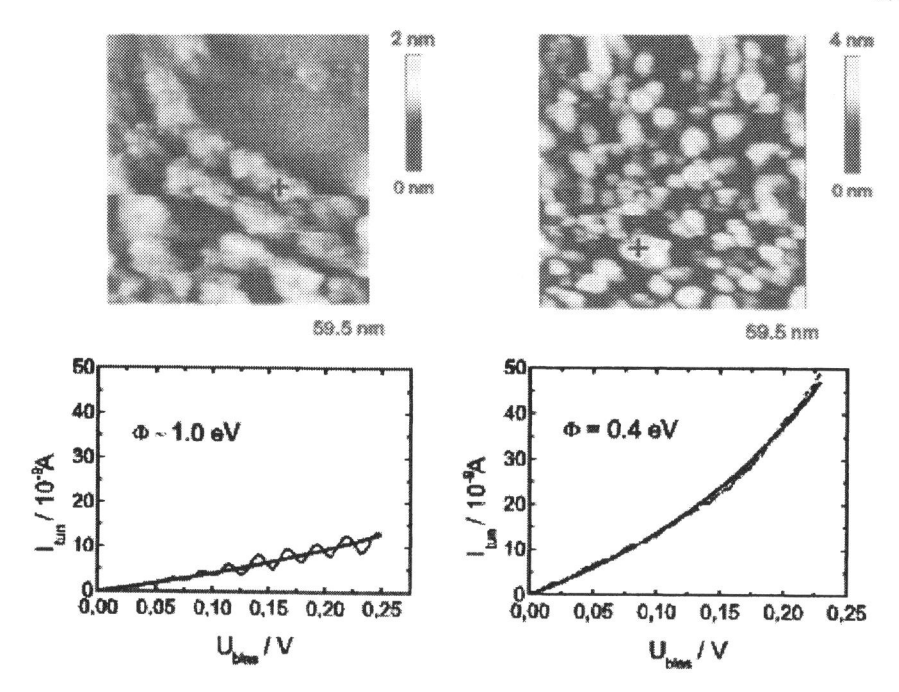

Figure 20: STM images (top row) and I-V curves (bottom row) of Ni and Ni-Al cluster deposition. Left: Ni deposition at –0.2 V, Right: Ni-Al Clusters deposited at –0.4 V. The effective work functions Φ have been determined from fits of the measured I – V curves by Equ. (28).

4. Concluding Remarks

Metal-molten salt interfaces have been considered from two different directions: wetting phenomena at the metal-molten salt/wall (sapphire) interface and electrocrystallization at the electrified molten salt/metal interface. Alkali metal –alkali halide melts with a liquid –liquid miscibility gap exhibit prewetting transitions at elevated temperatures in the homogeneous metalrich phase whereby a saltrich wetting film of mesoscopic thickness intrudes between the bulk metallic fluid and the wall. A transition to complete wetting accompanied by macroscopically thick wetting films seems to occur along the metastable extension of the liquid-liquid coexistence curve which is consistent with a tetra point wetting scenario. In comparison with fluid mixtures characterized by van der Waals type interactions the wetting behaviour of metal-molten salt solutions shows various peculiarities: strong undercooling of the saltrich wetting films by more than 200 K, unusual thickness of the prewetting films of the order of 10 – 100 nm, and a microscopic structure different from the bulk molten salts. Presumably surface charging plays an important role. An interesting question is, if and in how far these results of metal-molten salt solutions can be extended to purely ionic mixtures? It is difficult to give an answer at present as investigations of the wetting characteristics of molten salt mixtures, low or high temperature molten salts, are not known. Similarly, knowledge of the microscopic structure of the interface of ionic systems and its change as a function of the state variables is rather poor. This is also true for the respective electrified interfaces; the "double layer" problem of an ionic electrolyte/electrode interface is still unsolved.

Several of these fundamental problems have a strong impact on technological applications. A specific example is electrodeposition from molten salt electrolytes which are widely used in light and refractory metal production. For a better understanding of these processes and an improved control of deposits, knowledge of the initial stages of electrocrystallization on the nanometer scale is desirable. This topic is addressed in the second part of this lecture. For a few selected examples it is shown that *in situ* electrochemical scanning probe experiments are feasible also at the ionic electrolyte/electrode interface, both with ionic liquids like room temperature molten salts and the classical molten salts like NaCl/AlCl$_3$ mixtures. A specific advantage of these electrolytes is their relatively large electrochemical window which enables EC-SPM-studies of both metals and semiconductors with nanometer control.

Acknowledgement

Several of my present and former collaborators have made excellent contributions to the topics of this article which is appreciated. Expert assistance in preparing this manuscript by D. Rohmert-Hug and J. Szepessy is acknowledged. Special thanks are extended to C. A. Zell for critical reading of the manuscript and to the artist of Fig. 1. Part of the work reviewed here has been supported by DFG through SFB 195 and Project FR 299/17 as well as by the Fonds der Chemischen Industrie.

References

[1] Cahn, J. W., 1977, Critical Point Wetting, J. Chem. Phys., **66**, 3667
[2] Widom, B., 1978, Structure of the $\alpha\gamma$ Interface, J. Chem. Phys., **68**, 3878
[3] Dietrich, S., 1988, Wetting Phenomena, in C. Domb and J. Lebowitz (eds), Phase Transitions and Critical Phenomena, Academic Press, London
[4] Croxton, C. A. (ed.), 1986, Fluid Interfacial Phenomena, John Wiley and Sons, New York
[5] Charvolin, J., Joanny, J. F., and Zinn-Justin, J. (eds), 1990, Liquids at Interfaces, North-Holland, Amsterdam
[6] Dietrich, S., Findenegg, G., and Freyland, W. (eds), 1994, Phase Transitions at Interfaces, Proceeding of a Discussion Meeting, Ber. Bundsensges. Phys. Chem., **98** (3)
[7] Frenken, J. W. M., Marée, P. M. J., and van der Veen, J. F., 1986, Observation of Surface-initiated Melting, Phys. Rev. B, **34**, 7506
[8] Van der Veen, J. F., 1999, Melting and Freezing at Surfaces, Surface Science, **433 – 435**, 1
[9] Zhu, Da-Ming, and Dash, J. G., 1988, Surface Melting of Neon and Argon Films: Profile of the Crystal-Melt Interface, Phys. Rev. Lett., **60**, 432
[10] Dash, J. G., Fu, Haiying, and Ettläufer, J. S., 1995, The Premelting of Ice and its Environmental Consequences, Rep. Prog. Phys., **58**, 115

[11] Löwen, H., 1994, Melting, Freezing and Colloidal Suspensions, Physics Reports, **237**, 249
[12] Pershan, P. S. and Als-Nielsen, J., 1984, X-ray Reflectivity from the Surface of a Liquid Crystal: Surface Structure and Absolute Value of Critical Fluctuations, Phys. Rev. Lett, **52**, 759
[13] Ocko, B. M., Wu, X. Z., Sirota, E. B., Sinka, S. K., Gang, O., and Deutsch, M., 1997, Surface Freezing in Chain Molecules: Normal Alkanes, Phys. Rev. E, **55**, 3164
[14] Yang, B., Gidalevitz, D., Li, D., Hung, Z., and Rice, S. A., 1999, Two-dimensional Freezing in the Liquid-Vapour Interface of a Dilute Pb-Ga Alloy, Proc. Nat. Acad. Sci., USA, **96**, 13009
[15] Turchanin, A., Nattland, D., and Freyland, W., 2001, Surface Freezing in a Liquid Eutectic Ga-Bi Alloy, Chem. Phys. Letters, **357**, 5
[16] Lei, Ning, Huang, Zhengquing, and Rice, S. A., 1996, Surface Segregation and Layering in the Liquid/Vapour Interface of a Dilute Bismuth-Gallium Alloy, J. Chem. Phys. **104**, 4802
[17] Binnig, G., and Rohrer, H., 1981, Scanning Apparatus for Surface Investigation Using Vacuum Tunnel Effect at Cryogenic Temperatures, Eur. Pat. Appl., EP27517
[18] Sonnenfeld, R., and Hansma, P. K., 1986, Atomic-Resolution Microscopy in Water, Science, **232**, 211
[19] Wiechers, J., Twomey, T., Kolb, D. M., and Behm, R. J., 1988, An in-situ Scanning Tunneling Microscopy Study of Gold (111) with Atomic Resolution, J. Electroanal. Chem, **248**, 451
[20] Gewirth, A. A. and Sigenthaler, H., (eds), 1995, Nanoscale Probes of the Solid/Liquid Interface, Nato ASI Series E, Kluwer Academic Publishers, Dordrecht
[21] Lorenz, W. J., and Plieth, W. (eds), 1998, Electrochemical Nanotechnology, Wiley-VCH, Weinheim
[22] Kolb, D. M., Ullmann, R., and Will, T., 1997, Nanofabrication of Small Copper Clusters on Gold (111) Electrodes by a Scanning Tunneling Microscope, Science, **275**, 1097
[23] Penner, R. M., 2000, Hybrid Electrochemical/Chemical Synthesis of Quantum Dots, Acc. Chem. Res. **33**, 78
[24] Zell, Chr., Endres, F., and Freyland, W., 1999, Electrochemical *in situ* STM Study of Phase Formation During Ag and Al Electrodeposition on Au(111) from a Room Temperature Molten Salt, Phys. Chem. Chem. Phys. **1**, 697
[25] Zell, C. A. and Freyland, W., 2001, *In situ* STM and STS Study of Ni_xAl_{1-x} Alloy Formation on Au(111) by Electrodepositon from a Molten Salt Electrolyte, Chem. Phys. Letters, **337**, 293
[26] Defay, R., and Prigogine, I., 1966, Surface Tension and Adsorption, Longmans, Green and Co., Bristol
[27] Adamson, A. W., and Gast, A. P., 1997, Physical Chemistry of Surfaces, 6th ed., John Wiley and Sons, New York
[28] Rowlinson, J. S., and Widom, B., 1982, Molecular Theory of Capillarity, Clarendon Press, Oxford
[29] Israelachvili, J. N., 1995, Intermolecular and Surface Forces, 5th ed., Academic Press, London

[30] Somorjai, G. A., 1994, Introduction to Surface Chemistry and Catalysis, John Wiley and sons, New York
[31] Taborek, P. and Rutledge, J. E., 1994, Prewetting of Helium on Weak Binding Substrates, in Ref. [6], 361
[32] Christmann, K., 1991, Introduction to Surface Physical Chemistry, Springer-Verlag, New York
[33] Shen, Y. R., 1984, The Principles of Nonlinear Optics, Wiley, New York
[34] Mills, D. L. 1991, Nonlinear Optics, Springer, Heidelberg
[35] Tostmann, H., Nattland, D., and Freyland, W., 1996, Wetting and Prewetting Transition in Metallic Fluid K-KCl Solutions Studied by Second Harmonic Gerneration, J. Chem. Phys. **104**, 8777
[36] Azzam, R. M. A., Bashara, N. M., 1987, Ellispometry and Polarized Light, North-Holland, Amsterdam
[37] Staroske, S., 2000, Spektroskopische Charakterisierung von Benetzungs- und Vorbenetzungsfilmen im Fluiden K-KCl, Ph. D. Thesis, Universität Karlsruhe (TH)
[38] Staroske, S., Nattland, D., and Freyland, W., 2001, Tetra Point Wetting of Liquid K-KCl Mixtures: Spectroscopic Characterization of Mesoscopic Wetting and Prewetting Films, J. Chem. Phys. (submitted)
[39] Nattland, D., Poh, P. D., Müller, S. C., and Freyland, W., 1995, Interfacial Wetting in a Liquid Binary Alloy, J. Phys. Condens. Matter, **7**, L 457
[40] Wynblatt, P., and Chatain, D., 1998, Wetting and Prewetting Transitions in Ga-Pb Alloys, Ber. Bunsenges. Phys. Chem. **102**, 1142
[41] Tostmann, H., Di Masi, E., Shpyrko, O. G., Pershan, P. S., Ocko, B. M., and Deutsch, M. 2000, Microscopic Structure of the Wetting Film at the Surface of Liquid Ga-Bi Alloys, Phys. Rev. Lett, **84**, 4385
[42] Freyland, W., 1995, in: Metal-Insulator Transition Revisited, P. P. Edwards and C. N. R. Rao (eds), Taylor and Francis, London
[43] Bredig, M. A., 1964 in: Molten Salt Chemistry, Blander, M., (ed), Interscience, New York
[44] Nattland, D. and Freyland, W., 1988, Macroscopic Adsorption and Wetting Transition at the Fluid/Solid Interface of K_xKCl_{1-x} Solutions, Phys. Rev. Lett, **60**, 1142
[45] Heyers, D. M. and Clarke, J., 1981, Computer Simulation of Molten Salt Interfaces, J. Chem. Soc. Faraday Trans., **77**, 1089
[46] Dietrich, S., and Schick, M., 1997, Wetting at a Solid-Liquid-Vapour Tetra Point, Surf. Science, **382**, 178
[47] Juchem, R., Müller, S., Nattland, D., and Freyland, W., 1990, Interfacial Segregation and Wetting Transition in Fluid Metal-Salt Systems, J. Phys. Condens. Matter, **2**, SA 427
[48] Staroske, S., Nattland, D., and Freyland, W., 2000, Spectroscopic Evidence of a Wetting and Prewetting Transition in Liquid K-KCl Mixtures, Phys. Rev. Lett. **84**, 1736
[49] Welters, W. J. J. and Fokkink, L. G. J., 1998, Fast Electrically Switchable Capillary Effects, Langmuir, **14**, 1535
[50] Hamann, C. H. and Vielstich, W., 1981, Elektrochemie, I and II, Verlag Chemie, Weinheim

[51] Bard, A. J., and Faulkner, L. R., 2001, Electrochemical Methods, Fundamentals and Applications, 2nd ed, John Wiley and Sons, New York
[52] Greet, R., Peat, R. Peter, L. M., Pletcher, D., Robinson, J., 1990 Instrumental Methods in Electrochemistry, 2nd ed, Ellis Horwood, New York
[53] Fischer, H., 1954, Elektrolytische Abscheidung und Elektrokristallisation von Metallen, Springer-Verlag, Heidelberg
[54] Budevski, El, Staikov, G., and Lorenz, W. J., 1996, Electrochemical Phase Formation and Growth, Verlag Chemie, Weinheim
[55] Hamers, R. J., and Bonnell, D. A., (eds) 1993, Scanning Tunneling Microscopy and Spectroscopy: Theory, Techniques and Application, VCH, New York
[56] Chourankov, A., Endres, F., and Freyland, W., 2001, High Temperature EC-STM for Molten Salt Studies, Rev. Scient. Instr. (to be published)
[57] Ogaki, K., and Itaya, K. J., 1995, *In situ* Scanning Tunneling Microscopy of Underpotential and Bulk Deposition of Silver on Gold (111), Electrochimica Acta, **40**, 1249
[58] Endres, F., and Schrodt, C., 2000, *In situ* STM Studies on Germanium Tetraiodide Electroreduction on Au(111) in the Room Temperature Molten Salt 1-Butyl-3-Methylimidazolim Hexafluorophosphate, Phys. Chem. Chem. Phys., 2, 5517
[59] Pai, Wu Woli, and Wendelken, J. F., 2001, Evolution of Two-Dimensional Wormlike Nanoclusters on Metal Surfaces, Phys. Rev. Lett. **86**, 3088
[60] Morin, S., Lachewitzer, A., Möller, F., Magnusson, O. M., and Behm, R. J., 1999 Comparative *in situ* STM Studies on the Electrodeposition of Ultrathin Nickel Films on Ag(111) and Au(111) Electrodes, J. Electrochem. Soc. **146**, 1013; see also: 1999, Phys. Rev. Lett. **83**, 5066
[61] Zell, C. A., Freyland, W., and Endres, F., 2000, *In situ* STM Study of Ni Electrodeposition on Au(111) from the Room Temperature Molten Salt MBIC-ALCl$_3$, Progr. Molten Salt Chemistry 1, (eds) Berg, R. W., and Hjuler, H. A., Elsevier, Paris, 597

MODELLING OF THERMODYNAMIC DATA

HARALD A. ØYE
Department of Chemistry, Norwegian University of Science and Technology, N-7491 Trondheim, Norway

0. A General Introduction to Model Building and Thermodynamic Models in Particular

Model building is at best a dynamic process where model fitting and analysis may lead to new experiments and an improved model. The language of statistics is often used for evaluation of models. Statistics, however, is the science of random errors and random errors are not the main problem for model builders. Removal of systematic errors is generally the major issue. Hence, the given confidence limits are often many times too small. The non-statistical errors may be of three kinds:

Experimental errors.
Errors due to wrong model structure.
Errors due to wrong parameters.

Modelling will concern itself with minimizing these errors.
The following points should be observed:

1. Know the experimental set-up, evaluate whether the best methods are used and probe for systematic experimental errors.

2. Evaluate the data set critically and decide whether some data should be discarded.
    Suspicious data may not only complicate the model building, but also lead to wrong conclusions.

3. Check physical background.
    Physically based models generally needs fewer parameters and are more reliable outside the measured range than empirical models. It is advantageous if the model could be explained in a simple way (See 4.1).

4. Check thermodynamic consistency. (For thermodynamic models).
    For instance, the Gibbs-Duhem equation must always be obeyed for activity functions.

5. Check derivation of equations.

6. The model structure must fit the problems.
   The generally used function $y = a + bx + cx^2 + dx^3 + \cdots$ is often a very poor choice. Example: Accurate viscosity measurements of molten NaCl in range 824 - 920°C was fitted to the equation $\eta = A \exp(B/T)$ with a standard deviation 0.05 %. Use of the equation: $\eta = a + bt + ct^2$ gave a standard deviation 0.4 % and had one additional parameter.
   Power series with too many parameters will also lead to oscillations. If there are strong changes in activities mass actions based laws are useful, see later. If a property goes towards an asumptotic value, the model equation should reflect that.

7. Minimize the number of parameters.
   The model program should always have a test for the necessity of each parameter.

8. Check border case.
   Border cases and extrapolations may pinpoint inconsistencies in the model and give validity limitations.

9. Make a drawing of the model.
   Drawings pinpoint deficiencies and peculiarities much better than tables. Inconsistencies become clearer by derivation.

10. Present the model in a condensed and clear form and give a calculational example.
    The example gives the user a check whether he has understood the model properly. (For instance has a log function base 10 or e).

    The benefit of a good model can be summarized as follows:
    - The data are "canned".
    - The data can be manipulated further in a computer.
    - The data is in a form useful for process design.
    - Interpolation and extrapolation are easily performed.
    - The model may pinpoint experimental consistencies.
    - The model may predict properties not measured (See 3.1 and 4.1).
    - The model may predict properties which at first glance were unexpected (See 4.1).
    - The model may predict "unmeasurable" properties (See 5.1).

1. Models for Mixtures

1.1. IDEAL SOLUTION

The term ideal solution is used for binary mixtures where Raoult's law is valid over the whole concentration range:

$$a_i = X_i \text{): } \gamma_i = 1 \tag{1.1}$$

This means solvent standard states for all components:

$$d\mu_i = RTd \ln a_i = RTd \ln X_i \tag{1.2}$$

$$\mu_i - \mu_i^o = RT \ln X_i \tag{1.3}$$

For a two component system A, B, and one mole of mixture:

$$\Delta G_m = -T\Delta S_m = X_A RT \ln X_A + X_B RT \ln X_B \tag{1.4}$$

0.5 mole A, 0.5 mole B

$$\Delta G_m = -RT \ln 2 \tag{1.5}$$

ΔG is the driving force of a chemical reaction and we see here that the driving force is entirely caused by the entropy,): increase in disorder by mixing, the energetic effect being zero.

1.2 REGULAR SOLUTION

The regular solution model assumes complete interchange in a lattice with statistical distribution of A and B with co-ordination number z and a bond energy given by the equation

$$A - B = \frac{-\varepsilon_A - \varepsilon_B + w}{z} \tag{1.6}$$

where ε_A and ε_B are the A-A and B-B bond energies. Statistical derivation gives the following thermodynamic function

$$\Delta G_m = RT(X_A \ln X_A + X_B \ln X_B) + X_A X_B b \tag{1.7}$$

$b = N_o w$

where N_o is Avogadros number.

$$a_A = X_A \gamma_A = X_A \exp\left(\frac{bX_B^2}{RT}\right) \tag{1.8}$$

$$a_B = X_B \gamma_B = X_B \exp\left(\frac{bX_A^2}{RT}\right) \tag{1.9}$$

This model is inconsistent. Random distribution is assumed and ideal entropy even if there is a preferred tendency for formation of pairs ($b < 0$) or demixing ($b > 0$). Nevertheless it turns out that this approximation describes actual systems to an often surprisingly good degree. b is assumed temperature independent.

The parameter b is a measure of deviation from ideality as the term without b corresponds to ΔG_m^{id} or $\Delta \mu_A^{id}$.

$b > 0, \Delta H_m > 0$: positive deviation from ideality, endothermic process and tendency to unmixing.

$b < 0, \Delta H_m < 0$: negative deviation from ideality, exothermic process and tendency to compound formation.

The regular solution theory can also predict when the tendency to unmixing is so large that phase separation occurs.

$\Delta G_m /RT$ for different values of b/RT is plotted on Fig. 1.1.

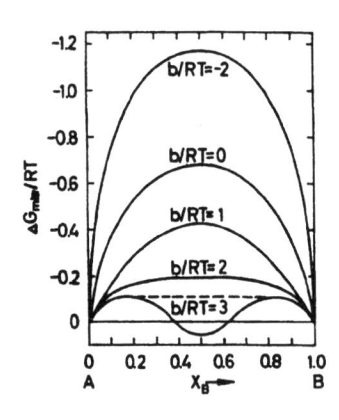

Figure 1.1. $\Delta G_m/RT$ as function of composition for different values

For $b/RT = 3$ unmixing must occur since the process

Mixture $(X_B = 0.5)$ → Mixture $(X_B \approx 0.2)$ + Mixture $(X_B \approx 0.8)$

is a voluntary process.

The temperature when $b/RT = 2$ and hence $T_c = b/2R$ (1.10)

is called the critical temperature of mixing.

1.3. ASYMMETRIC REGULAR SOLUTION

A variation of the regular solution model is the so-called asymmetric regular solution model which has been found useful when two components are unequal in size. The functions are as follow:

$$\Delta G_m = RT(X_A \ln X_A + X_B \ln X_B) + \frac{\alpha X_A X_B}{X_A + \beta X_B} \tag{1.11}$$

$$\ln \gamma_A = \frac{\alpha \beta}{RT} \left(\frac{X_B}{X_A + \beta X_B} \right)^2 \tag{1.12}$$

$$\ln \gamma_B = \frac{\alpha}{RT} \left(\frac{X_A}{X_A + \beta X_B} \right)^2 \tag{1.13}$$

1.4. GENERAL EQUATION FOR THE ACTIVITY OF LIQUID MIXTURES

As already mentioned regular solution is a surprisingly good approximation in spite of that the statistical condition of the model is not fulfilled, provided that the components are not too different chemically and not too close to the temperature of critical mixing.

There are, however, mixtures where a regular solution model is unsatisfactory. One approach is to modify the theory as, for instance, using the quasi-chemical model. Modification on the regular solution theory is, however, often marginal and when the deviation from ideality is substantial it is advisable to use a more empirical approach and, retaining regular solution term as the first term in the expansion.

Excess functions are used which is defined as:

$$\Delta G_m^E = \Delta G_m - \Delta G_m^{id} = \Delta G_m - RT \sum_i X_i \ln X_i \tag{1.14}$$

$$\Delta H_m^E = \Delta H_m \quad (as \ \Delta H_m^{id} = 0) \tag{1.15}$$

$$\Delta S_m^E = \Delta S_m - \Delta S_m^{id} = \Delta S_m + R \sum X_i \ln X_i \tag{1.16}$$

The equations are of a form which also can be used on multicomponent mixtures:

$$\Delta G_m = \sum_i X_i \Delta \mu_i = \sum_i X_i RT \ln a_i = RT \sum X_i \ln \gamma_i X_i \tag{1.17}$$

eqn. 1.14 can be alternately expressed as

$$\Delta G_m^E = RT \sum X_i \ln \gamma_i \tag{1.18}$$

The excess Gibb function is directly related to the activity coefficients.

Very often excess enthalpy and excess entropy will partly cancel rendering the ΔG_m^E curve less complicated than the two others.

In order to describe complex systems the Redlich-Kister expansion may be used. It incorporates ideal and regular solution as the 0th and 1st approximation.

$$\frac{\Delta G_m^E}{RT \, X_A X_B} = A + B(X_A - X_B) + C(X_A - X_B)^2 + D(X_A - X_B)^3 \quad (1.19)$$

By introduction of $X_A = 1 - X_B$ and partial differentiation is obtained:

$$\ln \gamma_A = X_B^2 \{A - B(4X_B - 3) + C(2X_B - 1)(6X_B - 5) + D(2X_B - 1)^2 (8X_B - 7)\} \quad (1.20)$$

and a symmetrical equivalent equation for $\ln \gamma_B$.

1.5. ACTIVITY OF MOLTEN SALT MIXTURES. THE TEMKIN MODEL

Due to Coulombic forces ionic crystals are ordered in a way that cations only are surrounded by anions and vice versa. An interchange of an anion and a cation in the lattice is energetically highly unfavourable. **An ionic melt can similarly be considered to consist of an anion lattice and a cation lattice with no interchanges between the two.** As there is no mixing between the two, the entropy will be given as a sum of a cation and an anion contribution.

$$S = S^+ + S^- \quad (1.21)$$

This is the general idea of the Temkin model which first was proposed by Temkin [1] and later applied by Flood, Førland and Grjotheim [2].

The ideal Temkin model then proposes that the mixture can be described by an ideal mixture of anions and an ideal mixture of cations. Ionic fractions are defined as:

$$X_{Na^+} = \frac{n_{Na^+}}{\Sigma n_+}, \quad X_{Cl^-} = \frac{n_{Cl^-}}{\Sigma n_-} \quad (1.22)$$

From the previous equations for ideal mixtures (1.4)

$$\Delta G_m = -T \Delta S_m = RT \left(\sum_{c_i} X_{c_i} \ln X_{c_i} + \sum_{a_i} X_{a_i} \ln X_{a_i} \right) \quad (1.23)$$

$$\Delta H_m = 0 \quad (1.24)$$

where X_{c_i} and X_{a_i} are cation fractions and anion fractions, respectively.

Using a mixture of NaCl and KBr as example:

$$\Delta G_m = -T\Delta S_m = RT\left[(X_{Na^+})(\ln X_{Na^+}) + (X_{K^+})(\ln X_{K^+}) + (X_{Cl^-})(\ln X_{Cl^-}) + (X_{Br^-})(\ln X_{Br^-})\right] \quad (1.25)$$

By partial differentiation:

$$\overline{\Delta S}_{NaCl} = -R(\ln X_{Na^+}) - R(\ln X_{Cl^-}) = -R\ln\left((X_{Na^+})(\ln X_{Cl^-})\right) \quad (1.26)$$

$$RT \ln a^{id}_{NaCl} = \overline{\Delta H} - T\overline{\Delta S} = 0 + RT \ln\left((X_{Na^+})(X_{Cl^-})\right) \quad (1.27)$$

$$a^{id}_{NaCl} = (X_{Na^+})(X_{Cl^-}) \quad (1.28)$$

The general equation is:

$$a^{id}_{A_m B_n} = X_A^m X_B^n \quad (1.29)$$

As an illustration of the formula, the activity of NaCl with 0.10 mol NaBr and 0.10 mol KBr is calculated.

0.90 NaCl - 0.10 NaBr: $a_{NaCl} = (X_{Na^+})(X_{Cl^-}) = 1 \cdot 0.90 = \mathbf{0.90}$

0.90 NaCl - 0.10 KBr: $a_{NaCl} = (X_{Na^+})(X_{Cl^-}) = 0.90 \cdot 0.90 = \mathbf{0.81}$

The Temkin model has some resemblance with the activity formulas used for aqueous electrolytes. Deviation from ideality will similarly be defined by an activity coefficient. For instance:

$$a_{NaCl} = (X_{Na^+})(X_{Cl^-})\gamma_{NaCl} \quad (1.30)$$

Deviation from ideality can, however, sometimes also be introduced by redefinition of the ionic species, see the following.

2. Ionic Models Assuming Ideal Mixing of Complexes

A useful concept for molten salts has been the assumption of defined structural species, so-called complexes. The term "complex" has a less defined meaning in molten salts, as the proposed entities are not isolated from each other by the solvent as for instance in water. Because of the somewhat unclear separation of possible complexes, a debate raged in the sixties on the justification of the concept for molten salts. This discussion has ceased due to the usefulness of such models and because several of the proposed complexes have been identified by vibration spectroscopy. Usefulness means that the thermodynamic properties can be explained in a simple and logical way by the assumption of complexes.

One should, however, always heed the warning Professor Håkon Flood gave us as students:

> "Thermodynamic experimental data can tell in principle only if a suggested structural model is possible or not, but they will frequently be unable to decide which one of several structures is correct or if a particular structure is the only one possible. It is also important to keep Hildebrand's warning in mind that principally an insufficient model may still fit the activity (ΔG) data well, however, when applied to ΔS and ΔH-data much more serious disagreement will be evident."

2.1. PHASE DIAGRAM KCl – MgCl$_2$ ON THE KCl SIDE

The interpretation of the phase diagram KCl – MgCl$_2$ on the KCl side by Flood and Urnes [3] is an early classical example of the usefulness of the complex concept.

An ideal Temkin model of K$^+$, Mg^{2+} and Cl will give the activity of KCl as

$$a_{KCl} = \left(X_{K^+}\right)\left(X_{Cl^-}\right) = 1 \cdot \frac{n_{K^+}}{n_{K^+} + n_{Mg^{2+}}} = \frac{n_{KCl}}{n_{KCl} + n_{MgCl_2}} = X^o_{KCl} \qquad (2.1)$$

This model did, however, not explain the phase diagram.

It was assumed that MgCl$_2$ reacted quantitatively with KCl forming the ion $MgCl_4^{2-}$ (which later also has been identified by Raman spectroscopy):

$$MgCl_2 + 2Cl^- \rightarrow MgCl_4^{2-} \qquad (2.2)$$

The ideal activity of KCl is then given as:

$$a_{KCl} = \left(X_{K^+}\right)\left(X_{Cl^-}\right) = 1 * \frac{n_{Cl^-}}{n_{Cl^-} + n_{MgCl_4^{2-}}} =$$

$$\frac{n_{KCl} - 2n_{MgCl_2}}{n_{KCl} - 2n_{MgCl_2} + n_{MgCl_2}} = \frac{3X^o_{KCl} - 2}{2X^o_{KCl} - 1} \qquad (2.3)$$

This activity function did explain the freezing point depression (Fig. 2.1) expressed as

$$\ln a_{KCl} = \frac{-\Delta H_f(KCl)}{R}\left(\frac{1}{T} - \frac{1}{T_f(KCl)}\right) \qquad (2.4)$$

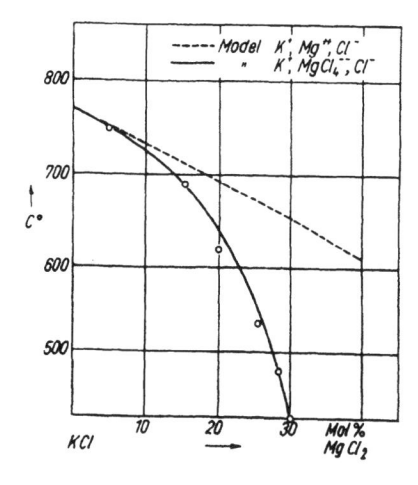

Figure 2.1. The KCl side of the KCl - MgCl$_2$ phase diagram [3].

Question: When does this model break up?

Answer: At $X^o_{KCl} > 0.33$ the activity thus becomes negative.

2.2. EXERCISE

Calculate liquidus temperatures for 20 mol % CuCl - 80 mol% KCl and compare with the phase diagram. KCl: Melting point 772°C, H°$_{fus}$ = 26.3 kJ/mol.

Model 1:   Ideal mixture of CuCl and KCl.
Model 2:   Quantitative formation of CuCl$_2^-$ and ideal mixture of CuCl$_2^-$ and Cl$^-$.

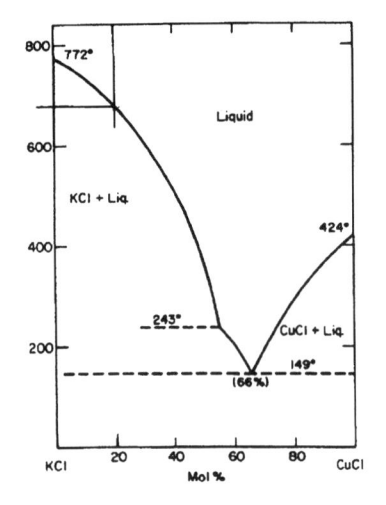

Figure 2.2. System CuCl-KCl [4].

Solution
Model 1

$$a_{KCl} = X^o_{KCl}$$

$$\ln a_{KCl} = \frac{-\Delta H_f(KCl)}{R}\left(\frac{1}{T} - \frac{1}{T_f(KCl)}\right)$$

$$\ln 0.8 = \frac{-27000}{8.314}\left(\frac{1}{T} - \frac{1}{772+273}\right) \quad T = 947, \ t = 702°C$$

Model 2

	KCl	+	CuCl	→	K⁺	+	CuCl₂⁻
Before reaction:	X^o_{KCl}		$+-X^o_{KCl}$		0		0
After reaction:	$X^o_{KCl}-1+X^o_{KCl}$		0		$1-X^o_{KCl}$		$1-X^o_{KCl}$

$$a_{KCl} = X_{K^+} \cdot X_{Cl^-} = 1 \cdot \frac{2X^o_{KCl}-1}{2X^o_{KCl}-1+1-X^o_{KCl}} = \frac{2X^o_{KCl}-1}{X^o_{KCl}}$$

$$\ln a_{KCl} = \frac{-\Delta H_f(KCl)}{R}\left(\frac{1}{T} - \frac{1}{T_f(KCl)}\right)$$

$$\ln\frac{2 \cdot 0.8-1}{0.8} = \frac{-27000}{8.314}\left(\frac{1}{T} - \frac{1}{772+273}\right) \quad T = 956, \ t = 683°C$$

Model 2 still gives a very good agreement with the experimental liquidus values ≈ 680°C.

2.3. IDEAL MIXING OF DISSOCIATED COMPLEXES, EXAMPLE Na$_3$AlF$_6$

Grjotheim [5] performed very accurate measurement of the liquidus temperature of the system NaF - AlF$_3$ around the cryolite composition Na$_3$AlF$_6$ (Fig. 2.3). He assumed the presence of AlF_6^{3-} which dissociated.

$$AlF_6^{3-} = AlF_4^- + 2F^- \tag{2.5}$$

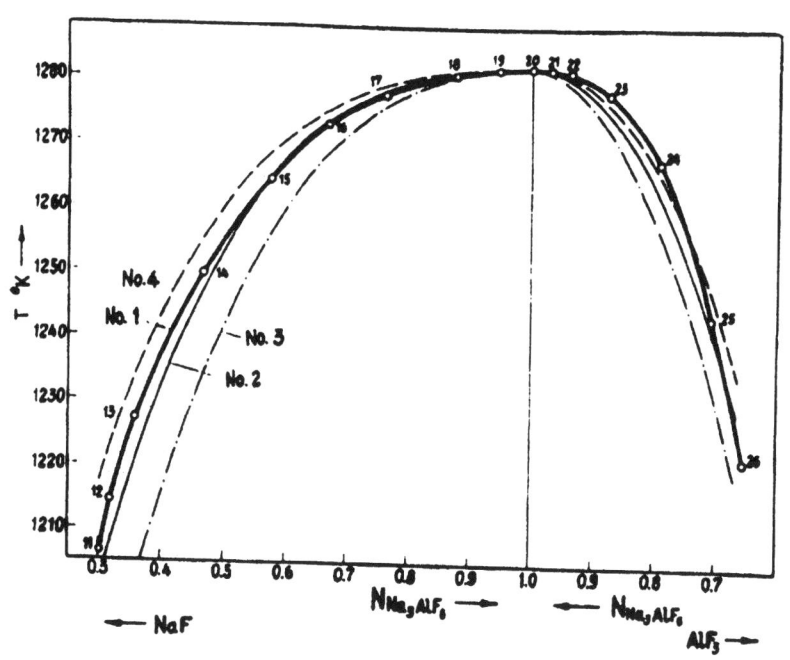

Figure 2.3. Liquidus line for $NaF-AlF_3$ compared with 3 models assuming the dissociation equilibrium $AlF_6^{3-} = AlF_4^- + 2F^-$ and different degree of dissociation α. 1: Experimental data, 2: α = 0.3, 3: α = 0.2, 4: α = 0.4. A dissociation scheme with α = 0.3 gives a quite good description of the liquidus curve. The ions AlF_6^{3-} and AlF_4^- have later been identified by Raman spectroscopy [5].

Newer Raman spectroscopy work of Gilbert et al. [6] shows the presence also of AlF_5^{2-}. The good fit without AlF_5^{2-} may, however, be taken as a confirmation of Flood's statement that a model that describes the properties is not necessary the correct structural model.

2.4. IDEAL MIXING OF DISSOCIATED COMPLEXES, EXAMPLE AlCl$_3$ – NaCl (ACIDIC MELT)

The properties of alkali chloride - aluminium chloride mixture spans from covalent with excess AlCl$_3$ to ionic with excess alkali chloride. As an example the phase diagram of NaCl - AlCl$_3$ is shown in Figure 2.4 [7]:

AlCl$_3$ is immiscible with NaCl between 100-80 mol %. Melts with excess AlCl$_3$ are covalent Lewis acids and low melting with a quite high vapour pressure. A dramatic change occurs around 50 mol % AlCl$_3$ with a sharp rise of the liquidus curve and transfer to a basic, ionic melt with low vapour pressure.

By analysis of vapour pressure data from Dewing [8] the activity of Al_2Cl_6 in the acidic region was calculated as follows, Table 2.1, [9].

TABLE 2.1. Activity of AlCl$_3$ –NaCl at 190°C

Mol fraction AlCl$_3$, $X^o_{AlCl_3}$:	0.658	0.640	0.581	0.536
Activity of Al_2Cl_6, a_6:	$2.89 \cdot 10^{-2}$	$1.21 \cdot 10^{-2}$	$4.41 \cdot 10^{-4}$	$1.6 \cdot 10^{-5}$

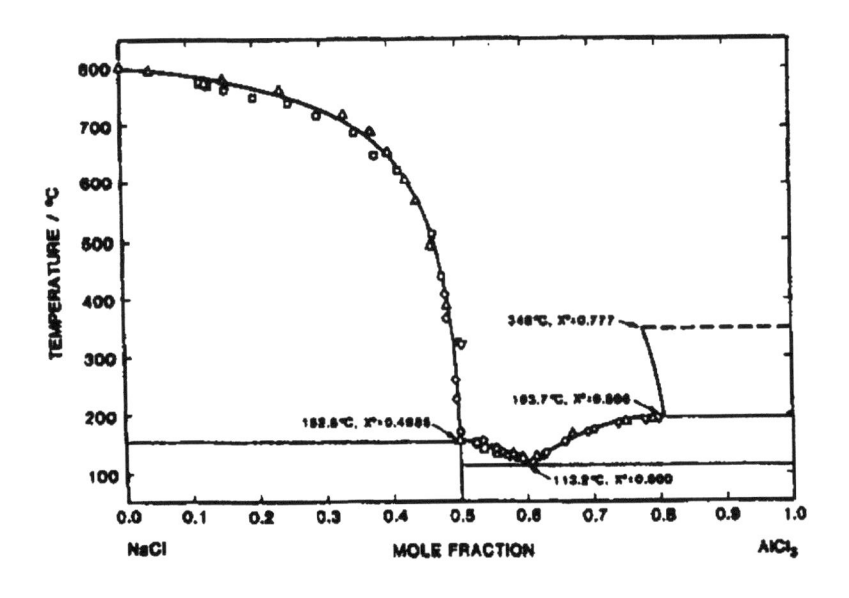

Figure 2.4. Phase diagram of the NaCl-AlCl$_3$ system [7].

It was known that liquid $NaAlCl_4$ probably existed as Na^+ and $AlCl_4^-$, pure $AlCl_3$ as Al_2Cl_6 and that the only vaporizing species at this low temperature were Al_2Cl_6 (and some $AlCl_3$).

The activity of $AlCl_3$ at for instance 65.8 mol % $AlCl_3$ is far too low to describe the system as a mixture of $AlCl_4^-$ and Al_2Cl_6. The immiscibility gap, (Figure 2.4), is also an indication that this is not the case. It is therefore reasonable to assume an intermediate species:

$$Al_2Cl_7^- : \begin{bmatrix} Cl & -Al \begin{matrix} Cl \\ \diagup \\ \diagdown \end{matrix} \begin{matrix} Cl \\ \diagup \end{matrix} Al \begin{matrix} Cl \\ - \\ \diagdown \end{matrix} Cl \\ Cl & & & Cl \end{bmatrix}^-$$

The following ions were assumed present: Al_2Cl_6, $Al_2Cl_7^-$, $AlCl_4^-$, Na^+ [9]. The ion $Al_2Cl_7^-$ is assumed to be partly dissociated in order to explain that the activity of $X^o_{AlCl_3}$ <0.67 is not zero:

$$2Al_2Cl_7^- = Al_2Cl_6 + 2AlCl_4^- \quad (2.6)$$

$$K = \frac{a_{Al_2Cl_6} \cdot a^2_{AlCl_4^-}}{a^2_{Al_2Cl_7^-}} \quad (2.7)$$

Both neutral and ionic species are present and a straight Temkin model cannot be applied. A modified Temkin model was proposed where Al_2Cl_6 was counted among the anions:

$$a_6 = X_6 = \frac{n_6}{n_6 + n_7 + n_4} \quad (2.8)$$

$$K = \frac{X_6 \cdot X_4^2}{X_7^2} = \frac{n_6 \cdot n_4^2}{n_7^2(n_6 + n_7 + n_4)} \quad (2.9)$$

where n denotes number of moles and the indexes the number of Cl atoms in the species. Introduction of mass and charge balance into the equilibrium equation 2.8 gives the following relationship between the equilibration constant and the activity of Al_2Cl_6:

$$K = \frac{a_6(2 - 3X^o_{AlCl_3} + a_6 \cdot X^o_{AlCl_3})^2}{(-1 + 2X^o_{AlCl_3} - a_6)^2} \qquad (2.10)$$

Least square fit of the experimental data by the MODFIT program [10] gave the following, best fit of the data for $K = 9.1 \cdot 10^{-4}$ (Figure 2.5). The ion $Al_2Cl_7^-$ and the equilibrium, Eq. 2.6 was later identified by Raman spectroscopy [11,12] for the system $AlCl_3$ - KCl.

NaAlCl$_4$ - AlCl$_3$ vapour pressure data
$K = 6.43 \cdot 10^{-4}$

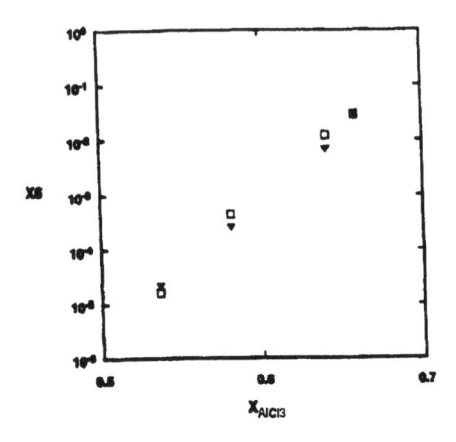

Fig. 2.5. Least square fit of the model given in Eq. 2.10: □ exp, and ▼ calc.

2.5. EXERCISE

a) Derive Eqn. 2.10. Assume one mole NaCl + AlCl$_3$. (Hint: Set up the two mass balances and the charge balance. Express all n's by $X^o_{AlCl_3}$ and a_6).

b) The following activity data are measured in the acidic range ($X^o_{AlF_3} > 0.25$) for the system LiF-AlF$_3$ and NaF-AlF$_3$, respectively:

LiF-AlF$_3$, 800°C		NaF-AlF$_3$, 1020°C	
$X^o_{AlF_3}$	a_{LiF}	$X^o_{AlF_3}$	a_{NaF}
0.320	0.220	0.420	0.0334
0.341	0.184	0.440	0.0230
0.362	0.158	0.459	0.0147
0.382	0.135	0.480	0.0091
0.403	0.116	0.500	0.0048
0.428	0.091		

Show that the LiF-AlF$_3$ system is best described by the equilibrium:

$$\text{LiAlF}_4 + 2\,\text{LiF} = \text{Li}_3\text{AlF}_6$$

and the NaF-AlF$_3$ system by the equilibrium:

$$\text{NaAlF}_4 + \text{NaF} = \text{Na}_2\text{AlF}_5$$

(E. Dewing, quoted in thesis by J.E. Olsen, NTNU, 1996 [13])

In both cases assume that the amount of F$^-$ ions is negligible and that the anions form an ideal Temkin mixture. The model is best proved by finding a model expression for a_{LiF} and a_{NaF} and plot this against the measured activity which then should give a straight line relationship, i.e. a constant K.

Solution:
a)

Alumina balance: $\qquad X = X^o_{AlCl_3} = 2n_6 + n_4 + 2n_7 \qquad (1)$

Charge balance: $\qquad 1 - X = n_4 + n_7 \qquad (2)$

$$a_6 = \frac{n_6}{n_6 + n_7 + n_4} \qquad (3)$$

$$K = \frac{n_6 \cdot n_4^2}{n_7^2 (n_6 + n_7 + n_4)} \qquad (4)$$

1 - 2: $n_7 = -1 + 2X - 2n_6 \qquad\qquad n_4 = 1 - X - n_7 = 2 - 3X + 2n_6$

3: $\quad a_6 = \dfrac{n_6}{n_6 - 1 + 2X - 2n_6 + 2 - 3X + 2n_6} \qquad n_6 = \dfrac{a_6(1 - X)}{1 - a_6}$

Now n_6, n_4 and n_7 can be expressed as function of X and a_6 and introduction into (4) gives the final answer (2.10).

b)
Model for LiF-AlF$_3$:

$$\text{LiAlF}_4 + 2\,\text{LiF} = \text{Li}_3\text{AlF}_6$$

$$\frac{a_{Li_3AlF_6}}{a_{LiAlF_4} \cdot a_{LiF}^2} = K$$

$$a_{LiF} = \sqrt{\frac{a_{Li_3AlF_6}}{K \cdot a_{LiAlF_4}}}$$

c)
Model for NaF-AlF$_3$:

$$\text{NaAlF}_4 + \text{NaF} = \text{Na}_2\text{AlF}_5$$

$$\frac{a_{Na_2AlF_5}}{a_{NaAlF_4} \cdot a_{NaF}} = K$$

$$a_{NaF} = \frac{a_{Na_2AlF_5}}{K \cdot a_{NaAlF_4}}$$

Ideal Temkin:

$$a_{LiF} = \sqrt{\frac{X_{AlF_6^{3-}}}{K \cdot X_{AlF_4^{2-}}}}$$

1 mole LiF + AlF$_3$

Al-balance: $X^o_{AlF_3} = n_{AlF_4^-} + n_{AlF_6^{3-}}$

n_{F^-} is assumed small.

Charge balance:

$1 - X^o_{AlF_3} = n_{AlF_4^-} + 3\, n_{AlF_6^{3-}}$

$n_{AlF_6^{3-}} = 0.5 - X^o_{AlF_3}$

$n_{AlF_4^-} = X_{AlF_4^-} = 2 X^o_{AlF_3} - 0.5$

$$a_{LiF} = \sqrt{\frac{0.5 - X^o_{AlF_3}}{K(2X^o_{AlF_3} - 0.5)}}$$

Ideal Temkin:

$$a_{NaF} = \frac{X_{AlF_5^{2-}}}{K \cdot X_{AlF_4^-}}$$

1 mole NaF + AlF$_3$

Al-balance: $X^o_{AlF_3} = n_{AlF_4^-} + n_{AlF_5^{2-}}$

n_{F^-} is assumed small.

Charge balance:

$1 - X^o_{AlF_3} = n_{AlF_4^-} + 2\, n_{AlF_5^{2-}}$

$n_{AlF_5^{2-}} = 1 - 2X^o_{AlF_3}$

$n_{AlF_4^-} = X_{AlF_4^-} = 3 X^o_{AlF_3} - 1$

$$a_{NaF} = \frac{1 - 2X^o_{AlF_3}}{K(3X^o_{AlF_3} - 1)}$$

LiF - AlF$_3$

$X^o_{AlF_3}$	a_{LiF}	$\sqrt{\dfrac{0.5 - X^o_{AlF_3}}{2X^o_{AlF_3} - 0.5}}$
0.320	0.220	1.134
0.241	0.184	0.935
0.362	0.158	0.785
0.382	0.135	0.669
0.403	0.116	0.563
0.428	0.091	0.449

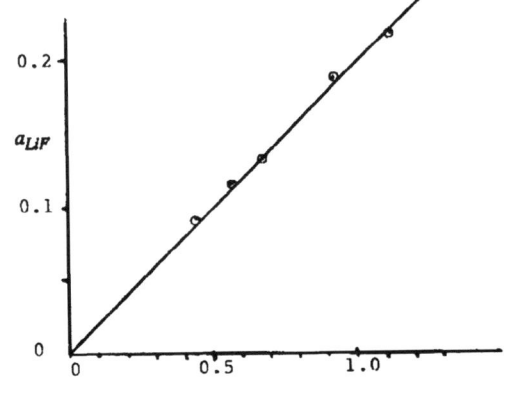

$\sqrt{\dfrac{0.5 - X^o_{AlF_3}}{2X^o_{AlF_3} - 0.5}}$ $0.25 < X^o_{AlF_3} < 0.50$

(Slope: 0.2)): $\frac{1}{\sqrt{K}} = 0.2$ $K = 2.23$)

NaF - AlF$_3$

$X^o_{AlF_3}$	a_{NaF}	$\frac{1-2X^o_{AlF_3}}{3X^o_{AlF_3}-1}$
0.420	0.0334	0.615
0.440	0.0230	0.375
0.459	0.0147	0.218
0.480	0.0091	0.091
0.500	0.0048	0

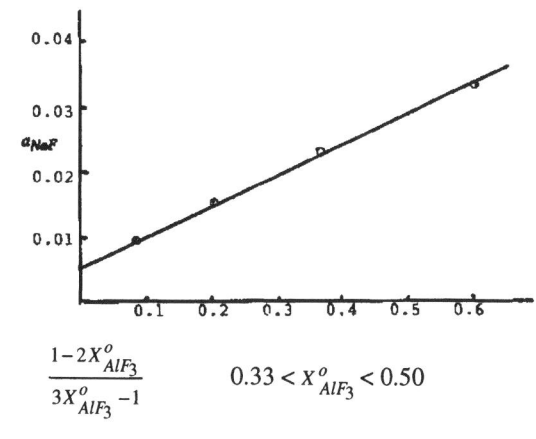

$\frac{1-2X^o_{AlF_3}}{3X^o_{AlF_3}-1}$ $0.33 < X^o_{AlF_3} < 0.50$

(Slope: $\frac{0.0232}{0.5} = 0.0464$ $K = 21.5$)

2.6. EXAMPLE. SOLUBILITY OF Nd IN CHLORIDE MELTS

The "solubility" of Nd metal in NdCl$_3$-LiCl-KCl melts was measured [14]. No real metal solubility was found but the reaction

NdCl$_3$ + Nd → 2 NdCl$_2$

The "solubility" was not dependent on temperature but the solubility measurements (a total of about 100 measurements) could all be characterized by an index, i, defined as the average stoichiometry of the Nd ions in solution: NdCl$_i$ assuming that all the dissolved Nd metal was converted to NdCl$_2$.

Hence

$$i = \frac{2X_{NdCl_2}}{X_{NdCl_2} + X_{NdCl_3}} + \frac{3X_{NdCl_3}}{X_{NdCl_2} + X_{NdCl_3}} = 2\,X'_{NdCl_3} + 3\,X'_{NdCl_3} = 3 - X'_{NdCl_2} \quad (2.11)$$

Fig. 2.6 shows a plot of i as function of KCl content.

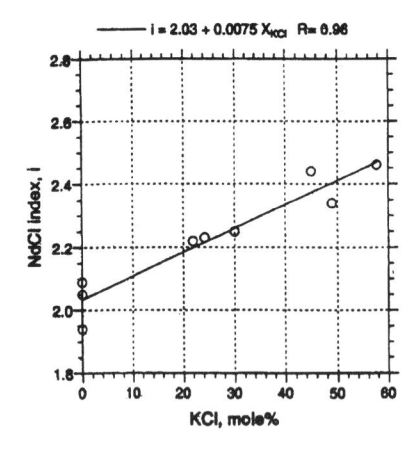

Figure 2.6. Solubility of neodymium expressed by the index i, as a function of the KCl content [14].

From the figure

$$i = 2.03 + 0.75\, X^o_{KCl} \approx 2 + 0.75\, X^o_{KCl}.$$

Hence all NdCl$_3$ is converted to NdCl$_2$ in LiCl melts while it is expected that $i = 2.75$ in a pure KCl melt,): only 25 % of the NdCl$_3$ is converted to NdCl$_2$. The behaviour is understandable. The weaker cation K^+ will stabilize highly charged anions as NdCl$_6^{3-}$. This is an empirical model but the advantage is that all the measurements could be represented by a very simple formulae.

3. Complex Formation with Deviation from Ideality

3.1. AlKCl – AlCl$_3$ (BASIC MELTS) (ASYMMETRIC REGULAR SOLUTION)

The power of thermodynamic models was demonstrated by modelling the total vapour pressure above basic AlkCl – AlkAlCl$_4$ melts [15]. To use NaCl – NaAlCl$_4$ as an example, the total vapour pressure was measured by the boiling point method and found to vary as follows:

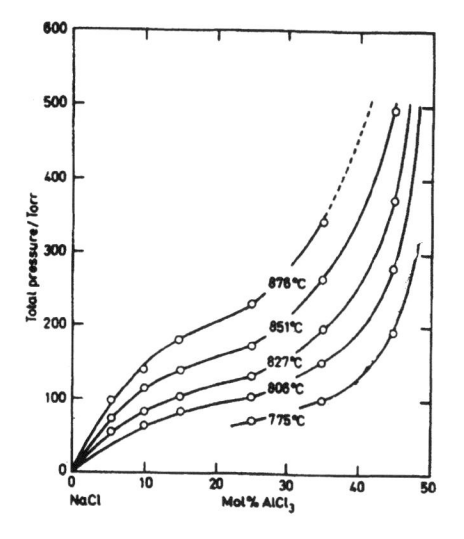

Figure 3.1. Total vapour pressures above molten mixtures of NaCl - AlCl$_3$ as function of composition [15].

The vapour pressure of the pure end member AlCl$_3$ at the temperatures in question is very high, e.g. about 100 atm at 800°C [16]. From the observed pressures it is obvious that very large negative deviations from ideality occur in the MCl - AlCl$_3$ systems in the molten state. These deviations are most readily accounted for by assuming formation of the complex $AlCl_4^-$ in the melt. This is consistent with the observation that the vapour pressure changes sharply as the composition approaches 50 mol % AlCl$_3$ (Fig. 3.1).

On the other hand, the near flatness of the vapour pressure curve around 20 mol% AlCl$_3$ indicates positive deviation. (A horizontal curve would have meant phase separation.)

The following model was constructed:

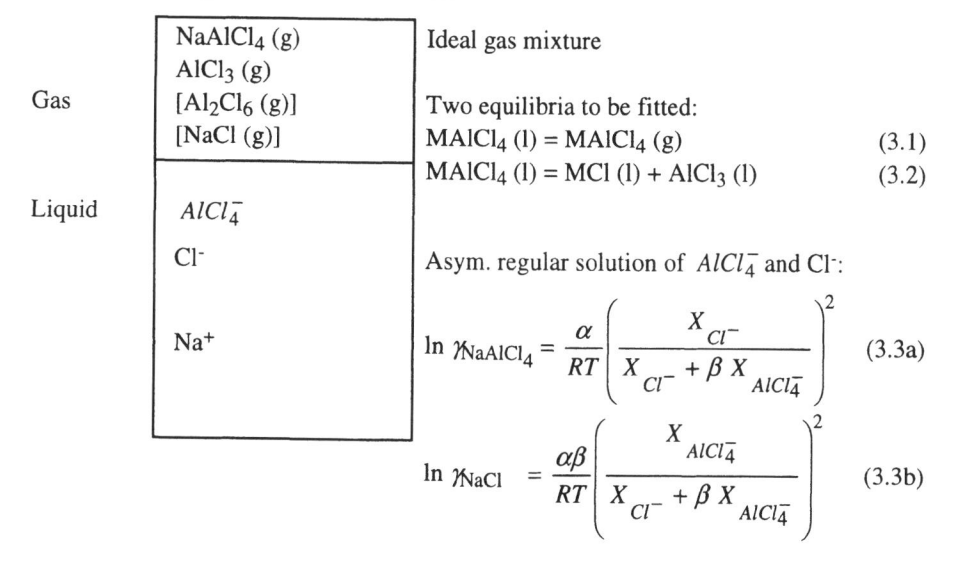

Gas	NaAlCl$_4$ (g) AlCl$_3$ (g) [Al$_2$Cl$_6$ (g)] [NaCl (g)]	Ideal gas mixture Two equilibria to be fitted: MAlCl$_4$ (l) = MAlCl$_4$ (g) (3.1) MAlCl$_4$ (l) = MCl (l) + AlCl$_3$ (l) (3.2)
Liquid	$AlCl_4^-$ Cl$^-$ Na$^+$	Asym. regular solution of $AlCl_4^-$ and Cl$^-$: $$\ln \gamma_{NaAlCl_4} = \frac{\alpha}{RT}\left(\frac{X_{Cl^-}}{X_{Cl^-} + \beta X_{AlCl_4^-}}\right)^2 \quad (3.3a)$$ $$\ln \gamma_{NaCl} = \frac{\alpha\beta}{RT}\left(\frac{X_{AlCl_4^-}}{X_{Cl^-} + \beta X_{AlCl_4^-}}\right)^2 \quad (3.3b)$$

In addition the known equilibria [16] are employed:

$$AlCl_3 \text{ (l)} = AlCl_3 \text{ (g)} \tag{3.4}$$

$$2AlCl_3 \text{ (g)} = Al_2Cl_6 \text{ (g)} \tag{3.5}$$

Equilibrium (3.1) has to be assumed in order to explain the presence of NaAlCl$_4$ in the gas phase and Equilibrium (3.2) to explain the presence of AlCl$_3$.

It was first tried assuming ideal and regular solution of Cl$^-$ and AlCl$_4^-$, but the asymmetric regular solution model was necessary to describe the data. This model is developed as follows

$$\Delta G^E = \alpha \frac{n_{MCl} \cdot n_{MAlCl_4}}{n_{MCl} + \beta n_{MAlCl_4}} \tag{3.6}$$

Simplification of notation:

$$n_{MCl} = n_1, \; n_{MAlCl_4} = n_4 \tag{3.7}$$

$$\Delta G^E = \alpha \frac{n_1 \cdot n_4}{n_1 + \beta n_4} \tag{3.8}$$

Partial differentiation:

$$RT \ln \gamma_1 = \frac{\delta \Delta G^E}{\delta n_1} = \alpha \frac{(n_1 + \beta n_4) n_4 - n_1 n_4 \cdot 1}{(n_1 + \beta n_4)^2} = \alpha \frac{n_1 n_4 + \beta n_4 - n_1 n_4}{(n_1 + \beta n_4)^2} \tag{3.9}$$

$$\ln \gamma_1 = \frac{\alpha \cdot \beta}{RT} \left(\frac{X_4}{X_1 + \beta X_4} \right)^2 \tag{3.10}$$

Similarly

$$\ln \gamma_4 = \frac{\alpha}{RT} \left(\frac{X_1}{X_1 + \beta X_4} \right)^2 \tag{3.11}$$

Assume: 1 mol of the mixture NaCl-AlCl$_3$.

Charge balance: $\quad X^o{}_{NaCl} = n_{Na^+} = n_{Cl^-} + n_{AlCl_4^-} = n_1 + n_4 \tag{3.12}$

Aluminium balance: $\quad X^o{}_{AlCl_3} = n^o{}_{AlCl_3} = n_{AlCl_4^-} = n_4 \tag{3.13}$

$$X_4 = \frac{n_4}{n_1 + n_4} = \frac{X^o{}_{AlCl_3}}{X^o{}_{NaCl}} = \frac{X^o{}_{AlCl_3}}{1 - X^o{}_{AlCl_3}} \qquad (3.14)$$

$$X_1 = 1 - X_4 = \frac{1 - X^o{}_{AlCl_3} - X^o{}_{AlCl_3}}{1 - X^o{}_{AlCl_3}} = \frac{1 - 2X^o{}_{AlCl_3}}{1 - X^o{}_{AlCl_3}} \qquad (3.15)$$

Introduction of the activities $a_{MCl} = X_1\gamma_1$; $a_{MAlCl_4} = X_4\gamma_4$ and the equilibrium constant K_2 for the dissociation reaction (3.2), the activity of $AlCl_3$ in the melt is given by:

$$a_{AlCl_3} = K_2(X_4\gamma_4)/(X_1\gamma_1) \qquad (3.16)$$

where X_1 and X_4 are given from Equations (3.14) and (3.15), γ_1 and γ_4 from (3.10) and (3.11).

The observed, total pressure is the sum of the partial pressures:

$$P_{tot} = P_{MCl} + P_{MAlCl_4} + P_{AlCl_3} + P_{Al_2Cl_6} \qquad (3.17)$$

The small partial pressure P_{MCl} may be safely neglected and by introduction of Eq. (3.5):

$$P_{Al_2Cl_6} = K_5 P^2{}_{AlCl_3} \qquad (3.18)$$

the total pressure is expressed as:

$$P_{tot} = P_{MAlCl_4} + P_{AlCl_3} + K_5 P^2{}_{AlCl_3} \qquad (3.19)$$

The partial pressure may be expressed as the product of the activity of the component in the melt and the pressure of the component in its standard state. $P^o{}_{MAlCl_4}$ then denotes the pressure of $MAlCl_4(g)$ over the hypotetical, un-dissociated pure liquid $MAlCl_4$, while $P^o{}_{AlCl_3}$ is the pressure of $AlCl_3(g)$ in its mixture with Al_2Cl_6 in equilibrium with pure liquid $AlCl_3$. Equation 3.19 then gives

$$P_{tot} = P^o{}_{MAlCl_4} X_4\gamma_4 + P^o{}_{AlCl_3} a_{AlCl_3} + K_5(P^o{}_{AlCl_3} a_{AlCl_3})^2 \qquad (3.20)$$

Combination gives the final equation:

$$P_{tot} = P^o_{MAlCl_4} \cdot X_4 \cdot \exp\left[\frac{\alpha}{RT}\left(\frac{X_1}{X_1 + \beta \cdot X_4}\right)^2\right]$$

$$+ P^o_{AlCl_3} \cdot K_2 \cdot \frac{X_4}{X_1} \exp\left[\frac{\alpha}{RT} \frac{X_1^2 - \beta \cdot X_4^2}{(X_1 + \beta \cdot X_4)^2}\right]$$

$$+ K_5 \cdot (P^o_{AlCl_3})^2 \cdot K_2^2 \cdot \left(\frac{X_4}{X_1}\right)^2 \exp\left[\frac{2\alpha (X_1^2 - \beta \cdot X_4^2)}{RT (X_1 + \beta \cdot X_4)^2}\right] \qquad (3.21)$$

In this equation, K_5 and $P^o_{AlCl_3}$ are known from the JANAF data [16] and are introduced in Equation 3.21 for each measured temperature. K_2 is unknown and is expressed as a function of temperature as follows:

$$K_2 = \exp\left[-\frac{\Delta H_2^o}{RT} + \frac{\Delta S_2^o}{R}\right] \qquad (3.22)$$

$P^o_{MAlCl_4}$ is likewise unknown and may be expressed by the equilibrium constant of the reaction

$$\text{MAlCl}_4 \text{ (hypotetical, undissociated liquid)} = \text{MAlCl}_4(g) \text{ }^{*)} \qquad (3.23)$$

i.e.

$$P^o_{MAlCl_4} = \exp\left[-\frac{\Delta H_1^o}{RT} + \frac{\Delta S_1^o}{R}\right] \qquad (3.24)$$

With introduction of Equations 3.22 and 3.23 in Equation 3.21, the complete equation contains altogether six unknown parameters. These are calculated by computer fitting of the equation to the complete set of experimental data (i.e., measured total pressures at all experimental temperatures and compositions) for one system at a time. The computer fitting was executed by means of the MODFIT program [10]. The standard deviations in fit were as follows. LiCl-AlCl$_3$: 6.5%, NaCl-AlCl$_3$: 3.7% and KCl-AlCl$_3$: 2.5%. The computed parameters are given in Table 3.1 and the results of the fitting are given graphically in Figs. 3.2 - 3.4.

---

*) or written as Eq. 3.1.

TABLE 3.1. Computed values for the parameters in the combined equations 3.21-3.24.

	$\dfrac{\Delta H_1^o}{kJmol^{-1}}$	$\dfrac{\Delta S_1^o}{JK^{-1}mol^{-1}}$	$\dfrac{\Delta H_2^o}{kJmol^{-1}}$	$\dfrac{\Delta S_2^o}{JK^{-1}mol^{-1}}$	$\dfrac{\alpha/R}{K}$	β	$\dfrac{SD}{\%}$
LiCl - AlCl$_3$	95.5	91.6	71.0	32.7	2128	3.64	6.5
NaCl - AlCl$_3$	117.4	96.2	85.1	26.2	1851	2.70	3.7
KCl - AlC$_3$	113.4	86.7	105.2	28.7	1391	2.83	2.5

The fully drawn lines represent the computer calculated values for the partial pressures and their sum, while points represent experimental values for total pressure. It is noted that the Al$_2$Cl$_6$ pressure is negligibly small in the KCl-AlCl$_3$ and NaCl-AlCl$_3$ systems, while in the LiCl-AlCl$_3$ system it appears as a noticeable contribution because of the lower temperatures employed for this system. A striking feature of the present result is the very high total pressure of the LiCl-containing system relative to the other alkali chloroaluminates, Fig. 3.5.

A general objection to the numerical treatment in the preceding section might be raised, in that an equation with as many as six adjustable parameters could be made to fit almost any set of experimental data. Thus it appears necessary to draw attention to some features which indicate the physical reality of the model.

The standard deviation in the fit of the vapour pressures is acceptably low (in the range 2 to 6 percent). This indicates that the parameters are essentially independent of temperature, as they should be according to the model.

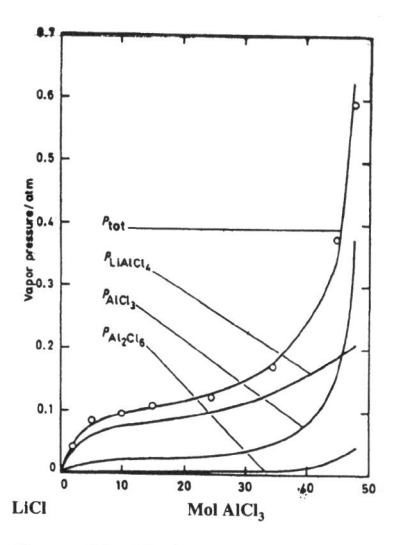

Figure 3.2. Total pressure (experimental and calculated) and partial pressures (calculated) in the LiCl-AlCl$_3$ system at 646°C

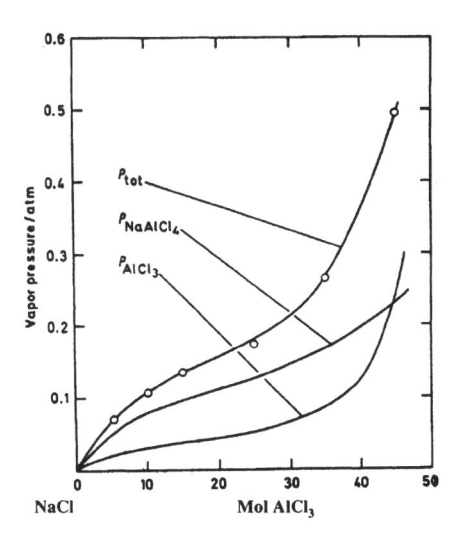

Figure 3.3. Total pressure (experimental and calculated) and partial pressures (calculated) in the NaCl - AlCl$_3$ system at 827°C.

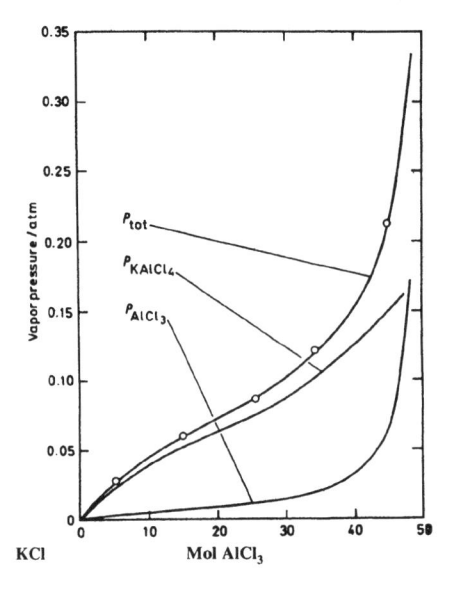

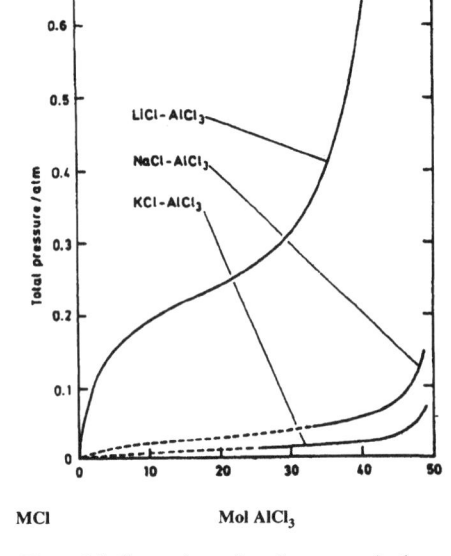

Figure 3.4. Total pressure (experimental and calculated) and partial pressures (calculated) in the KCl - AlCl$_3$ system at 851°C.

Figure 3.5. Comparison of total pressures in the three systems at a common temperature of 700°C. Dotted line: Extrapolated below freezing point.

The various parameters appear with physically reasonable values, and the trends in the values when going from the lithium to the potassium system are also reasonable. It is notable that the β is close to the volume ratio $AlCl_4^- / Cl^-$.

A final point in favour of the present interpretation may be had by comparison with the phase diagrams for the systems in question. The values obtained for α and β together with Equation (3.10), yield the activity of alkali chloride in the molten mixture for various compositions. This permits the calculation of the liquidus line on the MCl side of the MCl-AlCl$_3$ system by means of the equation

$$\ln a_{\text{MCl}} = \frac{-\Delta H_m^o}{R} \left[\frac{1}{T} - \frac{1}{T_m} \right] \tag{3.25}$$

where ΔH_m^o is the enthalpy of melting and T_m the melting point of pure MCl, and T is the liquidus temperature of the mixture. The results of such calculations are shown in Fig. 3.6 [17]. Fully drawn lines are calculated while the experimental points for the known systems NaCl-AlCl$_3$ and KCl-AlCl$_3$ are taken from the phase diagram investigations of Shvartsman [18] and of Fischer and Simon [19]. The agreement is almost perfect and can hardly be incidental.

Another confirmation was the following:

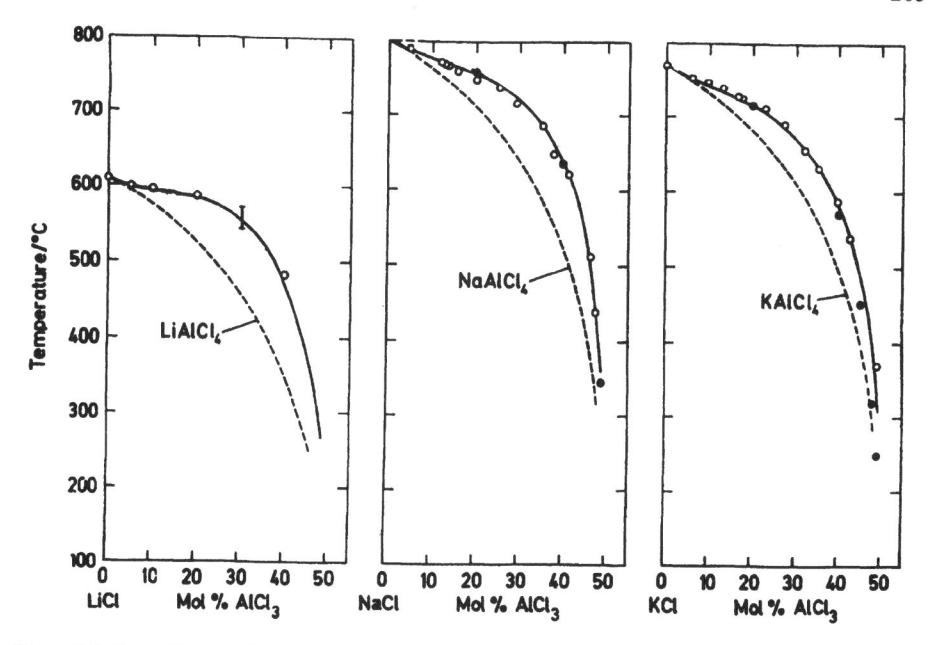

Figure 3.6. Phase diagrams for the alkali chloride side of the three systems LiCl-AlCl$_3$, NaCl-AlCl$_3$, and KCl-AlCl$_3$. Open circles: Experimental points for the NaCl-AlCl$_3$ and KCl-AlCl$_3$ systems from Shvartsman [18]. Filled circles: Do. from Fischer and Simon [19]. LiCl - AlCl$_3$: Our measurements: Dashed lines: Calculated by assuming ideal Temkin mixture of ions Cl$^-$ + AlCl$_4^-$. Fully drawn lines: Calculated from the present vapour pressure data using an asymmetric Temkin model.

Mr. Warren Haupin (Alcoa Technical Center, USA), receiving a copy of the manuscript for this paper, kindly responded by reporting some unpublished results of EMF measurements on cells of the type (Al(l)/LiCl-AlCl$_3$ (melt)/Cl$_2$(g) with the melt compositions in the range 1 to 10 mol% AlCl$_3$. For comparison he calculated the corresponding EMF values from the present model with parameters given in Table 3.1. His experimental and calculated values agree within 0.2 relative percent.

4. Modelling of Ternary Mixtures from Binary Data

4.1 LiCl – NaCl – AlCl$_3$

The published total vapour pressure [15] for the alkali chloride - aluminium chloride (Fig. 3.5) did, however, raise curiosity in ALCOA. ALCOA was at that time running industrial electrolysis of aluminium in a melt with approximate composition 4 wt % AlCl$_3$ in an equimolar mixture of LiCl and NaCl. The cell reaction was

$$\text{AlCl}_3 \text{ (LiCl-NaCl)} \overset{700°C}{=} \text{Al(l)} + 1.5 \text{ Cl}_2 \text{ (g)} \qquad (4.1)$$

The evolved gas carried out a saturated vapour from the electrolyte, which later was condensed. The condensate contained more NaAlCl$_4$ than LiAlCl$_4$, which at first glance appeared odd in view of the much higher total vapour pressure of LiCl - AlCl$_3$ compared with NaCl- AlCl$_3$. We were asked to carry out a study of the ternary system LiCl - NaCl - AlCl$_3$ and measure the total vapour pressure by the boiling point method as well as analyse the condensate by transportation studies with saturation of Argon-gas. Figures 4.1 and 4.2 give the total measured pressure and composition of the vapour phase [20].

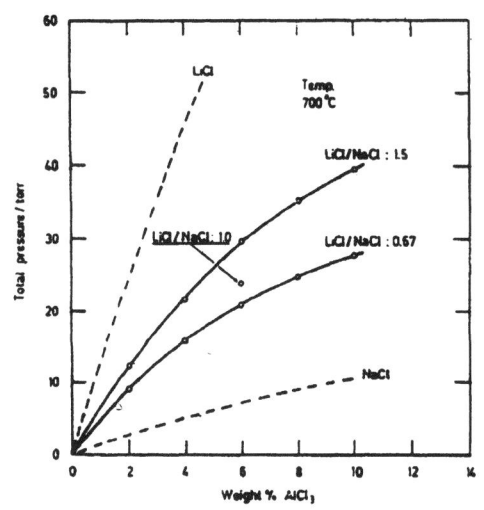

Figure 4.1. Total pressure of LiCl - NaCl - AlCl$_3$ melts at 700°C [20].

The study reconciled both the earlier binary data with the high vapour pressure of the binary LiCl - AlCl$_3$ system and the preponderance of NaAlCl$_4$ in the gas phase. When presenting the result and model to ALCOA I was asked to explain in a simple manner the preponderance of NaAlCl$_4$ in an equimolar LiCl - NaCl mixture and why **the pressure of NaAlCl$_4$ increases when NaCl is diluted with LiCl** (up to 60 mol-% LiCl).

The data was presented as a mathematical model not given here, but the work became a primary example how binary data can explain ternary behaviour which at first glance might seem surprising. The following binary data can be read from Fig. 4.2 and are given in Table 4.1.

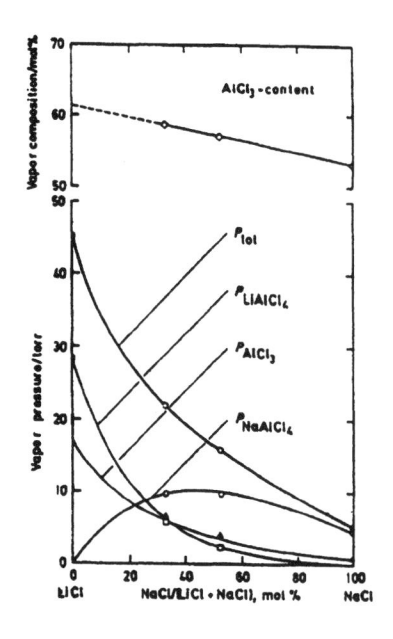

Figure 4.2. Upper part gives AlCl$_3$-content in the vapour based on condensation experiments and chemical analysis. Lower part gives total pressure and experimental and calculated pressure of LiAlCl$_4$, NaAlCl$_3$ and AlCl$_3$ at 700°C and 4 wt-% AlCl$_3$. [20]

TABLE 4.1. Total and Partial Pressures Read out from Figure 4.2.

		P_{LiAlCl_4} Torr	P_{NaAlCl_4} Torr	P_{AlCl_3} Torr	P_{tot} Torr
NaCl,	4 wt% AlCl$_3$	0	4.5	0.6	5.1
LiCl,	4 wt% AlCl$_3$	28.4	0	16.7	45.1

General Chemistry: For the binary systems:

$$LiCl(l) + AlCl_3(g) = LiAlCl_4(g): \quad \frac{P_{LiAlCl_4}}{P_{AlCl_3} \cdot a_{LiCl}} = K_{Li} \quad (4.2)$$

$$NaCl(l) + AlCl_3(g) = NaAlCl_4: \quad \frac{P_{LiAlCl_4}}{P_{AlCl_3} \cdot a_{NaCl}} = K_{Na} \quad (4.3)$$

$$K_{Li} = \frac{P_{LiAlCl_4}}{P_{AlCl_3} a_{LiCl}} = \frac{28.4}{16.7} = 1.70 \quad K_{Na} = \frac{P_{NaAlCl_4}}{P_{AlCl_3} a_{NaCl}} = \frac{4.5}{0.6} = 7.5 \quad (4.4)$$

The ternary equilibrium is obtained by combination:

$$LiAlCl_4(g) + NaCl(l) = NaAlCl_4(g) + LiCl(l): \quad \frac{K_{Na}}{K_{Li}} = 4.41 \quad (4.5)$$

$$\frac{P_{NaAlCl_4} \cdot a_{LiCl}}{P_{LiAlCl_4} \cdot a_{NaCl}} = \frac{K_{Na}}{K_{Li}} \tag{4.6}$$

Simple model:
$$\frac{a_{LiCl}}{a_{NaCl}} = \frac{X_{LiCl}}{X_{NaCl}} \tag{4.7}$$

$$\frac{P_{NaAlCl_4}}{P_{LiAlCl_4}} = \frac{K_{Na}}{K_{Li}} \frac{X_{LiCl}}{X_{NaCl}} = 4.41 \frac{X_{LiCl}}{X_{NaCl}} \tag{4.8}$$

Equimolar amounts of LiCl and NaCl:
$$\frac{P_{NaAlCl_4}}{P_{LiAlCl_4}} = 4.41$$

The "surprising fact" was explained.

The physical realities behind this reversal is a higher equilibrium constant for vaporization of $LiAlCl_4$ relative to $NaAlCl_4$ combined with a lower stability in the gas phase towards dissociation according to

$$LiAlCl_4(g) \rightarrow LiCl(l) + AlCl_3(g). \tag{4.9}$$

5. Modelling of "Unmeasurable Conditions"

5.1 $AlCl_3$ – NaCl – KCl (A TERNARY SYSTEM WITH STRONG CHANGES IN PROPERTIES)

The last model example is also taken from an industrial problem [21]. ARCO Metals produced $AlCl_3$ by carbochlorination in a fluidized bed reactor at about $600°C$:

$$Al_2O_3(s) + 1.5C(s) + 3Cl_2(g) = 2AlCl_3(g) + 1.5CO_2(CO)(g) \tag{5.1}$$

The $AlCl_3$ should then be used as a feed for chloride electrolysis of aluminium. A flow sheet of the process is shown in Figure 5.1.

Figure 5.1

```
A              B           C              D            E
Reactor        Cyclones    Demistor       Desublimer   Off-gas
                                                       treatment?
(500-700 °C)   (550-700°C) (200-350°C)    (50-100°C)   (25°C)
```

Stream compositions (as shown in the flow sheet):

- Feed to cyclones (and reactor outlet): (Na-K-Mg-Ca-P-V-S) $AlCl_3(g), Al_2Cl_6(g), CO_2, CO, (Cl_2)$
- After cyclones to demistor: (Na-K-Mg-Ca-P-V-S) $AlCl_3(g), Al_2Cl_6(g), CO_2, CO, (Cl_2)$
- From cyclones (solids): C, Al_2O_3 and $Al_2O_3 (AlCl_3), C$ (Na-K-Mg-Ca-P-V-S)
- Demistor outlet: $CO_2, CO, (Cl_2)$; liquid: $NaCl-KCl-AlCl_3$ melt (Other dissolved impurities?)
- Desublimer outlet: $CO_2, CO, (Cl_2)$, off-gas; solid: $AlCl_3$ (Which impurities will coprecipitate?)
- Off-gas treatment: H_2O in, CO_2, CO out, H_2O out
- Reactor inlet: $Cl_2, (O_2)$ (4 atm)

Reactor reaction:
$$Al_2O_3(s) + 1.5\,C(s) + 3\,Cl_2(g) = 2\,AlCl_3(g) + 1.5\,CO_2(g)$$

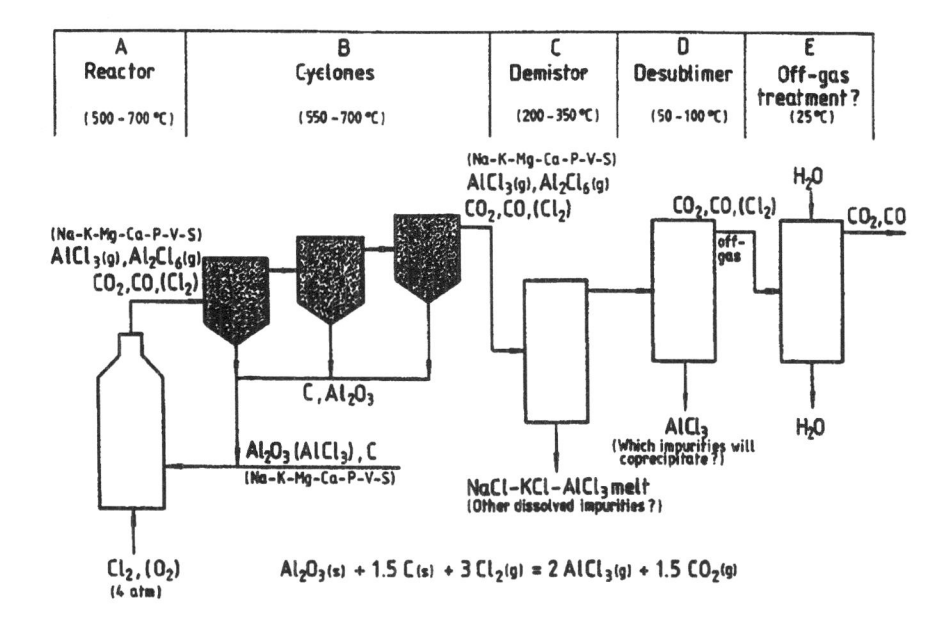

Figure 5.1. Flow sheet for production of $AlCl_3$ by carbochlorination.

Unreacted C and Al_2O_3 are separated in cyclones and returned. Impurities of NaCl and KCl are demisted at 200-350° and $AlCl_3$ desublimed at 50-100°C.

The aluminium oxide contained oxides of Na, K, Mg, Ca, P, V and S. The following problem should be addressed by a thermodynamic model:

a) What is the minimum reactor temperature to avoid enrichment of the impurities in the reactor forming a liquid phase?
b) What percent of Na and K will be demisted dependent on demistor temperature and gas composition?
c) What is the expected final purity of $AlCl_3$?

The main problem turned out to be the content of Na_2O and K_2O. Na_2O and K_2O are converted to NaCl and KCl in the reactor and can react with $AlCl_3$ forming a NaCl – KCl – $AlCl_3$ melt. Enrichment of this melt in the reactor will be catastrophic as the solid Al_2O_3 no longer will be fluidized. The problem is then to determine the minimum temperature for these impurities to leave the reactor as $NaAlCl_4$ and $KAlCl_4$. The vapour pressure will be low ($\approx 10^{-4}$ atm) in a gas with total pressure 2-4 atm. In addition the composition of the liquid will be close to 50 mol % $AlCl_3$ where small changes in composition give large changes in vapour pressure. Both facts make it practically

208

a)

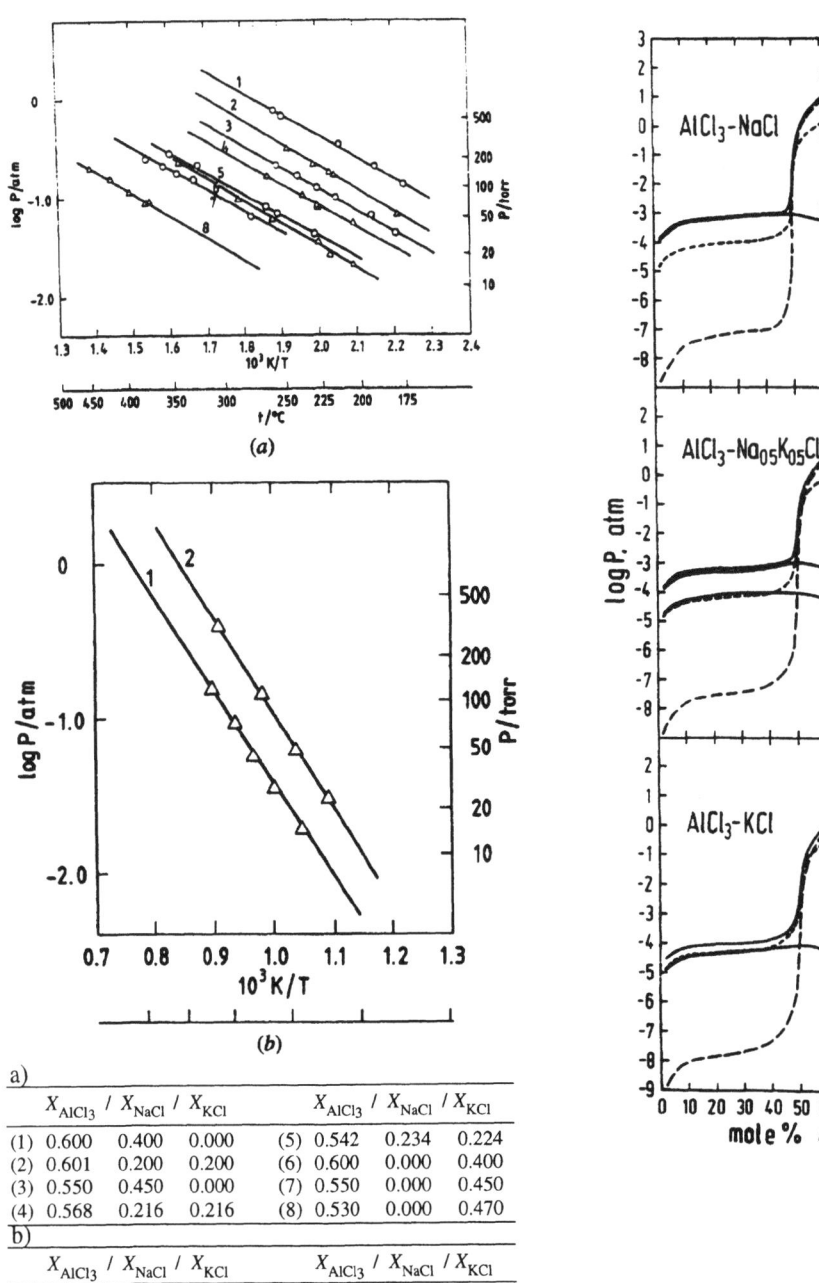

	X_{AlCl_3}	X_{NaCl}	X_{KCl}		X_{AlCl_3}	X_{NaCl}	X_{KCl}
(1)	0.600	0.400	0.000	(5)	0.542	0.234	0.224
(2)	0.601	0.200	0.200	(6)	0.600	0.000	0.400
(3)	0.550	0.450	0.000	(7)	0.550	0.000	0.450
(4)	0.568	0.216	0.216	(8)	0.530	0.000	0.470

b)

	X_{AlCl_3}	X_{NaCl}	X_{KCl}		X_{AlCl_3}	X_{NaCl}	X_{KCl}
(1)	0.300	0.350	0.350	(2)	0.450	0.275	0.275

Figure 5.2. Total vapour pressure for ternary mixtures of NaCl - KCl - $AlCl_3$ [21].

Figure 5.3. Vapour pressure above $AlCl_3$ – NaCl, $AlCl_3$ – $Na_{0.5}K_{0.5}Cl$, and $AlCl_3$ – KCl liquids at 773 K [21].

impossible to perform any kind of direct vapour pressure measurements. It was then decided to choose temperatures where the vapour pressure was easy to measure by the boiling point method, i.e. 10 – 600 Torr. At the acidic side (excess $AlCl_3$) a temperature around 250°C was chosen and at the basic side (excess alkali chloride) around 800°C. As log P versus 1/T is linear, (see Figure 5.2), these data could be extrapolated to the wanted temperature.

The ternary data were combined with earlier binary data, both in acidic and basic melts to a comprehensive model for the complete NaCl – KCl – $AlCl_3$ system in the temperature range 200-900°C. The model assumes the following species:

Gas phase: $Al_2Cl_6, AlCl_3, NaAlCl_4, KAlCl_4$

Liquid phase: $Al_2Cl_6, Al_2Cl_7^-, AlCl_4^-, Cl^-, Na^+, K^+$

The model becomes rather complex due to the wide composition and temperature range, and ref. [21] is referred to for details. The following equilibria were considered (Table 5.1).

TABLE 5.1. Considered equilibria with fitted parameters for the binary system $AlCl_3$ - NaCl and $AlCl_3$ - KCl.

Reaction		ΔH^o kJ / mol	ΔS^o J / mol K
$KAlCl_4$ (l)	= $KAlCl_4$ (g)	131.9	95.1
$NaAlCl_4$ (l)	= $NaAlCl_4$ (g)	101.5	74.8
$KAlCl_4$ (l)	= KCl (l) + $AlCl_3$ (l)	83.0	20.8
$NaAlCl_4$ (l)	= NaCl (l) + $AlCl_3$ (l)	99.0	44.6
KAl_2Cl_7 (l)	= $KAlCl_4$ (l) + $AlCl_3$ (l)	5.65	-10.1
$NaAl_2Cl_7$ (l)	= $NaAlCl_4$ (l) + $AlCl_3$ (l)	5.13	4.38

In addition 10 interaction parameters were fitted, which made a total of 22 parameters.

The model was used in an interactive program that can be used in two modes:

1) Specify: Liquid composition (Na/K as well as $X^o_{AlCl_3}$) and temperature.
   Calculate: Total pressure*) and vapour composition.

2) Specify: Gas composition and total pressure (P_{NaAlCl_4}, P_{KAlCl_4}, P_{tot}).
   Calculate: Condensation temperature and composition of condensing liquid.

---

*) Total pressure is $P_{NaAlCl_4} + P_{KAlCl_4} + P_{AlCl_3} + P_{Al_2Cl_6}$ excluding pressure of other gases as CO, CO_2, Cl_2 and inert gases.

The first option is used for generating vapour pressure tables. These tables can be used for manual prediction of condensation temperature and demistor efficiency. The second option is aimed specifically to determine the dew point of the gas, i.e., at which temperature the first liquid condenses out and the composition of this liquid.

As example of results Fig. 5.3 shows the calculated vapour pressures for $AlCl_3$ – $NaCl$, $AlCl_3$ – $Na_{0.5}K_{0.5}Cl$, and $AlCl_3$ – KCl, at 773K. Fig. 5.4 gives a calculation of the condensation temperature with composition of condensing liquid for the same two binary systems in the range

$$1 \cdot 10^{-4} atm < P_{MAlCl_4} < 10 \cdot 10^{-4} atm \qquad \text{and } P_{tot} = 2 \text{ atmospheres.}$$

Illustration: As an illustration the following simplified model is assumed. The alumina contains 0.10 mol% K_2O and no Na_2O. The vapour pressure of Al_2Cl_6 is negligible and the pressure of $AlCl_3$ is 2 atmospheres. (The real pressure may be closer to 4 atmospheres due to the presence of CO and CO_2.)

Hence
$$Al_2O_3 + 3C + 6Cl_2 \rightarrow 2 \cdot AlCl_3 + 3CO$$
$$K_2O + 2AlCl_3 + C + Cl_2 \rightarrow 2KAlCl_4 + CO$$

$$\frac{P_{KAlCl_4}}{P_{AlCl_3}} = \frac{0.0010}{2} = 5 \cdot 10^{-4} \text{ atm.}$$

From the figure the condensation temperature is 858 K and the condensing liquid contains 59.6 mol% $AlCl_3$ (In the real model no assumptions has to be made.)

A detailed discussion of the results is out of place here, but the results given in Fig. 5.4 illustrate the usefulness of the model. The vapour pressure that causes condensation is only a small fraction of the total pressure and would be near impossible to measure experimentally. On the other hand the phenomena of condensation of impurity elements are extremely important for the process design as with time unwanted condensation may cause build-up of impurities leading to an eventual shut-down of the process.

The final question in the physical reality and usefulness of such a model. Concerning usefulness the answer is yes. The model predicted that the reactor temperature had to be raised to 50°C from originally designed in order to eliminate the Na_2O - K_2O enrichment problem, and it was later confirmed by plant experience. The model is based on physical realities as all the species assumed can be identified by spectroscopy. By using an equilibrium type of description it is possible to describe the strong variation of vapour pressure with composition. A purely parametric model would never be able to perform in the same way.

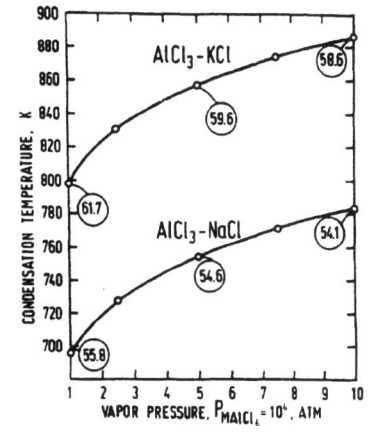

Figure 5.4. Condensation temperature of alkali chloride in a $MAlCl_4 - AlCl_3 - Al_2Cl_6$ gas mixture with total pressure 2 atmospheres. Mol % $AlCl_3$ in the condensing liquid is given in the circles.

References

1. Temkin, M (1945) *Acta Physicochim.*, URSS, **20,** pp. 411.
2. Flood, H, Førland, T. and K. Grjotheim, K. (1954): *Z. anorg. allgem. Chem.* **19**, p. 276.
3. Flood, H. and Urnes, J. (1955) *Zeitschrift f. Elektrochemie*, **59**, p. 834.
4. Levin, E.M., Robbins, C.R. and McMurdie, H.F. (1983) *Phase Diagram for Ceramists*. Am. Cer. Soc.
5. Grjotheim, K. (1956) *Det Kgl. Norske Videnskabers Selskabs Skrifter*, no. 5.
6. Gilbert, B., Robert, E., Tixhon, E., Olsen, J.E. and Østvold, T. (1996) Structure and Thermodynamics of $NaF-AlF_3$ Melts with Addition of CaF_2 and MgF_2. *Inorg. Chem.* **35**, 4198-4210.
7. Fannin, A.A., King, L.A. Seegmiller, D.W. and Øye, H.A. (1982) Densities and Phase Equilibria of Aluminium Chloride - Sodium Chloride Melts. 2. Two-Liquid-Phase Region, *J. Chem. Eng. Data*, **27**, 114-119.
8. Dewing, E. (1955) *J. Am. Chem. Soc*, **77**, 2639.
9. Øye, H.A., and Gruen, D.M. (1964) Octahedral Absorption Spectra of the Dipositive 3d Metal Ions in Molten Aluminium Chloride, *Inorg Chem.*, 3, 836-841.
10. Hertzberg, F. (1970) *MODFIT*, Department of Chemical Engineering, The Norwegian Institute of Technology, Trondheim, Norway.
11. Cyvin, S.J., Klæboe, P., Rytter, E. and Øye. H.A. (1970) Spectral Evidence for $Al_2Cl_7^-$ in Chloride Melts, *J. Chem. Phys.*, **52**, 2776-2778.
12. Øye, H.A., Rytter, E., Klæboe, P. and Cyvin, S.J. (1971) Raman Spectra of $KCl-AlCl_3$ Melts and Normal Coordinate Analysis of $Al_2Cl_7^-$, *Acta Chem. Scand.*, 1971, **25**, 559-576.
13. Olsen, E.J. (1996) *Thesis* no. 82, Institute of Inorganic Chemistry, NTNU.
14. Kvam, K.R., Bratland, D. and Øye, H.A. (1999) The Solubility of Neodymium in the Systems $NdCl_3$-LiCl and $NdCl_3$-LiCl-KCl. *Journal of Molecular Liquids* **83**, 111-118.
15. Linga, H., Motzfeldt, K. and Øye, H.A. (1978) Vapour Pressure of Molten Aluminium Chloride - Aluminium Chloride Mixtures, *Ber. Bunsenges. Phys. Chem.*, **82**, 568-576.
16. *JANAF, Thermochemical Tables 3. ed.*, J. of Physical and Chemical Reference Data, 1985, vol. 14, (Supplement no. 1.)
17. Linga, H. (1979) *Vapour pressure of basic alkalichlorid – aluminiumchlorid melts*. Thesis no. 36, Institute of Inorganic Chemistry, The Norwegian Institute of Technology, Trondheim, Norway.
18. Shvartsman, W.I. (1940) *Zhur. Fiz. Khim.* **14,** 254.
19. Fischer, W and Simon. A.I. (1960) *Z. anorg. allgem. Chem.* **1**, 306.
20. Knapstad, B., Linga, H. and Øye, H.A. (1981) Vapour Pressure of Ternary $LiCl-NaCl-AlCl_3$ Melts, *Ber. Bunsenges. Phys. Chem.* **85**, 1132-1139.
21. Grande, K., Hertzberg, T. and Øye, H.A. (1986) A Thermodynamic Computer Model for the System $AlCl_3$-NaCl-KCl, *Light Metals*, 431-436.

THERMODYNAMIC MODEL OF MOLTEN SILICATE SYSTEMS

V. DANĚK, Z. PÁNEK
*Institute of Inorganic Chemistry, Slovak Academy of Sciences
Dúbravská cesta 9, 842 36 Bratislava, Slovak Republic*

1. Introduction

The structure of glass-forming oxides SiO_2, GeO_2, P_2O_5, B_2O_3, etc., is based on linked AO_4 tetrahedrons (SiO_2, GeO_2, P_2O_5) or AO_3 triangles (B_2O_3). Førland [1] concluded that the structure of SiO_2 melt differs only little from that of cristobalite. The plausibility of this conclusion follows from the fact that the Si–O bonds are extremely strong. However, in spite of this, the melting enthalpy of SiO_2 is only 9.58 kJ.mol^{-1} and the melting entropy is 4.8 J.mol^{-1}.K^{-1} [2], which means that molten SiO_2 has an arrangement very similar to that of cristobalite. The densities of molten SiO_2 and crystobalite are nearly equal, the viscosity of molten SiO_2 is high [3]. SiO_2 belongs to tectosilicates, its structure consists of SiO_4 tetrahedrons linked by their apexes into the three-dimensional network. The structure contains only bridging oxygen atoms, i.e. those, bound to two neighboring central silicon atoms by covalent Si–O–Si bonds.

Let us consider what happens with the structure when adding another oxide (e.g. MeO) into SiO_2. The situation can be expressed by the scheme

$$-Si-O-Si- + MeO \rightarrow -Si-O^-\ Me^{2+}\ ^-O-Si- \qquad (A)$$

Now not all oxygen atoms are linked to two central silicon atoms. This is due to the addition of further oxygen atom, so that the $n(O)/n(Si)$ ratio exceeds 2. On principle, the process can be characterized as the decrease of the number of bridging oxygen atoms accompanied by simultaneous formation of the two-fold number of non-bridging oxygen atoms. With increasing MeO content the tectosilicate structural type changes to the inosilicates (e.g. pseudo-wollastonite) through sorosilicates (e.g. akermanite) to the structural type of nezosilicates (e.g. dicalcium silicate). The dependence of the number of bridging oxygen atoms on the content of MeO is linear from tectosilicates to nezosilicates.

Let us further consider the glass-forming ability of the MeO–SiO_2 melt, when understanding as glass the product formed by cooling the melt into the solid without crystallization [4]. The glass-forming ability of systems depends on the energy of bonds in the cooling melt, which disrupt in the course of crystallization. For the two limiting cases (tectosilicate–nezosilicate) there are therefore quite different conditions for glass formation. Crystallization of SiO_2 requires disruption of the very strong Si–O bonds and a new regular arrangement corresponding to the crystalline state. The crystallization process is a reconstructive one while in the case of nezosilicates the disruption of Si–O bonds is not necessary (the crystallization mechanism is not a reconstructive one). Practice shows that crystallization of SiO₂ does not virtually take place, while nezosilicates, on the other hand, crystallize very rapidly, so that one can assume a similar, even though irregular state of arrangement of the tetrahedrons in the melt when compared with the crystalline phase. Generally speaking, the ability to crystallize is indirectly proportional to the degree of polymerization, which in turn depends on the content and properties of the MeO oxides.

2. Theoretical

From the thermodynamic point of view solutions can be divided into specific groups, depending on the values of the enthalpy, $\Delta_{mix}H$, and entropy, $\Delta_{mix}S$, of mixing in the relation

$$\Delta_{mix}G = \Delta_{mix}H - T\Delta_{mix}S \tag{1}$$

For ideal solutions $\Delta_{mix}V = 0$, $\Delta_{mix}H = 0$, $\Delta_{mix}S = \Delta_{mix}S^{id}$, and $\Delta_{mix}G = \Delta_{mix}G^{id}$. To describe the behavior of real solutions we use different models. In the past there were many attempts to describe the behavior of molten systems, as a special kind of solutions, in terms of possible configuration of ions.

One of the most frequently used approaches is the model of regular solutions, which was originally developed for solutions of molten metals. In regular solutions $\Delta_{mix}V = 0$, $\Delta_{mix}H \neq 0$, $\Delta_{mix}S = \Delta_{mix}S^{id}$, and $\Delta_{mix}G \neq \Delta_{mix}G^{id}$, which means that the deviation from ideal behavior is ascribed to the change of the enthalpy at interaction of components at mixing. Different relations were proposed for the excess Gibbs energy of such systems, e.g. that of Redlich-Kister [5], Guggenheim [6], Margules [7], van Laar [8], etc. For the simple regular solution it holds that

$$\Delta_{ex}G = x_1 x_2 A \tag{2}$$

where A is the interaction coefficient. In simple regular solutions the entropy of mixing is equal to that in ideal solutions, what means that the excess entropy of mixing equals to zero and $\Delta_{ex}G = \Delta_{ex}H$.

According to the course of the functional dependence $a_i = f(x_i)$ in the limiting regions it is advantageous to distinguish real solutions of the I. and II. type. These limiting conditions are

$$\lim_{x_i \to 1} \left(\frac{da_i}{dx_i} \right)_{p,T} = k_{St} \tag{3}$$

and

$$\lim_{x_i \to 0} \left(\frac{d a_i}{d x_i} \right)_{p,T} = H_i^* \neq 0 \qquad (4)$$

where a_i is the activity of the component expressed in terms of mole fractions x_i according to any suitable model and k_{St} is the correction factor introduced by Stortenbeker [9], representing the number of foreign particles, which introduces the solute into the solvent at infinite dilution. If $k_{St} = 1$, the system obeys the Raoult's law and belongs to the I. type of solutions. If $k_{St} \neq 1$, the system belongs to the II. type of solutions, which do not obey the Raoult's law. H_i^* is proportional to the Henry's constant ($H_i^* = f_0 H$). However, the models of ideal and regular solutions can be applied only to systems of the I. type.

Each theoretical calculation of the solid-liquid phase equilibrium is based on the choice of a suitable thermodynamic model for the liquid phase, which also sufficiently considers its structural aspects. Owing to their polymeric character, silicate melts belong to the solutions of the II. type. The classic regular solution approach is not applicable, since the limiting laws (eqn (3) and (4)) are not obeyed. The Temkin's model of ideal ionic solution [10], which has been widely applied in molten salt systems, cannot be used, since the real anionic composition, owing to a broad polyanionic distribution, is not known a priori. The functional dependence $a_i = f(x_i)$ for the mentioned types of solutions is shown in Fig. 1.

Any structural model of silicate melts should be also in agreement with certain experimentally determined physico-chemical properties of the given silicate system:
- high equivalent conductivity, increasing with decreasing cation size, the conductivity being of ionic character, at least for the cations of I.A and II.A groups [11, 12];
- the activation energy of viscous flow is roughly constant within the range of 10 – 60 mole % M_2O (M = alkali metal) [13];
- the volume expansion is up to 10 mole % M_2O almost zero, approaching asymptotically for higher M_2O contents the value at about 50 mole % M_2O;
- the compressibility increases with increasing M_2O content up to 10 to 15 mole %, remaining further on constant value up to the concentration of about 50 mole % [14, 15];
- conductivity measurements have proved the presence of anions [16], studies of transport phenomena have shown that the O^{2-} anions do not contribute to the charge transfer [12].

It can be assumed that the course of the above relationships will be similar also in the systems $MeO-SiO_2$. On the basis of the above facts the structure of the $MeO-SiO_2$ melts can be imagined as to a certain degree polymerized lattice of SiO_4 tetrahedrons, where cations are situated in the free spaces between tetrahedrons. This concept has brought us to the formulation of the thermodynamic model of silicate melts [17].

In the thermodynamic model of silicate melts the chemical potential of component is expressed as the sum of chemical potentials of all energetically distinguishable atoms constituting the given component. Depending on the composition, oxygen atoms can be present as free oxygen anions, bridging, and non-bridging oxygen atoms. All oxygen atoms are thus not equivalent, i.e. their chemical potential is not equal. The material balance of available oxygen atoms and of Si–O bonds then gives the amounts of individual oxygen atoms. Silicon atoms are present exclusively in four-fold coordination, while other network forming atoms, e.g. B, Al, Fe, etc., may be present both as cations and/or central atoms in tetrahedral structural units. In the later case these atoms participate at least partially on the formation of the polyanionic network, which has to be taken into account in the calculation of the activity of component as well.

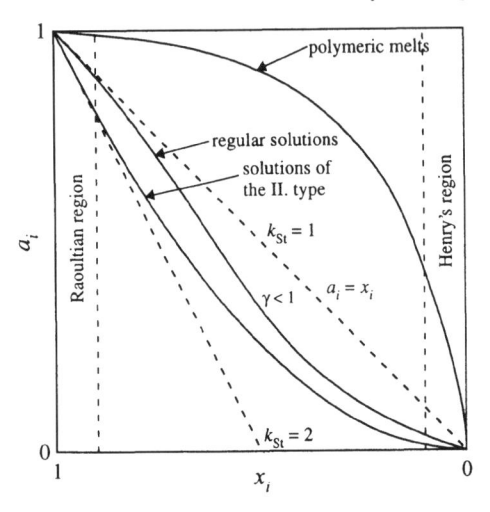

Fig. 1. The $a_i = f(x_i)$ plot for different types of solutions.

3. Thermodynamic model of silicate melts

The activities of individual components are calculated on the basis of the theory of conformal solutions [18]. This theory was derived for ionic systems without formation of complex ions, in which both anions and cations have identical charges. This theory was later applied to systems containing ions of various valences [19]. It should be emphasized that the application of this theory to silicate melts has only a formal character.

Let us define the polymerization degree of silicate melts as the fraction of bridging oxygen atoms in a single formula unit. The assumptions, on which basis the activities of components in the mixture are calculated, are as follows:
  a) linear dependence of the polymerization degree on MeO content throughout the tectosilicate–nezosilicate series is assumed,
  b) polymerization degree of SiO_4 tetrahedrons does not change during melting,

c) arrangement of polyanions in the melt in the vicinity of the melting point or the temperature of primary crystallization is identical at close distance to that in the crystalline state.

Let us consider an arbitrary melt in the MeO–SiO$_2$ system (Me = Mg, Ca, ...). The melt is composed of the following kinds of atoms: Me^{2+} cations, silicon atoms in tetrahedral coordination with oxygen atoms, and two kinds of oxygen atoms: the bridging ones, linking two neighboring SiO$_4$ tetrahedrons by means of Si–O–Si covalent bonds, and the non-bridging ones, bound to the silicon atom by one covalent bond and creating the coordination sphere of Me^{2+} cations. The two kinds of oxygen atoms have evidently different energetic state. Their mutual molar ratio defines the structure of the melt, i.e. its polymerization degree as well as the chemical potentials of the components. With regard to this structural aspect of silicate systems the chemical potential of an arbitrary component may be defined as the sum of chemical potentials of all atoms forming the considered component, when their particular energetic states are taken into account. The chemical potential of the i-th component in an arbitrary solution is defined by the relation

$$\mu_i = \sum_j n_{i,j} \mu_j \tag{5}$$

where $n_{i,j}$ is the amount of atoms of the j-th kind in the i-th component and μ_j is the chemical potential of of atoms of the j-th kind in the solution. For instance, the chemical potential of CaSiO$_3$, which is formed in the CaO–SiO$_2$ system, equals to the sum of the chemical potentials of calcium atoms, silicon atoms, and the bridging and non-bridging oxygen atoms. The activity of the i-th component in the solution also obeys the equation

$$\mu_i = \mu_i^o + RT \ln a_i \tag{6}$$

where μ_i^o is the chemical potential of the pure i-th component, defined similarly to the chemical potential of this component in the solution

$$\mu_i^o = \sum_j n_{i,j} \mu_{i,j}^o \tag{7}$$

where $\mu_{i,j}^o$ is the chemical potential of the pure j-th atoms in the pure i-th component. Substituting eqns (5) and (7) into eqn (6) we obtain

$$\sum_j n_{i,j} \mu_{i,j} = \sum_j n_{i,j} \mu_{i,j}^o + RT \ln a_i \tag{8}$$

The real mole fraction of j-th atoms in the i-th component and in the solution are given by relations

$$y_{i,j}^o = \frac{n_{i,j}}{\sum_j n_{i,j}} \tag{9}$$

$$y_j = \frac{\sum_i n_{i,j} x_i}{\sum_i x_i \sum_j n_{i,j}} \tag{10}$$

where x_i is the mole fraction of i-th component in the solution. The chemical potentials of j-th atoms in the pure i-th component and in the solution may also be expressed using eqns (9) and (10) in the following way

$$\mu_{i,j}^o = \mu_j^+ + RT \ln y_{i,j}^o \tag{11}$$

$$\mu_j = \mu_j^+ + RT \ln y_j \tag{12}$$

where μ_j^+ is the chemical potential of a hypothetical liquid composed exclusively of the j-th atoms. Inserting eqns (11) and (12) into eqn (8) we get for the activity of the i-th component in the solution the relation

$$\ln a_i = \sum_j n_{i,j} \ln y_j - \sum_j n_{i,j} \ln y_{i,j}^o \tag{13}$$

or after rearrangement

$$a_i = \prod_j \left(\frac{y_j}{y_{i,j}^o}\right)^{n_{i,j}} \tag{14}$$

Eqn (14) for the activity of a component derived in this way is wholly universal and may be used in any system. The calculation of the activity of a component in an actual system using this equation thus respects the characteristic features of silicate melts, e.g. the different energetic state of individual atoms of the same kind. Such cases may occur in the presence of three-valence atoms like B^{3+}, Al^{3+}, Fe^{3+}, etc., or four-valence atoms like Ti^{4+}.

In order to describe the structural aspect of the given silicate melt it is necessary to perform correctly the material balance of individual oxygen atoms as well as of the above mentioned double-acting atoms. The material balance is based on the following principles. Let us assume that in the system consisting of m oxides the polymeric network is formed by three- and four-fold coordinated atoms ^jA ($j = 1, 2, \ldots m$) linked to bridging (–O–) and non-bridging (–O$^-$) oxygen atoms. Designating the fraction of k-fold coordinated ^jA atoms as $\alpha_{k,j}$, the following inequality must hold

$$\alpha_{3,j} + \alpha_{4,j} \leq 1 \tag{15}$$

The distribution of the atoms according to their coordination, or to their participation in a covalent network, is then determined by the following material balance

$$n(^jA) = n_o(^jA) + n_3(^jA) + n_4(^jA) \tag{16}$$

where

$$n_o(^jA) = (1 - \alpha_{3,j} - \alpha_{4,j})n(^jA) \tag{17}$$
$$n_3(^jA) = \alpha_{3,j} n(^jA) \tag{18}$$
$$n_4(^jA) = \alpha_{4,j} n(^jA) \tag{19}$$

where $n_k(^jA)$ is the amount of the k-fold coordinated ^jA atoms and $n_o(^jA)$ is the amount of those ^jA atoms, which are not built into the polyanionic network. Then the amount of $n(^jA-O)$ bonds is given by the relation

$$n(^jA - O) = \sum_{j=1}^{m} [(3\alpha_{3,j} + 4\alpha_{4,j}) n(^jA)] \tag{20}$$

Assuming that the total amount of oxygen atoms, $n(O)$, equals to the sum of the bridging, $n(-O-)$, and non-bridging, $n(-O^-)$, oxygen atoms, we can calculate their amounts from the material balances of the total amounts of oxygen atoms and the ^jA-O bonds according to the equations

$$n(O) = n(-O-) + n(-O^-) \tag{21}$$
$$n(-O-) = n(^jA - O) - n(O) \tag{22}$$

If the solution does not have a physical sense (i.e. the values of $n(i)$ are negative), it is necessary to assume, that non-bridging oxygen atoms and oxide ions O^{2-} are present. The material balance is then given by the equations

$$n(O) = n(-O^-) + n(O^{2-}) \tag{23}$$
$$n(-O^-) = n(^jA - O) \tag{24}$$

For typical modifying atoms such as alkali metals and earth alkali metals it is assumed to holds

$$\alpha_{3,j} = \alpha_{4,j} = 0 \tag{25}$$

while typical network forming elements of the IV-th group of the periodic system like silicon and germanium $\alpha_{4,j} = 1$. In other cases the value of $\alpha_{k,j}$ can be chosen in order to fit the calculated and experimentally determined phase diagrams.

The formal polymerization degree P of the melt with given composition can be calculated from eqns (21) and (22). Since the polymerization degree was defined as the fraction of the bridging oxygen atoms in the total oxygen atoms, one can write

$$P = \frac{n(-O-)}{n(O)} = \frac{n(^jA-O) - n(O)}{n(O)} = \frac{\sum_{j=1}^{m}[n(^jA)(3\alpha_{3,j} + 4\alpha_{4,j})]}{n(O)} - 1 \qquad (26)$$

For instance, for the system MeO–SiO$_2$ with the composition $(1-x)$ MeO + x SiO$_2$ it holds that $n(^jA) = x$, $m = 2$, $^1A = $ Me, $^2A = $ Si, $\alpha_{3,1} = \alpha_{3,2} = \alpha_{4,1} = 0$ and $\alpha_{4,2} = 1$. The polymerization degree is then

$$P = \frac{4x}{1+x} - 1 \qquad (27)$$

In such a melt for $x = 0.333$, i.e. for the composition of orthosilicate, is $P = 0$, and for the pure SiO$_2$ melt ($x = 1$) $P = 1$.

4. Application of the model to different systems

The thermodynamic model of silicate melts has been applied in the calculation of the phase diagrams of a number of binary and ternary silicate, alumino-, ferro-, ferri-, titania-silicate, and borate melts. The calculation of liquidus temperature of individual components, $T_{i,\text{liq}}$, is performed using corresponding experimental values of the enthalpy and temperature of fusion according to the simplified and adapted LeChatelier-Shreder equation

$$T_{i,\text{liq}} = \frac{\Delta_{\text{fus}}H_i \cdot T_{\text{fus},i}}{\Delta_{\text{fus}}H_i - RT_{\text{fus},i} \ln a_i} \qquad (28)$$

where $T_{\text{fus},i}$ and $\Delta_{\text{fus}}H_i$ are the temperature and enthalpy of fusion of the i-th component, respectively, and a_i is its activity calculated according to eqn (14). In prevailing cases it was assumed that $\Delta_{\text{fus}}H_i = $ const. However, when necessary, i.e. at great difference between the melting and eutectic temperatures, the change of $\Delta_{\text{fus}}H_i$ with temperature was inserted in the form $\Delta_{\text{fus}}H_i(T) = \Delta_{\text{fus}}H_i(T_{\text{fus}}) - \Delta C_{P,s/l}(T_{\text{fus}} - T)$.

The first crystallizing phase at the given composition was determined according to the condition

TABLE 1. Temperatures and enthalpies of fusion of individual compounds.

Compound	T_{fus} / K	$\Delta_{fus}H$ / kJ mol^{-1}	Reference
Al$_2$O$_3$ (A)	2293	111.4	[7, 20]
3Al$_2$O$_3$.2SiO$_2$ (A$_3$S$_2$)	2123	188.3	[7, 20]
B$_2$O$_3$ (B)	723	22.2	[7]
CaO (C)	2843	52.0	[7, 20]
CaO.Al$_2$O$_3$ (CA)	1878	102.5	[7, 20]
CaO.2Al$_2$O$_3$ (CA$_2$)	2033	200.0	[7, 20]
12CaO.7Al$_2$O$_3$ (C$_{12}$A$_7$)	1728	209.3	[7, 20]
2CaO.Al$_2$O$_3$.SiO$_2$ (C$_2$AS)	1868	155.9	[7, 20]
CaO.Al$_2$O$_3$.2SiO$_2$ (CAS$_2$)	1826	166.8	[7, 20]
CaO.MgO.2SiO$_2$ (CMS$_2$)	1665	128.3	[21]
2CaO.MgO.2SiO$_2$ (C$_2$MS$_2$)	1727	85.7	[7, 20]
CaO.SiO$_2$ (CS)	1817	56.0	[7]
2CaO.SiO$_2$ (C$_2$S)	2403	55.4	[7, 18]
3CaO.2SiO$_2$ (C$_3$S$_2$)	1718	146.5	[7, 20]
CaO.TiO$_2$ (CT)	2243	127.3	estimated
CaO.TiO$_2$.SiO$_2$ (CTS)	1656	139.0	[22]
MgO.Al$_2$O$_3$ (MA)	2408	200.0	[7, 20]
MgO.SiO$_2$ (MS)	1850	75.2	[7]
2MgO.SiO$_2$ (M$_2$S)	2171	71.1	[7]
MnO.SiO$_2$ ($\overline{M}$S)	1564	66.9	[7, 20]
Na$_2$O (N)	1405	47.6	[7]
Na$_2$O.B$_2$O$_3$ (NB)	1239	72.4	[7]
Na$_2$O.2B$_2$O$_3$ (NB$_2$)	1016	81.1	[7]
Na$_2$O.3B$_2$O$_3$ (NB$_3$)	1045	105.7	estimated
Na$_2$O.4B$_2$O$_3$ (NB$_4$)	1088	133.4	estimated
K$_2$O (K)	1154	32.7	[23]
K$_2$O.B$_2$O$_3$ (KB)	1223	64.8	estimated
K$_2$O.2B$_2$O$_3$ (KB$_2$)	1088	104.1	[7]
Li$_2$O.B$_2$O$_3$ (LB)	1117	67.7	[7]
Li$_2$O.2B$_2$O$_3$ (LB$_2$)	1190	120.4	[7]
SiO$_2$ (S)	1996	9.6	[7]
TiO$_2$ (T)	2103	66.9	[7]

$$T_{pc} = \max_{i}(T_{i,\text{liq}}) \tag{29}$$

The liquidus curves in binary systems and liquidus surfaces in the ternary ones of individual phases were calculated on the basis of the calculated temperatures of primary crystallization using the method of multiple linear regression analysis. From these equations the isotherms of primary crystallization for the chosen temperature step were calculated. The boundary lines were determined as cross-sections of two neighboring liquidus surfaces with equal temperature of primary crystallization. The coordinates of invariant points were obtained as cross-sections of three boundary lines.

The basic thermodynamic data of individual components, i.e. the temperatures and enthalpies of fusion, were mainly taken from the literature and are summarized in Table 1. However, for some components the values of the enthalpy of fusion were not known. In such cases these data were estimated on the basis of thermodynamic analogy. More information on the estimation method the reader can find in the given reference.

4.1. SYSTEM CaO–MgO–SiO$_2$

In this system the calculation was focused to the phase diagrams of binary or pseudobinary systems SiO$_2$–CaO.SiO$_2$, CaO.MgO.2SiO$_2$–SiO$_2$, CaO.SiO$_2$–CaO.MgO.2SiO$_2$, CaO.SiO$_2$–2CaO.MgO.2SiO$_2$, CaO.MgO.2SiO$_2$–2CaO.MgO.2SiO$_2$, CaO.SiO$_2$–2CaO.SiO$_2$, and CaO.MgO.2SiO$_2$–2MgO.SiO$_2$ [23]. The experimental phase diagrams of these systems were taken in [24, 25]. The comparison of the calculated and experimentally determined phase diagrams is shown in Figs. 2–8. The change in the enthalpy of fusion with temperature was neglected in all cases. When solid solutions are formed in the systems, the activity of the component in the solid solution was taken to be equal to the mole fraction.

When comparing the experimentally determined liquidus curves with the calculated ones it can be seen that the model describes the course of the individual liquidus curves in these complex melts very well. The above explanation of the structure of silicate melts has been obviously simplified to a considerable degree. It is very probable that as the result of different charges and polarization effects of various cations the polymerization will not change linearly with composition. The proposed model also does not allow distinguishing a priori the way in which polyanions are arranged in the melt (circular, linear, spatial, etc.).

In the systems C–CS and S–CMS$_2$ the liquidus curves of SiO$_2$ were not calculated since two liquid phases are formed in this region. This phenomenon the proposed thermodynamic model does not consider. In the system CS–C$_2$S the congruently melting compound C$_3$S$_2$ is formed, which affects the activity of CS or C$_2$S in the melt. For this reason this fact should be considered in the calculation of the partial systems CS–C$_3$S$_2$ and C$_3$S$_2$–C$_2$S.

In the system CMS$_2$–M$_2$S no satisfactory agreement of calculated and experimental liquidus curve of forsterite was obtained [24] using the thermodynamic model. Good agreement was obtained when only the presence of non-bridging oxygen atoms was assumed at high M$_2$S concentrations. It may be therefore assumed that MgO present in

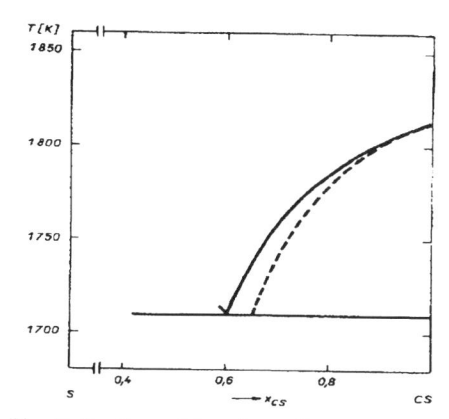

Fig. 2. Section of the phase diagram of the system SiO_2–$CaSiO_3$.

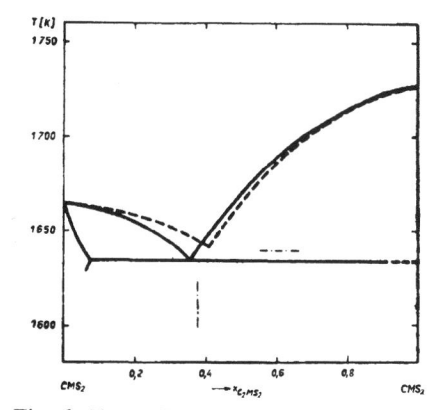

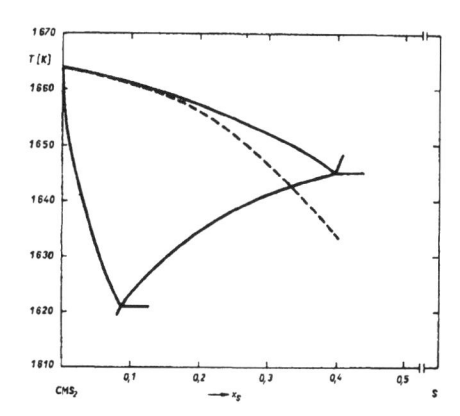

Fig. 3. Section of the phase diagram of the system $CaMgSi_2O_6$–SiO_2.

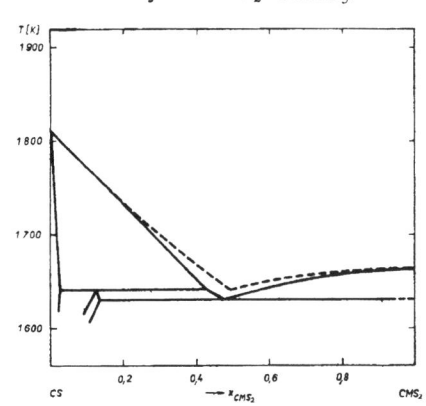

Fig. 4. Phase diagram of the system $CaSiO_3$–$CaMgSi_2O_6$.

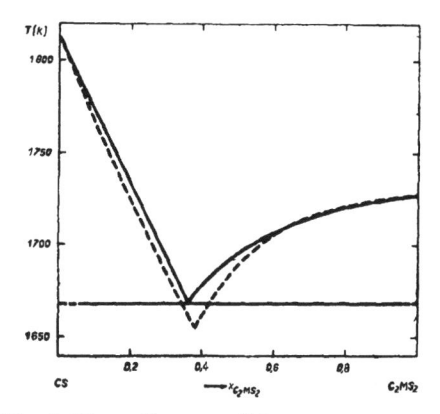

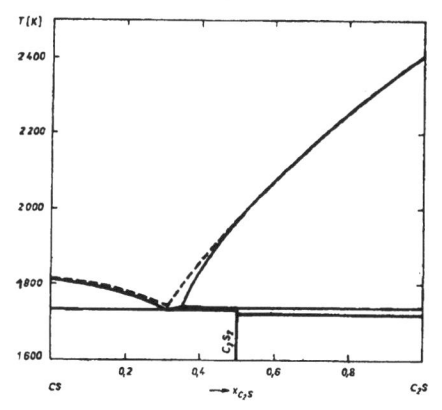

Fig. 5. Phase diagram of the system $CaSiO_3$–$Ca_2MgSi_2O_7$.

Fig. 6. Phase diagram of the system $CaMgSi_2O_6$–$Ca_2MgSi_2O_7$.

Fig. 7. Phase diagram of the system $CaSiO_3$–Ca_2SiO_4.

high concentration can cause increased disruption of the Si–O–Si bonds due to the formation of MgO$_4$ tetrahedrons.

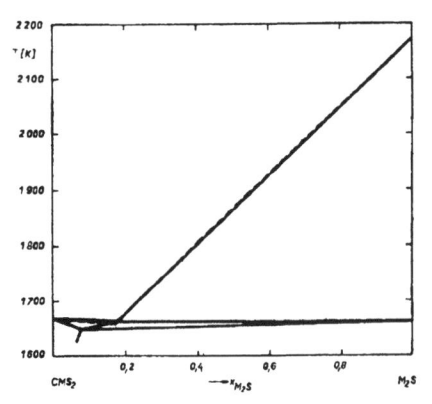

Fig. 8. Phase diagram of the system CaMgSi$_2$O$_6$–Mg$_2$SiO$_4$.

4.2. SYSTEM CaO–Fe$_x$O$_y$–SiO$_2$

4.2.1. *System CaO–FeO–SiO$_2$*

The melts of the system CaO–Fe$_x$O$_y$–SiO$_2$ are of great technological importance since they are the basis of metallurgical slags. In the melts of this system iron is present in two oxidation states, the amount of Fe^{2+} being dependent on temperature, oxygen partial pressure, and the composition of the melt. Phase equilibrium in the four-component system CaO–FeO–Fe$_2$O$_3$–SiO$_2$ is described in [26, 27]. Two phase diagrams corresponding to two oxygen partial pressures were considered: that of the CaO–"FeO"–SiO$_2$ system in equilibrium with metallic iron, and that of the CaO–"Fe$_2$O$_3$"–SiO$_2$ system in equilibrium with air ($P_{O_2} = 21$ kPa).

Relatively unambiguous is the situation in the CaO–"FeO"–SiO$_2$ system, where Fe^{II} atoms are placed in the interstitial sites with higher coordination that the fourfold one. However, according to the results of density measurement [28], these melts tend to microsegregation into regions richer in calcium oxide and regions richer on iron oxide.

On the other hand, Fe^{III} atoms in the CaO–"Fe$_2$O$_3$"–SiO$_2$ system can enter the polyanionic network being in the fourfold coordination (network former) or, similarly to the Fe^{II} atoms, be in higher coordination and behave as the network modifier [29, 30]. The dependence of interatomic distance Fe–O on composition is similar to that of Si–Si, which corresponds to the transition of Fe^{III} atoms from the octahedral sites to the tetrahedral ones.

The comparison of liquidus surfaces of CaSiO$_3$ calculated using the thermodynamic model of silicate melts with the experimental ones has been found a useful indirect method to study the real structure of the melts of the systems CaO–"FeO"–SiO$_2$ and

CaO–"Fe$_2$O$_3$"–SiO$_2$, i.e. the participation of individual Fe^{II} and Fe^{III} atoms in the polyanionic network. The calculation was made in [31].

The liquidus surface of CaSiO$_3$ in the CaO–"FeO"–SiO$_2$ system was calculated according to the LeChatelier-Shreder equation (28). Boundary curves of the CaSiO$_3$ liquidus surface were obtained by similar calculation of liquidus surfaces of Ca$_2$SiO$_4$ and Fe$_2$SiO$_4$. The boundary curve with tridymite could not be determined because of a large immiscibility gap in the region of high SiO$_2$ concentration, which the thermodynamic model does not take into account. The calculated part of the phase diagram is shown in Fig. 9. The similar part of the experimental phase diagram according to Muan and Osborn [27] is shown in Fig. 10 for comparison.

As it follows from the comparison of Figs. 9 and 10, a relatively good agreement between the experimental and calculated liquidus surface was obtained when all present Fe^{II} atoms are in higher coordination and behave as the network modifier. The dotted lines in Fig 9 represent isotherms in CaSiO$_3$ liquidus surface calculated for the case, when all Fe^{II} atoms behave as network former. It is evident, that such assumption is not correct. However, a little part of Fe^{II} atoms in tetrahedral coordination cannot be excluded. In this calculations the presence of rankinite was neglected, because of narrow region of its primary crystallization. The transition of pseudo-wollastonite to wollastonite was neglected too. These simplifications, however, could not essentially affect conclusions concerning the behavior of Fe^{II} atoms in the melts.

4.2.2. System CaO–Fe$_2$O$_3$–SiO$_2$

The calculation of the CaSiO$_3$ liquidus surface in the system CaO–"Fe$_2$O$_3$"–SiO$_2$ was carried out for two cases:
a) all Fe^{III} atoms are in tetrahedral coordination and together with the SiO$_4$ tetrahedrons participate on the polyanionic network formation,
b) only half of the Fe^{III} atoms is in tetrahedral coordination, while the other half behaves as the network modifier being in higher coordination.

The third possibility, that all Fe^{III} atoms behave as network modifier was as unlikely not considered.

The liquidus surface of CaSiO$_3$ in the CaO–"Fe$_2$O$_3$"–SiO$_2$ system was calculated according to the LeChatelier-Shreder equation. Boundary curves of the CaSiO$_3$ liquidus surface were obtained by similar calculation of liquidus surfaces of Ca$_2$SiO$_4$ and Fe$_2$O$_3$. Also the other limitations concerning the boundary limits and the immiscibility gap are similar with those of the system CaO–"FeO"–SiO$_2$.

The calculated part of the phase diagram of the CaO–"Fe$_2$O$_3$"–SiO$_2$ system is shown in Fig. 11. The similar part of the experimental phase diagram according to Muan and Osborn [27] is shown in Fig. 12 for comparison. The dotted lines in Fig. 11 represent isotherms of the CaSiO$_3$ liquidus surface for the first case (a). This assumption is obviously unsuitable. Very good agreement with the experimental phase diagram was attained in the second case. In Fig. 12 the full lines represent the right isotherms. It may be therefore stated that in the melts of the system CaO–"Fe$_2$O$_3$"–SiO$_2$ having the x_{CaO}/x_{SiO2} modulus in the range 0.6–1.5 roughly the half of the present Fe^{III} atoms is coordinated tetrahedrally and behaves as the network former, while the other one contributes to the polyanion destruction and behaves as the network modifier. Such arrangement does not exclude the possibility of the formation of anions of the $Fe_2O_4^{2-}$ and $Fe_2O_5^{4-}$ type as it is mentioned in the paper of Mori and Suzuki [33]. Also in the calculation of the $CaSiO_3$ liquidus surface in the CaO–"Fe_2O_3"–SiO_2 system the region of primary crystallization of rankinite was neglected. Because of the lack of the thermodynamic data, the calculation of the boundary curve with hematite was carried out using the experimentally determined coordinates of the cross section of compatibility join $CaSiO_3$–Fe_2O_3 and the boundary curve between the regions of the primary crystallization of hematite and magnetite (point P in Fig. 12).

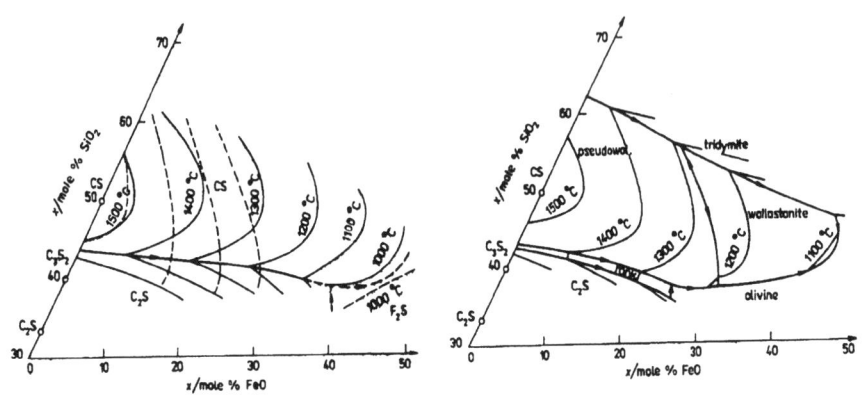

Fig. 9. Calculated part of the phase diagram of the system CaO–FeO–SiO_2.

Fig. 10. Part of the phase diagram of the CaO–FeO–SiO_2 system taken from [27].

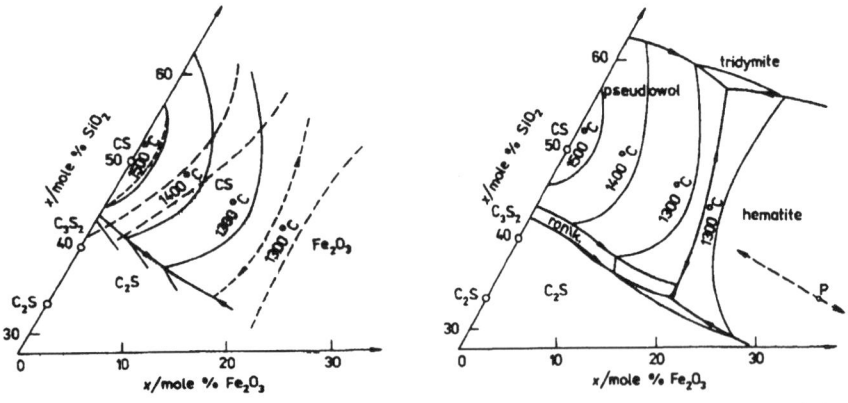

Fig. 11. Calculated part of the phase diagram of the system CaO–Fe_2O_3–SiO_2.

Fig. 12. Part of the phase diagram of the system CaO–Fe_2O_3–SiO_2 taken from [27].

It has to be concluded that in spite of many simplifications made in the calculation, using this method valuable information on the structure of these oxide melts on the semi-quantitative level has been obtained. However, the conclusions concerning the coordination of iron atoms in the silicate melts are obviously valid only in the concentration region with the modulus $x_{CaO}/x_{SiO2} \approx 1$.

4.3. SYSTEM CaO–Al$_2$O$_3$–SiO$_2$

The coordination of Al^{III} atoms in aluminosilicate melts is one of the important research directions in the structure of melts containing alumina. The interest is associated with extensive utilization of these melts in different sections of the silicate industry, such as glass, cement, porcelain manufacture, etc.

Studies of aluminosilicate melts were mainly based on the interpretation of physico-chemical properties, such as density and viscosity, in terms of composition, particularly of the $x(CaO)/x(Al_2O_3)$ ratio. The basic idea was the concept that in melts with the ratio $x(CaO)/x(Al_2O_3) \geq 1$, all the Al^{III} atoms are in tetrahedral coordination and that the octahedral coordination of Al^{III} atoms occurs only at ratios $x(Ca)/x(Al_2O_3) < 1$, i.e. at an excess of alumina with respect to CaO.

The actual coordination of Al^{III} atoms in the system CaO–Al$_2$O$_3$–SiO$_2$ melts was studied also in [34] using the calculation of the section of this phase diagram in the region of primary crystallization of wollastonite, anorthite, gehlenite, and dicalcium silicate using the thermodynamic model of silicate melts, taking into account the structural aspects of the coordination of Al^{III} atoms.

The values of the activity of components in question were inserted into the LeChatelier-Shreder equation (28) ($\Delta_{fus}H_i$ = const.). Fig. 13 shows the section of the phase diagram of the system CaO–Al$_2$O$_3$–SiO$_2$ in the region of the primary crystallization of wollastonite, anorthite, gehlenite, and dicalcium silicate, calculated on the assumption that one half of the Al^{III} atoms is in tetrahedral coordination over the entire concentration range in question, and that the other half of the Al^{III} atoms is in a higher coordination, obviously in the octahedral one, and this part do not participate in the formation of the polyanionic network. As follows from the comparison with the experimentally determined section of this phase diagram [27] (Fig. 14), the agreement is very satisfactory. This indicates that the accepted assumption is acceptable. The dashed lines in Fig. 13 represent the situation, when all Al^{III} atoms are in tetrahedral coordination. The existing idea of tetrahedral coordination of all the Al^{III} atoms in the given concentration range ($x(CaO)/x(Al_2O_3) \geq 1$) is therefore obviously incorrect, which also follows from the course of the boundary lines of the liquidus surfaces of the individual components and designated by the dashed lines in Fig. 13.

In any case, the finding on the partial tetrahedral coordination of Al^{III} atoms in the composition range for $x(CaO)/x(Al_2O_3) \geq 1$ is relatively surprising in view of the available information on the behavior of Al^{III} atoms in silicate melts. However, it is in agreement with the very similar behavior of Fe^{III} atoms in the melts of the system CaO–Fe$_2$O$_3$–SiO$_2$.

The success of the thermodynamic model in the calculation of a part of the system CaO–Al$_2$O$_3$–SiO$_2$ encouraged us to calculate the phase diagram of the whole system [35]. This system is rather complicated because of the formation of the number of binary and ternary compounds. The calculated phase diagram of this system is shown in Fig. 15, while the experimentally determined phase diagram according to Muan and Osborn [27] is shown in Fig. 16.

In this case the immiscibility region near the SiO$_2$ apex was neglected since such behavior is not considered in the thermodynamic model. Furthermore, because of the lack of thermodynamic data, the crystallization of rankinite, tricalcium silicate, tricalcium aluminate, and calcium hexaaluminate were not included into the calculation.

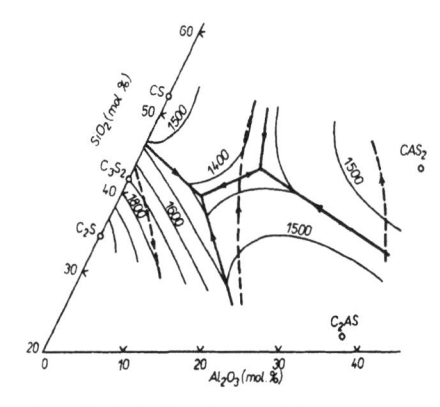

Fig. 13. Calculated part of the system CaO–Al$_2$O$_3$–SiO$_2$. Full lines – 0.5 Al^{III} in tetrahedrons, dashed lines – all Al^{III} in tetrahedrons.

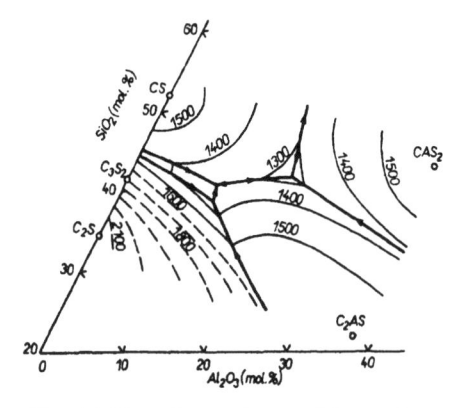

Fig. 14. Experimental part of the system CaO–Al$_2$O$_3$–SiO$_2$ according to Muan and Osborn [27].

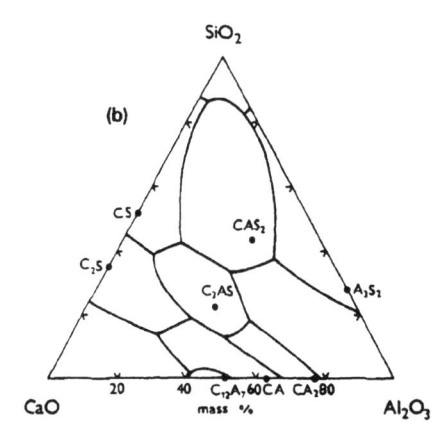

Fig. 15. Calculated phase diagram of the system CaO–Al$_2$O$_3$–SiO$_2$.

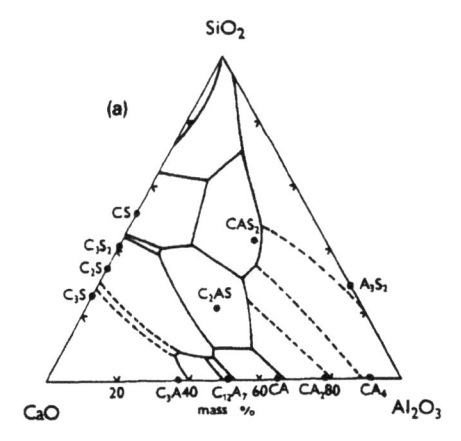

Fig. 16. Experimental phase diagram of the system CaO–Al$_2$O$_3$–SiO$_2$ [27].

From the comparison of the calculated and experimentally determined phase diagram of the system CaO–Al$_2$O$_3$–SiO$_2$ it follows that the chosen approach is very suitable for describing phase equilibrium also in aluminosilicate systems.

4.4. SYSTEM CaO–TiO$_2$–SiO$_2$

Silicate systems containing TiO$_2$ are of considerable technological and geochemical interest. Titanium dioxide is a common component of industrial glasses, enamels, pyroceramics, and of some metallurgical slags. The structural role of Ti^{IV} in silicate melts has been studied in many spectroscopic investigations, e.g. [36–40]. It is a complex function of several variables, namely TiO_2 and SiO_2 concentration, type and content of modifying cations and temperature [39, 41]. Despite the number of investigations neither consensus has been reached regarding the coordination state of Ti^{IV} atoms, nor how the structure of the melts is modified by their presence. The results obtained by various methods are often contradictory [38, 40].

In situ high temperature Raman spectroscopy of melts along the join $Na_2Si_2O_5$–$Na_2Ti_2O_5$ [39] have shown that the Raman spectra of Ti-bearing glasses and melts are consistent with Ti^{IV} in at least three different structural positions:

a) Ti^{IV} substitutes for Si^{IV} in tetrahedral coordination in the structural units in the melt (acts as network former),
b) Ti^{IV} forms TiO_2 like clusters with Ti^{IV} in tetrahedral coordination,
c) Ti^{IV} as network modifier, possibly occurring in octahedral or five-fold coordination.

The structural behavior of Ti^{IV} was determined also by means of calculation of the phase diagrams of some pseudobinary and one ternary systems using the thermodynamic model of the silicate melts, taking into account the participation of Ti^{IV} on the polyanionic network [42, 43].

The activities of considered compounds were derived applying equation (14) in the considered pseudo-binary and the ternary systems. It was assumed that all the present Ti^{IV} atoms are in tetrahedral coordination, i.e. they act as network formers. The calculation of phase diagrams and subsequent thermodynamic analysis were realized in the pseudo-binary systems $CaSiO_3$–$CaTiSiO_5$, $CaSiO_3$–$CaTiO_3$, Ca_2SiO_4–$CaTiO_3$, $CaTiSiO_5$–$CaTiO_3$, and $CaTiSiO_5$–TiO_2, in the binary system CaO–TiO_2, as well as in the whole ternary system CaO–TiO_2–SiO_2.

The comparison of experimental and calculated phase diagrams of the system $CaSiO_3$–$CaTiSiO_5$ is given in Fig. 17. In the calculation of the $CaSiO_3$ liquidus curve the solid solutions in the composition range $x(CS_{ss}) = 0.3$ at the eutectic temperature were assumed. However, this assumption was not confirmed experimentally. The course of liquidus curves calculated on the assumption that activities of both components are equal to mole fractions are also presented in Fig. 17 for comparison. It can be seen in from the figure, that the thermodynamic model fits the experimental data better than the assumption that $a_i = x_i$.

The phase diagram of the system $CaSiO_3$–$CaTiO_3$ is shown in Fig. 18. From the course of the liquidus curves it follows, that the thermodynamic model is almost identical with the assumption $a_i = x_i$. However, the course of the experimental liquidus curve of $CaTiO_3$ is not thermodynamically consistent, since in the vicinity of its melting point it exhibits dystectic mode of melting, which is not probable with regard to the nature of this compound.

The phase diagram of the system Ca_2SiO_4–$CaTiO_3$ is shown in Fig. 19. It can be seen from this figure that calculation using the $a_i = x_i$ assumption fits the experimental data best. However, this subsystem of the CaO–TiO_2–SiO_2 system is in the basic region of melts, where the thermodynamic model may not be valid.

The comparison of experimental and calculated phase diagrams of the system $CaTiSiO_5$–$CaTiO_3$ is shown in Fig. 20. With regard to the relatively big difference between the temperature of fusion of $CaTiO_3$ and the eutectic temperature the temperature dependence of the enthalpy of fusion of this compound has been taken into account introducing the difference of the heat capacity of solid and liquid $CaTiO_3$ $\Delta C_{p,sl}(CT) = 250$ J/mol.K. From the figure it follows, that the thermodynamic model of silicate melts describes the phase diagram of this system very good.

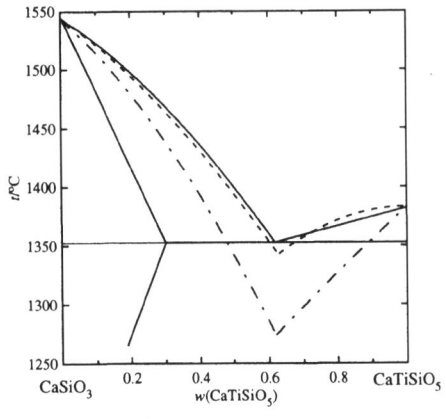

Fig. 17. Calculated and experimental phase diagram of the system $CaSiO_3$–$CaTiSiO_5$. full: [44], dashed: model, dott-dash: $a_i = x_i$

Fig. 18. Calculated and experimental phase diagram of the system $CaSiO_3$–$CaTiO_3$ full: [44], dashed: model, dott-dash: $a_i = x_i$

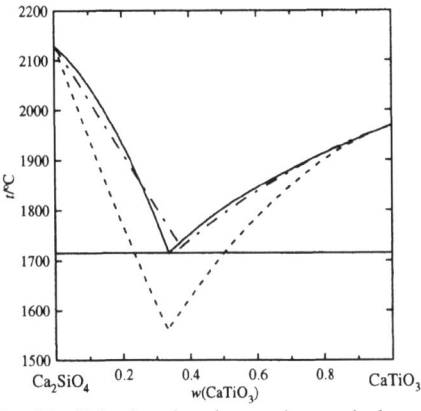

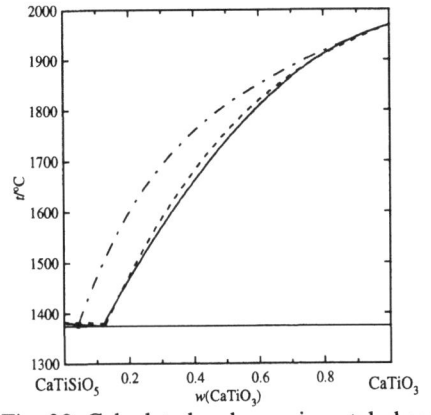

Fig. 19. Calculated and experimental phase diagram of the system Ca_2SiO_4–$CaTiO_3$. full: [45], dashed: model, dott-dash: $a_i = x_i$

Fig. 20. Calculated and experimental phase diagram of the system $CaTiSiO_5$–$CaTiO_3$ full: [44], dashed: model, dott-dash: $a_i = x_I$

The phase diagram of the system $CaTiSiO_5$–TiO_2 is shown in Fig. 21. As it can be seen, the liquidus curve of $CaTiSiO_5$ is best described by the thermodynamic model while the experimental liquidus curve of TiO_2 fits best the simple model $a_i = x_i$. The reason may again be the basic nature of the TiO_2 rich melts. More probable is that not all Ti^{IV} atoms are in tetrahedral coordination.

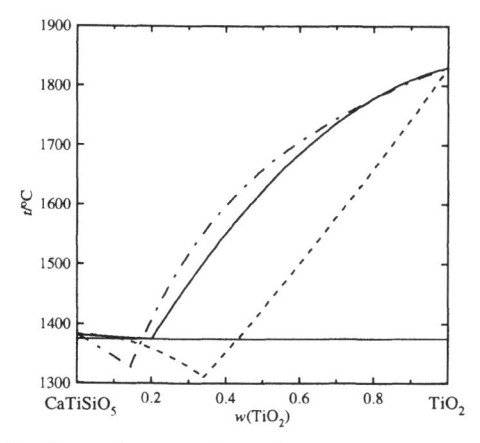

Fig. 21. Calculated and experimental phase diagram of the system $CaTiSiO_5$–TiO_2. Full line: [45], dashed line: model, dott-dashed line: $a_i = x_i$

The experimentally determined [44] and calculated phase diagrams of the system CaO–TiO_2 are shown in Figs. 22 and 23, respectively. The activities of individual phases, i.e. CaO, $Ca_3Ti_2O_7$, $CaTiO_3$, and TiO_2 were calculated according to the thermodynamic model. When all the present Ti^{IV} atoms are network forming, in the formula unit of $Ca_3Ti_2O_7$ there is one bridging oxygen atom and six oxygen atoms are the non-bridging ones. $CaTiO_3$, perovskite, is a mixed oxide with characteristic structure in solid state. In molten state, however, the structure depends on the behavior of Ti^{IV} atoms. When they participate on the formation of the network, the situation is similar like in pseudo-wollastonite, i.e. in the pure perovskite there will be one bridging oxygen atom and two non-bridging ones. In TiO_2 the situation is similar as in pure SiO_2 providing all Ti^{IV} atoms are network forming ones.

As follows from the comparison of Figs. 22 and 23, the agreement between the calculated and experimental phase diagram of the system CaO–TiO_2 is very good. Besides that, the liquidus curve of TiO_2 was calculated under the assumption that $a_i = x_i$. The total standard deviation between the calculated and experimentally determined liquidus temperatures is $SD = 64$ °C.

In the phase diagram of the system CaO–TiO_2–SiO_2 the following phases are present: CaO, TiO_2, SiO_2, Ca_3SiO_5, Ca_2SiO_4, $Ca_3Si_2O_7$, $CaSiO_3$, $CaTiO_3$, $Ca_3Ti_2O_7$, and $CaTiSiO_5$. Some existing phases in this ternary system, namely rankinite, tricalcium silicate, tricalcium aluminate, and calcium hexaaluminate, were not included into the calculation because of lack of relevant thermodynamic data. The expressions for the activity of individual phases were derived using the eqns (9), (10), and (14) of the thermodynamic model assuming that all the present Ti^{IV} atoms exhibit network forming character.

In the formula unit of Ca_2SiO_4 all the present oxygen atoms are definitely the non-bridging ones. In the calculation of the activity of Ca_2SiO_4, however, two different approaches were used depending on the acid-base character of the melt in question. In the region of basic melt it was assumed, that the amount of non-bridging oxygen atoms is equal to $2a$, while in the acidic region the amount of non-bridging oxygen atoms is equal to $4b + 4c$ (where a, b, and c are the amounts of individual oxides in the system CaO–TiO$_2$–SiO$_2$). The experimentally determined [44] and calculated phase diagrams of the system CaO–TiO$_2$–SiO$_2$ are shown in Figs. 24 and 25, respectively.

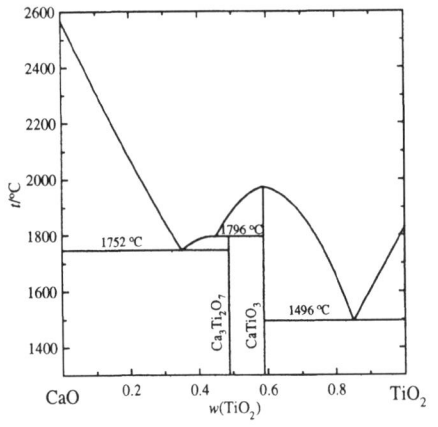

Fig. 22. Calculated phase diagram of the system CaO–TiO$_2$.

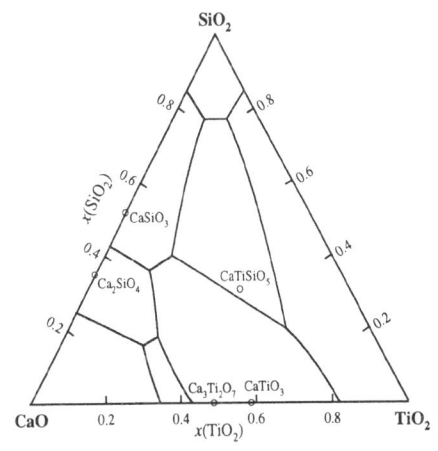

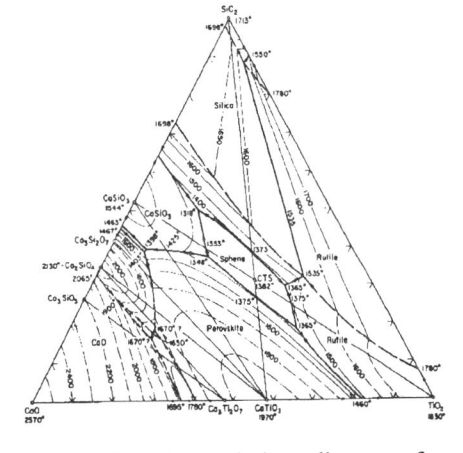

Fig. 23. Experimental phase diagram of the system CaO–TiO$_2$ [44].

Fig. 24. Calculated phase diagram of the system CaO–TiO$_2$–SiO$_2$.

Fig. 25. Experimental phase diagram of the system CaO–TiO$_2$–SiO$_2$ [44].

In the region of high content of SiO$_2$ the calculation of the phase equilibrium fails since the formation of two liquids is not considered in the thermodynamic model of silicate melts. This is also the reason for the enlarged liquidus surface of CaTiSiO$_5$ up to the high content of silica.

From the comparison of the calculated and experimental phase diagrams it follows that the thermodynamic model of silicate melts is suitable for describing the phase equilibrium also in titania-bearing silicate systems. The introduction of structural aspects into the thermodynamic model provides deeper information on the behavior of Ti^{IV} atoms. Discrepancies between some calculated and experimental phase diagrams are obviously caused by either inadequate structural assumptions or unreliable thermodynamic data. It was, however, shown that Ti^{IV} atoms behave in the silicate melts as network formers, excepting in the region of its high concentration, and in highly basic melts.

4.5. OTHER TERNARY SYSTEMS

Two types of ternary and pseudo-ternary subsystems were calculated [35]:
   a) simple eutectic systems,
   b) systems with four crystallization areas, where the figurative point of the fourth crystalline phase lies beyond the pseudo-ternary diagram.

As examples of the first case the systems $MgO.SiO_2$–$CaO.MgO.2SiO_2$–$CaO.Al_2O_3.SiO_2$, $CaO.MgO.2SiO_2$–$MnO.SiO_2$–$CaO.Al_2O_3.2SiO_2$, and $MgO.Al_2O_3$–$2CaO.SiO_2$–$2CaO.Al_2O_3.SiO_2$ were chosen, while the second case was represented by the system $2CaO.Al_2O_3.SiO_2$–$MgO.Al_2O_3$–$CaO.Al_2O_3.2SiO_2$. The experimentally determined phase diagrams were taken in [25]. The temperature dependence of the enthalpy of fusion was neglected and it was assumed that one half of the Al^{III} atoms are in tetrahedral coordination and the other half are in higher coordination and behave as network modifier. The comparison of the calculated and experimentally determined phase diagrams is shown in Figs. 26–33. All concentration data are in mass % and temperature is given in °C.

The calculated and experimentally determined [25] phase diagrams of the system $CaO.MgO.2SiO_2$–$MnO.SiO_2$–$CaO.Al_2O_3.2SiO_2$ are shown in Figs. 26 and 27, resp. There is a discrepancy in the temperature of fusion of rhodonite, $MnO.SiO_2$, given in [7] and in the experimental phase diagram [25]. The value of 1564 K was used in the calculation.

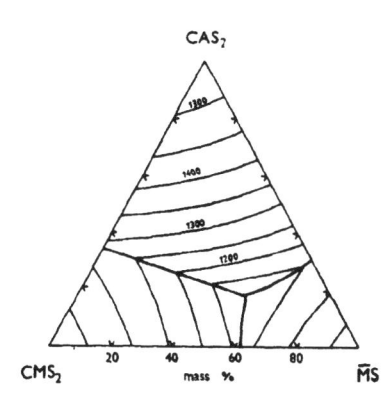

Fig. 26. Calculated phase diagram of the system CMS_2–$\overline{MS}$–CAS_2.

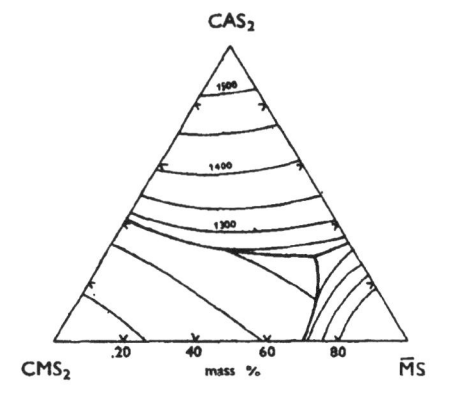

Fig. 27. Experimental phase diagram of the system CMS_2–$\overline{MS}$–CAS_2 [25].

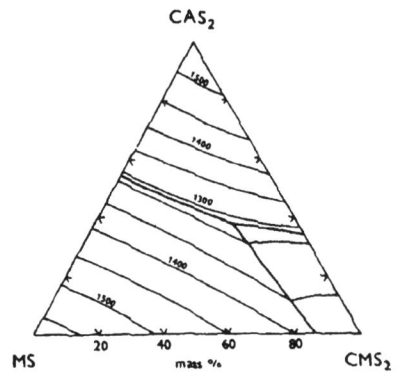

Fig. 28. Calculated phase diagram of the system MS–CMS$_2$–CAS$_2$.

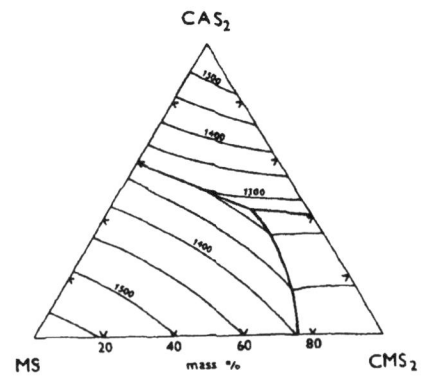

Fig. 29. Experimental phase diagram of the system MS–CMS$_2$–CAS$_2$ [25].

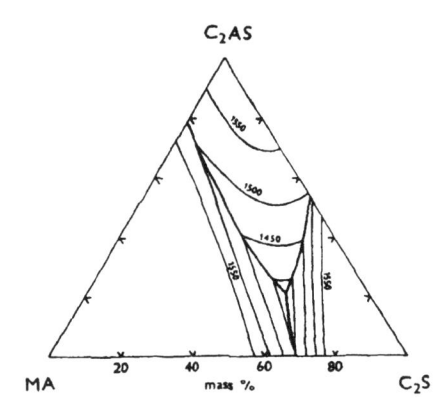

Fig. 30. Calculated phase diagram of the system MA–C$_2$S–C$_2$AS.

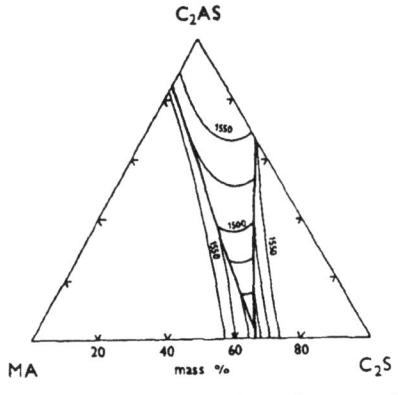

Fig. 31. Experimental phase diagram of the system MA–C$_2$S–C$_2$AS [25].

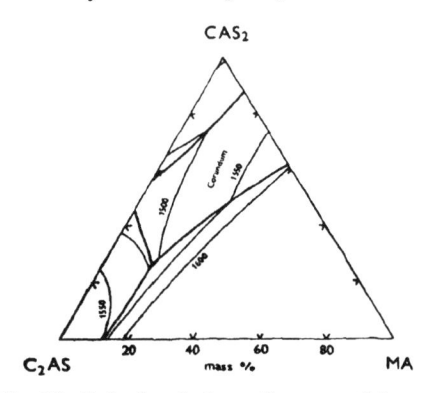

Fig. 32. Calculated phase diagram of the system C$_2$AS–MA–CAS$_2$.

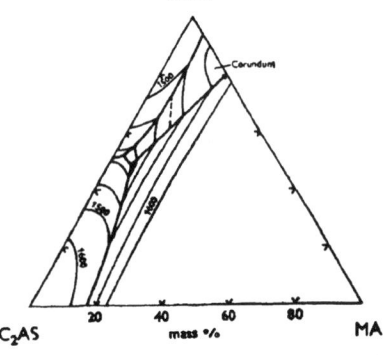

Fig. 33. Experimental phase diagram of the system C$_2$AS–MA–CAS$_2$ [25].

The comparison of the calculated and experimental phase diagrams of the system $MgO.SiO_2–CaO.MgO.2SiO_2–CaO.Al_2O_3.SiO_2$ is shown in Figs. 28 and 29, resp. For the last example of the simple eutectic systems, the $MgO.Al_2O_3–2CaO.SiO_2–2CaO.Al_2O_3.SiO_2$ one, the result of calculation is shown in Fig. 30, while the experimental phase diagram is in Fig. 31.

The systems with four crystallization areas, where the figurative point of the fourth crystalline phase lies outside the pseudo-ternary diagram, are represented by the system $2CaO.Al_2O_3.SiO_2–MgO.Al_2O_3–CaO.Al_2O_3.2SiO_2$. The calculated and experimentally determined phase diagrams of this system are shown in Figs. 32 and 33, respectively. Since the enthalpy of fusion of $MgO.Al_2O_3$ has not been found in the literature, this value was estimated on the basis of thermodynamic analogy.

4.6. SYSTEMS OF ALKALI METAL BORATES

It was shown using the X-ray diffraction techniques that vitreous and molten boron oxide is composed of BO_3 triangles linked by their apexes into an irregular three-dimensional network [46]. Grjotheim and Krogh-Moe [47] pointed out that the structure of boron oxide glass is similar to that of its hexagonal crystalline form and that it consists of two types of irregular BO_4 tetrahedrons. The first one is a hybrid between triangular and tetrahedral configuration with the boron located much closer to the three oxygen atoms. The other is a distorted tetrahedron with various B–O bond length. The mean coordination number of boron in the B_2O_3 melt, 3.1, as determined by Biscoe and Waren [48], does not provide explicit evidence of any change in boron coordination.

In the binary glass-forming systems of alkali metal borates it is possible to observe the change in the trend of a number of physico-chemical properties in the concentration range of approx. 20 mole % of alkali metal oxide. This phenomenon, called in the literature as "boric acid anomaly" are due to the change in the structure of the B_2O_3 melt caused by the alkali metal oxide addition and is related to the ability of boron to change its coordination number.

Krogh-Moe [49] concluded that the change in the coordination number of boron from 3 to 4 takes place up to a content of 33 mole % of alkali metal oxide, which corresponds to the maximum concentration of 50 % of four-coordinated boron. This assumption has been explicitly experimentally confirmed using the measurements of nuclear magnetic resonance carried out by Silver and Bray [50] and by Bray and O'Keefe [51]. These authors found out that within the concentration range of $x = 0–30$ mole % of alkali metal oxide the concentration of four-coordinated boron, N_4, may be quite accurately expressed by the relation

$$N_4 = \frac{x}{100-x} \qquad (30)$$

Equation (30) can be interpreted so that each oxygen atom added will change the coordination of two boron atoms from the triangular to the tetrahedral one. As a result of this there are no non-bridging oxygen atoms present within this concentration range. This fact has also been confirmed by X-ray structural analyses of various crystalline borates [49]. The comparison of results of NMR studies [50, 51] and of the glass forming ability of alkali metal borates indicates that the degree of polymerization in alkali metal borate systems is much higher than in silicate systems.

On the basis of the above mentioned facts one may reasonably expect the analogy between the structures of the crystalline phase and of the liquid one also in the systems of alkali metal borates. This assumption has been verified in the calculation of liquidus curves in alkali metal borate systems using the thermodynamic model of silicate melts adapted according to the characteristic feature of borate melts [52].

The structure of alkali metal borate melt can be imagined as the three-dimensional network of apex-joined BO_3 triangles and BO_4 tetrahedrons where the cations are located in the free spaces. The properties of such melts will obviously depend on the number of boron atoms in the individual coordination, on the amount of alkali metal cations and bridging oxygen atoms, and at higher concentrations of alkali metal oxide also on the amount of non-bridging oxygen atoms. In the region of low content of alkali metal oxide, each added oxygen atom of the alkali metal oxide changes the coordination of two boron atoms from 3 to 4. For instance, in the compound $Na_2O.4B_2O_3$ there are two boron atoms in tetrahedral coordination, six boron atoms in triangular coordination, and all oxygen atoms are the bridging ones.

Such trend is maintained up to approx. 30 mole % of alkali metal oxide. Above this concentration, non-bridging oxygen atoms arise as the result of the reverse transition of some boron atoms from tetrahedral coordination into the triangular one. The number of boron atoms in tetrahedral coordination decreases approaching zero at 70 mole % of alkali metal oxide [51]. It should be, however, noted that the picture on the structure of the melt is only approximate and differences between the liquid and crystalline phase may occur.

The liquidus curves were calculated in the binary systems B_2O_3–$Na_2O.4B_2O_3$, $Na_2O.4B_2O_3$–$Na_2O.2B_2O_3$, $Na_2O.2B_2O_3$–$Na_2O.B_2O_3$, $Li_2O.2B_2O_3$–$Li_2O.B_2O_3$, and $K_2O.2B_2O_3$–$K_2O.B_2O_3$. The necessary values of the enthalpy and temperature of fusion were taken in the literature [7, 22]. The numbers of individual kinds of atoms in the formula unit of pure components are listed in Table 2.

TABLE 2. Numbers of individual kinds of atoms in pure alkali metal borates.

Component*	Me^+	B(3)	B(4)	–O–	–O$^-$
B_2O_3 (B)	–	2	–	3	–
$M_2O.4B_2O_3$ (MB$_4$)	2	6	2	13	–
$M_2O.3B_2O_3$ (MB$_3$)	2	4	2	10	–
$M_2O.2B_2O_3$ (MB$_2$)	2	2	2	6.5	0.5
$M_2O.B_2O_3$ (MB)	2	1	1	2.5	1.5

* M = Li, Na, K

The number of three- and four-coordinated boron atoms in the formula unit was calculated according to eqn. (30). In the case of metaborate the presence of 50 % tetrahedrally coordinated boron atoms was assumed. The calculation of non-bridging oxygen atoms was based on experimental data given in [51]. For diborates the presence of 0.5 atom of non-bridging oxygen and 6.5 atoms of bridging ones was considered. For metaborates 2.5 bridging and 1.5 non-bridging oxygen atoms were accepted in the calculation. The experimental and calculated phase diagrams of the individual systems are shown in Figs. 34–38.

In the calculation of the liquidus curve of the system $Na_2O.4B_2O_3$–$Na_2O.2B_2O_3$ the presence of the incongruently melting compound $Na_2O.3B_2O_3$ has been taken into account. The primary crystallization curve for $Na_2O.3B_2O_3$ was calculated on the basis of its estimated hypothetical melting point of 1045 K and the enthalpy of fusion was estimated on the basis of the thermodynamic analogy.

From the comparison of the experimental and calculated liquidus curves it follows that the thermodynamic model of silicate melts describes satisfactorily the courses of liquidus curves in these complex glass-forming systems. The explanation of the structure of borate melts conforms the experimental data obtained by both the X-ray diffraction and NMR analyses and allows to characterize the thermodynamic behavior of the melt with satisfactory accuracy.

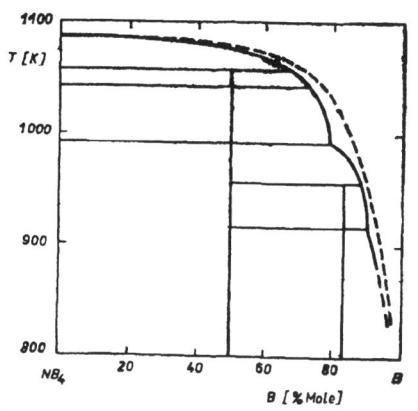

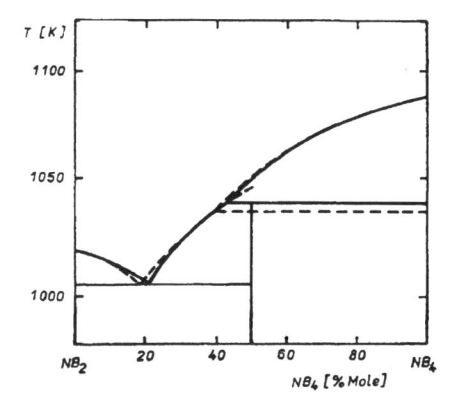

Fig. 34. Phase diagram of the system B_2O_3–$Na_2O.4B_2O_3$.
Dashed line – calculated.

Fig. 35. Phase diagram of the system $Na_2O.4B_2O_3$–$Na_2O.2B_2O_3$.
Dashed line – calculated.

In the system $Na_2O.2B_2O_3$–$Na_2O.B_2O_3$ the calculation has confirmed the course of the $Na_2O.B_2O_3$ liquidus curve determined experimentally by Miman and Bouazis [25] whereas the former results carried out by Morey and Merwin [24] can be considered as incorrect since the limiting relations in the vicinity of $Na_2O.B_2O_3$ are not fulfilled.

Significantly less agreement with experimental liquidus curves can be observed in the systems $Li_2O.2B_2O_3$–$Li_2O.B_2O_3$ and $K_2O.2B_2O_3$–$K_2O.B_2O_3$, where the assumption on the similarity of the structure of solid and liquid phases hold to a less degree. The effect of different polarizing ability if individual alkali metal cations plays evidently a more substantial role.

It can be concluded that the thermodynamic model of silicate melts can be successfully applied also in the systems of alkali metal borates. It can be expected that this model can be applied also in other inorganic glass-forming systems, like germanates, phosphates, etc.

Fig. 36. Phase diagram of the system Na$_2$O.2B$_2$O$_3$–Na$_2$O.B$_2$O$_3$.
Dashed line – calculated.

Fig. 37. Phase diagram of the system Li$_2$O.2B$_2$O$_3$–Li$_2$O.B$_2$O$_3$.
Dashed line – calculated.

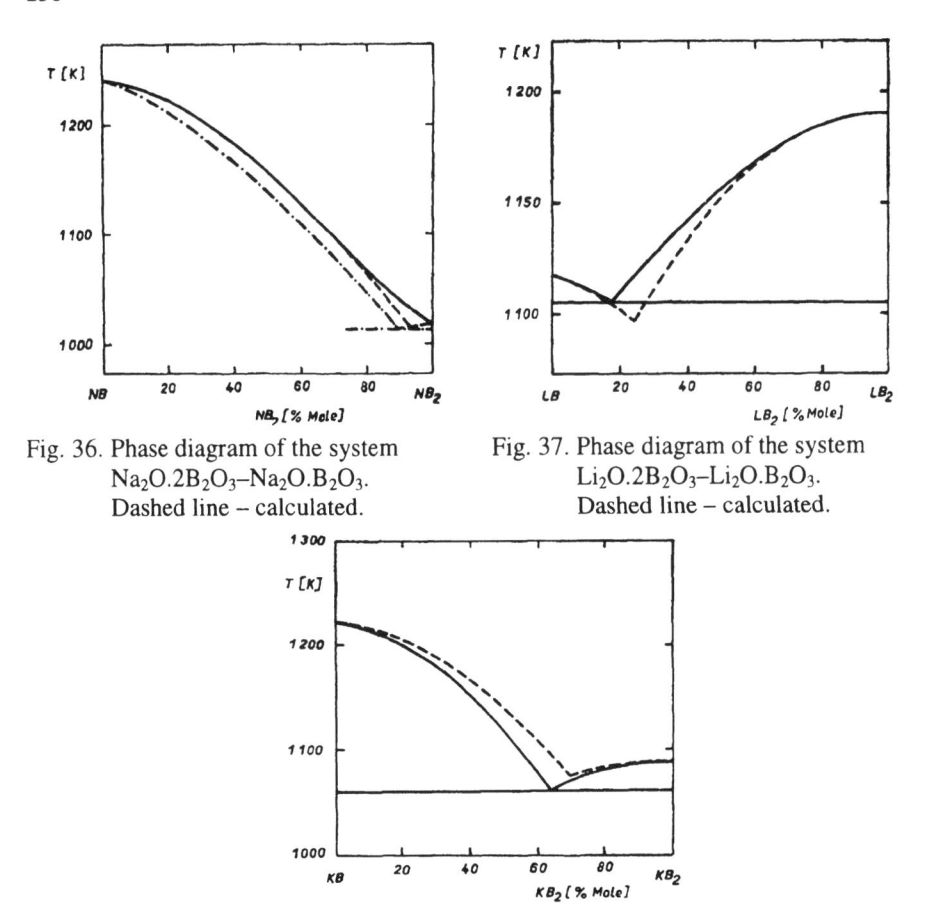

Fig. 38. Phase diagram of the system K$_2$O.2B$_2$O$_3$–K$_2$O.B$_2$O$_3$.
Dashed line – calculated.

5. Conclusions

The thermodynamic model of silicate melts, based on the analogy between the structure of the crystalline phase and that of the melt was used in the calculation of activities and subsequently of the liquidus curves in different types of binary, ternary, and quaternary melts.

From the comparison of the calculated and experimentally determined phase diagram it follows that the chosen approach of inserting characteristic structural aspects into the expression of the activity of components is very suitable for describing phase equilibrium in silicate systems. From the great number of successful applications of the proposed thermodynamic model of silicate melt it follows that it can be used as useful tool for planing experiments and thus the reduction of the number of tedious experimental measurements. In any case, more precise and correct values of the thermodynamic properties are therefore needed.

The thermodynamic model of silicate melts was applied also in the calculation of surface adsorption of SiO_2 in different silicate systems [53, 54], where it was used in the calculation of the activity of SiO_2. The thermodynamic model is consistent also with other properties of silicate melts, e.g. the electrical conductivity. The complex approach was published by Daněk and Ličko in [55].

Acknowledgement: This work was financially supported by the Scientific Grant Agency VEGA of the Ministry of Education of the Slovak Republic and the Slovak Academy of Sciences under No. 2/1033/2001.

6. References

1. Førland, T., ONR Tech. Rep. 63, Contract N6 on 269 Tash Order 8, The Pensylvania State University, University Park Pa (1955).
2. Barin, I. and Knacke, O., *"Thermochemical Properties of Inorganic Substances."* Springer Verlag, Berlin–Heidelberg–New York–Düsseldorf 1973.
3. Førland, T., in: *Fused Salts* (ed. B. R. Sundheim), p. 151. McGraw–Hill Book Company, New York 1964.
4. Rawson, H., *Inorganic Glass-Forming Systems,* p. 4. Academic Press, London, New York 1967.
5. Redlich, O. and Kister, A. T., *Ind. Eng. Chem.* **40**, 345 (1948).
6. Guggenheim, E. A., *Trans. Faraday Soc.* **33**, 151 (1937).
7. Margules, M., *S.-B. Akad. Wiss. Wien, math.-naturwiss. Kl.* **104**, 1243 (1895).
8. van Laar, J. J., *Z. physik. Chem.* **72**, 723 (1910).
9. Stortenbeker, W., *Z. physik. Chem.* **10**, 183 (1892).
10. Temkin, M., *Acta Physicochimica URSSR* **20**, 411 (1945).
11. Bockris, J. O'M., Kitchener, J. A., and Davies, A. E., *Trans. Faraday Soc.* **48**, 526 (1952).
12. Tomlinson, J. W., Heynes, M. S. R., and Bockris, J. O'M., *Trans. Faraday Soc.* **54**, 1822 (1958).
13. Bockris, J. O'M. and Lowe, D. C., *Proc. Roy. Soc. (London)* **A226**, 423 (1954).
14. Bloom, H. and Bockris, J. O'M., *J. Phys. Chem.* **61**, 515 (1957).
15. Bockris, J. O'M. and Kojonen, E., *J. Am. Chem. Soc.* **82**, 4493 (1960).
16. Bockris, J. O'M., Kitchener, J. A., Ignatowicz, S., and Tomlinson, J. W., *Trans. Faraday Soc.* **48**, 72 (1952).
17. Pánek, Z. and Daněk, V., *Silikáty* **21**, 97 (1977).
18. Reiss, H., Katz, J., and Kleppa, O. J., *J. Phys. Chem.* **46**, 144 (1962).
19. Saboungi, M. L. and Blander, M., *J. Am. Ceram. Soc.* **58**, 1 (1975).
20. Bottinga, Y. and Richet, P., *Earth Planet. Sci. Let.* **40**, 382 (1978).
21. Ferrier, A., *Rev. Int. Hautes Tempér. et Réfract.* **8**, 31 (1971).
22. *"Thermodynamic Properties of Inorganic Substances"* (in Russian). Atomizdat, Moscow 1965.
23. Nerád, I., Kosa, L., Mikšíková, E., Šaušová, S., and Adamkovičová, K., Proc. 6th Internat. Conf. Molten Slags, Fluxes and Salts, Stockholm–Helsinki, June 12–16, 2000, Poster Session 3. KTH Stockholm 2000, CD ROM.

24. Levin, E. M., Robbins, C. R., and McMurdie, H. F., *"Phase Diagrams for Ceramists"*. Am. Ceram. Soc., Columbus, Ohio 1964.
25. Levin, E. M., Robbins, C. R., and McMurdie, H. F., *"Phase Diagrams for Ceramists"*. Am. Ceram. Soc., Columbus, Ohio 1969, 1975.
26. Timucin, M. and Morris, A. E., *Met. Trans.* **1**, 3193 (1970).
27. Muan, A. and Osborn, E. F., *Phase Equilibria Among Oxides in Steelmaking*. Addison-Wesley Publishing Co., Reading, Mass., 1965.
28. Lee, Y. E. and Gaskell, D. R., *Met. Trans.* **5**, 853 (1974).
29. Bates, T., in *"Modern Aspects of the Vitreous State"*, (Mackenzie, J. D., Editor.) p. 195. London, 1962.
30. Henderson, J., *Trans. Met. Soc. AIME* **230**, 501 (1964).
31. Daněk, V., *Chem. Papers* **38**, 379–388 (1984).
32. Mori, K. and Suzuki, K., *Trans. Iron Steel Inst. Jap.* **8**, 382 (1968).
33. Engel, H.-J. and Vygen, P., *Ber. Bunsenges. Phys. Chem.* **72**, 5 (1968).
34. Daněk, V., *Silikáty* **30**, 97–102 (1986).
35. Liška, M. and Daněk, V., *Ceramics–Silikáty* **34**, 215–228 (1990).
36. Yarker, C. A., Johnson, P. A. V., and Wright, A. C., *J. Non-Cryst. Sol.* **79**, 117 (1986).
37. Abdrashitova, E. I., *J. Non-Cryst. Sol.* **38**, 75 (1980).
38. Schneider, E., Wong, J., and Thomas, J. M., *J. Non-Cryst. Solids* **136**, 1 (1991).
39. Mysen, B., and Neuville, D., *Geochim. Cosmochim. Acta* **59**, 325 (1995).
40. Liška, M., Hulínová, H., Šimurka, P., and Antalík, J., *Ceramics-Silikaty* **39**, 20 (1995).
41. Lange, R., and Navrotsky, A., *Geochim. Cosmochim. Acta* **57**, 3001 (1993).
42. Nerád, I. and Daněk, V., Thermodynamic analysis of pseudobinary subsystems of the system $CaO-TiO_2-SiO_2$. *Chem. Papers*, in press.
43. Daněk, V. and Nerád, I., Phase diagram and structure of melts of the system $CaO-TiO_2-SiO_2$. *Chem. Papers*, in press.
44. De Vries, R. C., Roy, R., and Osborn, E. F. J., *J. Amer. Ceram. Soc.* **38**, 158 (1955).
45. Pánek, Z., Kanclíř, E., Staroň, J., Kozlovský, M., and Palčo S., *Silikáty* **20**, 13 (1976).
46. Zarzycki, J., *Proc. 4th Int. Congress Glass, Paris* **6**, 323 (1956).
47. Grjotheim, K. and Krogh-Moe, J., *K. norske Vidensk. Selsk. Forh.* **27**, No. 18, 94 (1954).
48. Biscoe, J. and Waren, B. E., *J. Am. Ceram. Soc.* **21**, 287 (1938).
49. Krogh-Moe, J., *Arkiv Kemi* **12**, 475 (1958); *Phys. Chem. Glasses* **1**, 26 (1960).
50. Silver, A. H. and Bray, P. J., *J. Chem. Phys.* **29**, 984 (1958).
51. Bray, P. J. and O'Keefe, J. G., *Phys. Chem. Glasses* **4**, 37 (1963).
52. Daněk, V. and Pánek, Z., *Silikáty* **23**, 1–10 (1979).
53. Daněk, V. and Ličko, T., *Chem. Papers* **36**, 179–184 (1982).
54. Daněk, V., Ličko, T., and Pánek, Z., *Chem. Papers* **39**, 459–465 (1985).
55. Daněk, V. and Ličko, T., Thermodynamic model and physico-chemical properties of silicate melts. *Chem. Geology* **96**, 439–447 (1992).

DATA MINING AND MULTIVARIATE ANALYSIS IN MATERIALS SCIENCE:

Informatics Strategies for Materials Databases

> Krishna Rajan, A. Rajagopalan and C. Suh
> *Department of Materials Science and Engineering*
> *Rensselaer Polytechnic Institute*
> *Troy, NY 12180-3590 USA*

Abstract

Databases in materials science applications tend to be phenomenological in nature. In other words, they are built around a taxonomy of specific classes of properties and materials characteristics. In order for databases to serve as more than only a 'search and retrieve" infrastructure, and more for a tool for "knowledge discovery", data bases need to have functional capabilities. The recent advances in genomics and proteomics for instance provide a good example of the development of such "functional" databases. A first step to achieve this is to develop descriptors of materials properties that can be sorted and classified using appropriate data mining algorithms. In this paper we provide some examples of the use of some well established statistical tools to "prepare" such data especially when there is a multi-dimensional component associated with structure – chemistry-property relationships.

1. Introduction

The development, improvement or adaptation of materials is usually built on prior data. This is based on knowledge derived through experience, experiments and theory. However for each new application and engineering demand, the testing and modeling process has to be repeated or modified to account for the new conditions, and the knowledge building process repeats in an iterative manner. As many of the paradigms in materials science and engineering are in fact often derived through empirical correlations between observed behavior; theoretical frameworks of these observations have often been established or developed afterwards in order to achieve an explanation of existing data. The advent of high speed computing has opened the venue of simulation studies, which has effectively provided another source of data generation besides traditional experiments. However, this paradigm of research, namely the process of sequentially building on experiment / theory and simulation to iterate a search for a solution to a particular engineering problem has not changed. In this paper we are outline some of the requirements for the establishment of a "toolkit" which significantly challenges this paradigm by developing a methodology / information infrastructure which builds on the use of data to generate knowledge.

Our strategy is to build upon using prior data and the knowledge derived from it not only to retrieve information which we feel we need, but to use that data in a way that permits one to associate and anticipate structure-property correlations before conducting further experiments (either real or "virtual"). In this manner we can develop a methodology which will take advantage of prior knowledge as 'stored" in massive databases to search for potential links in structure-processing-property relationships. Integrating information in different types of databases using the appropriate data mining techniques permits one to link information across length and time scales in an useful manner.

The study of molten salts provides one of the best examples of the challenges in using databases. While for example vast amounts of property data exists, the question is how do we use it and what can we potentially gain in terms of new knowledge using prior data? The classic work of Janz and co-workers [1] for example provides an important testbed on which the complexity of data mining can be built. One of the important features of the molten salt database is that it covers a wide spectrum of chemistries and properties of a vast array of molten salts. Such a compilation of chemistry – property – structure information in one class of systems is rarely found in materials science. The sorting of that information to seek potentially new relationships is the purpose of data mining. The main challenge however is that there are numerous correlations across length and time scales to seek. The multi-variate nature of structure-property relationships is the major stumbling block in using data mining in the materials sciences.

The development of an informatics strategy to seek such information is complex (see Figure 1). In this paper we will outline some of the details involved in just one portion of this process, namely how we can find ways to seek patterns in a complex multi-dimensional or multivariate problem.

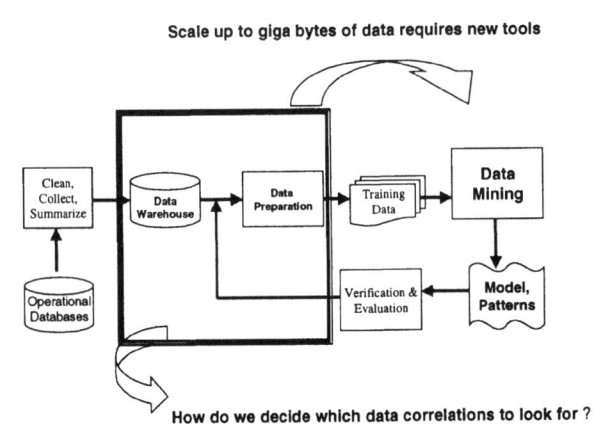

Figure 1: Schematic of infrastructure needed to apply data mining techniques

2. Reducing the Dimensionality of Multi-variate Data Sets

The challenge in linking length scales in materials science is that we do not necessarily have theories linking every aspect of materials characteristics in a unified manner. Much of materials design is based on phenomenological paradigms which provide guidelines for materials selection. An challenge that we are addressing in this proposal is how to integrate data at different length scales in such a way as to detect patterns of behavior (using statistical techniques) which could lead to (or suggest) new data or information (validated by experiments and theoretical formulations). The data preparation sequence outlined above has built into it a further level of detail that needs to be considered (Figure 2). As shown below there are numerous tools needed depending upon the nature of the data set that needs to be part of our proposed information infrastructure for a molten salt database.

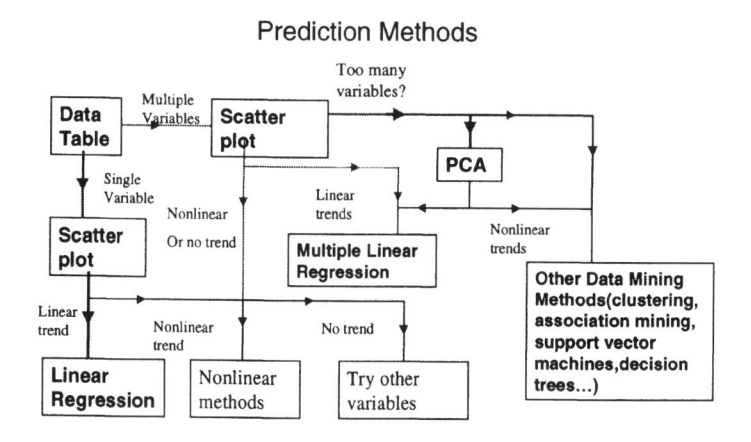

Figure 2: Data processing tools for different types of datasets

For the purposes of this discussion, we shall focus on situations where there exist a vast array of variables associated with single set of compounds or chemistry. A statistical evaluation to search for each descriptor is computationally expensive and most possibly ineffective. Principal Component Analysis (PCA) is a technique to reduce the information dimensionality that is often needed from the vast arrays of data as obtained from a combinatorial experiment, in a way so that there is minimal loss of information (Figure 3)

It relies on the fact that most of the descriptors are intercorrelated and these correlations in some instances are high [2]. From a set of N correlated descriptors, we can derive a set of N uncorrelated descriptors (the principal components). Each principal component (PC) is a suitable linear combination of all the original descriptors.

The first principal component accounts for the maximum variance (eigenvalue) in the original dataset. The second principal component is orthogonal (uncorrelated) to the first and accounts for most of the remaining variance. Thus the mth PC is orthogonal to all others and has the mth largest variance in the set of PCs. Once the N PCs have been calculated using eigenvalue/ eigenvector matrix operations, only PCs with variances above a critical level are retained. The M-dimensional principal component space has retained most of the information from the initial N-dimensional descriptor space, by projecting it into orthogonal axes of high variance. The complex tasks of prediction or classification is made easier in this compressed space.

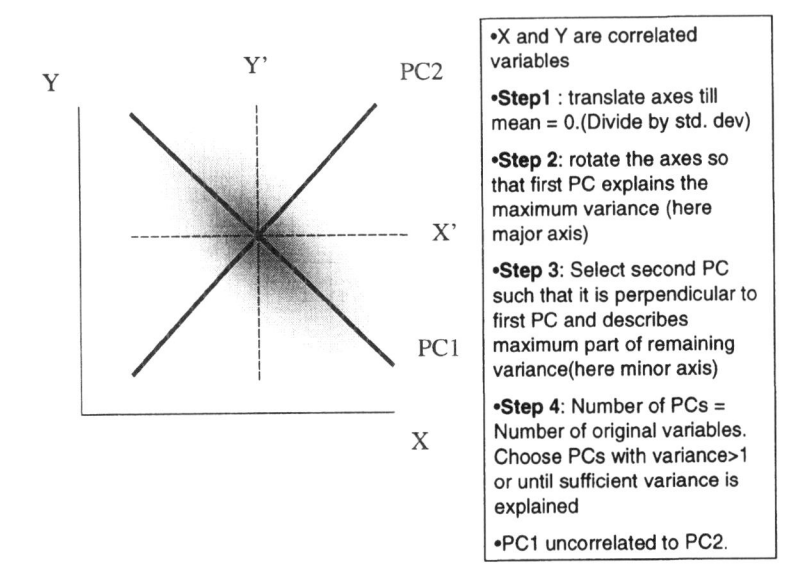

- X and Y are correlated variables
- **Step1** : translate axes till mean = 0.(Divide by std. dev)
- **Step 2**: rotate the axes so that first PC explains the maximum variance (here major axis)
- **Step 3**: Select second PC such that it is perpendicular to first PC and describes maximum part of remaining variance(here minor axis)
- **Step 4**: Number of PCs = Number of original variables. Choose PCs with variance>1 or until sufficient variance is explained
- PC1 uncorrelated to PC2.

Figure 3 : Schematic of PCA process

3. Application of PCA

As an example of the value of such data mining tools, we have conducted such an approach to reduce the dimensionality of the multivariate problem in developing descriptors for high temperature compound superconductors. We explored a wide array of descriptors based on "legacy" data of other inorganics possessing high temperature superconducting behavior including:
- Space group symmetry
- Stoichiometries of materials with iso-structural characteristics
- Diffraction data
- Stacking fault energy measurements and calculations

- Structure and property databases of other binary and multi-component borides, oxides and other chemistries
- A variety of classical "structure maps" including: for example "nearest – neighbor" coordination maps which map out possible bonding geometries

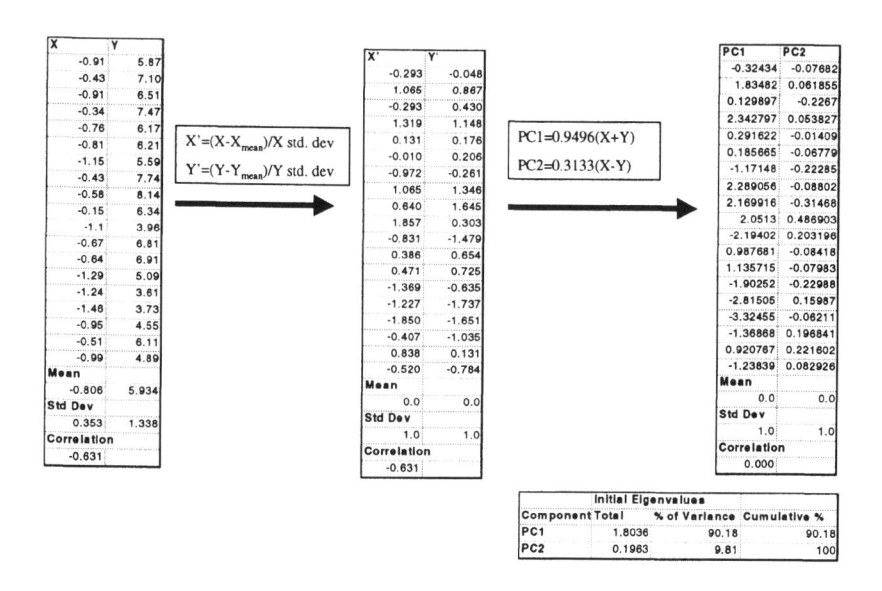

Figure 4: Illustration of eigenvalue calculations for PCA analysis of superconducting data

This multivariate information was analysed in terms of the PCA methodologies described above. An illustrative plot showing how correlations associated with numerous variables can be reduced to three dimensions which aids in visualization is shown below: The complexity of this information can be further reduced by data reduction techniques to explore and discover data trends that would never have been apparent by simply looking at individual data correlations. This specific example provides a demonstration of how by working with databases and integrating appropriate data mining tools, we can anticipate or provide potential predictive guidelines in identifying new materials.

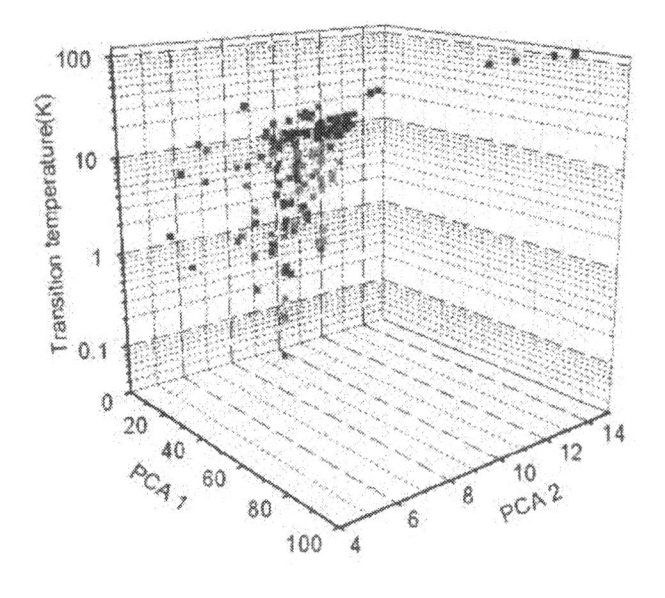

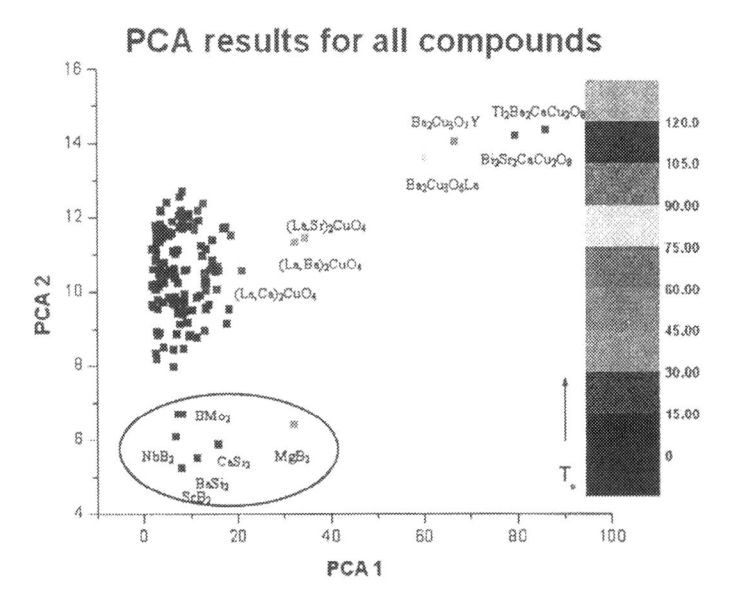

Figure 5: PCA plots in 3 and 2 dimensions of multivariate data showing MgB_2 clustered among other superconducting AB_2 stoichiometry type compounds

4. Conclusions

While the PCA techniques help to reduce the dimensionality of the multivariate problem, we still need to characterize the variability of the descriptors. It is proposed to so-called "Shannon or maximum entropy" methods. This process will help to identify the key descriptors and "rank" their influence on properties. Such an approach has been used in characterizing the variability of molecular descriptors for instance in combinatorial drug delivery problems . A central advantage of the maximum entropy technique is its ability to model overlapping data characteristics without increasing the number of parameters or fragmenting the initial data from the combinatorial experiments. Using maximum entropy techniques, it is possible to model a probability distribution on 2^n elements using only n free parameters. It is a powerful and efficient method for inducing statistical models from data [3]

A key aspect of developing an "informatics" approach to materials discovery is the need to establish the critical array of descriptors of materials attributes that may be subsequently input into a database. Having physically meaningful descriptors is key if one is to develop and search for associations between apparently disparate or disjointed datasets. Our initial work on this rapidly evolving field, has already provided potential descriptors which might indicate not only what materials may be worthwhile to investigate further, but also strategies of how new chemistries must influence the structure if superconductivity is to be promoted. This in turn of course provides possible insights into the mechanisms that govern high temperature superconductivity in these new classes of materials.

It is worthwhile to note that that the need for examining and searching for appropriate descriptors to build on prior materials discovery is getting further attention in the literature. For instance:
- Lacorre et.al [4] have suggested a stereochemical model for designing oxide ion conductors. Their approach involves the use of developing generic descriptors involving substitution of elements based on their oxidation state.
- MacGlashan et.al. [5] have proposed that the structures based on the structure $LiAsF_6$ may provide a basis of developing appropriate compositions / molecular structures of new conducting polymers. As in the case of ceramic systems, the incorporation of appropriate elemental species which alter the local oxidation state within a given coordination complex is suggested as a means of developing strategies for synthesizing new materials / molecular arrangements based on ionic conduction in polymeric systems. The molecular structure was expressed not as crystallographic coordinates but as internal stereochemical descriptors: bond lengths, bond angles and torsion angles. The characterization of structure provided clues as to potential properties and more specifically mechanisms (eg. Ionic conduction). This in turn establishes guidelines of designing new molecular chemistries for polymeric electrolytes.

Similar requirements for descriptors have been proposed in the designing of new semiconductors tailor optoelectronic properties, but none of these studies have built upon integrating these descriptors to data mining tools and combinatorial experiments to further develop the potentially new materials [6].

For the molten salt community, we have the challenge of a vast multivariate problem. As shown in the table below, for any given chemistry there are is a vast array of variables that exist to describe some type of attribute to that chemistry. The type of information infrastructure needed for this class of materials requires a complex array of tools. At this stage we suggest that the data preparation techniques such as PCA analysis could provide an invaluable first step in the virtual design of molten salts.

Table 1: Descriptors for molten salts

- For a specific chemistry....
  - Structural data
    - Atomic/ionic size
    - Coordination
    - Cation/cation & anion-anion distances
- Viscosity
- Heat capacity
- Conductivity
- Surface tension
- Refractive index
- Density
- Compressibility
- Zeta potential / ZPC
- Phase equilibria
- Cryoscopic behavior
- Heat conductance
- Solubility
- Melting point
- Electromigration
- Transport numbers
- Raman spectra
- Neutron / X-ray scattering
- NMR & EPR data

↓

References

1. Janz,G.J. (1967) *Molten Salts Handbook* Academic Press NY
2. Preisendorfer,R.W. (1988) *Principal Component Analysis in Meterology and Oceanography* Elsevier, Amsterdam
3. Godden, J., Stahura,F.L and Bajorath, (2000) Variability of Molecular Descriptors in Compound Databases Revealed by Shannon Entropy Calculations, *J. Chem. Inf. Comput. Sci.* **40** 796-800
4. .Lacorre,P., Goutenoire, F., Bohnke, O, Retoux, R and Laligant,Y : Designing fast ion conductors based on $La_2Mo_2O_9$ *Nature*, **404** 856-858 (2000)
5. MacGlashan, G.S., .Andreev,Y.G. and Bruce,P.R. : Structure of the polymer electrolyte poly(ethylene oxide)$_6$: $LiAsF_6$ *Nature* **398** 792-794 (1999)
6. Wang,T., Moll,N. Cho,K. and Joannopoulos,J.D. Deliberately designed materials for optoelectronic applications *Physical Review Letters* **82** 3304-3307 (1999)

PYROCHEMISTRY IN NUCLEAR INDUSTRY

Tadashi INOUE and Yoshiharu SAKAMURA
Central Research Institute of Electric Power Industry (CRIEPI)
2-11-1, Iwato-Kita, Komae-shi, Tokyo 201-8511, Japan
TEL: --81-3-3480-2111,   FAX: --81-3-3480-7956
E-mail: *inouet@criepi.denken.or.jp*

Abstract

The nuclear fission is a main device of energy production to sustain the development over the 21st century in the world. The recycling of fissile materials must be a key issue for keeping the use of nuclear energy. Pyrochemistry/pyrometallugy is one of potential devices for future nuclear fuel cycle. Not only economic advantage but also environmental safety and strong resistance for proliferation are required for the fuel cycle. In order to satisfy the requirement, actinides recycling used pyrochemistry with molten salt of LiCl-KCl, which could be implemented to light water reactor and fast breeder reactor fuel cycles, has been an issue of current interest. Electrorefining for U and Pu separation and reductive-extraction for TRU separation have been studied over decade. Reduction process of oxides in molten LiCl has been also examined by use of actinide oxides. The application of this technology on reprocessing of oxide and metal spent fuels and on separation of actinides in high level liquid wastes should improve the present nuclear fuel cycle system.

After looking the overview of pyrochemical programs in the nuclear field, the activity in CRIEPI is reported. The brief summary is follows,
- The thermodynamic properties of actinides, U, Np, Pu, and Am, and lanthanides, Y, La, Ce, Nd and Gd in LiCl/KCl have been evaluated by measuring electrochemical potential.
- The distribution coefficients of actinides and lanthanides have been measured in LiCl-KCl/Cd and LiCl-KCl/Bi systems.
- For technological implementation of pyrochemistry, electrorefining and reductive extraction to recover actinides have been basically developed by lots of experiments used actinide materials.
- The fuel dissolution into molten salt and the uranium recovery on solid cathode for electrorefining have been demonstrated by engineering scale facility with use of spent fuels in Argonne National Laboratory and by a kg scale of installation with use of uranium in CRIEPI.
- Concerning on actinide separation from high-level liquid waste, the conversion of nitrate solution to chlorides through oxides has been also established through uranium tests.
- It is confirmed that more than 99% of TRU nuclides can be recovered by

reductive extraction from simulated materials of high level liquid waste by TRU tests.
- Through these studies, the process flow sheets for reprocessing of metal and oxide fuels and for partitioning of TRU separation have been established.

The subjects to be emphasized for further development are classified into three categories, that is, process development (demonstration), technology for engineering development, and supplemental technology.

1. Introduction

Nuclear energy is one of the most potential sources for sustainable development in the world. The remarkable increase of energy demand is, especially in Asia, prospected by contribution of developing nations. Some of the nations are much concerned with nuclear energy as well as another resources. While, in Japan, the primary energy resources more than 80% is imported from overseas countries, in which more than 99% of oil is shipped mostly from the Middle East. Thus, the nuclear energy, which takes a role of basic load of electricity, has a non-substantial role as semi-domestic resources, depending not so much amount on the supply from the overseas. Another large contribution is expected on nuclear energy from the point of mitigation of global warming, because of much less emission of carbon dioxides, as expected from figure1[1].

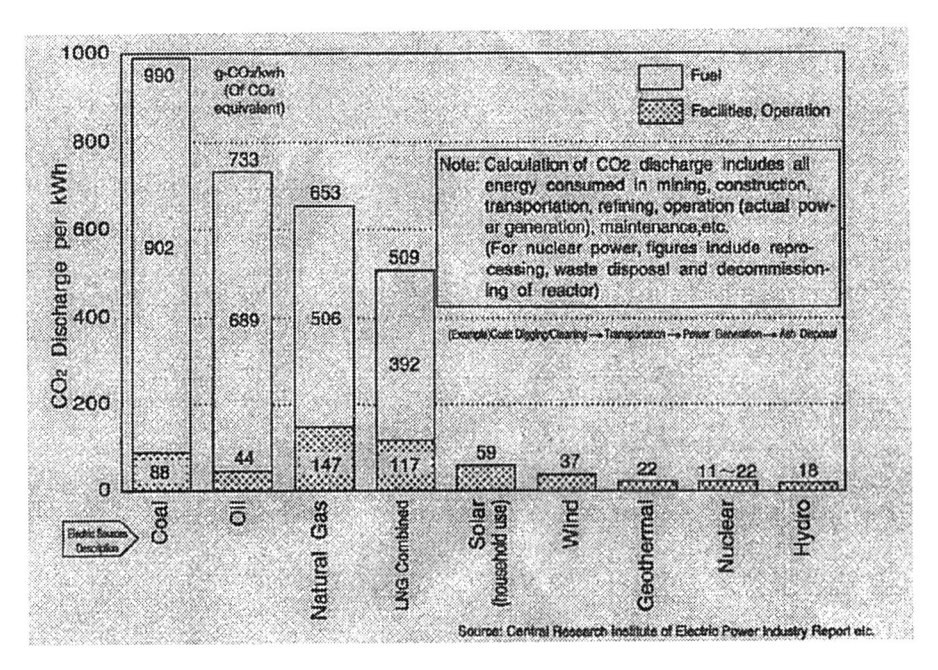

Figure 1. Life cycle CO_2 emissions of power generation devices

To say the least, the nuclear fission is a main device of energy production to sustain over 21st century. For keeping the use of nuclear energy, the recycling of fissile materials must be a key issue, though the transmutation of plutonium and minor actinides, MA: neptunium (Np), americium (Am) and curium (Cm) is, currently, more directed in some of front running nations in the nuclear world, because of the present surplus of Pu disposed of military use.

On the other hand, the future nuclear system has to be much concerned on the requirement of system safety, lower impact on environment and non-proliferation as well as economic advantage. A use of pyrochemistry is attractive to attain such potentials for establishing improved nuclear system. The pyrochemistry is on the programs in nations, such as United State, Europe, Japan, Russia and others. Among them, it is applied on the transmutation system used an accelerator-driven system in United State and Europe, and on the breeding system of fissile material in Japan, Russia and China.

CRIEPI has been proposing the "Actinide recycling by pyro-process with fast breeder reactor" in order to meet such requirement. The system is integrated into a metal fueled FBR cycle by means of pyrochemical reprocessing with reduction technology☐of U-, Pu- and MA- (Np, Am, Cm) oxides to metallic form, in which the pyrochemical separation of TRU from HLLW for burning in FBR (Partitioning and Transmutation) is accommodated. Partitioning and transmutation has another objectives to expand the public acceptance for high level liquid waste management. In proposed nuclear fuel cycle, recovered TRU nuclides will be mixed with U-Pu-Zr to make fresh fuels for FBR and will take a role of energy production. Consequently, only the trace amount of TRU will be disposed of into geologic formation [2].

2. Overview of activities on pyrochemistry in nuclear industry

The pyrochemistry is a potential device to separate actinides for recycling in the accelerator-driven system (ADS) and fast breeder reactor cycle, and to separate uranium for spent fuel treatment. The ADS in Europe and, in turn, accelerator transmutation of waste program (ATW) in United State, furnish the electrorefining in molten salt for separating transuranium elements from spent targets/fuels. The schematic process with electrorefining of ATW proposed by Los Alamos National Laboratory is shown in figure 2[3]. For treating oxides, uranium existing as the most quantity in fuels is removed by aqueous method prior to applying pyrochemistry. The transuranium elements are separated by electrorefining in chloride salt for introducing to ATW fuel cycle. The possibility of transmutation by ADS is also explored in France, Italy, Spain and others in Europe[4].

Another activity currently demonstrated the feasibility by a pilot scale is the program by use of the Fuel Conditioning Facility (FCF) in Argonne National Laboratory (ANL) for purposing uranium separation from spent metal fuels to minimize the volume of TRU waste at disposal in geologic formation. The 100 driver fuels and 25 blanket assemblies have been already treated successfully by electrorefining with solid cathode in a LiCl-KCl bath and, totally, ca.1.6 t of uranium has been separated[5].

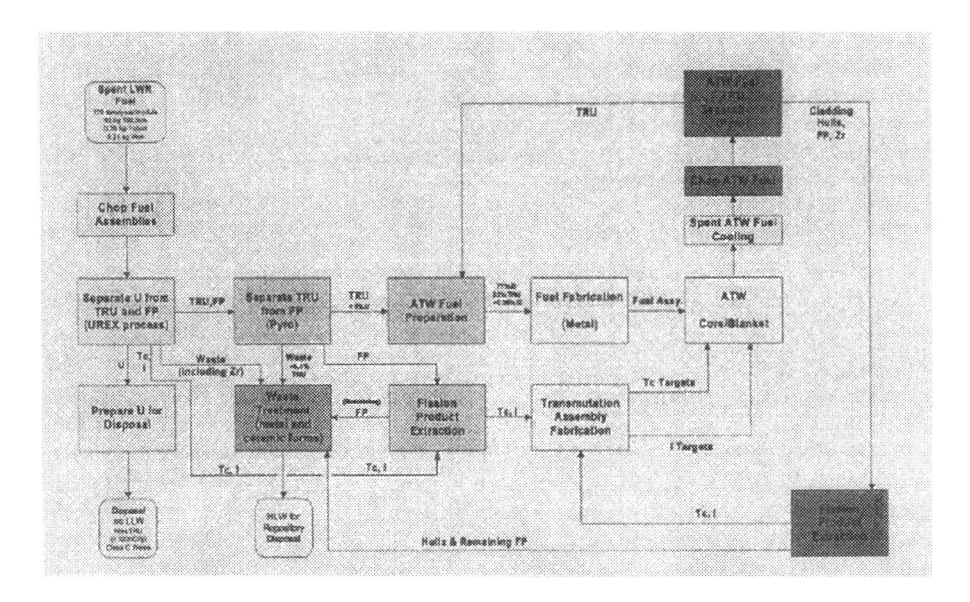

Figure 2. Process diagram of ATW system proposed by Los Alamos National Laboratory

The Pyrometallurgical Processing Research Programme, called "PYROREP", in the 5th framework program of the European Commission has started to evaluate the technical feasibility of potential devices of pyrochemistry, where CRIEPI is invited. The program covers electrorefining in chloride salt, fluoride volatilization, thorium cycle and related subjects on materials. The participants from EU are Commissariat a l'Energie Atomique (France) acting as a project coordinator, Centro de Investigaciones Energeticas Medio Ambientales Y Technologicas (Spain), Ente per le Nuove technologie, l'Energia E l'Ambiente (Italy), European Commission, Ustav Jaderneho Vyzkumu Rez a.s Nuclear Research Institute Rez plc (Czech Republic) and British Nuclear Fuels plc (GB).

The Research Institute of Advanced Reactor (RIAR) in Russia has been developing pyroelectrochemical reprocessing based on the electrorefining of oxides by combing vibro-pack fuel fabrication since 1960's [6]. Pyroelectrochemical reprocessing can be carried out by two ways: separation of UO_2 and PuO_2 during process and joint codeposition of $UPuO_2$. Figure 3 shows the flow diagram of the first pyroelectrochemical reprocessing option. After decladding of fuel pins, granulated or powdered oxides are chlorinated by pouring chlorine gas in order to dissolve uranium and plutonium as UO_2^{2+} and Pu^{4+} in molten NaCl-KCl. In the second stage, uranium oxychloride is electrolyzed to deposit UO_2 at cathode by using a potential not to codeposit PuO_2. In the third stage, crystallized PuO_2 is precipitated from the salt by flowing mixed gas of oxygen and chlorine. More than 99% of plutonium in fuels can be precipitated in this process. After recovering plutonium, UO_2 is deposited at the cathode by supplementary electrorefining.

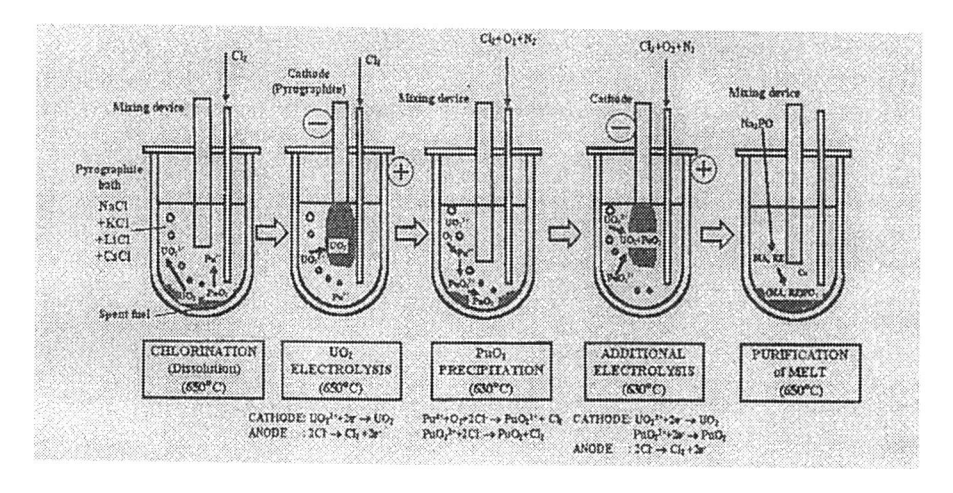

Figure 3. Process diagram of pyro-electrochemical reprocessing of oxide fuel in RIAR

In Japan, two kinds of activities for advanced nuclear system toward 21st century are put into national program. The OMEGA, a long-term program for research and development on partitioning and transmutation technology, started since 1988[7], in which Japan Atomic Energy Research Institute (JAERI), Japan Nuclear Fuel Cycle Development Cooperation (JNC) and CRIEPI are involved The another is the Feasibility Study by utilities, JNC, CRIEPI, JAERI and universities for making an advanced system of FBR and fuel cycle with potentials of effective utilization of fissile material, environmental safety and strong resistance of non-proliferation as well as economic advantage. In the framework of this study, CRIEPI is now installing a glove box system for pyrochemistry with the collaboration of JNC. JAERI, contributing to the OMEGA program and the Feasibility Study from point of transmutation of transuranium elements works on the nitride fuel cycle connected with accelerator-driven system. Figure 4 shows the schematic flow of the system[6], which consists of the first stratum of power reactor cycle and the second of transmutation cycle with nitride fuel.

The spent fuel is treated by means of electrorefining in LiCl-KCl for recycling actinides. Spent nitride target positioned in the anode basket are electrolyzed by producing nitrogen gas and a liquid cadmium cathode collects transuranium elements, which is converted to nitrides just by pouring nitrogen gas in cadmium. A Collaboration with CRIEPI has been continuing on a part of electrorefining study.

The actinide-recycle program has initiated at early 1980's in CRIEPI. The pyrochemistry has been studying for applying on separation of actinides to establish the advanced fuel cycle connected with metal fueled FBR. The participation to IFR (Integral Fast reactor) project of US DOE (ANL) from 1989 to 1995 [8] and the collaboration with domestic organizations, representing by universities and JAERI, and overseas institutes, Missouri University/Rockwell International in U.S.A and the Institute of Transuranium Elements in EU and AEA Technology in UK provide the effective expertise for data acquisition of actinides. The details of activity are described afterwards.

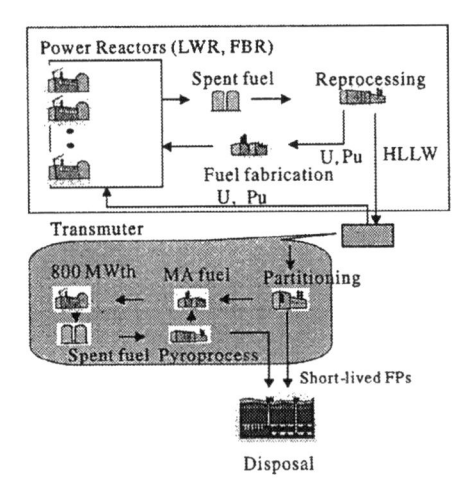

Figure 4. Process flow of transmutation cycle by the double strata system in JAERI.

3. Characteristics of the system with pyrochemical process and metal fueled FBR

3.1 PROCESS DESCRIPTION

The integral system of LWR and FBR with metal fuel for actinide-recycle by pyrometallurgy proposed by CRIEPI is shown in figure 5. The pyrometallurgical reprocessing consists of electrorefining with solid cathode for uranium recovery and liquid cadmium cathode for actinides collection in chloride salt of LiCl-KCl, reductive-extraction for recovering transuranium elements in LiCl-KCl/Cd and LiCl-KCL/Bi by use of lithium as reductant and fuel fabrication by casting after removing salt and cadmium by distillation for actinides products. The lithium reduction is applied to reduce actinide oxides by forming Li_2O, which has to be electrolyzed for recycling to reduce the process waste production, for oxide fuel treatment.

The conversion of HLLW to chloride is required for adopting the reductive-extraction to recover transuranium elements. To facilitate chlorination, HLLW has to be denitrated to obtain the oxides. The chlorination of oxides occurs in a LiCl-KCl bath with a same kind of chloride bath operating at a lower temperature for collecting the species with high evaporation. Metal fuels of U-Pu-Zr with minor actinides are prepared for FBR fuels, in which some amount of lanthanides separated together with actinides is contained as well, because of no large difference of free energy of formation of chlorides from actinides.

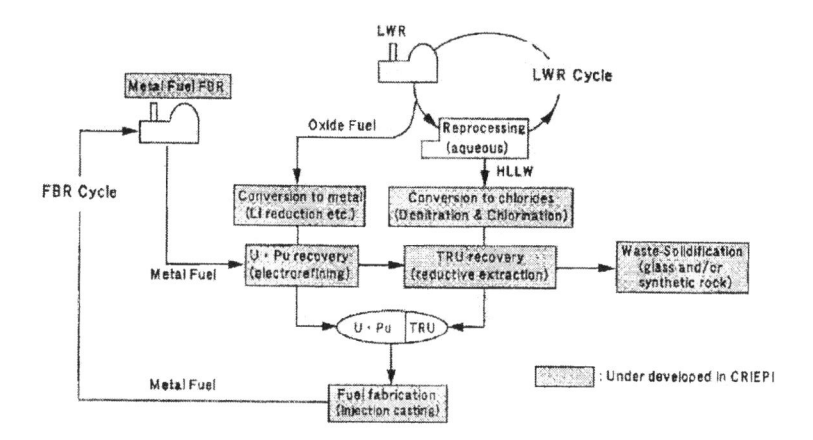

Figure 5. Integrated system of LWR and FBR cycles with pyro-metallurgical process for actinide-recycle proposed by CRIEPI

3.2 CHARACTERISTICS OF SYSTEM AND PROCESS TECHNOLOGY

The pyrometallurgy used in this system can be applied on any type and any composition of nuclear fuels, for instance, not only on metal fuel but also on fuel, like MOX, and nitride fuel, and is also applied on high burn-up fuels. It is rather preferable to apply on the short cooling fuels with high decay heat and radiation. The application of pyrometallurgy foresees that plutonium and neptunium cannot be separated with pure form in principle, which contributes to a high resistance for nuclear proliferation. In addition, a high potential on environmental safety is expected, because a trace amount of actinides goes into vitrification for disposal into geologic formation. The pyro-process facility, which can be compactly designed, is well fitted to the collocation with reactor(s), which fascinates to reduce a large number of transportation of nuclear materials. The process for reprocessing and recovering minor actinides is expected to be simple and consist of a compacted facility with relatively small size of equipments. Small amount of waste will be generated constantly under the normal operation, because neither organic solvent nor nitric acid solution is used. A large margin for criticality will be relatively expected, because of no aqueous phase exists. The proposed TRU recovery process can be applicable not only to high level liquid waste but also to the slurry, undissolved residue and solvent washing reagent in wastes coming from aqueous reprocessing.

The metal fuel has a high potential for FBR, because of high density of fissile material and high thermal conductivity, and the metal fuel core can be expected to be inherent safety. The reprocessing of metal fuels is well fitted with pyrometallurgical method. On the other hand, we have to note that, compared with Purex process and oxide fuel, less experiences has accumulated on both subjects. Large efforts shall be needed for commercial realization.

4. Description of the achievement

4.1 MEASUREMENT OF THERMODYNAMIC PROPERTIES

In order to evaluate the separation of actinides from other elements, electrochemical potential of chloride, that is free energy of formation of chlorides, is required. Especially precise measurement has to be done for actinides and lanthanides, because of the chemical similarity of element among those groups. Figure 6 shows the galvanic cells for measurements of equilibrium electrochemical potentials in a cell of $M/MCl_n,LiCl-KCl//AgCl,LiCl-KCl/Ag$. The potentials of lanthanides and actinides were measured as a function of molar fraction of MCl_n in LiCl-KCl at 450 C. Applying

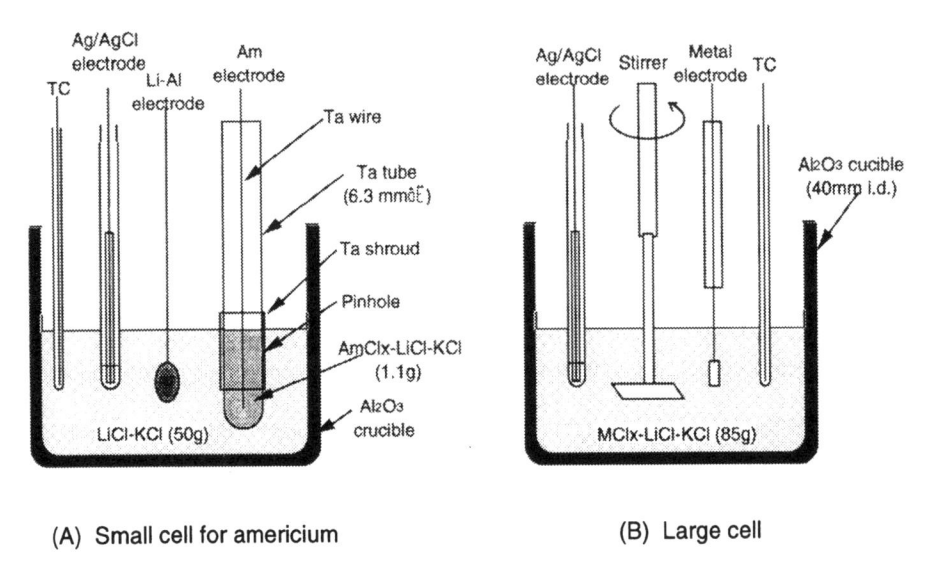

(A) Small cell for americium

(B) Large cell

Figure 6. Galvanic cells for measurements of equilibrium electrochemical potentials in a cell of $M/MCl_n,LiCl-KCl//AgCl,LiCl-KCl/Ag$

the Nernst equation on measured values gives the standard potentials of actinides and lanthanides, which is appeared in table 1. The potentials promise the precise prediction for separation of actinides from lanthanides, both of which have the chemical similarity. The electrochemical potentials of those metals and Cd or Bi alloys in LiCl-KCl at 500 C are calculated by taking into account of measured activity coefficients in those liquid metals. The potentials are given in figure 7 [9]. In addition, the distribution coefficients of actinides and lanthanides in LiCl-KCl/Cd and LiCl-KCl/Bi systems are essential for separation by reductive extraction. Experimental results show a higher separation factor among them in LiCl-KCl/Bi than in LiCl-KCl/Cd [5].

Table 1. Standard potentials vs. Ag/AgCl(1wt%-AgCl) or Li(I)/Li(0) reference electrode and standard free energies of formation for chlorides in LiCl-KCl eutectic at 450 C. The small cell and the large cell used for the measurements were made of a 6.3 mm diameter tantalum tube and a 50 mm alumina crucible, respectively

Couple	Cell	E_M^0 (Ag) V	E_M^0 (Li) V	ΔG_f^0 (MCl) kJ/mol
U(lll)/U(0)	Small	−1.283	1.158	−714.5
U(lll)/U(0)	Large	−1.283	1.158	−714.4
Np(lll)/Np(0)	Small	−1.484	0.957	−772.5
Pu(lll)/Pu(0)	Small	−1.593	0.847	−804.3
Am(ll)/Am(0)	Small	−1.642	0.799	−545.5
La(lll)/La(0)	Large	−1.918	0.523	−898.3
Nd(lll)/Nd(0)	Large	−1.862	0.579	−882.1

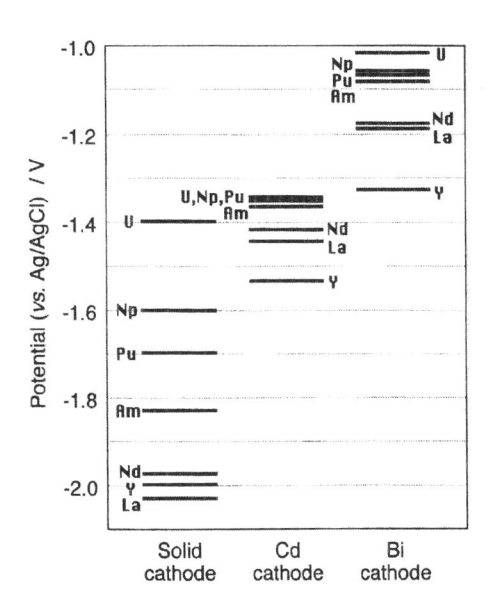

Figure 7. Reduction potentials of actinides and lanthanides for solid cathode, liquid cadmium cathode and liquid bismuth cathode in LiCl-KCl eutectic salt; T=500 C, X M in salt = X M in Cd = X M in Bi = 0.001

4.2 Process Technology

4.2.1 *U and Pu Electrorefining Process*

Development of electrorefining technology in ANL has attained to laboratory scale test and engineering scale demonstration test [11]. Due to the policy change of no use of Pu, the demonstration of the throughout process of pyro-reprocessing has not been yet realized. The technologies, which are demonstrated in engineering scale, are the spent fuel dissolution process, U collection process by electrorefining on the solid cathode and U separation process from adhered salt.

We, CRIPEI, are aiming to optimize the operating condition on electrorefining. Uranium test with 1 kg scale has been conducted successfully [12,13]. The electrorefining with solid cathode and liquid cathode has been progressing by a joint study with JAERI using Pu and Np [14]. Successful results are obtained by getting plutonium recovered in liquid cadmium cathode with 10.8 wt% by forming $PuCd_6$.

4.2.2 *Reduction Process of Oxide Fuel*

ANL and CRIEPI, independently, confirmed that more than 99% of UO_2 was converted into metal form [15]. The experiments used single elements of PuO_2, Am_2O_3 or NpO_2 and used MOX pellet revealed the feasibility of conversion to metals, which has been conducted CRIEPI and AEAT jointly [16]. Currently, the process feasibility using simulated spent fuel is under investigation.

4.2.3 *Chlorination Process of High Level Liquid Waste*

Demonstration test using simulated high level waste indicates that almost all of lanthanides, noble metals and actinides are denitrated, remaining alkali nitrates at 500 C [17]. In this step, most of alkali elements of FP with high heat emission can be separated by water rinsing after denitration. The oxides formed are converted to chlorides by pouring chlorine gas with carbon reductant over 700 C in a LiCl-KCl bath [17]. Most of all chlorides evaporated can be captured in a LiCl-KCl bath keeping at a lower temperature.

4.2.4 *Reductive Extraction Process for Recovery of TRUs*

A Multi-stage extraction, such as counter current extraction, is required to lower the lanthanide concentration in actinides recovered, when applied on waste salts concentrated fission products at electrorefining and on HLLW. Experiments for separation of TRUs used salts with simulated HLLW components were carried out and the recovery yields of all actinides are found to be over 99% with attaining the decontamination factor over 10 of lanthanides [18]. Figure 8 shows the experimental result attained over 99% recovery of each actinide from simulated composition of chloride of HLLW by multistage extraction following to electrorefining of uranium.

4.2.5 *Waste Treatment Process for Pyro-Process*

Two kinds of waste treatment process have been developed. Firstly, heated $NaAlO_2$ and SiO_2 with waste salts at high temperature, the sodalite structure is synthesized, in which waste chlorides are immobilized in an artificial mineral matrix [19]. The second process is to form borosilicate glass contained waste. Salt waste is converted to oxides through the step of electro-reduction in liquid lead and vitrified with B_2O_3 and SiO_2 at air atmosphere. Most of all solvent salt and solvent metal after electrorefining can be recycled, and Cl_2 gas generated in this process will be used at chlorination process [20].

Recovery ratio of uranium by electrorefining prior to reductive-extraction
Composition of the deposit collected on the cathode

Uranium	98.1%
Transuranics	1.2%
Rare earths	0.5%
Zirconium	0.2%

Recovery ratio of TRUs by the multistage extraction

Neptunium	99.8%
Plutonium	99.7%
Americium	99.4%

Results of TRU separation test at Missouri Univ.

Uranium and TRU metals collected by the electrorefining

Figure 8. Experimental result attained over 99% recovery of each actinide from simulated composition of chloride of HLLW by multistage extraction following to electrorefining of uranium

4.2.6 *Process Diagram*

The optimized process diagram with material balance for reprocessing of metal fuel and for TRU separation has been established based on the experimental evidences. Figure 9 shows an optimized TRU separation process flow sheet where more than 99% TRU could be separated, maintaining a large separation from lanthanides, and solvent salt, reductant and Cl_2 gas can be recycled in the process [21].

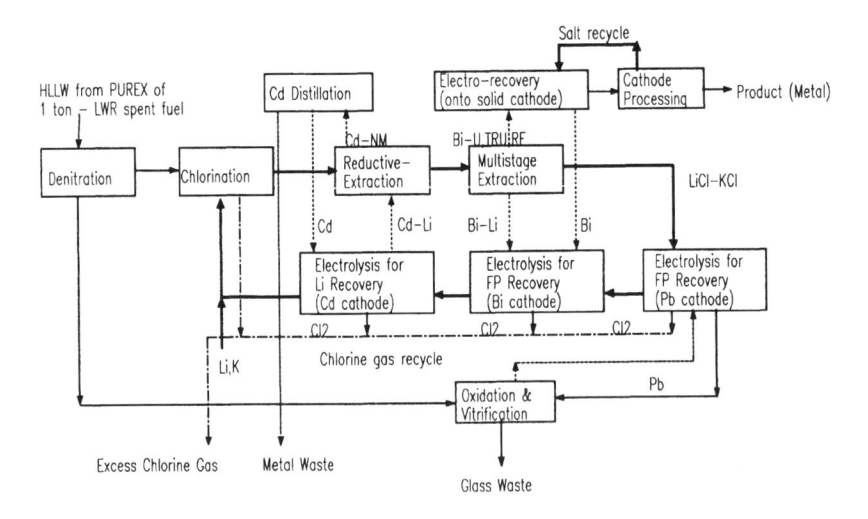

Figure 9 Process flow diagram of pyro-partitioning

5. Summary

The pyrochemistry is an attractive device to separate actinides for advanced nuclear fuel cycle. Related programs, in which the pyrochemistry takes a role to treat, or to reprocess fuels/targets, are in progress in United State, Europe, Russian Federation, Japan and the others. The electrorefining to separate transuranium elements in molten salt is employed in the accelerator-driven system (ATW, ADS) for burning them and in the fuel treatment program to separate uranium from spent metal fuels. The electrorefining is also implemented as a potential device to establish the advanced recycling technology for fast (breeder) reactors.

Actinide Recycling by Pyro-process integrated with FBR has been developing in CRIEPI since 1985 for aiming the next generation nuclear fuel cycle system with high potentials of environmental safety, non-proliferation as well as economic advantage. Through studies, elemental technologies, such as electrorefining, reductive extraction, and salt waste treatment and solidification, have been developed successfully. The fuel dissolution into molten salt and uranium recovery on solid cathode for electrorefining have been demonstrated by engineering scale facility in ANL by using spent fuels and on CRIEPI by uranium tests. Single element tests, using actinides, showed the Li reduction to be technically feasible, remaining the subjects of applicability on multi-elements system and effective recycle of Li by electrolysis of Li_2O. Concerning on the treatment of HLLW for actinide separation, the conversion to chlorides through oxides has been also established through uranium tests. It is confirmed that more than 99% of TRU nuclides can be recovered from the high level liquid waste by TRU tests. Through these studies, the process flow sheets for reprocessing of metal and oxide fuels and for partitioning of TRU separation have been established.

However, the pyro-process has much less experienced compared with aqueous process. Further efforts have to be done from engineering points of view, such as removal or dissolution of dross accumulated in equipments, handling of the cathode product and separation from salt, collection of dropped material from cathode product, transportation of the melt and remote operation technology. In addition, accounting measure of fissile material has to be developed, according to batch-wise operation. Material selection for high temperature operation is not the least issue.

References

[1] H. Hondo, Y. Uchiyama and Y. Moriizumu, "Evaluation of power generation technologies based on life cycle CO2 emissions", CRIEPI report Y99009, March 2000. (In Japanese)
[2] T. Inoue and H. Tanaka, "Recycling of actinides produced in LWR and FBR fuel cycles by applying pyrometallurgical process", Proc. on Future Nuclear Systems (GLOBAL'97), pp.646-652, Oct.5-10, 1997, Yokohama, Japan, (1997)
[3] A Report to Congress, "A roadmap for developning accelerator transmutation of waste (ATW) technology", October 1999.
[4] M.Hugon, "Overview of the EU research projects on partitioning and transmutation of long-lived radionuclides", EUR 19614 EN.
[5] Committee on electrometallurgical techniques for DOE spent fuel treatment, "Electrometallurgical techniques for DOE spent fuel treatment, Final report", National Academy Press, 2000.
[6] NEA/OECD, Proceedings of the workshop on Pyrochemical separation, Avignon, France, 14-16 March 2000.
[7] Japan Atomic Energy Commission, "State of art and future program of partitioning and transmutation in Japan", November, 2000
[8] Y.I.Chang, "The integral fast reactor", Nucl. Technol., **88**,129,(1989)
[9] Y.Sakamura, T.Hijikata, K.Kinoshita, T.Inoue, T.S.Storvick, C.L.Krueger, J.J.Roy, D.L.Grimmett, S.P.Fusselman, and R.L.Gay, "Measurement of standard potentials of actinides (U,Np,Pu,Am) in LiCl-KCl eutectic salt and separation of actinides from lanthanides by electrorefining", J. Alloy. Compound, **271-273**, 592, (1998)
[10] M.Kurata, Y.Sakamura, and T.Matsui, "Thermodynamic quantities of actinides and rare earth elements in liquid bismuth and cadmium", J. Alloy. Compound, **234**, 83, (1996)
[11] T.C.Totemeier and R.D.Mariani, "Morphologies of uranium and uranium-zirconium electrodeposits", J. Nucl. Mater. **250**, 131, (1997)
[12] T.Koyama, M.Iizuka, Y.Shoji, R.Fujita, H.Tanaka,T.Kobayashi and M.Tokiwai, "An experimental study of molten salt electrorefining of uranium using solid iron cathode and liquid cadmium cathode for development of pyrometallurgical reprocessing", J. Nucl. Sci. Technol., **34**, 384,(1997)
[13] T.Koyama, M.Iizuka, N.Kondo, R.Fujita and H.Tanaka, "Electrodeposition of uranium in stirred liquid cadmium cathode", J.Nucl.Mater., **247**, 227, (1997)
[14] M. Iizuka, K. Uozumi, T. Inoue, T. Iwai, O.shirai and Y. Arai, "Behavior of plutonium and americium at liquid cadmium cathode in molten LiCl-KCl electrolyte, To be submitted in J. Nucl. Mater.
[15] E.J.Karell, K.V.Gourishankar, L.S.Chow and R.E.Everhart, "Electrometallurgical treatment of oxide spent fuels", Int. Conf. On Future Nuclear Systems (GLOBAL'99), Aug.29 – Sep.3, 1999, Jackson Hole, Wyoming, (1999)
[16] T.Usami, M.Kurata, T. Inoue, J.Jenkins, H.Sims, S,Beetham and D.Browm, "Pyrometallurgical reduction of unirradiated TRU oxides by lithium in a lithium chloride medium", OECD Nuclear Energy Agency, Nuclear Science Committee, Workshop on Pyrochemical Separations, 14-15 March, 2000, Avignon, France, (2000)
[17] M.Kurata, T.kato, K.Kinoshita and T.Inoue, "Conversion of high level waste to chloride for pyrometallurgical partitioning of minor actinides", Proc. Of 7th Int. Conf. On Radioactive Waste Management and Environmental Remediation, ICEM'99, Sep.26-30, 1999, Nagoya, (1999)
[18] K.Kinoshita, T.Inoue, S.P.Fusselman, D.L.Grimmett, J.J.Roy, R.L.Gay C.L.Krueger, C.R.Nabelek and T.S.Storvick, "Separation of uranium and transuranic elements from rare earth elements by means of multistage extraction in LiCl-KCl/Bi system", J. Nucl. Sci. Technol., **36**, 189, (1999)
[19] T.Koyama, C.Seto, T.Yoshida, F.Kawamura and H.Tanaka, Evaluation of Emerging Nuclear Fuel Cycle Systems (GLOBAL'95), VOL.2, 1744, Sep.11-14, 1995, Versailles, (1995)
[20] Y.Sakamura, T.Inoue, T.Shimizu and K.kobayashi, "Development of pyrometallurgical partitioning technology for TRU in high level radioactive wastes –vitrification process for salt wastes-", , Int. Conf. On Future Nuclear Fuel Systems (GLOBAL'97), VOL.2, 1222, Oct.5-10, 1997, Yokohama, (1997)
[21] K.Kinoshita, M.Kurata and T.Inoue, "Estimation of materials balance in pyrometallurgical partitioning process of transuranic elements from high-level liquid waste", J. Nucl. Sci. Technol., **37**, 75, (2000)

MOLTEN SALTS FOR SAFE, LOW WASTE AND PROLIFERATION RESISTANT TREATMENT OF RADWASTE IN ACCELERATOR DRIVEN AND CRITICAL SYSTEMS

V.V. IGNATIEV
RRC - Kurchatov Institute 123182, Moscow, Russia

1. Introduction

At very beginning of the nuclear technology two approaches emerged in solving the question of what should a nuclear power be like. These approaches can be characterised as a «heat engineering» concept and a «physical-chemical» one. The first approach to nuclear energy system presumes that nuclear fuel has to be used in a maximum condensed form (and volume) that excludes reprocessing and management of the fuel while reactor operating.

Contrary to this « physical-chemical» concept represent a very different and radical approach. This concept is based on a philosophy that presuppose the use of fuel in the form that permits continuous management of nuclear-physical, chemical and heat transfer processes in fuel and regulation of its nuclides content. It means as if one includes an additional degree of freedom. This degree of freedom can be used for optimisation of nuclear energy system and gaining a maximal profits from physical potential of nuclear phenomenon.

A consistent implementation of the traditional philosophy has an advantage of technological simplicity, but it results in a loss of potential advantages of nuclear fission phenomenon and in deterioration of some nuclear plant indexes. The reactors with solid fuel elements belong to the «heat engineering» concept. The reactors with fuel in liquid or gas (plasma) phases are essentially ones of the «physical- chemical» concept. In reactor of such a concept not only are there possibilities of the general benefits such as unlimited burn-up, easy and relatively low cost of purifying and reconstituting the fuel, but also there are some more specific potential gains.

For example, expansion due to a rise in temperature in the reactor core reduces not only fluid density, but also the amount of fissile material in the core—thus reducing reactivity. The system offered the prospect therefore of being self-regulating, and the reactor experiments that were operated showed that the classical control rod absorber system was not necessary. Nor was it considered necessary to have a shutdown system. The combination of self-regulation and the ability to dump fuel to a previously prepared storage system, with a geometry designed to render criticality impossible and which also has built-in independent heat removal, was considered the better arrangement from a safety point of view. However, although using a self-heated fluid may perhaps overcome heat-transfer problems within the reactor core, the heat generated still has to be transferred through a heat-exchange system to the working fluid or a secondary coolant.

Hydrodynamics and thermodynamics of heat transport around the core are not principally different from the characteristics of heat transport in other solid fuelled systems. Again, although the inventory of fuel in the core itself can be small, it has to be remembered that the fluid fills the whole volume of the primary circuit, and if care is not taken to keep its volume to a minimum, fuel inventories can become excessive. Furthermore, fission products will be released in the fuel solution, and the materials of the primary circuit must be compatible with the resulting mixture. It may be necessary to process the fuel solution for this reason, in addition to any requirements imposed by neutron economy or other fuel cycle aspects. The whole of the circuit will become highly active owing to fission products and also as a result of some activation due to delayed neutrons. From the other side this approach provide the way significantly decrease the accumulation of the fission products in the reactor core. Therefore, although the total fission products inventory in the reactor plant integrated with processing facility remain at the same level, its significant part is moved in the conditions of the decreased specific power. Such a distribution for radionuclides additionally decrease the total risk of severe accidents in the integrated system if the danger 'separation' principle is realised.

Sometimes the critics of fluid fuel concept say, that the use of liquid fuel means the loss of two safety barrier (solid fuel matrix and fuel cladding) and thus diminishing the defence-in-depth principle. But it is a wrong way of understanding of defence-in-depth principle because:

- the use of fluid fuel does not mean that fission products are more movable. Indeed, some fission nuclides are soluble and thus are kept in the melt. Others, that are volatile, could been removed from fuel during operation. Meanwhile they are being accumulated in solid fuel and released while core melt accident.
- the numbers of barriers is not a magic number. What is of real importance is a reliability of barriers. And if one wants to keep a number of barriers to be constant, why shouldn't we to put an additional guard vessel for fluid fuel reactor, just as it is proposed for some advanced light water reactors?

Formidable though these problems look, the potential attractions of low fuel-cycle costs were sufficient to establish considerable activity, particularly during the 1950s, extending in one instance to the mid-1970s. Liquid solvents selected for reactor concepts at that time were: water (e.g. D_2O), liquid metals (e.g. Bi) and molten salts (fluorides and chlorides). The principal work was in the US, and a major incentive was provided by the interest in the use of thorium as fertile material in thermal reactors. In a thermal reactor the ^{233}U-Th cycle has little margin before it ceases to breed, and all the schemes have, of necessity, aimed for the utmost neutron economy. Provision has to be made for maintaining fission products at a low level by continuous processing where possible—xenon is particularly important in this respect. In some cases too the protactinium (Pa) product intermediate between thorium and ^{233}U (which has 27 day half-life) has to be removed, as neutron capture in Pa not only prevents ^{233}U being formed, but loses a further neutron from the cycle. This effect is proportional to neutron flux level, and as fluid-fuel reactors aim to have very high in-core fuel ratings, it is important to minimise the extent to which protactinium is present in these high-flux regions. This was one of the reasons why all the early concepts envisaged a two-zone reactor with fuel only in the core solution, which was fed from a thorium-containing blanket via a suitable processing system (a layout comparable with that in a fast reactor).

The fluid-fuel reactor which enjoyed the most extended run of investigatory life was the molten-salt system. This began with an aircraft reactor experiment, which operated successfully in 1954 in a 'proof-of-principle' short-term test at a power level of 2.5 MWt and at temperatures up to 860 °C. The fuel was a solution of UF_4 in other fluorides with Inconel-clad beryllium as the moderator. Those who took part in this program realised from the early stages that, potentially, the molten-salt system was a candidate as a power reactor, and a group at ORNL began work in 1957. A further reactor experiment (MSRE) was built and operated successfully at a power level of 7,3 MWt until the end of 1969. The fuel was the mixture of Li,Be,Zr,U/F with graphite as the moderator and Hastelloy N as the container material. The MSRE was very successful experiment, in that it answered many questions and posed a few new ones. Perhaps the most important result was the conclusion that it was quite practical reactor: Note, that in 1970-1980 US and others countries, including France, Japan and Russia, also placed some additional efforts in the U-Th. MSR power reactor developments (e.g. 1000MWe MSBR plant design and 1000MWe DMSR conceptual designs developed at ORNL, US).

The interest to explore molten salts, as a future option in nuclear power both to back end of the fuel cycle or for its overall simplification, currently is revisited. Its include using the molten salt fuel concept in order to manage plutonium and minor actinides, but also to envisage innovative systems producing less radioactive waste (e.g. by the use of thorium).

2. Choice of the fuel salt and coolant

In choosing a fuel for a given fluid fuel reactor design the following criteria are applied:
Low neutron cross section for the solvent components
Thermal stability of the salt components
Low vapour pressure
Radiation stability
Adequate solubility of fuel (including TRU's) and fission product components
Adequate heat transfer and hydrodynamic properties
Chemical compatibility with container and moderator materials
Low fuel and processing costs.

Fluorides of the metals other than U, Pu or Th are used as diluents and to keep the melting point low enough for practical use. Consideration of nuclear properties alone leads one to prefer as diluents the fluorides of Be, Bi, ^{7}Li, Pb, Zr, Na, and Ca, in that order. Salts which contain easily reducible cations (Bi^{2+} and Pb^{2+}, see Table 1) were rejected because they would not be stable in nickel – or iron base alloys of construction. Particularly, a disadvantage of ZrF_4 contained (more than 25 % mole) melts is associated with its condensable vapour, preponderantly ZrF_4 The «snow» that would form could block vent lines and cause problems in pumps that circulate the fuel. Note also, that use of Zr, instead e.g. the sodium in the basic solvent will lead to the increased generation of the long lived activation products in the system. This leaves BeF_2, ^{7}LiF and NaF as preferred major constituents. To achieve lower melting temperatures, two or more salts are combined to produce still lower melting mixtures. Since the freezing point is an important fluid fuel criterion, note that alternative composition $NaLiBeF_4$ (335°C) had the significantly lower liquidus temperature in this region of the phase diagram than that of Li$_2$BeF$_4$ (458 °C). For reasons of neutron economy in ORNL, the preferred solvent for prior MSR concepts have been LiF and BeF$_2$ with the lithium enriched to 99,995 in the ^{7}Li isotope. (At this level, in a typical MSBR configuration, the remaining ^{6}Li absorbs 40% as many neutrons as the ^{7}Li).

TABLE 1 Thermodynamic properties of fluorides [Ignatiev, 1999]

Compound (solid state)	$-\Delta G_{f,1000}$, kcal/mole	$-\Delta G_{f,298}$, kcal/mole	E$_{298}$, V (Me|F$_2$)
LiF	125	140	6.06
CaF$_2$	253	278	6.03
NaF	112	130	5.60
BeF$_2$	208	231	5.00
ZrF$_4$	376	432	4.70
PbF$_2$	124	148	3.20
BiF$_3$	159	200	2.85
NiF$_2$	123	147	3.20
FeF$_2$	138	158	3,43
CrF$_2$	151	172	3,73
UF$_3$ (UF$_4$)	300 (380)	330 (430)	4.75
PuF$_3$ (PuF$_4$)	320	360 (400)	5.20
ThF$_4$	428	465	5,05
AmF$_3$	325	365	5.30
CeF$_3$	345	386	5,58
LaF$_3$	348	389	5,63

As a class these salts satisfy also the characteristic properties of the thermal stability, radiation resistance, and low vapour pressure. For example, at RRC-KI in-reactor loop tests of radiation stability of fluoride molten-salt fuels were carried out. The radiolysis products (first of all, F$_2$) were detected in the gas phase over the melts. It was stated, that:

The radiolytic evolution of fluorine from molten fluoride mixtures when irradiating in a nuclear reactor is insignificant;

The measured values of radiation chemical yield G(F$_2$) (the number of F$_2$ molecules, evolving per 100 eV of absorbed energy) are in the range of 10^{-5}–10^{-6}, therefore the fluoride fluid fuel may be attributed to the category of radiation resistant materials in the temperature range up to 1200^{0}C;

The frozen fuel salt at 50^{0}C has the G(F$_2$) value about 10^{-2}, that is much better, as compared to water, which is used as the primary coolant in LWRs.

In considering the application of these molten salts to a MSR situation, three important and interrelated concepts characterising the chemical state must be controlled. These are solubility, Redox chemistry and chemical activity. These concepts are not completely independent, but each emphasises or correlates certain aspects of the chemical behaviour and physical properties.

One of these ways for characterisation arises from the relative proportions of the major constituents and related to their tendencies to form the complex ions. Monovalent salts are «basic» in that they supply fluoride ions. Polyvalent salts are «acidic» in that they form complexes with F$^-$. For example, BeF$_4^{2-}$, ZrF$_6^{2-}$, ThF$_7^{3-}$, AlF$_6^{3-}$ are typical complex ions.

The excess or deficit in the free fluoride-ion content can be placed on a rough scale, as follows:

$$Excess\ free\ fluoride = [MF] - 2[BeF_4^{2-}] - 3[AlF_6^{3-}] - 2[ZrF_6^{2-}] - 3[ThF_7^{3-}]\ etc. \quad (1)$$

This excess (or deficit) influences many properties of the salt. The chemical reactivities of this and other metal ions are higher when they are not sufficiently co-ordinated with fluoride ions. For example, in the absence of the extra fluoride ions supplied by the LiF or NaF component, BeF_2 and ZrF_4 would be volatile and distil from the system.
Also, excess / deficit free fluoride affect the solubility of actinides and lanthanides in the fuel. In case of the pure fluoride materials, solubility is determined by phase diagrams of the mixtures. These temperature versus composition determinations give the temperature at which components might fall from solution (as determined by the liquidus line on the plot).
In some MSR concepts ^{233}U and ^{235}U will be the primary fissile isotopes, while ^{232}Th, with important assistance from ^{238}U, is the fertile material, but some MSR concepts will derive its fission energy mainly from plutonium and minor actinide isotopes. Clearly, from neutronic point of view, the ^{232}Th concentration will need to be higher than total concentration of uranium and plutonium isotopes.
Uranium tetrafluoride and uranium trifluoride are only fluorides which appear useful as constituents of molten fluoride fuels. UF_4 is relatively stable, non-volatile, and nonhygroscopic. UF_3 disproportionate at temperatures above 1000°C by the reaction:

$$4UF_3 \leftrightarrow U^0 + 3UF_4 \quad (2)$$

Thorium tetrafluoride is the only known fluoride of thorium. It melts at 1111 C, but fortunately its freezing point is depressed by fluoride diluents which are also useful with UF_4.
The single fluid MSBR and DMSR designs developed at ORNL used Li,Be,Th,U/F composition with concentrations of ThF_4 much higher than that of UF_4. Accordingly the phase behaviour of the fuel will be dictated by that of Li,Be,Th/F system. Inspection of the diagram reveals that a considerable range of the compositions with > 10 mole % ThF_4 will be completely molten at or below 500°C.
Trivalent plutonium and minor actinides are only stable species in the various molten fluoride salts. Tetravalent plutonium could transiently exist if the salt Redox potential was high enough. But for practical purposes (stability of potential container material) salt Redox potential should be low enough and corresponds to the stability area of Pu(III).
PuF_3 solubility is maximum in pure LiF or NaF and decreases with addition of BeF_2 and ThF_4. Decrease is more for BeF_2 addition, because the PuF_3 is not soluble in pure BeF_2. The lanthanide trifluorides are also only moderately soluble in BeF_2–ThF_4 containing mixtures. If more than one such trifluoride (including UF_3) is present, they crystallize as solid solution of all the trifluorides on cooling of saturated melt so, that in effect, all the LnF_3 and AnF_3 act essentially as a single element. If so, the total (An+Ln) trifluorides in the end of life reactor might possibly exceed their combined solubility. The solubility of PuF_3 in fluoride based solvents is temperature and composition dependent.

According to ORNL data the solubility of PuF$_3$ and CeF$_3$ in the Li,Be,Th/F melt at 565°C (the minimum temperature anticipated within the MSBR and DMSR fuel circuit) is 1,3 mole %. For Li,Be/F and Na,Be/F systems the PuF$_3$ solubility is considerably smaller and seems to be minimal in the most established «neutral» melts. For these ones it reach only about 0,2 – 0,3 mole % at the temperature of 565°C. Note, that for some excess free fluoride ternary Na, Li, Be/F system (T$_{melt}$=480°C) solubility of PuF$_3$ will be about 1,1 mole % at 565 C. It is illustration of how effects of addition of third component by swelling the system structure may influence not only on lower melting temperatures, but also on increased solubility of plutonium trifluoride in flibe based system.

In addition to concerns related to those of the pure materials, solubility has significance of impurities such as moisture react with molten salts to produce metal oxides of much higher melting point and correspondingly lower solubility. For UF$_4$ fuelled systems this reaction is written as:

$$2H_2O + UF_4 \leftrightarrow 4HF + UO_2 \tag{3}$$

where the very insoluble UO$_2$ could precipitate out. To prevent the loss of uranium from solution and possible accumulation of UO$_2$ precipitate in amounts large enough to pose criticality concerns, five mole percents of ZrF$_4$ was added to the solution to getter any oxide impurities through the reaction:

$$2H_2O + ZrF_4 \leftrightarrow 4HF + ZrO_2 \tag{4}$$

which is more favourable to reaction with moister than the former. Plutonium as PuF$_3$ shows little tendency to precipitate as oxide even in the presence of excess BeO and ThO$_2$. The solubility of the oxides of Np, Am, Cm has not been examined. Some attention to this problem will be required.

Besides solubility concerns with respect to impurity reactions, one must prevent the liquid fuel component from reacting : with container materials and cause corrosion products plus structural weakening or with moderator graphite and leave behind insoluble fissionable materials. Reactions of the former like as:

$$UF_4 + Cr\,(solid) \leftrightarrow UF_3 + CrF_2 \tag{5}$$

and the later:

$$UF_3 + 2C \leftrightarrow UC_2 + 3UF_4 \tag{6}$$

were prevented by careful control of the solution Redox chemistry which was accomplished by setting the UF$_4$ / UF$_3$ ratio at approximately (50-60)/1. Additions of metallic Be to the fuel salt to effect the reduction of the UF$_4$ via:

$$2UF_4 + Be^0 \leftrightarrow 2UF_3 + BeF_2 \tag{7}$$

The significance of the redox control to MSR systems with uranium free fuel is that in some cases where the fuel is e.g. PuF$_3$, the Pu(III)/Pu(IV) Redox couple is to oxidising to present satisfactory Redox buffered system. In this case as it was proposed by ORNL Redox control could be accomplished by including an HF/H$_2$ mixture to the inert cover gas sparge which will not only set the Redox potential , but will also serve as the Redox indicator if the exit HF/H$_2$ stream is analysed relative to inlet.

In general, when make selection of the fuel composition for the MSR, one has the volatility and neutron capture problem with Zr, the some additional neutron capture problem with Na, the isotopic enrichment and tritium production problems with Li, and the Pu solubility and viscosity problems with Be.

The molten fluoride chemistry (solubility, Redox chemistry, chemical activity etc) for the 2LiF–BeF$_2$ system is well established and can be applied with great confidence, if PuF$_3$ fuels are to be used in the 2LiF–BeF$_2$ solvent. But the chemistry of other solvent systems are different and less understood (for example, in the basic Li,Na/F system) and requires a comprehensive studies. For MSR's transmuting system needs next more important is consideration of PuF$_3$ chemical behavior in these solvent systems: PuF$_3$ solubility in Li,Be/F and Na,Be/F or Na,Li,Be/F solvents; phase transition behavior of the ternary Na,Li,Pu/F or quaternary, e.g. Na,Li,Zr,Pu/F fuels, oxide tolerances of such mixtures and Redox effects of the fission products.

The secondary coolant is required to remove heat from the fuel in the primary heat exchanger and to transport it to the power generating. The coolant mixture chosen for the MSRE and shown to be satisfactory was Li,Be/F (T$_{melt}$= 458°C). The less expensive binary system NaF-NaBF$_4$ was the choice for the MSBR plant design (T$_{melt}$= 385°C). Very few fluorides or mixtures are known to melt at temperatures below 370°C. Coolant compositions which will meet the low liquidus temperature specification may also chosen from Na,Be/F or Na,Li,Be/F system. They are almost certainly compatible with structural material, low vapour pressure, and they are posses adequate transport properties (discussed in the next section). It is possible that substitution of ZrF$_4$ or even AlF$_3$ for some of the BeF$_2$ would provide liquidus of lower viscosity at no real expense in liquidus temperature

3. The heat and mass transport properties of the molten salts

The design of MSR requires detailed information on heat and mass transport properties of the proposed fuel and coolant fluoride melts. This criterion relates to the physical properties of the fuel (determinant factor for the design behaviour in normal and accidental conditions) , in particular the specific heat, density, thermal conductivity and viscosity.

Obviously a coolant capable of taking up a lot of heat per unit volume requires a lower flow rate to do its job. If it has a low viscosity it requires less pumping power to achieve that flow rate. An approximate figure of merit to express this factors in normal operation (fixed reactor power, fluid heat up in the core and pipe parameters) is:

$$\frac{C_p^{2.8} \rho^2}{\mu^{0.2}} \tag{8}$$

In connection with developments in safety philosophy (passive after heat removal) an approximate figure of merit for the natural convection quality of the coolant could be:

$$C_p \rho \beta^{0.5} \Delta T_c^{1.5} \qquad (9)$$

where C_p – specific heat (J /(kg K)), ρ -density(kg / m^3) , μ -viscosity (kg /(m s)), β - thermal expansion (1/K) and ΔT_c - fluid heat up in the core.

When compare different solvents for fluid fuelled reactors in term of merits given for physical properties, liquids are superior to gases and high pressure gases are superior to low pressure gases. Certainly, water demonstrate good transport potential as fluid fuel carrier. In term of merits given for physical properties molten salts look better than liquid metals.

One of principal problems for water and high pressure gases (e.g. UF$_6$) is the need to operate it at about 14.0 MPa to achieve reasonable efficiency, and any leaks that might arise from such a high-pressure active circuit could clearly cause difficulties.

Molten salts have a much higher heat capacity per unit volume than lead and sodium, so that the physical size (cost) of pumps and piping will be smaller. The molten salts have a much lower thermal conductivity than liquid metals , so that sudden coolant temperature changes which directly correlate with fluid heat up in the core will provide less thermal shock to system components.

Many of the physical properties dealing with heat and mass transport processes of molten fluoride salt have been obtained during the development of the MSR program in ORNL. The specific physical properties which were either measured within Russian MSR program include density, heat capacity, heat of fusion, viscosity, thermal conductivity and electrical conductivity. Particular emphasis has been placed for U/Th fuelled reactor cores.

Table 2 shows the composition and key properties of the fluids at the indicated temperatures. Most of the physical properties of Li,Be,Th/F (72-16-12 %mole) are known with reasonable accuracy, although several have been defined by interpolation from measurements on slightly different compositions. The liquidus temperature is well known, and density, viscosity, heat capacity are accurate within 3-7% . Thermal conductivity is the key property for predicting heat transfer coefficients of molten fluorides. ORNL measurements are probably accurate within 10-15% for this fuel composition.

TABLE 2 Physical properties of molten salts [Cantor,1963]

Composition, mole %	72LiF-16BeF$_2$-12ThF$_4$-0.3UF$_4$	46,5LiF-11,5NaF-42KF	92NaBF$_4$-8NaF	66LiF-34BeF$_2$	23LiF-41NaF-36BeF$_2$
At temperature, °C	700	600	621	700	600
Liquidus, °C	500	454	385	458	328
Heat capacity, kJ/kg/°C	1.34	1,86	1.51	2.34	1.97
Density of liquid, kg/m^3	3250	2090	1810	2050	2010
Thermal conductivity, W/m/°C	1.2	0.7	0.4	1.0	1,0
Viscosity, g /m/s	7	7,1	1,1	5.6	7.5
Prandtl modulus, $\eta \cdot C_p / \lambda$	8	7	5	13	15

Properties are in each case adequate to the proposal service. However, the tabulations available in some cases are such as to permit interpolation of reasonable accuracy for preliminary design purposes if melt contains small (< 1 mole %) addition of TRUF$_3$. In some cases careful remeasurement of some properties (in particular thermal conductivity for such solvents as Li,Na,K/F; Na,Be/F; Li,Na,Be/F) was reasonable and desirable. When the melt is complicated by the addition of large quantities of TRUF$_3$ the situation becomes considerably less favourable and need additional measurements of required physical properties.

The operation of the MSRE at ORNL represents the most large scale experience with heat transfer from fuel salt to coolant salt. The heat transfer correlations used for the MSRE primary heat exchanger were based on the previous development tests which showed that fluoride salts behave as normal fluids. When MSRE operation revealed the overall heat transfer coefficient to be less than predicted, a revaluation of the physical properties disclosed that the actual thermal conductivities of the fuel and coolant salts were below those used in the design calculations and accounted for the difference.

At RRC-KI heat transfer studies with different fuel and coolant salts (Prandtl modulus range from 2 to 35) flowing in forced and natural convection loops were made in a wide range of parameters typical for MSR designs.

Forced convection runs have covered a Reynolds modulus range from 5000 to 20000 and a heat flux q_s range up to 10^2 kW/m^2. The criterion dependencies generalising the heat transfer data were obtained for natural convection heat exchangers at separate or joint actions of surface and volume heat sources in the fluid up to q_s=30 kW/m^2 and q_v=7 MW/m^3.

No evidence of influence of corrosion and irradiation products on heat transfer was found (at temperatures up to 750^{0}C, neutron fluences up to $2 \cdot 10^{20}$ cm^{-2} and concentration of Cr, Fe, Ni metal impurities less than 10^{-1}% by mass.).

The conclusion is that the use of accurate physical property data with correlation's for normal fluids (Pr>1) is adequate for heat transfer for design with fluoride salts respect to forced and natural convection.

4. Container metal for primary and secondary circuits

The material used in constructing the primary circuit of an MSR will operate at temperatures up to 700 -750°C. The inside of the circuit will be exposed to salt containing fission products and will receive e.g. for MSBR case a maximum thermal fluence of about 5×10^{21} neutrons/cm^2 over the operating lifetime more than 30 years. This fluence will cause embrittlement due to helium formed by transmutation but will not cause swelling such as is noted at higher fast fluences. The outside of the primary circuit will be exposed to nitrogen containing sufficient air from inleakage to make it oxidising to the metal. Thus the metal must have moderate oxidation resistance, must resist corrosion by the salt, and must not be subject to severe embrittlement by thermal neutrons.

In the secondary circuit the metal will be exposed to the coolant salt under much the same conditions described for the primary circuit. The main difference will be the absence of fission products and uranium in the coolant salt and the much lower neutron fluences. This material must have moderate oxidation resistance and must resist corrosion by a salt not containing fission products or uranium.

The primary and secondary circuits involve numerous structural shapes ranging from a few inches thick to tubing having wall thickness of only a few thousandths of an inch. These shapes must be fabricated and joined, primarily by welding, into an integral engineering structure.

The success of an MSR is strongly dependent on the compatibility of the container materials with the molten salts used in primary and secondary circuits. Because the products of oxidation of metals by fluoride melts are quite soluble in corroding media, passivation is precluded, and the corrosion rate depends on other factors, including: Oxidants, Thermal gradients, Salt flow rate and Galvanic coupling.

Design of practicable system, therefore, demands the selection of salt constituents such as LiF, NaF, BeF_2, UF_4, ThF_4, PuF_3 etc., that are not appreciably reduced by available structural metals and alloys whose components Fe, Ni and Cr can be in near equilibrium with the salt. Equilibrium concentrations for these components will strongly depend on the solvent system.

Thermodynamic data clearly reveal that in reactions with structural metals (M):

$$2UF_4 + M\ (solid\ ) \leftrightarrow 2UF_3 + MF_2, \tag{10}$$

chromium is much more readily attacked than iron, nickel, or molybdenum.

Stainless steels, having more chromium than Ni-based alloy Hastelloy N, are more susceptible to corrosion by fluoride melts, but can be considered for some applications. Oxidation and selective attack may also result from impurities in the melt,

$$NiF_2 + M\ (solid\ ) \leftrightarrow MF_2 + Ni, \tag{11}$$
$$2HF + M\ (solid\ ) \leftrightarrow MF_2 + H_2$$

or oxide films on the metal,

$$NiO + BeF_2 \leftrightarrow NiF_2 + BeO, \tag{12}$$

followed by reaction of NiF_2 with M.

These reactions will proceed essentially to completion at all temperatures within the circuit. Accordingly, such reactions can lead (if system is poorly cleaned) to rapid initial corrosion. However, these reactions do not give a sustained corrosive attack. The impurity reactions can be minimised by maintaining low impurity concentrations in the salt and on the alloy surfaces.

Reaction with UF_4, on the other hand, may have an equilibrium constant which is strongly temperature dependent, hence, when the salt is forced to circulate through a temperature gradient, a possible mechanism exists for mass transfer and continued attack. This reaction is of significance mainly in the case of alloys containing relatively large amounts of chromium. Corrosion proceeds by the selective oxidation of Cr at the hotter loop surfaces and reduction and deposition of chromium at the cooler loop surfaces. In some solvents (Li,Na, K,U/F for example) the equilibrium constant for reaction (10) with Cr changes sufficiently as a function of temperature to cause formation of dendritic chromium crystals in the cold zone. For Li,Be,U/F mixtures the temperature dependence of the mass transfer reaction is small, and the equilibrium is satisfied at reactor temperature conditions without the formation of crystalline chromium.

In case of the MSR fuel salts (Li,Be,Th,U/F), the resultant maximum corrosion rate of Hastelloy N measured in extensive loop testing was below 5 μm/ yr, and similar rate was measured in RRC-KI corrosion tests on specimens of the nickel base alloy HN80MT. Higher Redox potential set in the system makes the salt more oxidising. At ORNL the dependence of the corrosion vs. flow rate was specially tested in the range of the velocities from 1 to 6 m/s. It was reported that the influence of the flow rate was significant only during first 1000-3000 hours. Later the corrosion rates as well as their difference are decreased.

If a 300-series stainless steels is exposed to uranium fuelled salt under the same closed system conditions the corrosion rate was manifested by surface voids of decreased Cr content to a depth of 60-70 μm at 600-650°C. Data on corrosion rates obtained in experiments with Li,Be,Th,U/F for the 304SS and 316SS at ORNL and later at RRC-KI for the Russian austenitic steels 12H18N10T and AP-164 agree well with each other. Note that for the Li,Be/F system not containing uranium, where the oxidation potential of the salt could be lowered by buffering with metallic Be without concerns for disproportionation of uranium trifluoride, the corrosion rate was decreased at 650°C from the 8 to the 2 μm/yr.

For the secondary coolant $NaF-NaBF_4$ studies show that the corrosion is also determined by the selective yield of Cr from alloy through reactions which proceed mainly due to the presence of moisture impurities:

$$H_2O + NaBF_4 \leftrightarrow NaBF_3OH + HF$$
$$NaBF_3OH \leftrightarrow NaBF_2O + HF \qquad (13)$$
$$6HF + 6NaF + Cr \leftrightarrow 2Na_3CrF_6 + 3H_2$$

ORNL data in thermal corrosion loops for Hastelloy N in the $NaF-NaBF_4$ lies in the interval of 5-20 μm/yr and is determined mostly by the degree of the salt purification. These data are in a good agreement with later RRC-KI corrosion studies with Russian nickel base alloy of the HN80MT type (about 10-15 μm/yr till to 600°C). At ORNL studies on occasions when leaks developed, the corrosion rate has increased and then decreased as the impurities were exhausted.

Early materials studies at ORNL led to the development of a nickel-base alloy, Hastelloy N, for use with fluoride salts. As shown in Table 3, the alloy contained 16% molybdenum for strengthening and chromium sufficient to impart moderate oxidation resistance in air but not enough to lead to high corrosion rates in salt. This alloy was the sole structural material used in the MSRE and contributed significantly to the success of the experiment. However, two problems were noted with Hastelloy N which needed further attention before more advanced reactors could be built. First, it was found that Hastelloy N was embrittled by helium produced from ^{10}B and directly from nickel by a two-step reaction. This type of radiation embrittlement is common to most iron— and nickel—base alloys. The second problem arose from the fission-product tellurium diffusing a short distance into the metal along the grain boundaries and embrittling the boundaries.

TABLE 3 Chemical composition of nickel based alloys [Engel, 1979; Ignatiev,1990]

Element	Content (% by weight)			
	Standard alloy Hastelloy N	Modified alloy US 1972	Modified alloy US 1976	HN80MTY RF 1984
Nickel	Base	Base	Base	Base
Molybdenum	15-18	11-13	11-13	11-12
Chromium	6-8	6-8	6-8	5-7
Iron	5	0.1	0.1	1,5
Manganese	1	0.15-0.25	0.15-0.25	0,5
Silicon	1	0.1	0.1	0,15
Phosphorus	0.015	0.01	0.01	0,015
Sulphur	0.020	0.01	0.01	0,012
Boron	0.01	0.001	0.001	0,001
Titanium		2		0,5-1
Niobium		0-2	1-2	
Aluminium				1,2

When ORNL studies were terminated in early 1973, considerable progress had been made in finding solutions to both problems. Since the two problems were discovered a few years apart, the research on the two problem areas appears to have proceeded independently. However, the work must be brought together for the production of a single material that would be resistant to both problems. It was found that the carbide precipitate that normally occurs in Hastelloy N could be modified to obtain resistance to the embrittlement by helium. The presence of 16% molybdenum and 0.5% silicon led to the formation of a coarse carbide that was of little benefit. Reduction of the molybdenum concentration to 12% and the silicon content to 0.1% and the addition of a reactive carbide former such as titanium led to the formation of a fine carbide precipitate and an alloy with good resistance to embrittlement by helium. The desired level of titanium was about 2%, and the phenomenon had been checked out through numerous small laboratory and commercial melts by 1972.

Because the intergranular embrittlement of Hastelloy N by tellurium was noted in 1970, ORNL understanding of the phenomenon was not very advanced at the conclusion of the program in 1973. Numerous parts of the MSRE were examined, and all surfaces exposed to fuel salt formed shallow intergranular cracks when strained. Some laboratory experiments had been performed in which Hastelloy N specimens had been exposed to low partial pressures of tellurium metal vapour and, when strained, formed intergranular cracks very similar to those noted in parts from the MSRE. Several findings indicated that tellurium was the likely cause of the intergranular embrittlement, and the selective diffusion of tellurium along the grain boundaries of Hastelloy N was demonstrated experimentally. One in-reactor fuel capsule was operated in which the grain boundaries of Hastelloy N were embrittled and those of Inconel 601 (Ni, 22% Cr, 12% Fe) were not. These findings were in agreement with laboratory experiments in which these same metals were exposed to low partial pressures of tellurium metal vapour. Thus, at the close of the program in early 1973, tellurium had been identified as the likely cause of the intergranular embrittlement, and several laboratory and in—reactor methods were devised for studying the phenomenon. Experimental results had been obtained which showed variations in sensitivity to embrittlement of various metals and offered encouragement that a structural material could be found which resisted embrittlement by tellurium.

The alloy composition favoured at the close of the ORNL program in 1973 is given in the Table 3 with the composition of standard Hastelloy N. The reasoning at that time was that the 2% titanium addition would impart good resistance to irradiation embrittlement and that the 0 to 2% niobium addition would impart good resistance to intergranular tellurium embrittlement. Neither of these chemical additions was expected to cause problems with respect to fabrication and welding.

When the ORNL program was restarted in 1974, top priority was given to the tellurium-embrittlement problem. A small piece of Hastelloy N foil from the MSRE had been preserved fur further study. Tellurium was found in abundance, and no other fission product was present in detectable quantities. This showed even more positively that tellurium was responsible for the embrittlement.

Considerable effort was spent in seeking better methods of exposing test specimens to tellurium. The most representative experimental system developed for exposing metal specimens to tellurium involved suspending the specimens in a stirred vessel of salt with granules of Cr_3Te_4 and Cr_5Te_6 lying on the bottom of the salt. Tellurium, at a very low partial pressure, was in equilibrium with the Cr_3Te_4 and Cr_5Te_6, and exposure of Hastelloy N specimens to this mixture resulted in crack severalties similar to those noted in samples from the MSRE.

Numerous samples were exposed to salt containing tellurium, and the most important finding was that modified Hastelloy N containing 1 to 2% niobium had good resistance to embrittlement by tellurium. An almost equally important finding was that the presence of titanium negated the beneficial effects of niobium. Thus, an alloy containing titanium, to impart resistance to irradiation embrittlement, and niobium, to impart resistance to tellurium embrittlement, did not have acceptable resistance to tellurium embrittlement, even though the mechanical properties in the irradiated condition were excellent As a result, it became necessary to determine whether alloys containing niobium (without titanium) had adequate resistance to irradiation embrittlement. One irradiation experiment showed that the niobium -modified alloy offered adequate resistance to irradiation embrittlement, but more detailed tests are needed. Several small melts containing up to 4.4% niobium were found to fabricate and weld well; so products containing 1 to 2%. niobium can probably be produced with a minimum of scale—up difficulties.

One series of ORNL experiments was carried out to investigate the effects of oxidation state on the tendency for cracks to be formed in tellurium-containing salt, on the supposition that the salt might be made reducing enough to tie the tellurium up in some innocuous metal complex. The salt was made more oxidising by adding NiF_2 and more reducing by adding beryllium. The experiment had electrochemical probes for determining the ratio of uranium in the +4 state (UF_4) to that in the +3 state (UF_3). Tensile specimens of standard Hastelloy N were suspended in the salt for about 260 hr at 700°C. The oxidation state of the salt was stabilised, and the specimens were inserted so that each set of specimens was exposed to one condition. After exposure, the specimens were strained to failure and were examined metallographically to determine the extent of cracking. At U^{4+}/U^{3+} ratios of 60 or less, there was very little cracking, and at ratios above 80 the cracking was very extensive. These observations offer encouragement that a reactor could be operated in a chemical regime where the tellurium would not be embrittling even to standard Hastelloy N. At least would need to be in the +3 oxidation state (UF_3), and this condition seems quite reasonable from chemical and practical considerations.

The development of Russian domestic structural material for MSR was substantiated by available experience accumulated in ORNL MSR program on nickel-base alloys for UF_4 –containing salts.

The alloy HN80MT was chosen as a base. Its composition (in wt.%) is Ni(base), Cr(6.9), C(0.02), Ti(1.6), Mo(12.2), Nb(2.6). The development and optimisation of HN80MT alloy was envisaged to be performed in two directions:

improvement of the alloy resistance to a selective chromium corrosion,

increase of the alloy resistance to tellurium intergranular corrosion and cracking.

About 70 differently alloyed specimens of the HN80MT were tested. Among alloying elements there were W, Nb, Re, V, Al and Cu. The main finding is that alloying by aluminium at a decrease of titanium down to 0.5% revealed the significant improvement of both the corrosion and mechanical properties of the alloy. The chromium corrosion and intergranular corrosion have reached the minimum value at Al content in the alloy ~2.5%. Irradiation effect on a corrosion activity of fuels was also studied. It was shown that at least up to the power density 10 W/cm^3 in fuel composition $LiF-BeF_2-ThF_4-UF_4$ there is no radiation induced corrosion.

Then the radiation study of 13 alloy modifications were carried out. Specimens (in nitrogen atmosphere) were exposed to the reactor neutron field up to the fluency of $3 \cdot 10^{20}$ n/cm^2. Experimental results of alloy mechanical properties at temperatures of 20, 400 and 650°C for nonirradiated and irradiated specimens permits to rest only four modifications. These alloys modified by Ti, Al and V have shown the best postirradiation properties.

At last, corrosion under the stressed condition was studied. It is known that tensile strain promotes an opening of intergrain boundaries and thus boosts intergranular corrosion and create prerequisites for an intergranular cracking. The studies did not reveal any dependence of intergranular corrosion on the stress up to the value 240 MPa, that is 0.8 of a tensile yield of the material and 5 times higher than typical stresses in MSR designs.

The results of combined investigation of mechanical, corrosion and radiation properties various alloys of HN80MT permitted us to suggest the Ti and Al–modified alloy as an optimum container material for the MSR. This alloy named HN80MTY (or EK–50) has the following composition given in Table 3. The comparison of our corrosion data with those obtained at ORNL for Hastelloy–N indicates that corrosion resistance of HN80MTY is higher and it's maximum working temperature could be up to 750°C. Nevertheless the weldability of the alloy deserves an improvement. To suppress crack formation during welding the metal penetration regime was set up and a maximum heat removal from the welded joint was ensured. These measures made it possible to increase significantly characteristics of the welded joints. The work on manufacturing of heat exchanger has confirmed once more that obtained HN80MTY alloy is technologically effective both in hot and cold process stages.

The high temperature, salt Redox potential, radiation fluence and energy spectrum poses a serious challenge for any structural alloy in a MSR. Data for UF_4 –containing salts provides a roadmap for establishing the corrosion properties for PuF_3 – containing salts, but additional corrosion testing under MSR transmuting system conditions will be required to quantify corrosion rates of candidate container materials.

NaF, LiF, BeF$_2$, CaF$_2$, ThF$_4$, PuF$_3$, AmF$_3$, CmF$_3$ and NpF$_3$ can not be oxidised in system considered and can be reduced only to the metals, and then only by reducing agents very much stronger than the constituents of Hastelloy–N. Mixtures of these materials would not be expected to be corrosive. Recent capsule experiments in Cheljbinsk-70 have demonstrated that PuF$_3$ addition in LiF–BeF$_2$ solvent system did not make the corrosion situation worse on both nickel- and iron-base alloys. The chemistry of PuF$_3$ needs further testing in corrosion loops studies for Redox control.

Also, included in further evaluation should be an assessment of lower salt Redox potentials from the stand point of allowing the use of stainless steels as structural materials and establishing the potentials that must be maintained to avoid intergranular cracking for nickel-based alloys. Techniques developed under other reactor programs to improve the resistance of stainless steels to helium emrittlement should be extended to include nickel-base alloys.

5. Graphite for molten-salt reactors

The graphite in a main part of MSR concepts serves no structural purpose (its primary function is, of course, to provide neutron moderation) other than to define the flow patterns of the salt and, of course, to support its own weight. Its moderating power is about twice that of the salt, while its effective macroscopic absorption cross section is about one-fifth that of the salt, exclusive of the actinides. Fuel salt will typically occupy 10-15% of the core volume. No metals are used in the core.

The requirements on the material are dictated most strongly by nuclear considerations, namely stability of the material against radiation induced distortion and nonpenetrability by the fuel bearing molten salt. The practical limitations of meeting these requirements, in turn, impose conditions on the core design, specifically the necessity to limit the cross-sectional area of the graphite prisms. The requirements of purity and impermeability to salt are easily net by several high quality, fine grained graphites, and the main problems arise from the requirement of stability against radiation induced distortion.

In MSBR for replacement cost to be reasonable (rate of expansion becomes quite rapid), graphite must survive a fluence of about 3×10^{22} neutron/cm^2 ($E > 50$ keV) or better. An approximate inverse relationship thus exists between the useful life of the graphite (T_g, year) and the core power density (q_c, w/cm^3):

$$T_g * q_c \leq 200 \qquad (14)$$

By the time the MSBR Program at ORNL was cancelled in early 1973, the dimensional changes of graphite during irradiation had been studied for a number of years. These changes depend largely on the degree of crystalline isotropy, but the volume changes fall into a rather consistent pattern.

In the MSBR concept the neutron flux is sufficiently high in the central region of the core to require that the graphite be replaced about every four years. It was further required that the graphite be surface sealed to prevent penetration of xenon into the graphite. Since replacement of the graphite would require considerable downtime, there was strong incentive to increase the fluence limit of the graphite. A considerable part of the ORNL graphite program was spent in irradiating commercial graphites and samples of special graphites with potentially improved irradiation resistance. The approach taken to sealing the graphite was surface sealing with pyrocarbon. Because of the neutronic requirements, other substances could not be introduced in sufficient quantity to seal the surface.

The pyrolytic sealing work was only partially successful. It was found that extreme care had to be taken to seal the material before irradiation. During irradiation the injected pyrocarbon actually caused expansion to begin at lower fluences than those at which it would occur in the absence of the coating. Thus the coating task was faced with a number of challenges.

The most detailed creep data exist on the US and German graphites fore the HTR plant designs . But these graphites, because their coarse granularity and large pore size, are unsatisfactory for molten salt applications. Generally much closer to the MSR requirements are the IG-100 graphite grade produced in Japan for their HTRs and GSP type graphite produced in Russia. Development of sealing techniques should be continuing, both with the «pulse-impregnation» technique and with isotropic pyrolitic coatings put on at a somewhat higher temperatures.

With the relaxed requirements for breeding performance in the new wave of the MSR concepts relative to the MSBR, the requirements for the graphite would be diminished. First, the lessened gas permeability requirements mean that the graphite damage limits can be raised. Secondly, if the salt flow rate through the core is reduced from the turbulent regime, and the salt film at the graphite surface may offer sufficient resistance to xenon diffusion so chat it will not be necessary to seal the graphite. Finally, the peak neutron flux in the place of graphite location can be reduced to levels such that the graphite will last for the lifetime of the reactor plant. The lifetime criterion adopted for the breeder was that the allowable fluence would be about 3×10^{22} neutrons/cm^2. This was estimated to be the fluence at which the structure in advanced graphites would contain sufficient cracks to be permeable to xenon.

Experience has shown that even at volume changes of about 10% the graphite is not cracked but is uniformly dilated. For some non-breeder devices where xenon permeability will not be of concern, the limit will be established by the formation of cracks sufficiently large for salt intrusion. It is likely that current technology graphites could be used to $3 * 10^{22}$ neutrons/cm^2 and that improved graphites with a limit of 4×10^{22} neutrons/cm^2 could be developed. Also, early efforts show promise that graphites with improved dimensional stability can be developed.

6. Fission product cleanup

For molten salt fuels, fission products could be grouped by the three broad classes: 1) the soluble at salt Redox potential fission products, 2) the noble metals and 3) the noble gases.

The MSR would manage the noble gas continuous removal from reactor circuit by purely physical means of stripping with helium. As it was mentioned before the problem here is to prevent the xenon from entering the porous graphite moderator. Such a stripping circuit would remove an appreciable (but not major) fraction of the tritium and a small fraction of the noble and seminoble fission products.

For the noble metals more experimental efforts is required in order to control their agglomeration, adhesion to surfaces and transport in purge gas. If they precipitate as adherent deposits on the heat exchanger, they would cause no particularly difficult problems (only additional decay heating). However, should they form only loosely adherent deposits that break away and circulate with the fuel, they would be responsible for appreciable parasitic neutron captures. If these species were to deposit on the core moderator graphite, they would constitute an even worse neutron situation. To the extent that they circulate as particulate material in the fuel, insoluble fission –product species could probably be usefully removed by a small bypass flow through a relatively simple Ni (or Hastelloy)- wool filter system with low pressure drop.

In MSRs, from which xenon and krypton are effectively removed, the most important fission products poisons are among lanthanides which are soluble in the fuel. Also, the trifluoride species of AnF_3 and the rare earth's are known to form solid solutions so, that in effect, all the LnF_3 and AnF_3 act essentially as a single element. In combination of all trifluorides, AnF_3 solubility in the melt is decreased by lanthanides accumulation. Since plutonium and minor actinides must be removed from the fuel solvent before rare earth's fission products the MSR must contain a system that provides for removal of TRUs from the fuel salt and their reintroduction to the fresh or purified solvent. A number of pyrochemical processes (reductive extraction, electrochemical deposition, precipitation by oxidation and their combinations) for removing the soluble fission products from the fluoride based salt have been explored in last years. Studies of the full scale fuel salt chemical processing system are not as far advanced, but small scale experiments lead to optimism, that a practicable system can be developed.

Reductive extraction of the fission products in bismuth is being developed most deep in ORNL. The Bi-Li alloy is used for reductive extraction from different melts according to the scheme:

$$MF_n \ (melt) + nLi^0(Bi) = M^0(Bi) + nLiF \ (melt) \tag{15}$$

In this process the rare earths are extracted from a fuel salt stream into lithiated bismuth after all the plutonium and minor actinides have been previously removed.

Taking in to account that under experimental conditions used LiF and individual activity coefficients were essentially constant, the equilibrium constant for this reaction can be written as:

$$K' = \frac{X_M}{X_{MF_n} X_{Li}^n} \tag{16}$$

in which X is a mole fraction. The distribution coefficients for component M is defined by:

$$D = \frac{mole\_fraction\_of\_M\_in\_bismuth\_phase}{mole\_fraction\_of\_MF_n\_in\_salt\_phase} = \frac{X_M}{X_{MF_n}} \tag{17}$$

Combination of (15) and (16) gives in logarithmic form,

$$\lg D = n \lg X_{Li} + \lg K' \qquad (18)$$

The easy with which one component can be separated from another is indicated by the ratio of their respective distribution coefficients, i.e., by separation factor Θ:

$$\Theta = (X_{An}/X_{AnFn})(X_{LnFn}/X_{Ln}) \qquad (19)$$

where X_{An}, X_{Ln} - mole fraction of An and Ln in alloy, X_{AnFn}, X_{LnFn} - mole fraction of AnF_n and LnF_n in a molten salt, The separation factors Θ of AnF_n and LnF_n are approximately equal to 10^3 for the Li,Be/F and Li,Na,K/F solvents. These values are very convenient for the lanthanide's separation by the reductive extraction. However, when the melt is complicated by the addition of large quantities of ThF_4 the situation becomes considerably less favourable (Θ_{Ln-Th} ~1-3,5 under the desired operating conditions $C_{Li} >$ 0,1 at.%). Rare earth removal unit based on Bi–Li reductive extraction flowsheet developed in ORNL for $LiF-BeF_2$ solvent system could provide negligible losses of TRU ($\approx 10^{-4}$) by use of several counter current stages.

Note the following drawbacks of reductive extraction as applied to MSR:
- less favourable scheme of An and Ln separation due to decreased difference in thermodynamic potentials of An and Ln (alloys with liquid metal);
- materials compatibility pose substantial problems;
- poor separation of thorium from rare earths for fluoride system; it can be made by use of LiCl;
- changes in fuel composition because of the significant amount of lithium required; rare earths are removed in separate contractors in order to minimise the amount of Li required.

Although the metal-transfer process appears to give the best fuel salt purification in case of processing system with relatively short cycle (10-30 days), there are other possibilities for rare earth removal that are perhaps worth keeping in mind. If bismuth containing system prove expensive or if unseen engineering difficulties (e.g. material required) develop, the other methods may be applicable, especially at longer processing cycle times. If treatment of a MSR fuel on a cycle-time of 100 days or more is practicable, such an oxide precipitation might be used for periodic fuel cleanup.

For example, in Russian experiments a successful attempt was made to precipitate mixed uranium, plutonium, minor actinides and rare earths from LiF–NaF molten salt solution at temperatures 700–900°C by CaO (Al_2O_3) oxidation. Particularly, it was shown that La and Ce precipitate from the melt by reactions:

$$2LnF_3 + 3CaO = Ln_2O_3 + 3CaF_2 \qquad (20)$$

The rare earths concentration in the molten salt solution was about 5–10 mole %. It was found the following order of precipitation in the system U–TRU–Al–Ln–Ca. Essentially all U and TRU were recovered from the molten salt till to rest concentration $5 \cdot 10^{-4}$ %, when rare earths still concentrated in solution. For U/Th fuelled salts this suggests that, after the Pa and U have been precipitated as oxides, the thorium may also be separated out in this way, leaving the carrier salt containing TRU and rare-earths for next processing steps.

Main advantage of a method of processing of fuel composition by a sequential sedimentation of its components by oxides, nonsoluble in fluoride melts, is the simplicity of the equipment for processing unit and more acceptable corrosion of structural materials in comparison with reductive extraction. Note, that recovery of oxide precipitates from a molten salt need further development.

Main advantage of electrochemical methods is a possibility for the fuel clean up without introduction to the melt of additional reagents, which could change salt composition and influence its chemical behaviour and properties. Some processing flowsheets with electrochemical deposition on solid electrodes, at first of all TRU elements, and after that the fission products, by the same way, or as alternative, for example, by oxides precipitation are possible. The reintroduction TRU in molten salt could be carried out electrochemically, or by simple dissolution, for example, with HF use.

In principle, An / Ln separation on solid electrodes could be more attractive in comparison with liquid electrodes. For the first case the overall reaction like as:

$$Ln + AnF_3 = An + LnF_3 \qquad (21)$$

and the later:

$$Bi(Ln) + AnF_3 = Bi(An) + LnF_3 \qquad (22)$$

In first case equilibrium in the separating process is carried out at the maximum value of a difference of thermodynamic potentials for actinides and lanthanides.

Regarding technological aspect the electrochemical method has the important advantage consisting in possibility of a continuous quantitative control of the process. During the process its rate and also depth of fuel processing are set and controlled by the value of a potential on electrodes of an electrolytic bath.

Note the following drawbacks of this method:
- Space limitation on processes area (bath electrodes), decreased capacity of units as contrasted to chemical processes in volume, especially for the end phase of the process.
- Necessity of dendrites control when use solid electrodes.

Several combinations of the preferred process with some of the alternatives are possible to be used for MSR fuel cleanup unit. A drawback with electrochemical deposition and precipitation by oxidation is that they are intrinsically of batch type, and so likely to be more expensive on a large scale than that the continuous mode preferred for industrial purpose.

With discussed above and other possibilities for chemical processing relatively unexplored there seems to be much room left for interesting studies in the units when TRU are the fuel.

Summary.

The use of the molten salts as the fuel material has been proposed for many different reactor types and applications. Particularly in US, Russia, France and Japan molten salt fuelled reactor concepts have been prepared for fast breeders and thermal reactors. Though, molten salt nuclear fuel concept has been proven by successful operation experience of MSRE experimental reactor at ORNL, this approach has not been implemented in industry.

It is obvious from the discussion above that use of molten fluorides as fuel and coolant for a reactor system of energy production and incinerator type faces a large number of formidable problems. Several of these have been solved, and some seem to be well on the way to solution. But it is also clear that some still remain to be solved. The molten salts have many desirable properties for such applications, and it seems likely that – given sufficient intellectual effort, development time and money - a successful TRU free or transmuting system could be developed. Performing of some additional experimental work give us possibility to understand the practicability of operating in MSR transmuting system.

In contrast with well-established $2LiF-BeF_2$ solvent system, which could not be optimal for MSR transmuting system case, for the other alternative solvents there is essential uncertainty in fundamental data and it is necessary to estimate their potential for MSR fueled by plutonium and minor actinide trifluorides. The general uncertainties in the MSR transmuting system conceptual design are in the areas of $TRUF_3$ chemistry, Redox control for the U free fuel matrix, tritium confinement, fuel salt processing, maintenance procedures, choice of the intermediate salt and behavior of some fission products.

7. Acknowledgement

The author would like to thank L. Mac Toth former from ORNL (now retired), as well as R. Zakirov from RRC-Kurchatov Institute for permission to use in this lecture some their findings and illustrations.

8. References

1. Afonichkin V.F.e.a. (1996) Interaction of actinide and rare-earth element fluorides with molten fluoride salts and possibility of their separation for ADTT fuel reprocessing, *In Proc. of the second international conference on ADTTA*, Kalmar, Sweden, June 3, pp.1144- 1155
2. Barton C.J. (1960) Solubility of plutonium trifluoride in fused alkali fluoride beryllium mixture, *J.Phys. Chem.*, **64**, pp.306-312
3. Cantor S. (1963) Physical properties of molten-salt reactor fuel, coolant, and flush salts, *ORNL-TM-2316*
4. De Van H., e.a. (1993) Materials considerations for molten salt accelerator based plutonium conversion system, *In Proc. of the Global'93 international conference*, Las Vegas, USA
5. Engel J.R., e.a. (1979) Development status and potential program for development of proliferation-resistant molten salt reactors, *ORNL/TM-6415*
6. Gorbunov V.F. e.a. (1976) Experimental studies on interaction of plutonium, uranium and rare earth fluorides with some metal oxides in molten fluoride mixtures, *Radiochimija*, **17**, pp.109–114,
7. Haubenreich P. (1973) Molten salt reactors concepts and technology, *J. Brit. Nucl. Energy Soc.*, **12** (2), p.147
8. .Ignatiev V.V., e.a. (1999) Experimental study of molten salt reactor technology for safe, low-waste and proliferation resistant treatment of radioactive waste and plutonium in accelerator-driven and critical systems, *In Proc. of the Global'99 international conference,* Jackson hole, USA,.
9. Mac Pherson H. (1985) The Molten Salt Reactor Adventure. *Nuclear Science Engineering*, **90** (4), pp.374-380,
10. Ignatiev V.V., e.a. (1990) Molten salt reactors: perspectives and problems, *Energoatomizdat*, Moscow

ELECTROCHEMICAL TECHNIQUES
Some Aspects of Electrochemical Behaviour of Refractory Metal Complexes

S.A. KUZNETSOV

Institute of Chemistry Kola Science Centre, 14 Fersman Str., Apatity, 184200 Russia

1. Introduction

Molten salts have been used in industrial metallurgical processes for many years. For example, the production of aluminium metal by electrolysis of alumina-cryolite melt. Now the considerable attention is devoted to the use of molten salts in such applications as fuel cells, nuclear reactors, as well as the processes of catalysis, electrosynthesis, electrowinning and electrorefining. Indeed, electrowinning and electrorefining in molten salts is a promising method for high purity refractory and rare-earth metals production. Electrochemical synthesis is an effective method for obtaining alloys of these metals in melts. Another important problem concerning molten salts is recycling of spent fuel. Not withstanding the active research, especially in electrowinning and electrorefining the important fundamental problems of electrochemistry in molten salts remains outside the field of vision or have not been addressed in depth. At the same time the electrochemical techniques offer information not only of electrochemical character, but it is powerful method to study complex formation, chemical reactions in melts and to determine thermodynamic properties of molten salts. Electroanalytical techniques provide unique possibility to monitor the concentration and control the technological processes in situ. There are several excellent reviews that deal with electrochemical techniques [1-3]. The purpose of this paper is to focus on the problems of electrochemical techniques, which require a special attention and to describe some aspects of electrochemical studies in molten halides.

2. Methodical aspects

2.1. PREPARATION OF HALIDE ELECTROLYTES

Impurities in molten salts can influence significantly the mechanism and kinetics of electrode processes. Considerable efforts including chemical, physical methods, application of gaseous reagents and pre-electrolysis are required to prepare the high quality salts. Some specific methods of halide melts preparation can be found in review [1]. In our opinion now the main method of high quality preparation alkali halide salts is the gaseous pre-treatment of melt with further application process of recrystallization or zone melting or using only the last processes, at least twice.

2.2. DETERMINATION OF IMPURITIES IN MOLTEN HALIDES

The main impurities in molten salts the up-to-date level of its preparation are oxide, hydroxide ions and water. There are some good methods for their determination: electrochemical, spectroscopic, thermal measurements and LECO-analysis. Electrochemical methods are more flexible by comparison with others.

2.2.1. Electrochemical Methods
Probably, the first indication on a possibility of polarographic oxide ions determination in molten salts was published in the paper [4]. Among the number of electroanalytical techniques cyclic voltammetry, square wave voltammetry, differential voltammetry and in some cases stripping voltammetry can be recommended for analysis. Cyclic voltammetry is most suitable and has received the most extensive application in molten salts [5-10]. Cyclic voltammetry can be applied to estimate the melt quality before purification procedures, and significant peak current densities>1 mA cm^{-2} at 100 mV s^{-1} scan rate are indicative of the need purification. White [1] suggested to use standard scan rate 100 mV s^{-1}, to mention a working electrode material, its area, type of reference electrode and temperature for comparison of melts quality especially for chloride, bromide and iodides.

In many papers [4-10] it was found that the height of the oxide-ions oxidation peak, measured on a glassy carbon working electrode was directly proportional to the amount of oxide added to the melt. The oxide and hydroxide ions concentration, as well as water was determined in some studies for common melts such as FLiNaK, LiCl-KCl, NaCl-KCl, MgCl$_2$-KCl etc. [1]. For example, the next equations for the impurities detection on the base of cyclic voltammetry measurements were obtained in LiCl-KCl melt [5]:

$$C_{oxide} = 9.7314 \cdot 10^{-7} j_p^a T^{1/2} v^{-1/2} exp2540/T \quad (1)$$

$$C_{hydroxide} = 6.5383 \cdot 10^{-7} j_p^c T^{1/2} v^{-1/2} exp2629/T \quad (2)$$

$$C_{water} = 7.2331 \cdot 10^{-7} j_p^c T^{1/2} v^{-1/2} exp2629/T \quad (3)$$

where T is the temperature, K; v is the scan rate, V s^{-1}; j_p is the observed peak current density, mA cm^{-2}; C is the concentration, mol cm^{-3}.

However, the voltammograms usually were rather noisy, making it difficult to determine precisely the peak current. Therefore, instead of the measurements the peak current, the area below the voltammetric curve has been used for determination of the oxide content [11-12].

2.2.2. Drawbacks of electrochemical methods
It is known [13] that the anodic reaction of the oxide on carbon electrode can be described by two-stage scheme:

$$O^{2-} + xC - 2e^- \rightarrow C_xO_{(ads)} \quad (4)$$

$$O^{2-} + C_xO_{(ads)} - 2e^- \rightarrow CO_{2(g)} + (x-1)C \qquad (5)$$

The process (4) reflects the reaction of electrochemical adsorption, and (5)- electrochemical desorption. The electrochemical process (5) is accompanied by reaction of thermal decomposition of compounds $C_xO_{(ads)}$ due to reaction:

$$C_xO_{(ads)} \rightarrow nCO_{(g)} + mCO_{2(g)} \qquad (6)$$

It is clear from reaction (4-5) that the electrode surface area is changed after measurement and it is necessary to renew the glassy carbon electrode regularly. After each experiment the electrode must be replaced by a new one and such procedure leads to error in the surface area determination. Thus for the estimation of oxide-ions concentration in the melt the alternating current (A.C.) impedance techniques are preferable.

Another error can be obtained as a result of interaction of $CO_{2(g)}$ which's evolution takes place during anodic polarization with the flow of oxide species O^{2-} moving to the electrode with the formation of carbonate ions:

$$CO_{2(g)} + O^{2-} \Leftrightarrow CO_3^{2-} \qquad (7)$$

Carbonate ions oxidize at more positive potentials by comparison with oxide ions [13].

In chloride melts in some cases the determination of oxide ions by electrochemical methods is impossible because the potential of $TaOF_6^{3-}$ complexes, for example, has more positive value than the potential of chlorine evolution [14].

The certain difficulties appear when the two different type of oxide-containing complexes coexist in the melt. In this case the calibration line of peak current or integrated current versus oxygen concentration must change the slope, because the kinetic parameters of electroreduction of these complexes are different. The linear dependence of the peak current on the amount of Na_2O added to the $FLiNaK-K_2TaF_7$ melt until the ratio O/Ta=2, which was found in the paper [15] is questionable. Later [11] for the same system it was found the deviation from the straight line at the ratio O/Ta=0.4 due to existence of $K_xTa_2OF_{8+x}$ and K_2TaOF_5 complexes.

2.3. CONTAINER MATERIALS

In review [1] it is summarized the compatibility of molten salts with glasses, ceramics, different metals and refractory materials available. It should be of the certain care to take these recommendations into account. In our opinion oxide materials are not advisable to use not only for fluoride melts, but for some chloride melts also.

The reaction between titanium chlorides and oxide refractory materials with the formation of lower titanium oxides is mentioned in [16]. The following schemes of the processes are proposed:

$$MeO + TiCl_2 \Leftrightarrow MeCl_2 + TiO \qquad (8)$$
$$3MeO + 2TiCl_2 \Leftrightarrow 3MeCl_2 + Ti_2O_3 \qquad (9)$$

However, in the electrolysis of melts containing $TiCl_2$, a metallic coating is formed on the walls of the container (SiO_2), and the cathodic metal has high purity with respect to silicon. This metal formation is stated [16] to be due to the disproportionation (DPP):

$$2TiCl_2 \Leftrightarrow TiCl_4 + Ti \qquad (10)$$

however the equilibrium constant (10) in NaCl-KCl melt at 1000 K has the order of 10^{-5} and even providing the nonequilibrium conditions does not lead to significant metallization of crucibles. These contradictions can be eliminated if we assume that coating formation on oxide materials in the case of titanium (zirconium) occurs mainly because of the reactions of the type:

$$3TiCl_2 + MeO_2 \Leftrightarrow 2TiOCl \downarrow + MeCl_4 + Ti \qquad (11)$$

The conclusion is also in agreement with the results of [17], where it was determined that the reaction of lower titanium and zirconium halides with quartz is accompanied by the reaction of disproportionation:

$$19/8 TiCl_2 + 3/8 SiO_2 \Leftrightarrow 3/4 TiOCl + 1/8 Ti_5Si_3 + TiCl_4 \qquad (12)$$
$$2ZrCl_2 + 1/2 SiO_2 \Leftrightarrow ZrSi + 1/2 ZrO_2 + ZrCl_4 \qquad (13)$$

As the results of the silicides formation the disproportionation equilibrium of $TiCl_2$ and $ZrCl_2$ shifts sharply toward their decomposition. The interaction of silica with lowest chloride of refractory metals was also established in [18-19]. The contamination of the melts containing $AlCl_3$ due to the contact with quartz or glass was observed at temperature 473 K [20]. The silica was corroded by the melt $NaCl$-KCl-$CsCl$-$NbCl_5$ [21] and tantalum pentachloride interacted with quartz at temperature higher than 773 K [22]:

$$2TaCl_5 + SiO_2 \Leftrightarrow 2TaOCl_3 + SiCl_4 \qquad (14)$$

In paper [23] the interaction of the melt NaCl-KCl K_2NbF_7(10 w/o), which was in equilibrium with niobium metal with different oxides compounds was investigated. The sample with the highest energy of Me-O bond rupture showed the lowest corrosion.

Boron nitride crucibles are widely used for study of molten salts especially for fluoride melts.

The comparison of the sintered and pyrolytic boron nitride behaviour was studied [24] in the following molten salts: 1) KCl-NaCl-K_3AlF_6 (15w/o); 2) LiF-BaF_2-LaF_3 (40w/o)-La_2O_3 (2w/o); 3) KCl-NaCl-K_2HfF_6 (15w/o); 4) KCl-NaCl-K_2ZrF_6 (15w/o); 5) KCl-NaCl-K_2NbF_7 (10w/o); and 6) KCl-NaCl-K_2TaF_7 (10w/o).

A monitoring of sample weight changes while holding in molten salts showed sintered boron nitride to gain weight in all melts (1-6). The increase was the difference between the weight gain by electrolyte penetration and losses due to interaction of boric oxide in the boron nitride with fluoride salts to form oxofluoride complexes. As the result of interaction of the sintered boron nitride with melts (3) and (4) its surface after the experiment contained HfO_2 and ZrO_2, respectively. In the pyrolytic boron nitride studied no changes in sample weights were found in melts (1)-(4) and no interaction products therewith. At the same time when the pyrolytic and sintered boron nitride samples were in contact with melts (5) and (6) containing K_2NbF_7 and K_2TaF_7, the electrically conductive films of NbN and TaN formed on the sample surfaces. Niobium and tantalum formation occurs directly as a result of interaction between boron nitride and potassium fluoroniobate and fluorotantalate by the reactions:

$$5BN + 3\ K_2NbF_7 \rightarrow 3NbN + 5KBF_4 + KF + N_2\uparrow \quad (15)$$
$$5BN + 3\ K_2TaF_7 \rightarrow 3TaN + 5KBF_4 + KF + N_2\uparrow \quad (16)$$

The formation of boron fluoride complexes was confirmed by voltammetric studies and the nitrogen evolution by gas chromatography. The Gibbs energy for these reaction turned out to be -41.77 and -43.64 kJ mol^{-1} respectively. The interaction of sintered boron nitride is further complicated by the reaction of boric oxide therein with potassium fluoroniobate and fluorotantalate to form niobium and tantalum oxofluoride complexes.

Recently the formation of conductive NdB_6 film was found on the surface of the sintered boron nitride (0.2-0.4 w/o of oxygen) after its interaction with the $NdCl_3$-$NdCl_2$ melt at temperature 1133 K. At the same time the surface of pyrolytic boron nitride had not changes after the same test. The reason for such a different behaviour is boric oxide impurity in sintered boron nitride which led to the disproportionation:

$$3NdCl_4^{2-} + 2O^{2-} \rightarrow 2NdOCl + Nd + 10Cl^- \quad (17)$$

The metallic containers are not universal material also. For example, the molybdenum crucible is not indifferent to chlorides melts containing $ReCl_4$ [25] or $EuCl_3$ [26] due to reaction of the type:

$$2Eu^{3+} + Mo \Leftrightarrow Mo^{2+} + 2Eu^{2+} \quad (18)$$

The nickel crucibles application is more limited than molybdenum containers because the potential of nickel electrooxidation is more negative by comparison with molybdenum. Probably the best material for crucibles is a glassy carbon, but in some cases the formation of carbides on their surface is possible.

Thus the above discussed results should be taken into account for the choice of container materials.

2.4. ELECTRODES

2.4.1. *Working Electrodes*

A great attention should be paid to the selection of working electrode. The most frequently used materials for the indicator electrode are platinum, molybdenum, tungsten, silver, nickel, pyrographite and glassy carbon. The alloy formation is common process during the metal deposition from molten salts [27-32]. The process of formation of intermetallic compounds proceeds with depolarization and is characterized by the appearance of additional waves on the voltammetric curves, which complicates investigation of the electroreduction mechanism. The number of waves on voltammograms due to alloy formation depends on the experimental conditions: the nature of substrate and deposited metal, temperature, sweep rate, reverse potential value, electrolyte composition, etc. [30]. It is necessary to note that the process of electron transfer and alloy formation is one elementary act. In paper [28] it has been shown that alloying process can be studied by electrochemical techniques such as cyclic voltammetry and current reversal chronopotentiometry. In contrast to one peak of HfF_6^{2-} complexes discharge on molybdenum electrode [33], a number of peaks or potential plateaux were obtained on copper electrode [28]. Chronopotentiometry was used to study the process of alloy formation in the system Cu-Hf. A constant current electrolysis was used to deposit a certain amount of hafnium at the surface of a copper electrode and then the current switched off. The chronopotentiogram recorded shows a five potentials plateaux corresponding to two-phase equilibrium. The Gibbs energy of formation of the alloys $HfCu_4$, $HfCu_3$, Hf_2Cu_5, Hf_2Cu_3 and Hf_2Cu were calculated from the potentials of the two-phase systems [28].

The ignorance of alloy formation process led to numerous mistakes in interpretation of electrode reactions mechanism as well as in determination of kinetic parameters.

Thus two remarks should be pointed out. From the one side a care must be taken to the nature of working electrode, but on the other hand the analysis of chronopotentiograms provides the thermodynamic information on the alloy formation.

The material of the working electrode is selected mainly on the base of phase equilibrium diagrams: "deposited metal-substrate" and the values of the interdiffusion coefficients in this couple. The working electrode, which forms solid solution with deposited metal with a low or insignificant solubility is preferable.

At the same time, while using metallic indicator electrodes, the fact that the recharge processes of metal complexes can occur in the region of potentials more positive than potential of the metal dissolution [25, 34-35] is not always taken into account. The study of chromium trichloride electroreduction in the NaCl-KCl melt conducted in [34] showed that the equilibrium potential of platinum in the melt partially overlaps the region of the recharge potentials of electrode reaction:

$$Cr(III) + e^- \to Cr(II) \qquad (19)$$

As a result, platinum can dissolve via the reaction:

$$2Cr(III) + Pt \Leftrightarrow 2Cr(II) + Pt(II) \qquad (20)$$

Thus platinum electrode in the NaCl-KCl melt containing Cr(III) is not indifferent. The reversible process (19) is not observed at the platinum electrode, and is clearly recorded at the cathode from glassy carbon of SU-2000 type.

The voltammetric curve obtained at the glassy carbon electrode in the CsCl-ReCl$_4$ melt is characterized by two peaks. The first peak corresponds to diffusion-controlled recharge process:

$$Re(IV) + e^- \rightarrow Re(III) \qquad (21)$$

while the second one is associated with the Re(III) complexes discharge to the rhenium metal, controlled by charge transfer [25, 36]:

$$Re(III) + 3e^- \rightarrow Re \qquad (22)$$

The voltammetric curve obtained at the molybdenum electrode displays only one cathodic wave (22) and dissolution wave of molybdenum. The electroreduction-electrooxidation process Re(IV) + e$^-$ $\Leftrightarrow$ Re(III) is not observed at the molybdenum electrode, because it occurs in the region of more positive potentials than potentials of molybdenum dissolution.

The examples given demonstrate clearly that in some cases the use of metallic electrodes may results in an erroneous interpretation of the electrode processes. Therefore, one of the electrodes used in a study of the electrode processes should be a glassy carbon (graphite), because its use in the region of positive potentials against the halide melts background is confined to the processes of halide ions oxidation.

In the cathodic region, the application of glassy carbon (graphite) electrodes in halides of alkali metals is restricted by the formation of intercalation compounds and alkali metal carbides [37]. The use of glassy carbon electrode for studying of electroreduction refractory metal complexes in details was discussed in [38]. It was shown that in some cases glassy carbon electrode allows to determine correctly the kinetic parameters of the electroreduction process. This is possible despite the refractory metal carbides formation, because their influence on the reduction process involving deposition of a refractory metal at the cathode is insignificant [38].

It is necessary to note that alloy formation process can be observed even prior to the polarization. This is due to phenomenon of currentless transfer.

When the metallic niobium is immersed in the NaCl-KCl-K$_2$NbF$_7$ melt, the metal-salt reaction takes place spontaneously with the formation of the reduced form of niobium [39]:

$$4Nb(V) + Nb \rightarrow 5Nb(IV) \qquad (23)$$

Niobium (IV) complexes diffuse, for example, to the glassy carbon electrode and disproportionate on its surface with the niobium carbide formation [40-41]:

$$5Nb(IV) + C \rightarrow NbC + 4Nb(V) \qquad (24)$$

The energy of carbide formation ΔG_{NbC} is the driving force of reaction (24). Niobium (V) complexes forming in the melt owing reaction to (24) diffuse to the niobium and react with it according to equation (23). Thus the niobium transfer to the surface of carbon electrode becomes a cyclic process, and resulting reaction can be written as:

$$Nb + C \to NbC \quad (25)$$

In the electrolyte NaCl-KCl-K$_2$HfF$_6$ (10 w/o) -NaF(10 w/o) in equilibrium with hafnium on the different substrates Ni, Co or Fe intermetallic compounds HfNi$_5$, HfCo$_6$, Fe$_2$Hf. were obtained. For example, the formation of intermetallic compound HfNi$_5$ can be written as [42]:

$$Hf + 4NaF \Leftrightarrow HfF_4 + Na^0 \quad (26)$$
$$Hf^{4+} + 4Na^0 + 5Ni \Leftrightarrow HfNi_5 + 4Na^+ \quad (27)$$

The transfer of electrons in this case being by cations of alkali metals, which dissolved in the melt.

The phenomenon of catalytic dissolution of copper in the melt, containing complexes of refractory metals occurs by the reactions of type:

$$\to 4Cu(I) + NbF_5^- \Leftrightarrow Nb + 4 Cu(II) + 5F^- \quad (28)$$
$$Cu(II) + Cu \Leftrightarrow Cu(I) \quad (29)$$

with formation of a porous niobium (tantalum, hafnium) coating with a bad adhesion to substrate [43].

So the reactions of currentless transfer change the nature of substrate yet before electrochemical study.

2.4.2. Determination of the Working Electrode Surface Area

It is very difficult to define the surface area of the working electrode in electrochemical methods. This is due to wetting between the electrode and the fused salt. Another reason is the lack of suitable insulator materials, which have enough corrosion resistance and approximately the same thermal expansion coefficient with the electrode material. The surface area of working electrode is generally deduced from immersion length measured through the transparent window with a cathetometer (when it is possible). The result after removing of the electrode from the melt are controlled by visual determination of the surface wetted by melt. However, it is hard to estimate correctly the capillary effects. The error in the immersion length due to capillary effects does not depend on the immersion length and the increment of the electrode area for different immersion lengths can be used for calculation of the kinetic parameters [44].

It is necessary to mark difficulties in the determination of surface area of microelectrodes ($A \leq 1 \cdot 10^{-2}$ cm^2) for the processes with the formation of insoluble product due to the partial coverage of the electrode by metal deposit. For a such type of the processes macroelectrodes have a higher accuracy in determination of electrochemical parameters by comparison with microelectrodes [45].

2.4.3. Reference Electrode

Electrochemical studies as a role require the use of suitable reference electrode to provide a stable and accurate potential. The reactivity of most molten salts is too high that problems arise mainly because of the corrosion action on glass or ceramics used as containers or diaphragms in construction of reference electrode. It is the same high temperature corrosion problem, which was discussed in section 2.3. Undetermined liquid junction, thermoelectric and thermal diffusion potentials give uncertainties in the measurements also. However, despite of these difficulties many reversible reference electrodes have been designed for molten chlorides [1-3, 46].

It is traditionally difficult to find the suitable reference electrode for chloride-fluoride and fluoride melts. Some reference systems for molten fluoride have been suggested [1-3]. The nickel-nickel fluoride electrode is contained in porous boron nitride envelope [47-48]. The electrode was developed for potential measurements in molten LiF-NaF-KF at a temperature 773-823 K. At higher temperatures, the boric oxide binder contained in the boron nitride would be dissolved in the melt and contaminate the electrolyte, thus changing the electrode potential. The Ni(II)/Ni electrode utilizes single crystal of LaF_3 as the membrane [49]. This system was tested for temperature up to 873 K. The electrode is difficult to assemble, and the crystal cracks after a few experiments. In some works the graphite was used as a diaphragm [50] or protective shell [51]. It is not a good idea because the graphite has electronic conductivity and a redox potential of the molten salt system are formed on it.

In many studies that to avoid these difficulties platinum, nickel wires or glassy carbon rods are used as a quasi-reference electrode in molten fluoride. Surely their potential is not define thermodynamically, but the value of electrode potential remains fairly time independent in the melt of certain composition [52]. The electrode potential of such electrodes can be changed both with fluctuations in the melt composition and temperature.

Another idea consists in generating an in situ redox couple by a short time electrolysis and using this system as an internal reference electrode [53-56]. The amount of foreign species introduced into the electrolyte by this technique must be very small to avoid contamination and possible reactions of electrogenerated species with fused salts, disproportionation and product accumulation at the monitoring electrode. In paper [55] the internal reference systems Pt^{2+}/Pt and Cl_2/Cl^- were studied in NaCl-KCl containing increasing amount of NaF. A complex formation with fluoride ions induced a positive shift in the standard potentials of Pt^{2+}/Pt and Cl_2/Cl^- couples. The Na^+/Na, Ni^{2+}/Ni, Mo^{3+}/Mo and Fe^{2+}/Fe couples were examined using convolution voltammetry as internal reference systems in molten sodium fluoride at 1298 K [56]. The Na^+/Na internal reference system not showed reproducibility and was abandoned. The mechanism of Ni and Mo oxidation do not appear to be totally reversible and therefore the systems Ni^{2+}/Ni and Mo^{3+}/Mo cannot be recommended. Only Fe^{2+}/Fe system was found suitable as internal reference electrode for molten sodium fluoride. The Fe^{2+}/Fe system is not universal and usage of this couple is problematic for fluoride melts containing some compounds. This is due to a possible strong interaction of iron electrode with melts because the Fe^{2+}/Fe standard potential is not too positive.

Thus a short survey showed that the problem of proper reference electrode for chloride-fluoride and fluoride melts remains open and further studies are necessary in this field.

2.5. INERT ATMOSPHERE

Most of halide melts show reactivity to water and oxygen, which results in significant changes of electrochemical behaviour. So it is advisable to conduct experiments with molten halide in an inert atmosphere. At present time it is not significant problem, because the furnace with electrochemical cell can be disposed inside of glove box with controlled atmosphere, usually the concentration of oxygen and water in box is less than 2 ppm.

It is clear that only a high quality molten salt, indifferent materials for electrochemical cell and an inert atmosphere permit to get correct electrochemical signal from studying system.

3. Diagnostic criteria for electrochemical processes

For determination of the electrochemical processes nature diagnostic criteria are used. The criteria for different electrochemical methods such as: steady-state voltammetry, linear sweep and cyclic voltammetry, square-wave and staircase voltammetry, chronopotentiometry, chronoamperometry and A.C. techniques are described in the books and papers [57-68]. Whole set of criteria should be used to understand the process nature, because in many publications conclusion on diffusion control was made only on the base of the dependence I_p-$V^{1/2}$ for linear sweep voltammetry and on the basis of transition time measurements for chronopotentiometry. Using only these criteria it is not possible to distinguish reversible and irreversible processes. Correct diagnostics of electrode reaction is essential condition for the choice of proper equations for determination electrochemical reaction parameters.

4. Some aspects of electrochemical behaviour of refractory metal complexes

By varying the anionic and cationic composition of the melt, concentration, and temperature, one is in a position to control the electroreduction of refractory metals [69-70]. Therefore, it is important to study how these factors affect both the number of steps and the mechanism involved in the electroreduction.

4.1. THE ANIONIC COMPOSITION

The number of steps and the mechanism involved in the electroreduction of complexes is affected most strongly, by the anionic composition of the melt, as it determines the composition of the first coordination sphere. The relevant data on the effect of the anionic composition of the melt on the mechanism involved in the electroreduction of refractory-metal complexes in a NaCl-KCl melt are listed in Table 1.

The influence exerted by the anionic composition of the melt depends on element, but a common feature is that an increase in the basicity of the melt reduces the number of discharge steps and causes the discharge of complexes to the metal to change from a reversible to an irreversible process.

4.2. THE CATIONIC COMPOSITION

The effect of the cationic composition of the melt on the number of steps involved in electroreduction for hafnium has been observed on changing from NaCl-KCl to CsCl. When about 70 m/o cesium chloride was added to the NaCl-KCl-HfCl$_4$ melt, the two-step mechanism of electroreduction with a two-electron discharge at each step gave way to a one-step mechanism with a four-electron discharge of hafnium complexes to the metal [73, 82]. This change in the electroreduction mechanism is traceable to the replacement of potassium and sodium cations by cesium in the second coordination sphere of the $HfCl_6^{2-}$ complexes. One-step irreversible discharge of HfF_6^{2-} complexes has also been reported in an equimolar NaCl-KCl mixture [33, 83]. Drawing upon Lux's view on the acidic-basic properties of ionic melts, one may conclude that the increase in the basicity of the melt, due to a change in the anionic or cationic composition, is responsible for the transition from the two-step mechanism with a two-electron discharge at each step in the $NaCl$-KCl-$HfCl_4$ melt [71-72] to the irreversible one-step mechanism in the $NaCl$-KCl-K_2HfF_6 [33] and $CsCl$-$HfCl_4$ [73, 82].

A change in the second coordination sphere can give rise to additional waves on voltammetric curves. It has been established that the NaCl melt contains solely Re(III) complexes, which are characterized by a single wave corresponding to Re(III)+3e$^-$ →Re [25, 36]. When 60 m/o potassium chloride or 70 m/o cesium chloride is added to NaCl, the sodium cations are displaced from the second coordination sphere by potassium or cesium. In the melt, Re(IV) complexes are formed. They are characterized by the electroreduction wave Re(IV)+e$^-$→Re(III) [25]. A higher concentration of CsCl is required because cesium has a lower ionic potential than potassium. The second coordination sphere likewise affects the electroreduction mechanism: the process Re(III)+3e$^-$→Re is reversible in sodium and potassium chlorides, but irreversible in the CsCl melt [25].

TABLE 1. Effect of the anionic composition of the melt on the steps and mechanism involved in the electroreduction of refractory-metal complexes in the NaCl-KCl melt

Element	Mechanism of electroreduction in melts		
	chloride	chloride-fluoride	oxofluoride
Hf [33,71-73]	I Hf(IV)+2e → Hf(II) (R); at v≤0.66 V/s complicated by DPP reaction II Hf(II)+2e → Hf(IR)	I Hf(IV) + 4e → Hf(IR)	
Nb [74-77]	I Nb(V)+e→Nb(IV)(R) II Nb(IV)+ 4e → Nb(R)	I Nb(V)+e→Nb(IV) (R) II Nb(IV)+4e→Nb(IR)	I Nb(V)+5e→Nb(Q); at v≤2.0 V/s complicated by preceding chemical reaction
Cr [35,78-79]	I Cr(III)+e→Cr(II) (R) II Cr(II)+2e→Cr(R)	I Cr(III)+e→Cr(II) (R) II Cr(II)+2e→Cr(IR)	I Cr(VI)+3e→Cr(III) end product Cr_2O_3
Re [25, 36]	I Re(IV)+e→Re(III) (R) II Re(III)+3e→Re(R)	I Re(IV)+e→Re(III) (R) II Re(III)+3e→Re(IR)	I Re(VII)+7e→Re (IR)
Ir [80-81]	I Ir(IV)+e→Ir(III) (R) II Ir(III)+3e→Ir(R)	I Ir(VI) + 4e→Ir (IR)	I Ir(VI)+6e→Ir(IR)

Note: (R), reversible process; (IR), irreversible process controlled by charge transfer rate; (Q) quasi-reversible process.

4.3. TEMPERATURE

The temperature affect on the mechanism of electroreduction niobium complexes has been observed in the LiCl-KCl-NbCl$_5$ melt [74]. At a temperature below 903-923 K the electroreduction of niobium proceeds in a three-step mechanism:

I	Nb(V) + e$^-$ → Nb(IV) (reversible),	(30)
II	Nb(IV) + 2e$^-$ → Nb(II) (irreversible),	(31)
III	Nb(II) + 2e$^-$ → Nb (reversible).	(32)

It should be added that the second step of electroreduction is complicated by the formation of insoluble cluster compounds owing to the reaction of Nb(IV) and Nb(II) complexes. At the second step, NbCl$_{2.33}$ cluster compound and a large assortment of nonstoichiometric compounds in the composition range NbCl$_{2.67}$-NbCl$_{3.13}$ have been obtained by potentiostatic electrolysis. The electrolysis at more negative potentials leads to an increased amount of niobium in the cathodic deposit. At the third step, the Nb(II) complexes still not reacted with Nb(IV), or reduced from cluster compounds, are reduced to the metal. At temperatures above 923 K, the second and third waves merge into a single four-electron wave, and the formation of cluster compounds of the NbCl$_x$, type becomes thermodynamically unfeasible. The voltammograms display only two reduction waves:

Nb(V) + e$^-$ → Nb(IV)	(reversible),	(33)
Nb(IV) + 4e$^-$ → Nb	(reversible).	(34)

At a temperature higher than 923 K, the deposition of niobium is not complicated by the formation of subchloride impurities.

According to [81], when in the cathodic reduction of iridium, Ir(III)+3e$^-$→Ir, in a NaCl-KCl-CsCl eutectic melt the temperature is raised to 873 K, the process changes from irreversible to reversible, and this in turn causes the structure of the deposits to change from coherent to dendritic.

4.4. CONCENTRATION

With an increasing of the metal-containing component concentration, one finds it somewhat difficult to analyze voltammetric curves and to cancel out the ohmic component in obtaining voltammograms.

During the electrodeposition of rhenium coatings from the NaCl-KCl melt at a rhenium concentration of 5.0 w/o and a current density of higher than 0.05 to 0.07 A cm^{-2}, rhenium metal powder and the salt are deposited at the cathode. Voltammograms for the NaCl-KCl electrolyte containing 5.0 w/o of rhenium show an extra wave at more negative potentials. Potentiostatic electrolysis at the potentials of the first wave results in the formation of a rhenium coating. Electrolysis at the potentials of the second wave yields rhenium powder or rhenium sponge. In addition to the metal, a dark-green salt was discovered on the cathode. XRD-analysis and chemical analyses identify this salt as K$_2$Re$_2$Cl$_8$ whose formation can be explained only by the participation in the second discharge process of the polynuclear rhenium complexes that are likely to appear in melts of high rhenium concentration [25, 36]. Strubinger et al. [84] reported the following polynuclear complexes of rhenium: $[Re_2Cl_8]^{2-}$, $[Re_2Cl_8]^{3-}$, $[Re_3Cl_{12}]^{3-}$ and $[Re_3Cl_{11}]^{3-}$. The formation of $K_2Re_2Cl_8$ at the cathode indicates that trinuclear rhenium complexes must be participating in the discharge process along with the formation of binuclear complexes. The reduction of trinuclear rhenium complexes may be assumed to proceed according to:

$$4[Re_3Cl_{12}]^{3-} + 2K^+ + 5e^- \rightarrow K_2Re_2Cl_8 + 5[Re_2Cl_8]^{3-} \qquad (35)$$

4.5. POLARIZATION RATE

An increase in polarization rate, (v) in the $NaCl$-KCl-$NbCl_5$ system, leads to a change in the mechanism responsible for $Nb(V)+e^- \rightarrow Nb(IV)$ recharge successively from reversible ($v<0.2$ Vs^{-1}) to quasi-reversible ($0.2 \leq v \leq 1.0$ V/s) and irreversible ($v>1.0$ Vs^{-1}) [85]. Quite often, high polarization rates offer a way to avoid the effect of preceding or succeeding chemical reactions on the electrode process. For example, a polarization rate of $v \geq 0.66$ Vs^{-1} makes it possible to avoid the effect of the DPP reaction in the Hf(IV) reduction in a NaCl-KCl melt [71], and escape the preceding chemical reaction on the electroreduction of oxofluoride complex $NbOF_6^{3-}$ [75-76]. Different polarization rates can produce different numbers of waves on voltammograms. In the electroreduction of hafnium ($C_{HfCl_4} \geq 6.09 \cdot 10^{-5}$ mol cm^{-3}), the recharge process is complicated by the subsequent DPP reaction:

$$HfCl_6^{2-} + 2e^- \rightarrow HfCl_4^{2-} + 2Cl^- \qquad (36)$$

$$2HfCl_4^{2-} \Leftrightarrow Hf + HfCl_6^{2-} + 2Cl^- \qquad (37)$$

As could be seen on a steady-state voltammetric curve, step (37) had time to reach completion. Therefore, in the region of the diffusion limited current density the process becomes equivalent [71]:

$$HfCl_6^{2-} + 4e^- \rightarrow Hf + 6Cl^-. \qquad (38)$$

At the same time, two peaks are observed on the unsteady-state voltammograms.

4.6. CONNECTION BETWEEN THE NUMBER OF ELECTROREDUCTION STAGES AND THERMODYNAMIC STABILITY OF COMPLEXES

The number of electroreduction steps is connected with the thermodynamic stability of refractory metal complexes of various oxidation states at given acid-base properties of the melt. By voltammetry in the NaCl-KCl melts at a ratio of $[F^-]/[Hf^{4+}] < 5$, three cathodic peaks are associated with the reduction of the Hf-IV species in the chloride-fluoride Hf(IV) complex to Hf(II), the discharge of Hf(II) to the metal and the electroreduction of Hf(IV) fluoride complex to hafnium in one four-electron step [83, 86]. The formation of Hf(IV) fluoride complexes is due to generation of fluoride anions in the vicinity of the cathode as the result of the Hf(II) chloride-fluoride complexes discharge and following ligand exchange reaction. At a ratio of $[F^-]/[Hf^{4+}] \geq 5$ only one cathodic four-electron wave is observed. Thus an increase in the number of fluoride anions in hafnium ligand shell results in Hf(IV) stabilization and a change in the discharge mechanism [83].

The conditions for stabilization of the highest oxidation states of refractory metals in NaCl-KCl melts have been determined in [87]. The changes of the most stable oxidation states of various d-elements in NaCl-KCl molten salt from purely chloride to oxochloride, chloride-fluoride and oxofluoride complexes are given in Table 2.

TABLE 2. Oxidation states of refractory metals in NaCl-KCl molten salt

3d *)	Composition of complexes				4d *)	Composition of complexes				5d *)	Composition of complexes			
	Cl	O-Cl	Cl-F	O-F		Cl	O-Cl	Cl-F	O-F		Cl	O-Cl	Cl-F	O-F
Ti	II-IV	III, IV	III, IV	IV	Zr	II, IV	IV	IV	IV	Hf	II, IV	IV	IV	IV
Cr	II,III	III,V,VI	II, III	VI	Nb	IV-V	V	IV,V	V	Ta	IV-V	V	V	V
										Re	III, IV	V-VII	III, IV	VI, VII
										Ir	III, IV	V, VI	IV	VI

*) - elements; Cl-chloride; O-Cl - oxochloride; Cl-F - chloride-fluoride; O-F- oxofluoride

The stabilization of higher oxidation states is often observed when passing from pure chloride, oxochloride, and chloride-fluoride to oxofluoride melt. This trend revealed for the alkali-halide melts containing d-elements is supported by numerous data on stabilization of the highest oxidation states in solutions and in solid phase. These data show that anionic ligands as F^- and O^{2-} are preferable for stabilizing highest oxidation states. The highest valencies obtained in chloride-fluoride or oxofluoride melts, are often the same. But oxofluoride melts are preferable for stabilizing the highest oxidation states of chromium, niobium, rhenium, and iridium. Obviously in these melts the effect of fluorine electronegativity and the stabilizing effect of electron delocalization typical for oxygen are acting synergistically. For instance, the highest oxidation state +7 was obtained for rhenium in the oxofluoride melt [25]. By addition of oxygen anions to NaCl-NaF melts at various ratios of $[O^{2-}]/[Re^{4+}]$ the Re(VI) and Re(VII) complexes can be stabilized. This stabilization is accompanied by disproportionation reactions [42].

4.7. FORMATION OF COMPLEXES OF DIFFERENT COMPOSITION IN THE VICINITY OF THE ELECTRODE DIFFERING FROM THE COMPLEXES IN BULK OF THE MELT

Voltammetric curves for the NaCl-KCl-NaF (5 w/o)-K$_3$NbOF$_6$ melt at different polarization rates allowed us to discover that, in the voltammograms recorded at $v \leq 0.1$

V·s⁻¹, a second peak appears at more negative potentials [76, 86]. As the polarization rate was decreased, the ratio between the heights of the first and second peak change in favour of the latter peak. Such behaviour observed in the presence of solely monoxofluoride complexes in the melt can be explained by the generation of free oxygen anions by discharge of $NbOF_6^{3-}$ in the vicinity the electrode and by occurrence of the exchange reaction:

$$NbOF_6^{3-} + O^{2-} \Leftrightarrow NbO_2F_4^{3-} + 2F^- \quad (39)$$

Dioxofluoride complexes that are formed by reaction (39) discharge at more negative potentials than monoxofluoride ones [76]. The appearance of the second peak and its height are controlled by the ratio between the rates of polarization and exchange reaction (39).

Earlier [25] the same behaviour was found for monoxofluoride complexes of rhenium in NaCl-NaF eutectic melt.

4.8. REVERSIBILITY-IRREVERSIBILITY OF ELECTROREDUCTION REFRACTORRY METAL COMPLEXES AND THE STRUCTURE OF CATHODIC DEPOSITS

Let us consider how reversibility-irreversibility of the discharge of the refractory metal complexes influences on the formation of coherent coating. One of the most important characteristics for formation coherent coating is the number of crystals Z. Thus, the coating growth on a heterogeneous substrate proceeds on three stages [69]: (1) nucleation of separate crystals on the substrate surface; (2) growth of these crystals to their consolidation in a coherent layer, and (3) collective growth of crystals in the coherent deposit.

The second stage is completed with the consolidation of crystals in a coherent layer only if the separate crystals before their consolidation do not reach such a size that their becomes unstable and, as a result, dendrites are formed. The size of crystals to their consolidation is determined by the following equation:

$$L = 1/Z^{1/2} \quad (40)$$

where Z is the number of crystals on the unit cathode area. One of the conditions for preparation a coherent coating on the substrate is the definite number of nuclei on it.

If in electrolysis of dilute solution salt of deposited metal in solvent melt the growth rate of the crystal nuclei is determined by the rate of diffusion to its surface, then at low phase overpotential Z is determined by the following equation [69]:

$$Z = \frac{2.3 \cdot 10^{-12} K_3^{1/2} T^{3/2} i^{3/2}}{n^{5/2} V^{1/2} D^{3/2} C_o^{3/2} C_c^{1/2} \eta_{max}^{3/2}} \quad (41)$$

whereas at fairly high overpotential it is calculated as follows:

$$Z = \frac{2.8 \cdot 10^{-6} K_3^{1/2} T^{3/2} i^{3/2}}{n V^{1/2} D^{3/2} C_0^{3/2} C_c^{1/2} \eta_{max}^{3/2}} \qquad (42)$$

where n is the ion charge, V is the molar volume of deposited metal, D is the diffusion coefficient of ion in the melt, C_0 is the volume concentration, C_c is the total capacity of the cathode, T is the temperature (K), η_{max} is the maximal value of overpotential, and K_3 is the factor characterizing the substrate reactivity.

From the equations (41) and (42) it follows that the number of crystal nuclei formed will grow up significantly with the decrease in the concentration of deposited metal and diffusion coefficient. In turn, the diffusion coefficient falls with the decrease in temperature and increase in strength of complexes.

The discharge of Cr(II) complexes to metal in NaCl-KCl melt is reversible [88], but a coherent chromium coating can be prepared by electrolysis of the NaCl-KCl-CrCl$_2$ melt. It should be noted that deposition of a coherent chromium coating is possible only at low concentrations of chromium dichloride (2-5 m/o) and cathodic density of 0.025-0.05 A cm^{-2} [89]. At a concentration $\geq$20 m/o the deposit is composed of separate crystals. Under these conditions the coherent coatings can be obtained only if the initial current pulse (0.5-5.0 A cm^{-2}) is applied for 1 s to increase the number of crystallization centres [89].

As shown in [90], the Cr(II) +2 e$^-$ →Cr discharge in the LiBr-KBr-CsBr melt at 543-723 K is limited by diffusion; however, electrolysis of this melt yielded coherent coatings. From the authors' point of view this is due to the low diffusion coefficient of Cr(II) (2.0·10^{-6} cm^2 s^{-1} at 543 K). In the NaCl-KCl-NbCl$_5$ (2 w/o) melt in equilibrium with niobium metal coherent coatings were also obtained, owing to the low concentration of Nb(IV) complexes, though the Nb(IV) + 4 e$^-$→Nb discharge is reversible [74, 85, 88].

The above examples suggest that it is possible to control the number of crystals required to form a coherent coating even in the melts in which metal electroreduction is reversible.

Transition from the reversible discharge to the irreversible process is characterized by decrease in the heterogeneous rate constants of the charge transfer k^0_{fh} and in the exchange current [91]:

$$i_0 = k^0_{fh} n F C_0^{1-\alpha} \qquad (43)$$

The transfer at the electrolyte-nucleus boundary is described by the following equation [92]:

$$i = i_0 \{C_n/C_0 exp[\alpha n F/RT(\eta - \eta_p) - exp[-\beta n F/RT(\eta - \eta_p)]\} \qquad (44)$$

where i is the current of nucleus growth, C_0 is the concentration of ions in the melt, C_n is the concentration of ions near nucleus, η is the cathode overpotential, α and β are the transfer coefficients ($\alpha + \beta = 1$), i_0 is the exchange current density at the melt-nucleus boundary, and η_p is the phase overpotential:

$$\eta_p = \frac{2\sigma V}{nF r_0} \qquad (45)$$

where σ is the surface tension of the melt-nucleus boundary, r_o is the nucleus radius, and V is the molar volume of the metal deposited.

Analysis of this equation [93] shows that the decrease in the exchange current at the melt–nucleus boundary increases the number of forming crystals. This is due to the fact that the growth current reduces with the decrease in the exchange current, and thus the system occurs in supersaturated state for a longer period, which provides a larger number of the nuclei.

Unfortunately, experimental data on the changes in the number of crystals on passing from the reversible to the irreversible reaction are virtually lacking. Only Tarasova et al. [94] found that on passing from the NaCl-KCl-CrCl$_2$ (5.5 w/o) system to the chloride-fluoride melt (F/Cr=10) the number of crystals grows up by an order of magnitude.

Thus the reversible discharge of complexes to the metals imposes certain hinders on electrodeposition of coherent coatings (for some metals this hindrance is probably insurmountable). At the same time, in spite of the reversible discharge of chloride complexes of Nb(IV), Cr(II) and Re(III) to metal, coherent coatings of these metals were obtained [95].

So the irreversibility of electroreduction is desirable, but not mandatory condition for the deposition of coherent coatings at least for niobium, chromium and rhenium.

4.9. MULTI-ELECTRON TRANSFER IN NARROW INTERVAL POTENTIALS AT HETERONUCLEAR COMPLEXES DISCHARGE WITH FORMATION OF COMPOUNDS

The formation of heteronuclear complexes is important not only for high temperature coordination chemistry but also for electrochemistry and electrolytic alloys. Indeed, during the discharge of heteronuclear complexes, conditions are created for the direct formation of an alloy. The refractory boride synthesis by joint deposition of components from molten salts becomes energetically more favourable if the components are complexed in heteronuclear complex ions, consisting of boron–refractory metal bonds [96-100]. The suggestion of the existence of heteronuclear complexes originates from the results of electrochemical experiments rather than from straight-forward structural studies. At the present time spectroscopic data on heteronuclear complexes formation were obtained for low temperature melts on the AlCl$_3$ base [101-105]. The following experimental facts based on voltammograms can be explained by the existence of heteronuclear complexes:

-although the individual components (B and Me) are deposited in irreversible electrochemical reaction, the electrochemical synthesis of some borides takes place as a reversible reaction. For example, for the NaCl-KCl-NaF-KBF$_4$-K$_2$TiF$_6$ system the synthesis wave of TiB$_2$ is reversible and is limited by diffusion of ions in the melt [106], although the deposition of boron in the NaCl-KCl-NaF-KBF$_4$, and that of titanium in the NaCl-KCl-NaF-K$_2$TiF$_6$ system take place irreversibly and are limited by the charge transfer;

-in some cases an initial chemical reaction takes place prior to the charge transfer during electrochemical synthesis of borides; these preceding reactions can be interpreted as formation or dissociation of heteronuclear complexes. For example, in the NaCl-KCl-NaF-KBF$_4$-K$_2$HfF$_6$ [97] and NaCl-KCl-NaF-KBF$_4$-K$_2$TaF$_7$ [96, 98] systems the existence of the initial chemical reaction prior to the charge transfer was found based on electrochemical criteria [67];

diffusion coefficients of electrochemically active species during the electrochemical synthesis of borides (calculated from voltammograms) appear to be much smaller, as compared with those obtained in melts consisting only of boron ions, or ions of the metallic component of the boride. For example, the diffusion coefficients of electrochemically active species (T=973 K) used for synthesis of TiB$_2$, ZrB$_2$ and HfB$_2$ equal $8.9 \cdot 10^{-6}$, $9.9 \cdot 10^{-6}$, $4.0 \cdot 10^{-6}$ cm^2 s^{-1} correspondingly, being 2-3 times lower than the diffusion coefficients of "pure" Ti, Zr, Hf or B-complexes [100]. This is obviously due to the larger size of the electrochemically active species during synthesis of borides compared to those during deposition of metals or boron; indicating the formation of heteronuclear species.

One of the most challenging phenomena in borides electrochemical synthesis is the formation of periodically layered structure of borides, as in the case of tantalum borides [96]. The coating with a total thickness of 30-50 μm was obtained using constant current density and consisting of numerous sub-layers each with thickness less than 1 μm. The formation of such micro-laminated structure was explained by concentration fluctuation near the cathode surface, leading to the appearance of the ever-changing situation of over-saturation and under-saturation of heteronuclear complexes [96].

In the NaCl-KCl melt, containing chromium and aluminium chlorides the formation of heteronuclear complexes was determined, which during discharge formed chromium-aluminium alloy [107]. So in this case it is possible to obtain the alloy of components with significantly different values of standard potentials. It was shown [108-109] that in the AlCl$_3$-NaCl (2:1) electrolyte containing Ti(II) complexes at a certain titanium concentration the TiAl$_3$ intermetallic compound was deposited on the cathode in a wide range of cathodic current density. The authors [108] explained this fact by the [Ti(AlCl$_4$)$_3$]$^-$ electroactive-species formation proved by spectroscopy data [105]. In some studies correlation between composition of heteronuclear complexes and composition alloy was found [110].

It is also very challenging for general electrochemical kinetics a multi-electron transfer in one stage. During electrochemical synthesis of refractory borides 10, or even 11 electrons are transferred in one step in the very narrow potential interval. According to classical electrochemical kinetics, multi-electron electrochemical reactions are interpreted as the consecutive sequence of 1-electron transfer. In the case of electrochemical synthesis, even the application of modern electrochemical technique does not allow the division of the 10-electron process into 10 steps of 1-electron transfer. Modern quantum-chemical calculations proved the energetic possibility of one-step multi-electron processes against the multi-step one-electron processes in high temperature molten salts [111-112].

4.10. FORMATION OF IONIC-ELECTRONIC MELTS DURING DISCHARGE OF ELECTRONEGATIVE METALS

The dicharge of complexes of such negative metals as Hf, Zr, Th, Ti, rare-earth metals and etc., especially, in chloride-fluoride and fluoride melts is accompanied by electroreduction of alkali metals ions. Formation of alkali metals solution results in the appearance of electronic conductivity in the melt, which brings to the shift of the electrochemical interface into the bulk of the melt. Probably, this shift is quite large because the diffusion coefficients of alkali metals have values of 10^{-3}-10^{-2} cm^2 s^{-1} [113] that at least two orders of magnitude higher than diffusion coefficients of electronegative metal complexes. So the process occurs not only on the surface substrate, but in the bulk of the melt also. The appearance of electronic conductivity in the melt leads to the uncertain values of the electrode area and the reactant concentration. For the melt with electronic conductivity the usual equations of voltammetry and other methods are not applicable.

5. Acknowledgements

The author is grateful to the University of Provence and Prof. M. Gaune-Escard (I.U.S.T.I., 13453 Marseille, France) as well as Russian Academy of Sciences (project "High temperature chemical and electrochemical synthesis of new compounds on the base of rare and rare-earth metals) for the financial support of this study.

6. References

1. White, S.H. (1983) *Molten Salt Techniques*, Plenum Press, New York **1**, 19.
2. Lantelme, F., Inman, D., and Lovering, D.G. (1984) *Molten Salt Techniques*, Plenum Press, New York **2**, 137.
3. Minh, N.Q., and Redey, L. (1987) *Molten Salt Techniques*, Plenum Press, New York **3**, 105.
4. Grachev, K.Ya., and Grebennik, V.Z. (1968) *Z. Analyt. Khim. (Rus.)* **23**, 186.
5. White, S.H. (1981) *Ionic Liquids*, Chap.12, Plenum Press, New York.
6. Shapoval, V.I., Taranenko, V.I., Uskova, N.N., and Lugovoi, V.P. (1982) *Ukr. Khim. Zh. (Rus.)* **48**, 835.
7. Yto, Y., Takenaka, T., Ema, K., and Oishi, J. (1983) *Proceedings of the First International Symposium On Molten Salt Chemistry and Technolog.*, Kyoto, Japan, 45.
8. Davianacos, D., Lantelme, F., and Chemla, M. (1983) *Electrochimica Acta* **28**, 217.
9. Brookes, H.C., and Inman, D. (1988) *J. Electrochem. Soc.* **135**, 373.
10. Polyakova, L.P., Elizarova, I.R., Kononova, Z.A., and Polyakov, E.G. (1994) *Zh. Analyt. Khim.* **49**, 1228.
11. Robert, E., Christensen, E., Gilbert, B., and Bjerrum, N.J. (1999) *Electrochimica Acta* **44**, 1689.
12. Rosenkilde, C., Vik, A., Ostvold, T., Christensen, E., and Bjerrum, N.J. (2000) *J. Electrochem. Soc.* **147**, 3790.
13. Nekrasov, V.N., Barbin, N.M., and Ivanovskiy, L.E. (1989) *Rasplavy (Melts)* **3**, 51.

14. Kuznetsov, S.A., Kuznetsova, S.V., and Stangrit, P.T. (1998) *Molten Salts Forum* **5-6**, 379.
15. Polyakova, L.P., Polyakov, E.G., Matthiesen, F., Christensen, E., and Bjerrum, N.J. (1994) *J. Electrochem. Soc.* **141**, 2982.
16. Sukhodskii, V.A., and Gopoienko, V.G. (1960) *Trudy VAMI (Rus.)* **46**, 146.
17. Polyachenok, L.D., Sonin, V.G., Novikov, V.G., and Polyachenok, O.G. (1969) *Zh. Fiz. Khim. (Rus.)* **43**, 2407.
18. Smirnov, M.V. (1973) *Electrode Potentials in Molten Chlorides (Rus.)*, Nauka, Moscow.
19. Pint, P., and Frengas, S.N. (1978) *Trans. IMM* **87**, 29.
20. Berg, R.M., Hjuler, H.A., and Bjerrum, N.J. (1984) *Inorg. Chem.* **23**, 557.
21. Elizarova, I.R., Polyakov, E.G., and Polyakova, L.P. (1991) *Russ. J. Electrochemistry* **27**, 755.
22. Puzynina, N.I., Kulyukin, V.N., and Kamburg, V.G. (1982) *Zh. Fiz. Khim. (Rus.)* **56**, 2205.
23. Kuznetsov, S.A., Polyakov, E.G., and Stangrit, P.T. (1982) *Izv. Vyssh. Uchebn. Zaved. (Rus.)* **4**, 76.
24. Kuznetsov, S.A., Polyakov, E.G., and Stangrit, P.T. (1985) *Russ. J. Appl. Chem.* **58**, 2016.
25. Kuznetsov, S.A. (1994) *Russ. J. Electrochemistry* **30**, 1462.
26. Kuznetsov, S.A., and Gaune-Escard, M. (2001) *Electrochimica Acta* **46**, 1101.
27. Lepinay, J., Bouteillon, J., Traore, S., and et al. (1987) *J. Appl. Electrochem.* **17**, 294.
28. Kuznetsov, S.A., Kuznetsova, S.V., Polyakov, E.G., and Stangrit, P.T. (1990) *Russ. J. Electrochemistry* **26**, 815.
29. Xie, G., Ema, K., Yto, Y., and Minshou, Z. (1993) *J. Appl. Electrochem.* **23**, 753.
30. Kononov, A.I., Elizarov, D.V., Kuznetsov, S.A., and et al. (1995) *J. Alloys and Comp.* **219**, 149.
31. Lantelme, F., and Salmi, A. (1996) *J. Electrochem. Soc.* **143**, 3662.
32. Okano, Y., and Katagiri, A. (1997) *J. Electrochem. Soc.* **144**, 1927.
33. Kuznetsov, S.A., Kuznetsova, S.A., Polyakov, E.G., and Stangrit, P.T. (1988) *Rasplavy (Melts)* **2**, 110.
34. Kuznetsov, S.A., and Stangrit, P.T. (1990) *Rasplavy (Melts)* **4**, 100.
35. Kuznetsov, S.A., and Stangrit, P.T. (1998) *Refractory Metals in Molten Salts*, Kluwer Academic Publisher, Dordrecht **53/3**, 251.
36. Kuznetsov, S.A., Smirnov, A.B., Shchetkovsky, A.N., and Etenko, A.L. (1998) *Refractory Metals in Molten Salts*, Kluwer Academic Publisher, Dordrecht **53/3**, 219.
37. Popov, B.N., Kimble, M.C., White, R.E., and Wendt, H. (1991) *J. Appl. Electrochem.* **21**, 351.
38. Kuznetsov, S.A. (1996) *Russ. J. Electrochemistry* **32**, 1215.
39. Kuznetsov, S.A., and Grinevitch, V.V. (1994) *Russ. J. Appl. Chem.* **67**, 1423.
40. Kuznetsov, S.A., Glagolevskaya, A.L., Kuznetsova, S.V., and et al. (1990) *Russ. J. Appl. Chem.* **63**, 2078.
41. Kuznetsov, S.A. (1999) *Russ. J. Appl. Chem.* **72**, 1187.
42. Kuznetsov, S.A. (1998) *Refractory Metals in Molten Salts*, Kluwer Academic Publisher, Dordrecht **53/3**, 189.
43. Kuznetsov, S.A., Kuznetsova S.V., and Glagolevskaya, A.L. (1994) *Rasplavy (Melts)* **8**, 38.
44. Iizuka, M. (1998) *J. Electrochem. Soc.* **145**, 84.
45. Kuznetsov, S.A., and Stangrit, P.T. (1991) *Rasplavy (Melts)* **5**, 42.
46. Alabyshev, A.F., Lantratov, M.F., and Morachevskii, A.G. (1965) *Reference Electrode for Fused Salts*, Sigma Press, Washington.
47. Jenkins, H.W., Mamantov, G., and Manning, D.L. (1968) *J. Electroanal. Chem.* **19**, 385.
48. Jenkins, H.W., Mamantov, G., and Manning, D.L. (1970) *J. Electrochem. Soc.* **117**, 183.
49. Bronstein, H.R., and Manning, D.L. (1972) *J. Electrochem. Soc.* **119**, 125.
50. Kontoyanis, C.G. (1995) *Electrochimica Acta* **40**, 2547.
51. Polyakova, L.P., Polyakov, E.G., Sorokin, A.I., and Stangrit, P.T. (1992) *J. Appl. Electrochem.* **22**, 628.
52. Mamantov, G. (1969) *Molten Salts: Characterization and Analysis*, Dekker, New York.
53. Wendt, H., Reuhl, K., and Schwarz, V. (1992) *Electrochimica Acta* **37**, 237.
54. Robin, A., and Lepinay, J. (1992) *Electrochimica Acta* **37**, 2433.
55. Berghout, Y., Salmi, A., and Lantelme, F. (1994) *Electroanal. Chem.* **365**, 171.
56. Adhoum, N., Bouteillon, J., Dumas, D., and Poignet, J.C. (1993) *Electroanal. Chem.* **391**, 63.
57. Delahay, P. (1954) *New Instrumental Methods in Electrochemistry*, Interscience, New York.
58. Geirovskiy, Ya., and Kuta, Ya. (1966) *Principles of Polarography*, Academic Press, New York.
59. Bard, A.J., and Faulkner, L.R., (1980) *Electrochemical Methods: Fundamentals and Applications*, Wiley, New York.
60. Galuz, Z. (1994) *Fundamentals of Electrochemical Analysis*, Ellis, Harwood Ltd., Chichester.
61. Christensen, P.A., and Hamnett, A. (1994) *Techniques and Mechanisms in Electrochemistry*, Chapman and Hall, London.
62. David, K., and Gosser, Ir. (1993) *Cyclic voltammetry. Simulation and Analysis of Reaction Mechanisms*. VCH Publishers, Inc., New York.

63. Southampton Electrochemistry Group (1990) *Instrumental Methods in Electrochemistry*. Ellis Horwood, West Sussex, England.
64. Macdonald, D.D. (1997) *Transient Techniques in Electrochemistry*, Plenum Press, New York.
65. Macdonald, J.R. (Ed.) (1987) *Impedance Spectroscopy*, Wiley, New York.
66. Matsuda, H., Ayabe, Y. (1955) *Z. Elektrochem.* **59**, 494.
67. Nicholson, R.S., and Shain, I. (1964) *Anal. Chem.* **36**, 706.
68. Martins, M.A.G., and Sequeira, C.A.C. (1984) *Tecnica (Port.)* **1**, 47.
69. Baraboshkin, A.N. (1976) *Electrocrystallization of Metals from Molten Salts. (Rus.)*, Nauka, Moscow.
70. Inman, D., and Williams, D.I. (1977) *Electrochemistry: The Past and the Future Thirty Years*, Plenum Press, New York.
71. Kuznetsov, S.A., Kuznetsova, S.V., and Stangrit, P.T. (1990) *Russ. J. Electrochemistry* **26**, 63.
72. Kuznetsov, S.A., Kuznetsova, S.V., and Stangrit, P.T. (1990) *Russ. J. Electrochemistry* **26**, 102.
73. Kuznetsov, S.A. (1990) *Russ. J. Electrochemistry* **32**, 1209.
74. Kuznetsov, S.A., Morachevskii, A.G., and Stangrit, P.T. (1982) *Russ. J. Electrochemistry* **18**, 1522.
75. Kuznetsov, S.A., Glagolevskaya, A.L., and Grinevitch, V.V. (1992) *Russ. J. Electrochemistry* **28**, 1344.
76. Kuznetsov, S.A., Kuznetsova, S.V., and Grinevitch, V.V. (1997) *Russ. J. Electrochemistry* **33**, 259.
77. Grinevitch, V.V., Reznichenko, V.A., Kuznetsov, S.A., and et al. (1999) *J. Appl. Electrochem.* **29**, 693.
78. Baimakov, A.N., and Kuznetsov, S.A. (1983) *Izv. Vyssh. Uchebn. Zaved. Tsvetn. Metall. (Rus.)* **1**, 79.
79. Kuznetsov, S.A., and Stangrit, P.T. (1990) *Rasplavy, (Melts)* **4**, 44.
80. Kuznetsov, S.A., Smirnov, A.B., and Shchetkovskii, A.N. (1987) *Russ. J. Electrochemistry* **23**, 690.
81. Kuznetsov, S.A., Smirnov, A.B., and Shchetkovskii, A.N. (1987) *Russ. J. Appl.Chem.* **60**, 1730.
82. Kuznetsov, S.A., Kuznetsova, S.V., and Stangrit, P.T. (1991) *Rasplavy (Melts)* **5**, 19.
83. Kuznetsov, S.A., Kuznetsova, S.V., and Stangrit, P.T. (1992) *Russ. J. Electrochemistry* **28**, 291.
84. Strubinger, S.K.D., Sun, I-W., Cleland, W.E., and Hussey, C.L. (1990) *Inorg. Chem.* **29**, 4246.
85. Kuznetsov, S.A., Kuznetsova, S.V., and Glagolevskaya, A.L. (1996) *Russ. J. Electrochemistry* **32**, 981.
86. Kuznetsov, S.A. (1998) *Molten Salts Forum* **5-6**, 505.
87. Kuznetsov, S. A., and Stangrit, P.T. (1993) *Russ. J. Appl. Chem.* **66**, 1993.
88. Kuznetsov, S.A., (1993) *Russ. J. Electrochemistry* **29**, 1323.
89. Tarasova, K.P., Baroboshkin, A.N., and Martem'yanova, Z.S. (1970) *Tr. Inst. Elektrohim. Ural. Nauch. Tsentra Akad. Nauk SSSR* **18**, 94.
90. White, S.H., and Twardoch, U.M. (1987) *J. Appl. Electrochem.* **17**, 225.
91. Vetter, K.J. (1961) *Elektrochemische Kinetik*, Springer, Berlin.
92. Isaev, V.A., and Baraboshkin, A.N. (1988) *Rasplavy (Melts)* **2**, 112.
93. Isaev, V.A. (1981) *PhD Thesis*. Institute Electrochemistry, Sverdlovsk.
94. Tarasova, K.P., Baroboshkin, A.N., and Martem'yanova, Z.S. (1970) *Tr. Inst. Elektrohim. Ural. Nauch. Tsentra Akad. Nauk SSSR* **17**, 118.
95. Kuznetsov, S.A. (1997) *Russ. J. Appl. Chem.* **70**, 59.
96. Kuznetsov, S.A., Glagolevskaya, A.L., and Belyaevskiy, A.T. (1994) *Russ. J. Appl.Chem.* **67**, 1093.
97. Kuznetsov, S.A., Devyatkin, S.V., and Glagolevskaya, A.L. (1992) *Rasplavy (Melts)* **6**, 67.
98. Kuznetsov, S.A., and Glagolevskaya, A.L. (1996) *Russ. J. Electrochemistry* **32**, 344.
99. Novoselova, I.A. Malyshev, V.V., and Shapoval V.I. (1997) *Teoreticheskie Osnovu Khimicheskoi Technologii (Rus.)* **31**, 286.
100. Devyatkin, S.V, Kaptay, G., Shapoval,V.I., and et al. *Refractory Metals in Molten Salts*, Kluwer Academic Publisher, Dordrecht **53/3**, 73.
101. Oye, H.A., and Gruen, D.M. (1964) *Inorg. Chem.* **3**, 836.
102. Volkov, S.V. (1989) *Koord. Khim. (Rus.)* **5**, 723.
103. Volkov, S.V., Babushkina, O.B., and Buryak, N.I. (1990) *Zh. Neorg. Khim.(Rus.)* **35**, 2881.
104. Volkov, S.V.,Buryak, N.I., and Babushkina, O.B. (1990) *Ukr. Khim. Zh. (Rus.)* **57**, 339.
105. Dent, A., Seddon, K., and Welton, T. (1990) *J. Chem.Soc. Chem. Commun.* **4**, 315.
106. Zarutskiy, I.V., Malyshev, V.V., and Shapoval, V.I. (1997) *Russ. J. Appl. Chem.* **70**, 1475.
107. Kuznetsov, S.A., and Glagolevskaya, A.L. (1995) *Russ. J. Electrochemistry* **31**, 1389.
108. Stafford, G.R. (1994) *J. Electrochem.Soc.* **141**, 945.
109. Stafford, G.R., and Moffat, T.P. (1995) *J. Electrochem.Soc.* **142**, 3289.
110. Kuznetsov, S.A. (1999) *Russ. J. Electrochemistry* **35**, 1301.
111. Gorodyskii, A.V., Karasevskii, A.I., and Matyushov, D.V. (1991) *J. Electroanalyt. Chem.* **315**, 9.
112. Karasevskii, A.I., and Karnaukhov, I.N. (1993) *J. Electroanalyt. Chem.* **348**, 49.
113. Haaberg, G.M., Osen, K.S., Egan, J.J., Heyer, H., and Freyland, W. (1988) *Ber. Bunsen.-Ges. Phys. Chem.* **92**, 139.

ORIGIN AND CONTROL OF LOW-MELTING BEHAVIOR IN SALTS, POLYSALTS, SALT SOLVATES, AND GLASSFORMERS

C. A. ANGELL
*Department of Chemistry, Arizona State University,
Tempe, AZ 85287-1604*

Abstract In this chapter we first analyze the reasons that some substances, ionic or otherwise, have melting points that are low relative to others of their class. In the case of ionic liquids such systems then may remain liquid at ambient temperature, and provide the basis for novel electrolytes and synthetic reaction media, or alternatively may form glassy solids during cooling. We then discuss a variety of examples of systems that may be useful in either of the above ambient temperature liquid categories. We analyse the factors that decide the fluidities of ambient temperature ionic liquids and give data on a class of such liquids that may have superior properties for "green" chemistry and photoelectrochemical cell purposes.

1. Introduction

In this chapter we will address various aspects of the subclass of molten salts that exhibit low melting points relative to the strength of the interactions between the ions. When ions under consideration are small and of high charge, and therefore the interaction strengths intrinsically large, the consequence is the formation of glassy solid phase. In these, the interest content lies in such properties as the transmission of light and communications signals, and perhaps the support of high ionic currents, as in solid electrolytes. On the other hand, when the ions are large and/or of low charge, then the interactions are weak and the result is an ionic liquid which is stable and fluid at room temperature. Such liquids are currently attracting great interest as solvents for low temperature synthetic chemistry purposes ("green" solvents), as well as for electrolyte, and even absorption refrigeration, purposes.

Apart from their applications possibilities, these low-melting systems are of interest for a number of academic reasons. They provide a challenge to theories of melting. The challenge is to understand the relations between molecular-ion interaction strengths and particle shapes, and the competition between energy and entropy that determines the melting point through the ratio of their changes on fusion,

$$T_m = \Delta H_m \Delta S_m \tag{1}$$

If the melting process gives the system access to an unusually large number of additional states relative to the energy increase which occurs when the crystal lattice breaks down, then Eq. (1) tells us that the change will occur at a low temperature. If the forces between the particles are weak then the system will be highly fluid when it melts, and the kinetics of processes that depend on molecular encounters will not be controlled by the solvent viscosity. Such a liquid is a desirable synthetic processes medium. It also satisfies the first requirement of an electrolyte for use in solar cells. On the other hand it may not be of any use as a glass-former because it will quickly crystallize during cooling. To have glassforming properties. the liquid must be rather sluggish at its thermodynamic melting point, so that the fundamental step of crystallization, which is the generation of the first seed or nucleus of the crystal, will be suppressed. This requires that the melting occurs at a temperature which is low relative to t he strength of the forces between the particles.

While the Coulomb energy is an important component of the energy of any ionic system, hence of the forces between its particles, its importance decreases rapidly as the size of the ions increases, and the role of van der Waals forces consequently tends to increase. Understanding well the connection between ion constitution and the van der Waals interaction is a major challenge here, particularly where the fluoride atom contribution comes into play. It is no accident that most of the "green" ionic liquid solvents involve heavily fluorinated anionic species. Unfortunately this is also the source of much of the financial penalty that must be paid in order to reap the advantages of these solvents.

For systems that fail to crystallize on cooling from high temperatures and, therefore, yield glasses, the challenge is one of understanding not only the low melting points, but also the nature, and kinetics, of crystallization and growth of crystal phases from their melts. In addition, some novel and fascinating aspects of liquid state behavior emerge in considering the detailed behavior of the network liquids, BeF_2 and $ZnCl_2$. In this chapter we will give equal weight to consideration to "low melters" of the glassforming variety, and "low melters" of the ambient temperature, ionic liquid-forming, variety.

2. Origin Of Low Melting Points And Glassforming Properties

2.1 SINGLE COMPONENT SYSTEMS

Here we look into the question of why glasses exist and how they can be formed when needed [1]. After all, they shouldn't exist. In almost every case, they are *metastable* with respect to the crystal state of the same composition (if it is a one-component system) or a combination of crystals (if it is a multi-component system). This means they only exist because the system did not have time to reach the lowest available free energy surface. "Not having time" means that crystals don't have time to nucleate and grow during the process of cooling the liquid to below the glass transition temperature. For crystals to nucleate and grow, fluctuations in order must occur, and these will occur more slowly the more viscous the liquid. As noted earlier, liquids which vitrify easily are almost always rather viscous at their freezing temperatures.

Let us try to understand why this happens, so we can know how to *make* it happen if we want to.

Figure 1 shows the free energy vs. temperature relations for a hypothetical system which has a variety of possible crystal states of different lattice energy (the lattice energy is G at T = OK when the reference state is the dilute gas state). For simplicity, we assume the same heat capacity-temperature relations, hence the same curvature of G. Melting occurs when the liquid free energy curve cuts across the free energy curve.

The liquid phase, which has a higher entropy, is drawn to cut across each of the crystal curves. The intersection of G(liquid) with each G(crystal) determines the melting point each would have if transitions between crystal polymorphs were excluded. Clearly the polymorph with the poorest crystal packing (smallest lattice energy) has the lowest melting point, and would accordingly melt to the liquid of highest viscosity.

From this "thought" example, we would conclude that systems which vitrify easily should be those which lack any large lattice energy crystal packing arrangements. We see, by this argument, that the ability of a substance to vitrify is determined in the crystalline state, *not* in the liquid state as is often supposed.

To support this line of thought, we consider the case of the three isomers of the disubstituted benzene, xylene. The liquids all boil within a few degrees of each other, hence we know that the attractive forces between the molecules are roughly the same, and consistent with this, the viscosities of the three isomers measured at the same temperature are very similar. However, only the meta-isomer is observed to form a glass on fast cooling [2]. At a simple level, the explanation follows immediately on examination of the melting points of the three isomers – the meta-isomer has quite the lowest fusion point (185K compared with 211 K for the o-isomer and 216 K for the p-isomer). This means that, consistent with our earlier generalization, the meta-isomer has

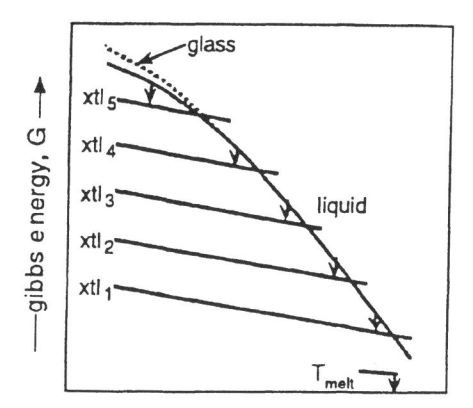

Figure 1. Schematic variation of Gibbs free energy (G) with temperature, showing access to different metastable crystals from the supercooled liquid. The crystal phase with the largest lattice energy (i.e. the one with energy most negative with respect to the ideal gas reference state) has the highest temperature intersection with the liquid curve, hence has the highest melting point. Since liquid viscosity increases super-exponentially with decreasing temperature, the crystal with the lowest lattice energy is the one which melts to the liquid with the highest viscosity. It is this crystal phase, then, that would experience the most difficulty in nucleating crystals during steady cooling.

the largest viscosity at the temperature at which it becomes thermodynamically metastable. Our problem, therefore, becomes one of explaining why the meta-isomer should have the lowest melting point of the three.

The explanation according to our argument above should lie in the difficulty of packing the meta-isomer molecules in long range three-dimensional order, and in this case, we can prove it by calculating that there is a smaller crystal lattice energy for the meta-isomer than for either of the others. This is done in Fig. 2, from ref [3] using precise thermochemical data available for these substances due to Pitzer and Scott [4]. The difference in lattice energy responsible for the ~70 K difference in melting points of m- and p-isomers is only ~1 kcal mol^{-1}.

In summary, therefore, we attribute the stability of certain materials in the amorphous state to the failure of nature to find suitable solutions to the three-dimensional long-range order packing problem in the case of these substances.

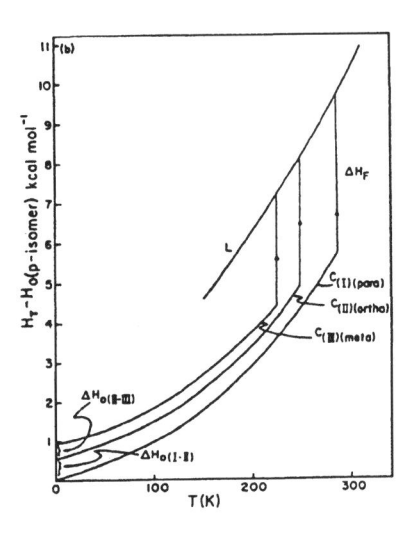

Figure 2. Enthalpy vs. temperature relations relative to enthalpy of p- xylene at *0K* for crystal and liquid isomers of xylene, showing how smaller lattice energy (less negative H$_o$) due to packing problems in case of *m*-isomer leads to low fusion temperature, hence greater glassforming ability for this isomer. (From ref. 3).

2.2 BINARY AND MULTICOMPONENT GLASSFORMERS

The probability of glass formation during cooling of a pure liquid, during cooling at a given rate will usually be enhanced when it incorporated in a liquid mixture. In a mixture, the free energy of the liquid phase is decreased, while that of the phase which is to crystallize usually remains unchanged (the general case in which the crystallizing solid is pure). In Fig. 3, the thermodynamic relations are shown, in relation to the phase diagram, for the case of the binary mixture of o- and p- xylenes [3], which should be almost ideal. The lowering of the m-xylene melting point, simply by the ideal mixing entropy decrease in free energy in the liquid, is enough to confer glass-forming properties on a range of compositions near the eutectic composition.

If the components in the mixture interact attractively, rather than just mixing ideally, then the freezing point of the solvent will be depressed more rapidly than in the case of ideal mixing illustrated in Fig. 3.

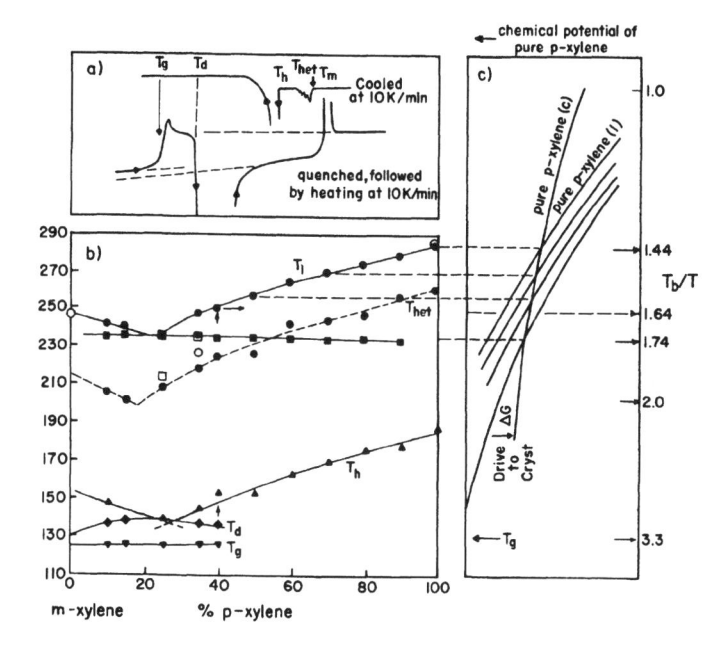

Figure 3. Enthalpy vs. temperature relations relative to enthalpy of p-xylene at O K for crystal and liquid isomers of xylene, showing how smaller lattice energy (less negative H_O) due to packing problems in case of m-isomer leads to low fusion temperature, hence greater glassforming ability for this isomer. (From ref.)

The attractive interaction lowers the free energy of the solvent in the mixture more rapidly than is done by the entropy of mixing alone, hence the effect or the freezing point seen in Fig. 2 becomes stronger. The more the freezing point is depressed, the more viscous becomes the liquid at the point (the liquidus) where thermodynamic instability of the liquid phase is reached. The more viscous the liquid, the slower the growth of nucleogenic fluctuations, the less likely that crystals will form, and accordingly the more likely that the liquid will instead become trapped in the glassy state.

On the other hand, if attractive interactions become too great, a new crystal structure will become favored. The more stable the new compound formed, the higher in temperature one must heat it before the increased entropy of the liquid state leads to free energy crossover of Fig. 1, i.e. fusion. The higher the fusion temperature, the less viscous the liquid at the liquidus temperature and the more probable the crystallization of the *compound* on cooling. Thus very strong interaction between the components of the mixture are not favorable for glass formation. The most favorable situation for low liquids temperatures and glass-forming propensity will therefore be the intermediate case where the solvent-solute interactions are strong enough to depress the liquidus much further than calculated for ideal solutions, but not strong enough to generate a new and competitively crystallizing compound.

It is usually found that in the glassforming region of binary systems, there are one or more crystalline compounds with very low melting points, implying poor structures. The optimum situation for glass formation is when the non-ideal (exothermic) mixing is strong enough to produce a compound, but only just. This maximizes the melting point lowering – thus we name this approach to finding glassforming compositions as the "barely stable compound" approach. This implies that, rather than multicomponent glasses having some especially stable amorphous state arrangement of particles (to be evaluated by scattering techniques, and appreciated for its stability - as is the common approach to interpretation of glass structure), our line of thought would suggest the opposite. Glasses form as a weak, disordered reflection of a three- dimensional crystal structure that is so energetically incompetent that it can barely compete with the disordered form at 0 K.

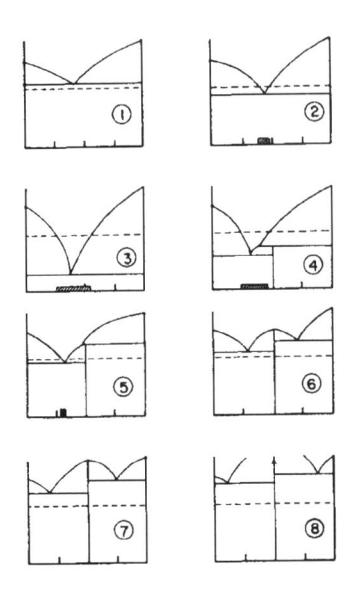

Figure 4. Illustration of the effects of increasing strength of attractive interactions between the components of a binary system, on the form of the phase diagram. Very strong attractions lead to compound formation with high melting points. Intermediate attractions lead to barely stable compounds but strong melting point lowerings due to the stabilization of the liquid over crystal states. Glass formation is most probable in the vicinity of such deep eutectics. A leading source of attractive interactions in inorganic systems is the Lewis acid base interaction. Panel 1 represents ideal mixing. The dashed line represents a temperature common to all systems, as a reference for the liquidus temperature behavior from panel to panel. Panel 4 bears a strong resemblance to the phase diagram for the system $WO_3 + ZrO_2$, in which glassforming ability has recently been found for rapidly cooled samples [5]. (Fig. 4 is reproduced from ref. 6 by permission of Academic Press)

It is quite consistent with the above line of thought that, in metallic systems, those which vitrify most easily are those which do not show the existence of any stable binary compounds, but which have strong solvent-solute interactions, hence deep eutectic temperatures. A good example is the system Au-Si [7]. It is furthermore consistent that, during warm-up of glasses formed during rapid quenching of the binary liquids, crystalline *compounds* can ge generated which are metastable, have no thermodynamically stable range, and later (at higher T) decompose to a combination of the components (or of one component plus a binary solution). stability within that range.

We illustrate the above argument by showing in Fig. 4a a representative series of possible phase diagrams for the system A + B in which the strength of the A - B interaction increases systematically in the sequence 1-8. In the unlikely event that the viscosity were to be independent of composition in all these cases, hence a function only of temperature, then the glassforming composition ranges in the above systems could be related straightforwardly to the position of the liquidus curves, and for a fixed cooling rate would be those indicated in the diagram. Although the glassforming range would be similar in cases 3 and 4, system 3 would have the most stable glass composition (near the deep eutectic). A dependence of viscosity on composition would modify these considerations quantitatively. Nevertheless the basic ideas suffice to qualitatively explain the glassforming ranges observed in many binary system families of limited glassforming ability, *e.g.* the BiCl3-alkali chloride glasses described by Topol *et al.* [8], nitrate glasses [9] and calcium aluminate glasses [10]; in the latter case, melting point lowerings of ~1000 K are involved. The principle has recently [5] been used with some success in the search for glass compositions with large ZrO_2 contents that can serve as oxide ionic conductors.

A recent example relevant to the search for non-crystallizing molten salt systems is that of $LiSCN + AlCl_3$ solutions [11], in which the melting point of the LiSCN is lowered by $AlCl_3$ complexation and the barely stable compound $LiAlCl_3SCN$ is formed. A wide range of compositions around that of the compound is glassforming

The case of the nitrate glassformers, which has become well known. In particular, the composition $40Ca(NO_3)_2.60KNO_3$ now known generally as CKN, has become a model composition for testing theories of liquids. It was, for instance, the system of choice for the first tests [12] of the now-famous mode coupling theory of glassy dynamics [13]. For this reason, and because these systems illustrate so well the principles we have discussed here, we reproduce in Fig. 5 the original set of Group I + Group II (A or B) nitrate phase diagrams published by Thilo. *et al.* [9].

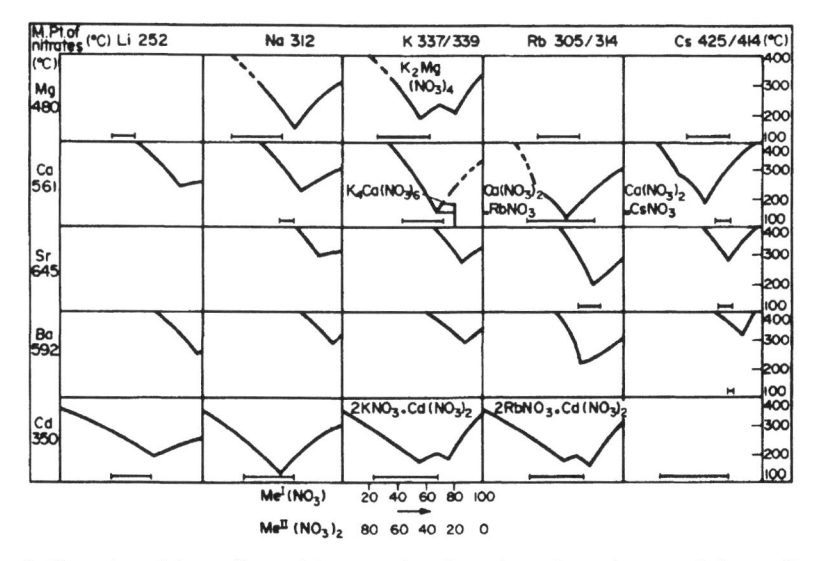

Figure 5. Illustration of how effects of increases in solute-solvent interaction may influence liquidus temperatures in binary systems, hence glassforming composition ranges (hatched bars) and glass. (from ref. 9, by permission)

To the extent that the above line of reasoning is valid, theories for stable glassforming ranges in multi-component systems can be regarded as theories for free-energy lowering of given solid phases on dissolution of solutes, as a function of solute character. Even the famous Zachariasen rules for glassforming substances may be regarded as rules for predicting structures which have lattice energies which are weakly competitive with their amorphous equivalents, hence which have low melting points relative to the forces acting between the particles. There are now known a great many inorganic glassforming systems in which the "glassformer," *e.g.* ZrF4 [12] does not conform to the Zachariasen rules. These are usually mixtures which satisfy the above-discussed "barely stable compound" principle.

3. Low-Viscosity, Low Vapor Pressure, Ionic Liquids

The above line of thought is helpful for understanding the origin of glass-forming ability and accounts for the existence of some systems that are stable at room temperature in the liquid state when this would otherwise not be expected. However, it does not deal adequately with the many cases of molten salts that are now being studied for their high *fluidity* at low temperatures. High fluidity in molten salts at ambient temperature is a key requirement for the use of molten salts as low vapor pressure reaction media. These are cases in which the alkali cations of interest to battery technology have been replaced by large organic cations, which are combined with anions of special character to ensure high fluidity, see below. They have become the subject of widespread interest due to their combination of the low vapor pressures typical of systems with long range attractive potentials (cf. liquid gallium) with low melting and, in many cases, also hydrophobic character.

To obtain the low viscosity ambient temperature molten salts of current interest, these cations must be combined with anions of low coulomb potential (low negative charge density z/r), and/or low polarizability. For a long time the only cases under study were isolated nitrates of alkylammonium cations (e.g ethylammonium nitrate which has been known since early in the last century) and organic cation salts of the tetrahaloaluminate anion $AlCl_4^-$ or its dimeric cousin $Al_2Cl_7^-$ elaborated in particular by Wilkes and co-workers [15]. These were exploited for their utility in room temperature solvent-free battery systems, although their instability against reduction to polymeric forms by the action of alkali metals has prevented them from finding widespread application.

In 1983 Cooper and the author [16] discovered that replacement of the iodide anion in low-melting tetrra-alkyl ammonium iodide by the tetrafluoroborate anion, resulted in a stable ambient temperature ionic liquid of low viscosity and high electrical conductivity. Although these authors remarked that " This room temperature molten salt has a remarkable conductivity of 1.7×10^{-3} (ohm.cm)$^{-1}$ at 25°C and may in fact represent an important development in non-aqueous supporting electrolytes for ambient temperature applications", there was no follow-up until Cooper and Sullivan in 1992 [17] made a systematic examination of large cation salts of perfluorinated anions, and set the stage for the now-burgeoning field of ambient temperature molten salt chemistry.

These liquids are of interest to several diverse groups of scientists. Firstly there is a branch of low temperature molten salt electrochemistry for which the existence of solvent-free but ambient temperature-mobile liquids is of interest. The interest content lies mainly in their potential use as alkali battery electrolytes [18], but there are other possible electrochemical applications. Secondly there is the related field of photoelectrochemistry in which the non-volatility, and the ability of redox-active species to diffuse rapidly between electrodes, are of equal importance (and the conductivity is of lesser significance) [19]. Finally there is the field of synthetic chemistry [20] to which the electrical conductivity of these salts is irrelevant but the combination of low viscosity with low vapor pressure is paramount.

To understand the low melting behavior of this family of salts, some of which fuse well below 0°C, we need to implement a line of thought that is different from that given above for mere glass-forming ability (though many of the new low-melting systems indeed have glass-forming ability). The requirement is that the energy needed to move particles past one another is small, whereas the energy needed to separate the particles completely (into gas phase molecules) is large. This requires a special combination of short and long range forces.

Fig. 6 shows the components of the anion-cation interaction that are important for understanding the existence of room temperature molten salts. Firstly there is the coulomb interaction which varies slowly with ionic separation and acts as if the charges are localized at the centers of the ions. Provided that the anions and cations arrange themselves into a quasilattice rather than forming quasi-molecular species, then the equivalent of the crystal Madelung energy ensures that a amount of large energy must be accumulated locally before an ion pair can escape into the vapor phase. By the Boltzmann factor, this represents a very improbable fluctuation at ambient temperature and so the vapor pressure of molecules (ion pairs) over any such composition that remains liquid
at ambient temperature, is very low. This is primary requirement for most applications of ambient temperature molten salts.

By contrast with the energetics of escape from the liquid surface,
the energetic requirements for local rearrangements can be small if the short range attractions are weak. The short range interactions in a molten salt are the repulsive interaction due to the overlap of electron clouds (which tends to violate the Pauli principle) and the atttractive exchange, or van der Waals, interaction. The strength of the van der Waals attraction depends strongly on the polarizability of the electron clouds on nearest neighbors and is determined only by the electrons in orbitals belonging to the outermost atoms in the molecular ion species. i.e. they do not vary as the inverse of the center-of mass separations as do the coulomb interactions. What is special about the perfluorinated salts, then, is that although they often form a true quasi-lattice of coulomb points, hence occupy a deep free energy minimum with respect to the isolated pair of ions in the vapor state, they remain mobile at low temperatures. This is ensured by the combination of large ion sizes (which keeps the lattice energy low relative to other salts) and the weak short range van der Waals attractions (which ensures that the decreased coulomb energy is not compensated by a large short range van der Waals attraction). Under these circumstances it is possible for the energy needed to move ions past each other to remain small, while the energy necessary to extract an ion pair from the salt into the gas phase, remains large. The only rival to the perfluorinated ions in this role is the perchlorate anion in which the oxygens are so strongly bonded to (charge clouds polarized by) the Cl(VII) species, that the van der Waals interaction with neighboring species is very weak. So far, perfluorination of the cations has not been exploited, but the effect might be expected to further enhance the performance of the salts.

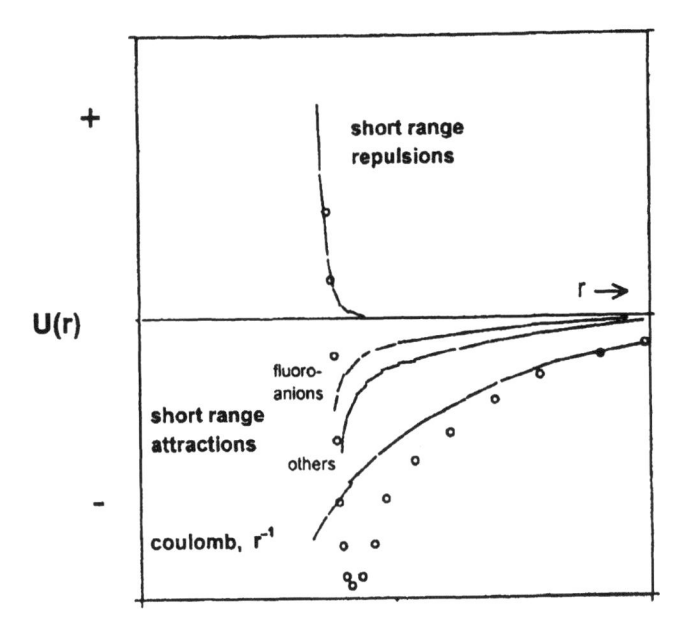

Figure 6. Components of the potential of interaction of cation and anion species in low temperature molten salts. The coulomb potential, in the form of a Madelung potential for the system, is of long range and guarantees that it is difficult for an ion pair to escape from the liquid surface. Locally, due to overlap of the long range interactions from different pairs, the potential is fairly flat. By contrast, the short range interactions do not act from the centers of the molecule ions, but rather between the centers of the outermost atoms. The Van der Waals attractions depend on the polarizability of the outermost atoms and act like a "sticky" component on the otherwise flat background potential. Because of the low polarizablity of the contacting species in perfluorinated anions, this component is much weaker than for non-fluorinated species. The total potential is indicated by open circles for the case of the weaker short range attraction. The local well of the total potential will be deeper in the case of the other short range attraction.

4. Low-Melting Salt Systems for Various Applications

In this section we will give some examples of molten salt systems which have been designed for low temperature operations with different applications in mind. The relevance of the principles discussed in the foregoing sections will be evident. We will consider briefly the areas of battery electrolytes, photoelectrochemical cells, and low temperature synthetic chemistry media.

4.1. AMBIENT TEMPERATURE MOLTEN SALT SYSTEMS FOR LITHIUM BATTERY OPERATION

The possibility that room temperature molten salts based on lithium cations might exist at room temperature and might offer the possibility of decoupled (single ion) conducting liquids of high interest to lithium battery operation, was raised in 1993 [21]. In that report it was noted that compositions containing lithium salts like $LiClO_4$ and

LiClO$_3$, and particularly those containing AlCl$_3$, could exhibit conductivities of 10^{-2} Scm^{-1}, values that rival those of the most fluid non-aqueous solutions.

Unfortunately none of these salts, or their combinations, is acceptable for practical battery applications, and it does not seem possible to find alternative more stable anions that have the same conductivity performance. Lithium salts of anions that have asymmetry, like FSO$_3^-$ and PO$_2$F$_2^-$, melt at higher temperatures than does the salt of the symmetrical ClO$_4^-$, which shows that mere symmetry does not control the phenomenon. Among symmetrical salts, the combination of LiAlCl$_4$ + LiAlI$_4$ is found to melt at 70°C, and rapid cooling produces a glassy state with conductivity at T$_g$ (–45°C) of 10^{-8}Scm^{-1}[22]. Since the value predicted from the viscosity increase would be 10^{-15}Scm^{-1}, it is clear that the Li$^+$ ions can move rather freely in this liquid structure.

The only salts found to melt at lower temperatures than the perchlorate and nitrate have been salts in which the Li cation seems to be quite strongly bound to the anion. For instance the lithium salt of the chlorinated bis-sulfonyl imide [(ClSO$_2$)$_2$N$^-$] does not have a freezing point as far as can be told [23]. However, despite the fact that its glass transition temperature is as low as that of LiBF$_4$ (the latter obtained by extrapolation of binary solution data [24]) its conductivity is extremely low. Clearly the lithium ion is firmly coupled to the anion, probably due to its interaction with the imide nitrogen. This is only partly alleviated by complexing the imide nitrogen with a Lewis acid stronger than the Li$^+$ cation e.g. AlCl$_3$ [23].

The decoupling of Li$^+$ to anion is enhanced if the imide nitrogens are placed in close proximity, as in polymeric anions [25] but the reason for this is not understood. Indeed, the formation of polymeric anions not only enhances the decoupling but seems to be the only way of suppressing the crystallization phenomenon. Unfortunately it also tends to raise the glass transition temperature, and this tends to negate the beneficial effects of cation decoupling. Nevertheless, polymeric forms plasticized with LiAlCl$_4$ have been obtained with conductivites as high as $10^{-2.9}$Scm^{-1}[25].

The most recent setback, and lesson, in low-melting lithium salt phenomenology has been that sustained in measuring the conductivity of a newly-synthesized salt of a very large perfluorinated anion, bis-perfluoropinnacolato orthoborate [26]. This salt has a massive anion that has very weak van der Waals interactions with its neighbors because of its almost complete coverage with fluorine atoms. With a total of 24 fluorine atoms per anion and a charge of only –1e, a very weak coulomb energy might be anticipated. Indeed, the salt melts at only 120°C, and the melt is very fluid at its melting point. Despite all these promising signs, the conductivity proves to very low, only 10^{-8}Scm^{-1} at 120°C. Furthermore the salt does not become glassforming nor conductive when fused with other salts, e.g. LiTFSI. The salt is easily distilled at temperatures of about 150-180°C indicating that it has a low boiling point not much above its decomposition temperature [26]. Clearly this melt does not establish a quasi-lattice with well-developed Madelung energy, but rather behaves more like an inorganic molecular liquid more than a molten salt. The corresponding salt in which all carbons are protonated, rather than fluorinated, is infusible, as indicated by its very low solubility in the common salt solvents [27].

The conclusion to be drawn from these observations is that the phenomenology of lithium salts is very complex and difficult to anticipate. The chances of successfully developing a class of stable room temperature ionic liquids of simple ion character would appear to be remote. On the other hand the existence of polymeric, ambient-temperature-amorphous, versions of such liquids seems established [25], and these will merit further investigation.

4.2. SOLVATE SYSTEMS

Another way of maintaining a low vapor pressure over an ionic liquid, while keeping the coulomb energy low enough to permit reasonable fluidity at ambient temperature, is to incorporate an involatile or chelating solvent in sufficient quantity to just satisfy the first coordination shell of the cation. Under the right conditions, a relatively fluid, highly conducting, molten solvate can result, in which the total vapor pressure over the solution is very low. Examples are cited in ref. 28. The most effective cation sheaths are the crown ethers and cryptands, but these are very expensive. Compromise solvate sheaths may comprise chelating agents like EDTA.

4.3. ORGANIC CATION IONIC MELTS FOR ELECTROCHEMISTRY, PHOTO-ELECTROCHEMICAL CELLS, AND SYNTHETIC CHEMISTRY

This segment of the field now has a considerable literature. The purpose of this section is primarily to expand its range with additional examples, rather than to review the area as a whole.

The early example of the non-chloroaluminate ambient temperature molten salt quoted earlier [16], remains one of the most conducting salts reported. As its cation was not of the familiar substitiuted imidazolium type, it merits some attention here. It contained an ethoxy group on a tetraXammonium cation, where X is any group attached to the central nitrogen. Its precise formula will be given below but first it is appropriate to place its development in context.

The problem of determining the relation between melting point and structure of organic cation molten salts was tackled by Cooper and the author [29, 30] using the test case of tetrapropylammonium bromide as a starting point. Based on the idea that poor packing was a primary cause of low melting point and high fluidity, these workers synthesized and characterized all of the 14 isomers of this salt, and studied their properties. The different isomers are obtained by redistribution of the 12 carbon-hydrogen "beads" around the central nitrogen. The different cations were named according to the number of carbons in the individual sidegroups. Thus the reference compound contains the "3333 cation" and isomers include cases like 4332, 5511, 6411 and 9111, each set summing to 12. As the symmetry of the cation decreased the melting point rapidly decreased from the 270-300°C (with decomposition) of the 3333 isomer to values like 114°C for the 5421 compound 115°C for the 6222 compound and 100°C for the 5511 compound.

The latter melts could be vitrified on rapid cooling and the glasses obtained exhibited glass transition temperatures that were all in the same range, -36 to –40°C. The higher-melting cases tended to yield plastic crystal phases which had "glassy crystal" analogs of the glass transition phenomenon, the transition temperatures for which fell at lower and more scattered temperatures, -50 to –100°C. A table summarizing these findings is provided in ref. 29 where the results were briefly reported. A more detailed account of the very complex behavior observed [30] was unfortunately twice rejected by reviewers, and has remained unpublished to this time. Copies are available from the author on request.

Apparently what prevents the melting points of the less symmetric cation-containing salts from being depressed below 100°C is the formation of a high symmetry, but highly disordered, polymorph, in which some of the configurational entropy of the melt is taken up pre-emptively in the crystal. To avoid this effect it was found effective to replace one or more of the alkyl sidegroups with an ether-containing group [31]. The fluidity of melts of this type is increased if the molecular weight can be decreased without simultaneously increasing the melting points.

The most fluid case was found to be the salt with the methoxyethyl,ethyl,dimethyl ammonium cation (MeOEt,Et,diMeN$^+$). As the iodide, this salt melted at 47°C and had a glass temperature of –52°C [31]. It was the substitution of the iodide anion in this salt with the tetrafluoroborate anion, that lead to the first example of the water stable, ambient temperature, molten salt, mentioned previously [16]. In ref. 31 the melting point of this salt is given as 13°C and the glass temperature as –98°C. These data seem to have been overlooked by other workers in the field, perhaps because they were only reported in an article focussed on the use of these salts in solid state ionic conductors. They were used as components in a double salt with LiBF$_4$, which yields a plastic crystal fast ion conductor (PLICFIC [31]).

Another salt with even better properties cited in ref. 31 is MeOMe,Et,diMeN$^+$BF$_4^-$ which melts at –16°C and has a glass temperature of –115°C. This is the lowest T_g on record for an ionic liquid.

The full possibilities of this type of cation for room temperature ionic liquid applications have never been fully explored. When the perfluorinated anion salts of organic cations were first properly examined for their ambient temperature liquid properties [17], it was the substituted imidazolium moities that were used as cations. The electrochemical stabilities of the tetraXammonium cations deserve to be properly evaluated since they may well be more favorable than those of the imidazole-based cations. Their fluidities may in some cases also become competitive. The article by Ohno and co-workers [32] lists cases with T_g values as low as –109°C, but these have conductivities up to an order of magnitude higher than measured for the MeOEt,Et,diMeN$^+$. BF$_4^-$ melt, cited in ref. 16. The factors that determine the differences in conductivity for these cases deserve to be examined in detail since, with molecular liquids, it is always the aromatic liquids that have the higher glass transition temperatures. The fluidities of the salts reported in ref. 31 may therefore be higher than those of the salts based on the imadazolium-type cations, and their properties as "green" solvents may therefore be superior.

Some electrical conductivity, and glass transition temperature, data for this type of salt, and mixtures with lithium salts, are shown in Fig. 7.

4.4. IMMISCIBILITY AND HYDROPHOBICITY OF LOW TEMPERATURE IONIC LIQUIDS

One of the advantageous aspects of the perfluorinated anion ambient temperature molten salts is their insensitivity to water, which is a reflection of their relatively hydrophobic character. The origin of this characteristic has not been discussed in detail, but is clearly a consequence of the same low polarizability characteristics responsible for the high fluidities of these salts. An interesting example of this effect is

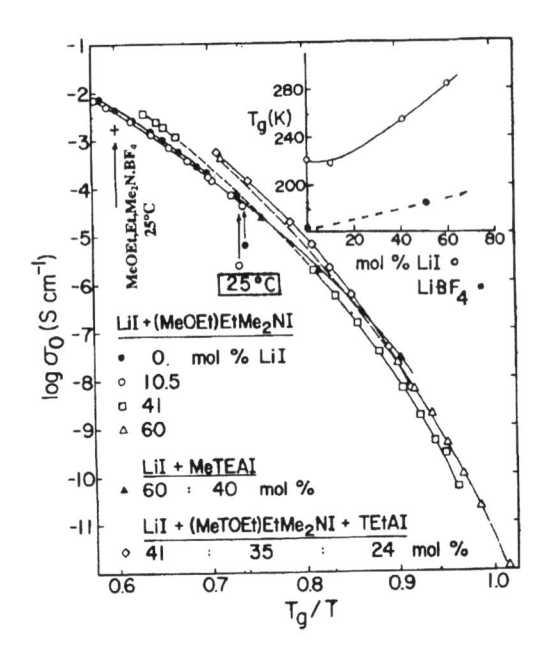

Figure 7. The electrical conductivities and glass transition temperatures of some substituted ammonium cation salts and their mixtures with alklai salts. The use of a Tg-scaled temperature coordinate allows judgment of the relative fragilities of the different salt types. **Insert**: T_g vs lithium salt content. The strong effect on the glass temperature (hence on the fluidity) of substituting unpolarizable (BF_4^-) for polarizable (I⁻) anions, is obvious.

provided by the behavior of aqueous solutions containing both perfluorinated and other anions, as follows.

When an aqueous solution of lithium bistrifluoromethanesulfonyl imide is mixed with lithium bromide solution containing the same mole ratio of salt to water, an unexpected phenomenon is encountered [33]. The mixture splits into two layers. The lower (denser) phase is rich in the imide salt, and it occupies less volume than the upper layer because the water has been concentrated in the lithium bromide-containing layer. The layers remain immiscible on addition of water until the equivalent of 14 moles of water per mole of Li⁺ has been added. This is the composition of the ambient temperature plait point for the ternary system.

Homogeneous solutions can be obtained at somewhat lower water contents if the temperature is raised. Then the meniscus separating the two layers will vanish at a critical solution point. Why should two aqueous solutions of strong electrolytes not be miscible? It must be related to the unfavorable energy associated with the mixing of perfluorinated and nonfluorinated anions. The difference between such anions lies primarily in their polarizabilities.

It might therefore be expected that anion polarizability-difference-driven immiscibility will be found between perfluorinated anion low-melters, and other low melting systems in which the anions are not perfluorinated and the low-melting characteristics arise from other sources. An example of such alternative low-melters would be the reduced coulomb energy in the solvate systems of subsection 2 above.

Accordingly it is expected that ionic liquids of the solvated or chelated cation type which utilize nitrate, halide or thiocyanate anions will be immiscible with ionic liquids of the organic cation-perfluorinated anion type, if the temperature is high enough to avoid a metathetical crystallization. It is quite likely that a redistribution of the cation anion combinations would occur. Irrespective of this outcome, the coexistence of immiscible ionic liquid phases might offer additional possibilities for product solvent separations.

5. CONCLUSIONS

The field of ambient temperature molten salts offers many challenges to our state of understanding of the liquid state, and many opportunities to apply that understanding in the quest for improved energy utilization and environmental quality. The full range of possible fluorinated anions and cations has not yet been explored.

ACKNOWLEDGEMENTS

This work was supported by the National Science Foundation under Solid State Chemistry grant Nos. DMR-9614531 and 0082535.

REFERENCES

1. This section is taken almost without modification from an article prepared by the present author for the Enrico Fermi series of summer schools in physics, and is incorporated here by permission of the Italian Physical Society. The original article containing this section, was entitled "The Glassy State Problem: Failure to Crystallize, and Vitrification," and it may be found in: Proc. Int. School of Physics, "Enrico Fermi", Course CXXXIV, edited by F. Mallamace and H. E. Stanley, IOS Press Amsterdam, 1997, p. 571.
2. Alba C., Busse L. E., and Angell C. A., *J. Chem. Phys.* **92** (1990) 617.
3. 3.Wong J. and Angell C. A., *Glass: Structure by Spectroscopy* (Marcel Dekker, New York) (1976) Chap. 1.
4. R. S Pitzer. and D. W. Scott, *J. Amer. Chem. Soc.* **65** (1943) 803.
5. S. Jacob, J. Javornizky, G. H. Wolf and C. A. Angell, Intern. J. Inorg. Mater. 3(3) 241-251,(2001).
6. C. A. Angell, in *Preparation and Characterization of Materials,* ed. C. N. R. Rao and J. M. Honig, Academic Press (1981), p. 449.
7. H. S. Chen. and D. Turnbull, *J. Chem. Phys.* **48** (1968) 2560; *J. Appl. Phys.* **38** (1967) 3646; *Acta. Metall.* **17** (1969) 1021.
8. Topol L. E., Mayer S. W. and Ransom L. D., *J. Phys. Chem.* **64** (1960) 862.
8. (b) C. A Angell. and D. C Ziegler., *Mat. Res. Bull.* **16** (1981) 269.
9. E Thilo, C. Wieker and W. Wieker, *Silic. Tech.* **15** (1964) 109.
10. *Phase Diagrams for Ceramicists,* edited by E. M. Levin, C. R. Robbins and H. F. McMurdie, (American Ceramic Society) Diagram No 391.
11. Changle Liu and C. A. Angell *Solid State Ionics* (Proc. 7th Int. Conf. Sol. State Ionics), **86-88**, 467-473 (1996).
12. F. Mezei, W Knaak and B Farago, *Phys. Rev. Lett.* **58**, 571 (1987)
13. Götze, W. in *Liquids, Freezing, and the Glass Transition,* Eds. Hansen, J.-P. and Levesque, D., NATO-ASI, North Holland (Amsterdam) (Les Houches 1989) 287-503.
14. Poulain M., Chantanasingh M. and Lucas J., *Mater. Res. Bull.*, **12** (1977) 151.
15. (a) Wilkes, J.S., Levisky, J.A., Wilson, R.A., and Hussey, C.L., *Inorg. Chem.*, **21**, 1263 (1982). (b) C. L. Hussey, *Adv. Molten Salt Chem.*, **5**, 185, (1983)

16. E. I. Cooper and C. A. Angell, *Solid State Ionics*, **9 & 10**, 617 (1983). (see "note added in proof").
17. E. I. Cooper and E. J. M Sullivan, *Proc. 8th. Intern. Symp. Molten Salts*, The Electrochem. Soc., Pennington NJ, 1992, Proc. Vol. 92-16, pp.386-396
18. R. T. Carlin, and J. S. Wilkes, in *Chemistry of Nonaqueous Solutions*, G. Mamantov and A. I. Popov, Editors, Ch. 5, VCH Publishers, New York (1994) and the article by R. Carlin in this issue.
19. P. Bonhote, A.-P. Dias, M. Armand, N. Papageorgiou, K. Kalyanasundaram, and M.Graetzel, *Inorganic Chem.*, **35**, 1168 (1996)
20. (a) C. J. Bowles, D. W. Bruce and K. R. Seddon, *Chem. Commun.* 1625 (1996),
    (b)J. D. Holbrey and K. R Seddon, *J. Chem. Soc.,Dalton Trans.*, 2133, (1999).
21. C. A. Angell, C. Liu and E. Sanchez, *Nature*, **362**, 137-139, (1993).
22. M. Videa and P. Lucas (unpublished work)
23. K. Xu and C.A. Angell, *Symp. Mat. Res. Soc.*, **369**, 505 (1995).
24. M. Videa and C. A. Angell, *J. Phys. Chem.* **103**, 4185-4190 (1999).
25. S.-S. Zhang, Z. Chang., K. Xu and C. A. Angell, *Electrochimica Acta* **45**, 1229 (2000).
26. W. Xu and C. A. Angell, *Electrochem. and Solid State Lett.*, **3**, 366 (2000).
27. W. Xu, unpublished work.
28. "Lithium ion conducting electrolytes" C. A. Angell and C. Liu, US Patent No. **5,506,073**,April 9, 1996.
29. C. A. Angell, L. E. Busse, E. E. Cooper, R. K. Kadiyala, A. Dworkin, M. Ghelfenstein, H. Szwarc, and A. Vassal, *J. de Chim. Phys.*, **82**, 267 (1985).
30. Liquids, Rotator Phases, and Glass Transitions in Relation to Cation Symmetry in C12 Tetra-alkylammoniuma Bromides," E. I. Cooper and C. A. Angell, *J. Phys. Chem.* (rejected), copies may be downloaded from the website at www.public.asu.edu/~caangell/.
31. E. I. Cooper and C. A. Angell, *Solid State Ionics*, **18 & 19**, 570 (1986).
32. M. Hirao, H. Sugimoto, and H. Ohno, J. Electrochem. Soc., **147**, 4168 (2000)
33. V. Velikov and C. A. Angell, (unpublished work)

ELECTRODES AND ELECTROLYTES FOR MOLTEN SALT BATTERIES: EXPANDING THE TEMPERATURE REGIMES

R. T. Carlin
Office of Naval Research
800 North Quincy Street
Arlington, VA 22217-5660
USA

J. Fuller
Air Force Office of Scientific Research
801 North Randolph Street
Arlington, VA 22217-7230
USA

1. Introduction

Molten salts have been employed in energy storage and generation technologies for numerous applications. Examples include thermal and rechargeable batteries employing LiCl-KCl and NaCl-AlCl$_3$ electrolytes, molten carbonate fuel cells, and thermal energy storage using molten nitrates. For electrochemical applications, molten salt electrolytes offer many benefits over conventional aqueous and organic electrolytes including minimal vapor pressure, wide electrochemical window, inherently high ion concentration, and low to minimal toxicity. Additionally, the high temperature operation of molten salt electrolytes is attractive for fast kinetics and high power requirements (and is required for thermally activated batteries and for thermal energy storage); however, it also poses materials difficulties in some instances. Room temperature molten salts, also termed ionic liquids, retain many of the benefits of high temperature electrolytes, but without the thermal handling problems.

Our work over the past 15 years has evaluated the applicability of ionic liquids as electrolytes for electrochemical energy storage (primarily batteries), as well as considered these unique electrolytes for electrochemical energy generation (*e.g.*, fuel cells and photovoltaic devices. This lecture series will summarize our work in the area and will provide insight for current and future researchers wishing to pursue this promising energy technology field. The lecture also provides basic information on battery principles and terminology, as well as, limited background material on relevant research and technologies. However, it should be emphasized that this is not a review, so excellent research by many scientists in the field may not be included, and readers will be referred to appropriate review articles whenever possible.

2. Battery Basics

Due to the specialized nature of battery technologies, the general principles and terms in the field are not familiar to many, and the terminology is often confusing. Therefore, this section provides a short description of the important terms and concepts needed to follow the discussions to follow. It is assumed that the reader has some basic understanding of electrochemistry. For more detailed discussions of battery technologies, the reader should refer to books on the topic including *Modern Battery Technology* by Clive Tuck [1] and Linden's *Handbook of Batteries* [2].

The negative and positive electrodes of an electrochemical cell are called the anode and cathode, respectively. An electrolyte is required to provide ionic conductivity between the electrodes. In a practical cell, active material for each electrode are usually attached to metallic current collectors to provide an efficient path for electronic conductivity. In addition, electronic conductors (*e.g.*, carbon blacks) and polymer binders are usually mixed with the active electrode materials to enhance electronic conductivity and physical stability.

A battery is a collection of single cells (sometimes only one cell) wired and packaged to deliver specific voltage and current levels. When cells are wired in series, their voltage is additive, producing a battery with a higher voltage. When cells are wired in parallel, the voltage will be the same as a single cell. In both cases, the overall battery electrical energy (*vide infra*) will be the sum of the individual cells; however, a series arrangement provides a higher voltage and may be more applicable to high power applications. A bipolar cell arrangement is one in which individual cells are directly stacked in series without wiring. In this structure, the current collector for the anode of one cell serves as the current collector for the cathode of the next cell in the stack, and so forth. A bipolar arrangement eliminates some of the non-active materials in the battery design. Drawbacks to this design include the need to assemble and adequately seal the individual cells to avoid leakage currents between the cells.

Batteries fall into two principle categories: primary (single-use) and secondary (rechargeable) designs. Reserve batteries are specialized forms of primary batteries that require physical or thermal activation, usually by injection of electrolyte into the electrode compartment (physical activation) or rapid melting of a solid salt already positioned between the electrodes to a molten salt (thermal activation). Reserve batteries possess exceptionally long shelf-life because they essentially eliminate the opportunity for self-discharge, and they avoid any detrimental long-term interactions between the electrolyte and electrodes. With rechargeable batteries, some scientists become confused with the terms "anode" and "cathode." In either discharge or charge mode, the electrode designation for a rechargeable battery always remains the same;

therefore, the anode undergoes *oxidation* during discharge and *reduction* during recharge, and vice versa for the cathode.

Electrical capacity is the amount of electrical charge a cell or battery can provide. This quantity is usually reported in ampere-hours (Ah), which corresponds to 1 Ah = 1 coulomb/s x 3600 s = 3600 coulombs. The Ah capacity provides a convenient indicator of the size of a cell or battery. Unlike energy units discussed next, connecting cells in parallel increases the capacity of the battery, while series arrangements increase voltage, but not Ah rating. Electrical capacity may be reported on a weight or volume basis and may then be referred to as specific charge or charge density, respectively.

Energy density is the electrical energy stored in per unit volume of a single cell or battery pack and is usually reported in watt-hours per liter (Wh/l). A Wh is simply an energy unit that corresponds to 3.6 kJ/kg and is calculated from 1 watt x 1 hour = 1 joule/s x 3600 s = 3600 joules = 3.6 kJ.

Specific energy is the electrical energy stored per unit weight and is usually reported in Wh/kg. Depending upon the application, a high specific energy or a high energy density may be more desirable. For example, space applications tend to emphasis specific energy due to the cost of launching payload weight, while energy density becomes a concern when an application requires packing as much energy as possible into a confined space, such as a torpedo.

Power density and specific power are the power equivalents of energy density and specific energy and carry units of W/l and W/kg, respectively. It is important to properly distinguish between energy content and power capability when considering battery performance.

Electrochemical testing of single electrodes (in a laboratory test cell), cells (single anode and cathode), and complete batteries are typically done in a chronopotentiometric mode. That is, a constant current is applied to the test unit, and the potential (voltage) is followed as a function of time. Time is easily converted to electrical capacity of the electrode, cell, or battery through Electrical Capacity = Current x Time. By plotting potential *vs.* electrical capacity, it is possible to see "drops" in potential at capacities corresponding to thermodynamic phase changes and to maximum electrical capacity. The maximum capacity of an electrode material is often converted to Ah/kg to provide a convenient ranking scale for energy storage. Other electrochemical techniques such as cyclic voltammetry and chronoamperometry are also used to evaluate electrode materials and cells; however, these techniques are less common and are more often used in the academic community.

3. Lithium Metal and Lithium-Ion Rechargeable Batteries

This section provides only a brief overview of lithium metal and lithium-ion batteries. Numerous reviews and books are available on these topics and should be consulted for more details on the range of research and technology in these areas [1-5]. Reference 5 provides a good overview of insertion electrode materials for rechargeable lithium batteries.

3.1 LITHIUM METAL ANODES

Lithium offers the highest electrical capacity of any metal at 3,862 Ah/kg and is highly desirable as an anode material. Furthermore, by combining the very negative potential of the Li$^+$/Li couple (-3.0 V *vs.* NHE) with an appropriate cathode, cell voltages in excess of 4 V can be achieved, making possible very high energy density batteries. However, lithium metal is highly reactive to essentially all electrolytes, and it is only by formation of a protective film at the electrode-electrolyte interface...termed the solid electrolyte interface (SEI)...that lithium batteries are able to operate. This SEI must conduct lithium ions, yet be electrically insulating to prevent continual electrochemical "corrosion" reactions between the lithium metal and the electrolyte. For rechargeable lithium metal cells, the SEI must also be stable to charge-discharge cycles or be self-healing in case the thin interface should crack or rupture.

While lithium metal rechargeable batteries have been introduced to the marketplace in the past, these were withdrawn due to instability between the lithium metal and the electrolyte. During recharging, the lithium metal electrodeposited as dendrites, which would disconnect from the electrode to form fine lithium metal particles. These highly reactive particles eventually reacted with the electrolyte leading to undesirable battery "events." More recent work with lithium ion conducting polymer and inorganic electrolytes indicate that these solvent-free, solid-state electrolytes may eliminate or reduce dendrite growth and prevent unwanted lithium metal reactions. However, conductivities of these solid-state electrolytes at ambient and sub-ambient temperatures need to be improved before true lithium metal batteries can be highly successful in the commercial marketplace. Research in lithium-polymer batteries is a dynamic research area and is receiving considerable support by various funding agencies.

3.2 LITHIUM-ION ANODES

To avoid the problems with lithium metal, battery researchers and developers have turned to lithium intercalation anodes, principally carbon-based coke or graphite. By also employing a lithium intercalation cathode (*e.g.*, lithiated transition metal oxides), a cell undergoing charge and discharge simply shuttles lithium ions between anode and cathode. In this design, the electrolyte serves only as a transport medium for the lithium ion and is not consumed or generated during the electrochemical processes. Such cells have been termed "rocking-chair" batteries and are commonly referred to as lithium-ion batteries. Figure 1 below illustrates the operation of a lithium-ion cell.

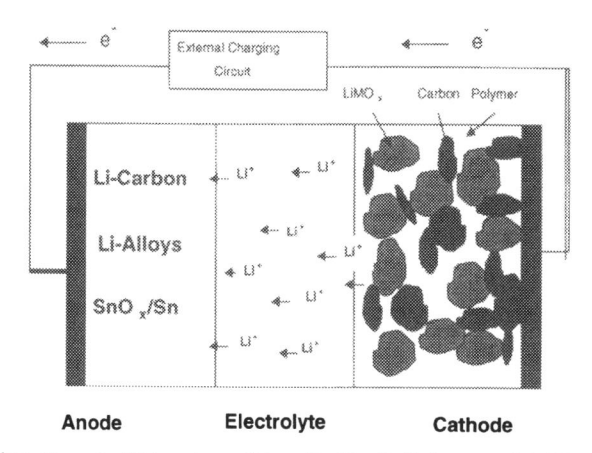

Figure 1. Illustration of a lithium-ion cell (provided by D. Sadoway and A. Mayes at MIT)

By eliminating lithium metal as the anode, the lithium-ion rechargeable cell increases overall battery safety; however, the design also requires the development of high capacity anodes and cathodes. In addition, it is critical to produce a stable SEI at the carbon-electrolyte interface to maintain stability of the charged lithiated carbon anodes and to prevent co-intercalation of electrolyte organic solvent, which exfoliate and destroy the carbon structure. Most lithium-ions batteries today employ a graphitic anode, which has a limiting lithium composition of LiC_6 corresponding to a theoretical specific charge of 372 Ah/kg and a theoretical charge density of 837 Ah/l, based on carbon weight and volume only [5].

Lithium alloys are also viable anodes, with LiAl being particularly advantageous: 993 Ah/kg and 2681 Ah/l based on aluminum. While lithium alloys are promising anodes from a capacity standpoint, most of the desirable alloying metals undergo large dimensional changes upon lithium alloying that cause fragmentation and rapid capacity loss during cycling. Very recently, researchers have discovered that nanoscale tin (Sn) particles encapsulated in an inert inorganic matrix can undergo reversible lithium alloying over many cycles with minimal losses in capacity [6]. Some examples of these nanocomposite anodes include $Sn-B_{0.56}P_{0.4}Al_{0.42}O_{3.6}$ glass (600 Ah/kg, 2200 Ah/l) [6], $Sn_2Fe-SnFe_3C$ (620 Ah/kg, 1600 Ah/l) [7], and $Sn-Li_2O$ (600 Ah/kg) [8]. Many other nanoscale metals are anticipated to provide performance in excess of lithium-carbon anodes, if appropriate nanocomposite structures can be identified. This is a highly active area of research and commercial interest.

3.3. METAL OXIDE CATHODES

Numerous cathode materials have been studied for lithium metal and lithium-ion batteries, including metal chalogenides, metal oxides, conducting polymers, and polysulfides. To date, the most successful of these remains the transition metal oxides.

Table 1 summarizes the performance of current state-of-the-art metal oxide cathode materials available today. In the table, the theoretical capacity assumes all the lithium in the structure can be cycled reversibly; however, in reality only a portion of the lithium is electrochemically reversible due to phase changes at various lithium contents. The practical capacity corresponds to reversible lithium intercalation, and the cell voltages assume a lithium-graphite anode.

Cathode Material	Theoretical Capacity (Ah/kg)	Practical Capacity (Ah/kg)	Average Cell Voltage (V)
$LiCoO_2$	274	142	3.6
$LiNiO_2$	275	145	3.6
$LiNi_{0.8}Co_{0.2}O_2$	274	180	3.6
$LiMn_2O_4$	148	120	3.8

Table 1. Transition metal cathode materials

In addition to the materials listed in Table 1, vanadium oxides have drawn a great deal of interests. The vanadium oxides display a range of V:O molar ratios (*e.g.*, V_2O_5, V_3O_8, and V_6O_{13}) and display good lithium capacity; however the voltage of the VO_x is lower than the other metal oxides (*ca.* 2 – 3 V). Recently, nanophase V_2O_5 cathode materials have been prepared with capacities in excess of 450 Ah/kg and a corresponding specific energy of approximately 1500 Wh/kg when coupled with a lithium metal anode [9]. Other nanophase cathode materials, including some in Table 1, show increased capacities over their micron-scale particles. This has been attributed to facile Li^+ diffusion into the interior of the nanoparticles with concomitant reduction in electrode polarization. Nanophase cathode materials and nanocomposite cathodes are a promising and active area of research.

3.4. ORGANIC ELECTROLYTES

The ideal electrolyte for lithium metal and lithium-ion rechargeable batteries would (1) provide high lithium-ion conductivity, even at sub-ambient temperatures; (2) possess a wide electrochemical window; (3) exhibit chemical and electrochemical stability at elevated temperatures; and (4) be nontoxic, nonvolatile, nonflammable, and low cost. In addition, the electrolyte (both organic solvent and lithium salt) should not be reactive towards the cathode and should form a stable SEI at the anode.

While many organic solvents have been studied for lithium batteries, the solvents of choice are organic carbonates. These solvents readily solvate lithium salts and provide some of the ideal properties listed above. The lithium salt of choice for the electrolyte tends to be $LiPF_6$ for performance, cost, and toxicity reasons; although, other inorganic lithium salts are used, as well. Because PF_6^- tends to be unstable at elevated temperatures, other more-stable anions are under consideration, including bis(trifluoromethylsulfonyl)imide (($CF_3SO_2N^-$)) and similar perfluoro organic anions, but the cost of these salts may be prohibitive. The exact composition of a lithium battery electrolyte will typically be a mixture of organic solvents and additives to provide the desired physical properties, while maintaining a stable SEI.

The most recent commercial lithium-ion batteries employ a "gelled" electrolyte in which the organic electrolyte is retained in a microscopic polymer matrix such as polyacrylonitrile (PAN) or poly(vinylidene fluoride) hexafluoropropylene copolymer (PVdF(HFP)). These lithium-ion polymer batteries possess the same chemistries and electrochemistries as liquid-electrolyte batteries; however, the polymer gel electrolyte prevents the liquid electrolyte from readily flowing. Without the free-flowing liquid, lithium-ion polymer batteries are easier to manufacture and can be packaged in lightweight plastic pouches, increasing the overall energy density of the battery. In standard metal casing batteries, packaging and other non-active materials of construction can easily account for 60-70% of the total weight of the battery; therefore, methods to reduce the non-active and non-essential materials are major drivers in battery design.

4. Ionic Liquid Electrolytes for Battery Systems

4.1 IONIC LIQUID ELECTROLYTE PROPERTIES

The ionic liquids, or room-temperature molten salts, used in our investigations are illustrated in Figure 2 and are composed of a 1,3-dialkylimidazolium cation, particular 1-ethyl-3-methylimidazolium (EMI$^+$), and chloroaluminate (AlCl$_4^-$) or perfluorinated anions (*e.g.*, BF$_4^-$, CF$_3$SO$_3^-$, and (CF$_3$SO$_3$)$_2$N$^-$). The 1,2,3-trialkyl imidazolium cations are more chemically stable and electrochemically stable towards reduction than the 1,3-dialkylimidazolium due to the lack of the C-2 proton, which is the most acidic proton on the 1,2-dialkylimidazolium cation [10].

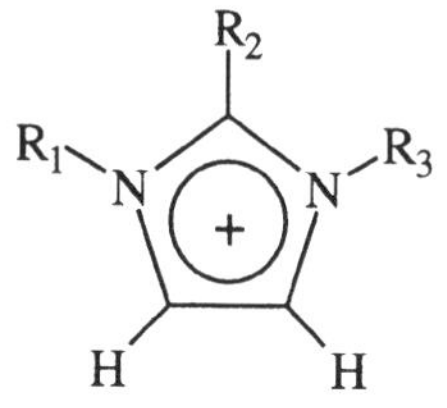

Anion = AlCl$_4^-$, Al$_2$Cl$_7^-$, BF$_4^-$, PF$_6^-$, CF$_3$SO$_3^-$, (CF$_3$SO$_2$)$_2$N$^-$, (CF$_3$SO$_2$)$_3$C$^-$

R$_1$ = Et, Bu, Pr; R$_2$ = H, Me; R$_3$ = Me

Figure 2. Ionic liquid imidazolium cations and various anions for battery electrolytes.

Some of the important properties of the ionic liquids as battery electrolytes are their nonvolatility, nonflammability, wide electrochemical windows, high inherent ionic conductivities, chemical inertness, and thermal stability. The benefits of these properties for battery technology are briefly described below.

- Nonflammable and Nonvolatile: The ionic liquid electrolytes are inherently nonflammable, thus eliminating fire hazards, even with highly reactive electrode materials. The negligible vapor pressure allows the ionic liquid electrolytes to be operated in sealed cells at elevated temperatures without venting. In addition, the lack of organic vapors eliminates the possibility of vapor ignition, increasing the overall safety of ionic liquid batteries. Moreover, the construction of ionic liquid cells is simplified since the negligible evaporation rate makes it possible to process the electrolyte as a laminate by impregnating it in a porous polymer matrix.

The nonflammability of the ionic liquid electrolytes was tested by soaking a piece of glass filter paper in 1.1:1.0:1.0 $AlCl_3$:EMIC:LiCl, $EMIBF_4$, or $EMI((CF_3SO_3)_2N)$ and then placing the saturated paper directly into a yellow flame. After several seconds, the ionic liquid began to carbonize and produced a small yellow ($EMI(CF_3SO_3)_2N$) and chloroaluminate) or green ($EMIBF_4$) flame; however, upon removing the flame source, the ionic liquid immediately extinguished, leaving the remaining ionic liquid unaltered. Therefore, the ionic liquid does not burn unless a direct flame is continually applied. This is in stark contrast to common organic electrolyte solvents used in lithium batteries, such as propylene carbonate, which burn to completion with a hot blue flame after ignition.

- Wide Electrochemical Windows: The ionic cations and anions are extremely resistant to electrochemical reduction and oxidation, respectively; therefore, high voltage anode and cathode materials can be employed in the ionic liquid electrolytes to create cells that operate in excess of 4 V [11]. While windows in excess of 6 V have been reported [12], such ultra-wide windows are due to slow kinetics of the anodic and cathodic limiting processes at test electrodes.

- High Inherent Conductivities: The ionic liquids are single-component electrolytes that possess ionic conductivities equivalent to the best, current organic battery electrolytes; therefore, high-power battery systems can be achieved with the ionic liquid electrolytes. Also, the manufacture of these electrolytes is easier than conventional organic electrolytes, which require both an organic solvent and an lithium salt, because only a single component must be synthesized and purified.

- Wide Thermal Operating Range: The room temperature molten salts are thermally stable to greater than 400 °C (perfluorinated anions) and greater than 200 °C (chloroaluminate anions) and, in some cases, maintain liquid electrolyte properties at temperatures well below -40 °C. Therefore, these electrolytes can be safely and reliably employed in batteries intended for operation in harsh, fluctuating temperature environments.

- <u>Chemically Inert</u>: The ionic liquid electrolytes are compatible with metals and polymers used in common battery devices; therefore, they do not require any specialized materials of construction or manufacturing technologies.

The chloroaluminate melts received attention initially as battery electrolytes because of the possibility of using aluminum as reversible anodes [13]. This required the use of acidic chloroaluminate melts where the molar ratio of $AlCl_3$:EMIC (EMIC = 1-ethyl-3-methylimidazolium chloride) is greater than 1. In such acidic melts, $Al_2Cl_7^-$ is readily reduced to aluminum metal. In basic melts ($AlCl_3$:EMIC > 1), where $AlCl_4^-$ and Cl^- are the only anions, aluminum cannot be electrodeposited from the kinetically stable $AlCl_4^-$ anion. Despite the promise of utilizing a reversible aluminum battery anode, the electrochemical window of the acidic melt is limited by oxidation of $AlCl_4^-$ to only *ca.* 2.5 V. Also, identification of an appropriate cathode is difficult, although metal chlorides and chlorine have been examined.

A particular class of chloroaluminate ionic liquids, termed buffered neutral, has drawn much attention for lithium and sodium anode batteries because high concentrations of alkali metal ions (*ca.* 2 M) can be obtained with these liquids, while still maintaining ambient temperature operation [6-8]. The buffered neutral ionic liquids are prepared by adding solid alkali metal chloride to a room-temperature molten salt initially containing an excess of $Al_2Cl_7^-$. The solid alkali metal chloride reacts with the Lewis acid $Al_2Cl_7^-$ species in the acidic melts, forming the Lewis neutral $AlCl_4^-$ anion and introducing a high concentration of the corresponding alkali melt cation into the melt.

$$Al_2Cl_7^- + MCl(s) \Leftrightarrow M^+ + 2\,AlCl_4^- \quad (M^+ = \text{alkali metal cation}) \tag{1}$$

Because the alkali metal chloride is not soluble in the resulting melt, the only anion present is the Lewis-neutral $AlCl_4^-$ species. The cations are alkali metal and the imidazolium ions. In these melts, any $Al_2Cl_7^-$ produced by an electrochemical process are neutralized by excess solid MCl, and any chloride ions formed are precipitated as MCl. Therefore, the ionic liquid is maintained in the Lewis-neutral state, and the system is termed a buffered neutral room-temperature molten salt. The buffering process maintains the wide electrochemical window (>4 V) available only in the neutral chloroaluminate ionic liquids. We have examined a number of ternary room-temperature melts with compositions $AlCl_3$-EMIC-MCl (MCl = LiCl, NaCl, KCl, RbCl, and CsCl) and have found that only lithium and sodium can be electrodeposited and anodized in their respective buffered neutral melts under appropriate conditions.

4.2. LITHIUM AND SODIUM ANODES IN BUFFERED CHLOROALUMINATES

As mentioned above, high concentrations of alkali metal ions can be introduced into the chloroaluminate room-temperature molten salts by adding an alkali metal chloride to a melt initially containing an excess of $Al_2Cl_7^-$. We have examined the alkali metal couples in the buffered neutral melts and have found that only lithium and sodium, with respective standard reduction potentials of -2.066 (± 0.005) and -2.097 (± 0.050) V *vs.* Al/Al(III), can be electrodeposited before reduction of the organic cation [14]. Despite these thermodynamic indications, elemental lithium and sodium cannot be electrodeposited from the pure buffered neutral melts. Instead, early researchers, including ourselves, found it necessary to add a source of hydrogen chloride – HCl_2^- [14-16], gaseous HCl [17], or a hydrogen chloride salt [18] – to the melt before elemental lithium and sodium could be deposited and stripped at an inert substrate electrode. However, because hydrogen chloride volatilizes from the melts, the desired deposition-stripping behavior can only be maintained by adding HCl_2^- at regular intervals, controlling the gaseous HCl partial pressure, or introducing a large excess of a hydrogen chloride salt. Therefore, addition of hydrogen chloride is not an ideal method for achieving rechargeable alkali metal anodes in these melts. In addition, the cycling efficiency (*i.e.*, charge for anodization ÷ charge for electrodeposition) was poor due to instability of the alkali metal deposits.

Due to poor performance with HCl, we undertook an effort to identify new "additives" that would provide better performance. Our search was based on the established understanding that lithium and lithium-ion anodes in organic-electrolyte batteries rely on the presence of a solid-electrolyte interface (SEI) to stabilize the anode against parasitic reactions. We selected thionyl chloride ($SOCl_2$) as a likely candidate based on its SEI-forming abilities in Li-$SOCl_2$ primary batteries and that fact that LiCl (the SEI in Li-$SOCl_2$ primary batteries) is insoluble in the buffered neutral melt. We immediately achieved reversible lithium and sodium deposition-stripping behavior in chloroaluminate ionic liquids by adding small quantities (<100 mM) of thionyl chloride, $SOCl_2$, to the electrolyte [19,20]. The cycling efficiencies in the presence of the thionyl chloride solute were >90%, and the deposited alkali metals exhibited improved stability under open-circuit conditions. In addition, the low volatility of $SOCl_2$ (b.p. = 75.8 °C) maintained the high cycling efficiencies, even in an open vessel under inert atmosphere, for extended times. A cyclic voltammogram for lithium deposition-stripping is shown in Figure 3.

Observations using *in situ* optical microscopic of the growth of the lithium film from a buffered chloroaluminate melt containing 48 mM $SOCl_2$ revealed the formation of a rough and dendritic surface. The presence of the dendritic lithium was prevalent and seemed independent of $SOCl_2$ concentration. During discharge, the lithium deposit appeared to disconnect from the tungsten electrode and seemed to "float" near the surface. Stripping efficiencies remained reasonably constant through all $SOCl_2$ concentrations. Repeated cycling of the deposition/stripping experiments without cleaning the electrode resulted in a steady decay in efficiencies; 85.6%, 80.7%, 70.6%, and 69% for four runs, respectively. This decrease in efficiency is further evidence of "dead" lithium on or near the electrode surface.

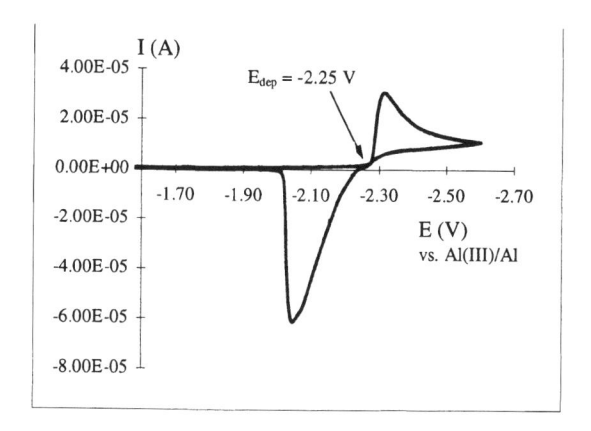

Figure 3. Cyclic voltammogram for Li/Li$^+$ from 1.1:1.0:0.1 AlCl$_3$:EMIC:LiCl containing 21 molal SOCl$_2$ at a 250-μm tungsten microelectrode. Scan rate = 500 mV s^{-1}.

Similar *in situ* observations of the deposition of sodium from a NaCl buffered melt containing 48 mM SOCl$_2$ revealed the growth of the sodium deposit to be slightly roughened, but free of dendrites. Stripping efficiencies remained reasonably constant through all SOCl$_2$ concentrations. Efficiencies obtained from repeated cycling of the deposition/stripping experiments without cleaning the electrode were relatively constant at 85% for four sequential runs.

As expected, the dramatic improvements in the cycling of lithium and sodium in the presence of SOCl$_2$ have been attributed to the formation of a thin, stabilizing SEI at the electrode-electrolyte interface. This SEI has been extensively studied for lithium and lithium-ion electrodes in organic solvents [21], but less so for sodium [22,23]. The SEI is formed by reaction of the alkali metal with components in the electrolyte to produce inorganic salts, such as LiCl, Li$_2$O, and Li$_2$CO$_3$. To achieve a stable rechargeable system, the SEI must be a good ionic conductor for the ion of the alkali metal comprising the active electrode. In chloroaluminate ionic liquids containing SOCl$_2$, we believe the SEI is composed of LiCl or NaCl, depending upon the alkali metal ion present in the melt.

In addition, SOCl$_2$ is electrochemically reduced to form an insoluble LiCl film at a bare, inert electrode (see AC-impedance study below). We believe this thin LiCl film impedes the direct electrochemical reduction of the imidazolium cation and shifts its irreversible reduction potential to more negative values. This kinetically induced negative shift of the imidazolium potential allows the clean electrodeposition of lithium and sodium metal within a widened electrochemical window. The electrodeposited alkali metal reacts immediately with additional SOCl$_2$ to produce an SEI that protects it from parasitic reaction with the organic cation (and anion).

To further study the SEI in these systems, AC-impedance was performed on the lithium and sodium deposits to evaluate the presence, growth, and stability of the SEI. Interpretation of Nyquist plots of the AC-impedance data collected at a tungsten disk electrode for the clean electrode, after reduction of SOCl$_2$ only, and after alkali metal electrodeposition indicated the following: (1) the electrode before and after SOCl$_2$ reduction showed simple capacitance behavior; (2) both lithium and sodium exhibited capacitance loops consistent with the presence of an SEI on the alkali metal surface; (3) the resistance of the SEI increased with time for both metals; and (4) based on the capacitance loops at higher frequencies, the sodium SEI possessed a higher resistance (*ca.* ten times greater) and higher capacitance than the lithium SEI. Importantly, these results were direct confirmation of SEI formation on the alkali metal surface in these ionic liquid electrolytes. Also, the latter result was consistent with a thinner (higher capacitance), higher resistive compact SEI for sodium than for lithium and may account for the higher over-potentials observed for sodium electrodeposition and stripping in these electrolytes. The lithium SEI also exhibited a large capacitance loop at low frequencies, indicative of a thick porous layer in series with the compact SEI layer.

In addition to SOCl$_2$, we also found that other reducible materials could be used to form a stabilizing SEI. For example, the electrochemistry of the Li/Li$^+$ couple was examined in a buffered neutral 1.1:1.0:0.1 AlCl$_3$:EMIC:LiCl melt containing 0.09 molal sodium 4-styrenesulfonate to promote reversible lithium electrodeposition. This system relies on the reductive polymerization of 4-styrenesulfonate to a protective film of poly(4-styrenesulfonate) on the electrodeposited lithium metal. We refer to this organic solid electrolyte interface (SEI) as a self-healing anionic polymer electrolyte (SHAPE) film. Figure 4 below shows three successive (no cleaning steps between runs) cyclic voltammograms for lithium electrodeposition and stripping at a 250-µm tungsten microelectrode.

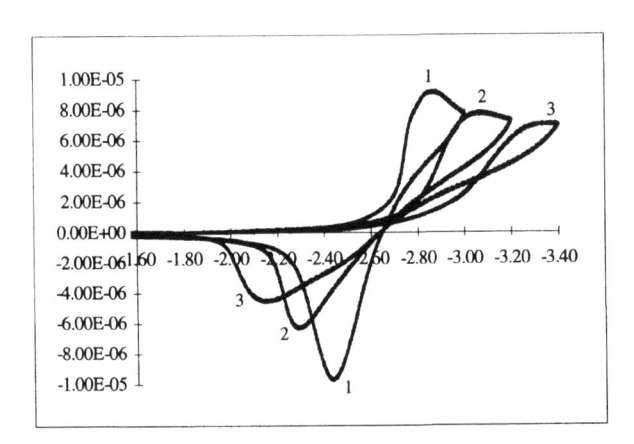

Figure 4. Successive cyclic voltammograms for lithium in 1.1:1.0:0.1 AlCl$_3$:EMIC:LiCl containing 0.09 molal sodium 4-styrenesulfonate at a 250-µm tungsten microelectrode. Scan rate = 200 mV s^{-1}.

With each scan, the lithium deposition and stripping waves become broader and more separated. The coulombic efficiencies are 62, 43 and 39% for scans 1, 2 and 3, respectively. This behavior is consistent with an increase in the polymer film thickness and a concomitant rise in the interfacial resistance, so that it is more difficult to electrodeposit and strip lithium at or below the polymer film. Additional electrochemical studies showed that 4-styrenesulfonate is electrochemically reduced at the inert electrode prior to the electrodeposition of lithium. This has been confirmed by *in situ* optical observations. The polymer forms as a yellow-orange film on the 250-μm tungsten electrode at potentials positive of lithium reduction. In addition, the lithium appears to initially electrodeposit as nodules below the pre-formed polymer film, and the lithium maintains a nodular morphology as it continues to deposit below, or possibly on top of, the polymer film. Additional work with this and other SHAPE systems is warranted.

4.3. LITHIUM ANODE IN EMIBF$_4$-LIBF$_4$

While reversible anode behavior can be obtained in the neutral buffered choroaluminate melts, these ionic liquids are moisture sensitive and generate HCl slowly upon exposure to humid air. Therefore, it is desirable to develop reversible electrode couples for the air- and water-stable perfluoroanion ionic liquids.

We found that LiBF$_4$ exhibits good solubility (>0.3 M) in the ionic liquid EMIBF$_4$, thus making possible a lithium or lithium-ion battery based on this chemically inert, nonflammable ionic liquid. However, because alkali metal chlorides are partially soluble in EMIBF$_4$, SOCl$_2$ is a poor choice to form the required SEI and is highly sensitive to moisture. During initial trials, we discovered that trace amounts of water promoted deposition of metallic lithium from a EMIBF$_4$-LiBF$_4$ binary system. We postulate that the water reacts with the lithium to form a Li$_2$O and/or LiOH SEI, similar to the interface films formed in Li-water primary batteries. For the system EMIBF$_4$ + 0.2 M LiBF$_4$ + <100 mM H$_2$O, the cycling efficiency was 60%; however, this is expected to improve for an optimized system [24].

4.4. LITHIUM-ALUMINUM ANODE IN IONIC LIQUIDS

A promising approach to improving the lithium cycling efficiency is to employ an aluminum collector electrode so that lithium reduction proceeds by formation of the β-LiAl alloy. However, expansion of the aluminum electrode upon alloying with lithium possesses problems with physical stability during cycling; however, these structural problems are likely to be overcome with the nanoparticle alloys being developed for conventional lithium-ion batteries.

We evaluated an aluminum wire electrode in the EMIBF$_4$-LiBF$_4$ (0.2M) containing trace amounts of water. The cycling efficiency for the β-LiAl anode was measured consistently at ≥ 90%, even in unoptimized studies. The reduction potential of the alloy is 0.35 V more positive than that for elemental lithium, and its electrical capacity is 790 Ah/kg. At this time, it is not clear if water is needed for alloy formation; however, preliminary data indicates that water is required only to shift the EMI$^+$ reduction to more negative potentials (as discussed above) and not to form an SEI on the alloy surface [24].

The lithium-aluminum alloy anode has also been demonstrated in buffered neutral LiCl-EMIC-AlCl$_3$ melts [25]. In this work, it was necessary to add either lithium metal or SOCl$_2$ to the buffered neutral melt to achieve clean Li-Al alloy formation. In our own work, we have shown that an aluminum-rich Li-Al alloy phase is electrodeposited slowly at potentials negative of -1.3 V (*vs.* Al/Al(III)) [26]. This process cannot be a result of direct electroreduction of Al$_2$Cl$_7^-$ because of its extremely low concentrations. Instead, the deposition process is believed to result from the polarizing nature of the Li$^+$, which promotes reduction of the kinetically stable AlCl$_4^-$ anion - effectively increasing the *activity* of Al$_2$Cl$_7^-$ in the buffered neutral melt compared to neutral 1:1 AlCl$_3$:EMIC melt. The addition of lithium metal should reduce the activity of the aluminum anions in the melt. While SOCl$_2$ (a base in these melts) may similarly reduce the aluminum anion activity, or it also may produce a LiCl passive film at the aluminum electrode surface that impedes aluminum anion reduction, but allows Li-Al alloy formation to proceed.

4.5. CATHODES FOR IONIC LIQUIDS

While much research has been performed on quantifying the properties of alkali metal anodes in the ionic liquid electrolytes, less work has been devoted to the cathode. Most cathode work in the chloroaluminates has dealt with halogen (chlorine or bromine) and transition metal chlorides. The halogens suffer from reaction with the organic cation and are difficult to engineer into a practical cell. While the metal chlorides hold promise as reversible cathodes, they suffer from large phase changes during cycling, and they function best in chloride-rich systems. Instead, intercalation or insertion cathodes would be more practical systems. In this vein, two primary approaches have been taken: (1) conducting organic polymers that insert or expel ions during charge and discharge and (2) transition metal oxides that undergo intercalation reactions during cycling.

Recent results in LiCl-EMIC-AlCl$_3$ buffered neutral melts show promise for transition metal oxides as reversible cathodes [27]. The results are summarized in the table below.

Cathode	Capacity (Ah/kg)	Cycling Efficiency (%)	Discharge Voltage (*vs.* Al)
LiCoO$_2$	124	91.9	1.9-0.9
LiNiO$_2$	100	74.0	1.8-0.8
LiMn$_2$O$_4$	92	65.7	2.0-1.0
V$_2$O$_5$ (crystalline)	146	100	1.4-0.4

Table 2. Transition metal cathode performance in buffered neutral LiCl-EMIC-AlCl$_3$ melt

Importantly, the solubilities of the metal oxides were low in these melts: LiCoO$_2$, 0.336 mg/l; LiNiO$_2$, 0.185 mg/l; LiMn$_2$O$_4$, 0.3921 mg/l; V$_2$O$_5$, 4.372 mg/l. Clearly additional work is need in this area to improve the cycling efficiency and to optimize capacity; however the initial results are very promising.

In our own work, we have identified a cathode material for acidic melts formed by the oxidation of ClSiPh$_3$ (Ph = phenyl) in an acidic AlCl$_3$-EMIC molten salt [28]. The exact nature of the material remains elusive; however, x-ray photoelectron spectroscopy indicates the material is composed of polymerized silane and imidazole units in an approximately 1:1 molar ratio. From the XPS analyses of charged and discharged materials, we estimated that the silane-imidazole material should have a specific electrical capacity of 46 - 107 Ah/kg. Subsequently, we prepared thin films of this material in 1.5:1.0 AlCl$_3$:EMIC and measured capacities in the range of 32 to 57 Ah/kg; however, this capacity is probably low since it was impossible to completely remove the melt from the film prior to weighing.

The silane-imidazole cathode can be combined with an Al anode to achieve a cell voltage of 1.65 V. In addition, we have found that these cathodes can be discharged at rates in excess of 100 mA/cm^2. The silane-imidazole cathode is compared to other polymeric materials in the table below, in which predicted performance parameters for a cell with an Al anode are also listed. Values for the other polymers are estimated from studies performed primarily by Robert Osteryoung and co-workers [29-35] as well as some data from other work [36].

Polymer	Capacity (Ah/kg)	E_{cell} (V)	Specific Energy (Wh/kg)
Polyaniline	37 - 133	1.6	59-213
Polyfluorene	54 - 82	1.4-1.6	74-129
Silane-imidazole	32 - 57	1.65	52-92

Table 3. Performance of conducting polymers cathode against a reversible Al anode in acidic melts.

Although the theoretical energy storage densities of the Al/polymer cells are not exceptionally high compared to other battery technologies, their safety, low cost, and excellent recharging characteristics make them viable battery systems where high-voltage. More importantly, these conducting polymer should function in the neutral buffered melts and provide highly reversible cathode couples for alkali metal-based anodes.

4.6. DUAL INTERCALATING MOLTEN ELECTROLYTE (DIME) BATTERIES

The interest in Li-carbon anodes for Li-ion batteries with organic electrolytes prompted us to examine the possibility of using an intercalating lithium-graphite anode in an AlCl$_3$-EMIC-LiCl room-temperature molten salt. Initial results appeared to show a lithium ion intercalation at -1.4 V; however, subsequent studies showed this process to be the lithium-ion promoted reduction of AlCl$_4^-$, as discussed above.

Although we were not been able to intercalate lithium into carbon electrodes, we did discover that the organic cation and melt anion could be effectively intercalated into graphite or coke electrodes through the processes illustrated by Equations 2 and 3 for a neutral 1:1 AlCl$_3$:EMIC melt [37].

$$EMI^+ + xC + e^- \Leftrightarrow (EMI)C_x \quad (2)$$
$$AlCl_4^- + yC \Leftrightarrow C_y(AlCl_4) + e^- \quad (3)$$

The graphite intercalation-deintercalation of EMI$^+$ is found at -1.6 V while the irreversible oxidation of EMI$^+$ begins at -2.0 V at GC. Tetrachloroaluminate undergoes graphite intercalation-deintercalation at +1.8 V while AlCl$_4^-$ is oxidized to chlorine gas at +2.5 V. Therefore, the electrochemical window of this melt at GC is seen to be 4.5 V, and the electrochemical intercalations at graphite occur well within this window. The anodic and cathodic intercalation couples at graphite can be used to create a rechargeable cell with a voltage of >3 V. Because the two electrochemical couples in such a cell involve intercalation of only the molten salt cations and anions, we have termed these systems Dual Intercalating Molten Electrolyte (DIME) batteries. We have demonstrated the DIME battery concept for a number of ionic liquids, as illustrated by selected data provided in Table 4 on the next page. Importantly, we have found that the 1,2-dimethyl-3-propylimidazolium (DMPI$^+$) cation performs much better than the EMI$^+$ cation as a carbon intercalation ion. This is attributed to the relatively high reactivity of the acidic C-2 proton in EMI$^+$, which is replaced with a methyl group in DMPI$^+$. Experiments have demonstrated that DMPI$^+$ can be reversibly intercalated (>90% efficiency) into a thin graphite electrode at least to a composition of (DMPI)C$_{36}$, with staging phenomena observed at compositions corresponding to (DMPI)C$_{141}$ and (DMPI)C$_{69}$ (Figure 5).

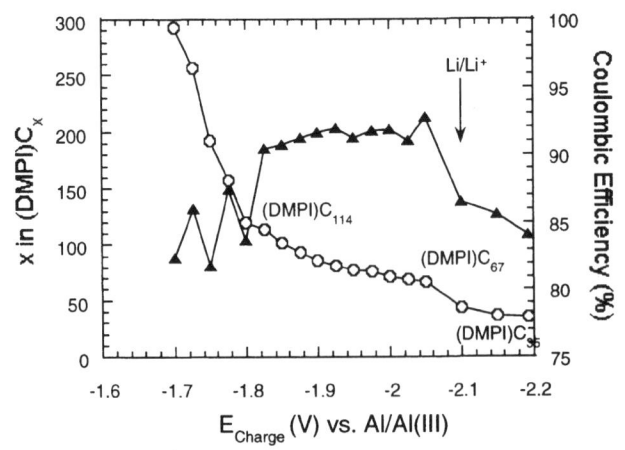

Figure 5. Intercatalation of DMPI$^+$ into lithium-ion graphite anode. Charge:Discharge = 0.1:0.2 mA/cm^2.

Molten Salt Electrolyte	Intercalated Ion	T (°C)	I_{charge} (mAcm^{-2})	E_{charge} (V)	$I_{discharge}$ (mAcm^{-2})	$E_{discharge}$ (V)	Efficiency %
(EMI)(AlCl$_4$)	EMI$^+$	24	1	-1.64	1	-1.47	59
	EMI^{+a}	24	2	-1.69	2	-1.50	65
	AlCl$_4^-$	24	1	+1.93	1	+1.6	41
	AlCl$_4^-$	24	2	+2.01	2	+1.6	64
(DMPI)(AlCl$_4$)	DMPI$^+$	30	1	-1.70	1	-1.25	60
	DMPI$^+$	30	1	-1.70	0.5	-1.36	100
	DMPI$^+$	30	1	-1.70	0.25	-1.35	94
	AlCl$_4^-$	30	1	+2.15	1	+1.50	64
	AlCl$_4^-$	30	1	+2.15	0.5	+1.60	86
	AlCl$_4^-$	30	1	+2.15	0.25	+1.60	76
(EMI)(BF$_4$)	EMI^{+b}	24	6	-1.89	6	-1.30	33
	BF$_4^{-b}$	24	1	+2.33	1	+1.50	55
(EMI)(PF$_6$)	EMI$^+$	90	5	-2.4	5	-1.4	21
	PF$_6^-$	90	5	+2.0	5	+1.3	23

^{a}Charging time = 60 min; ^{b}Charging time = 10 min.

Table 4. Anion and cation graphite intercalation in room-temperature molten salt electrolytes. Data collected with chronopotentiometry using charging times of 30 min, unless stated otherwise.

The DIME battery concept offers several practical advantages over conventional battery systems including (*i*) inexpensive graphite electrodes are used, (*ii*) only a single molten salt needs to be synthesized and purified, (*iii*) no organic solvents are required, (*iv*) the battery can be assembled in the discharged state, and (*v*) all components have low toxicity and are nonflammable. Despite these advantages, a DIME cell suffers from an inherently low energy storage capacity; for example, in (DMPI)(AlCl$_4$) the theoretical specific energy is only 91 Wh/kg, assuming a fully charged stoichiometry of [(DMPI)(C$_{24}$)][(AlCl$_4$)(C$_{24}$)]. Even so, it is possible that the low cost, ease of preparation, and safety of the DIME system will make it a practical choice for certain battery applications. The DIME concept is currently being pursued by researchers at the Naval Research Laboratory under the direction of Hugh De Long and Paul Trulove.

5. Ionic Liquid-Polymer Gel Electrolytes

While the ionic liquid electrolytes provide many benefits over conventional organic and aqueous electrolytes, they still must be contained in liquid-tight containers to prevent leakage. Therefore, a means to convert these free-flowing liquids into a solid-state form is highly desirable. To this end we have developed novel ionic liquid-polymer gels, composed of a perfluoroanion ionic liquid and a poly(vinylidene fluoride) hexafluoropropylene copolymer (PVdF(HFP)) and have evaluated them as solid-state electrolytes [38,39].

5.1. IONIC LIQUID AND IONIC LIQUID-POLYMER GEL PREPARATIONS

5.1.1. *EMIBF$_4$ Preparation in Acetonitrile*

In a typical acetonitrile preparation of EMIBF$_4$, 7.78 g NH$_4$BF$_4$ (Alfa, 99.5%) was added to 10.91 g of EMICl in 250 ml of acetonitrile. The mixture was stirred for 3 days, and the NH$_4$Cl filtered from the reaction mixture. The acetonitrile filtrate containing the soluble EMIBF$_4$ was treated with neutral alumina. After filtering the alumina and rotoevaporation of the solvent, the EMIBF$_4$ ionic liquid product was dried under vacuum for 3 h at 110 °C, giving a final yield of 94%. While we have also reported preparation of this hydrophilic ionic liquid in other solvents, detrimental side-reactions of acid in the NH$_4$BF$_4$ can lead to impure ionic liquid products; therefore, we **strongly** recommend the use of acetonitrile in this preparation.

The hydrophilic triflate ionic liquid, EMITrif can also be prepared by this route using NH$_4$(triflate) (99%, Aldrich) in place of the NH$_4$BF$_4$.

5.1.2. *1-n-Butyl-3-Methylimidazolium Hexafluorophospate (BMIPF$_6$)*

The preparation of this material followed an aqueous metathesis route similar to those described for other hydrophobic ionic liquids. Aqueous solutions of *ca.* 50 ml each containing 10.22 g (55.52 mmol) KPF$_6$ (98%, Aldrich) and 12.26 g (55.75 mmol) BMIBr were combined to produce a turbid solution. Upon sitting, hydrophobic globules of BMIPF$_6$ ($\rho \approx 1.36$ g cm^3) settled to the bottom of the flask and were recovered by removing the upper aqueous solution with a pipette. The ionic liquid product was extracted three times with 50 ml aliquots of deionized water to remove residual KBr. The final ionic liquid product was dried *in vacuo* at *ca.* 80 °C for several hours. Yield BMIPF$_6$: 10.19 g (35.76 mmol, 64.1 %). Although BMIPF$_6$ readily supercools to a highly viscous liquid or glass, even down to dry ice temperatures, we have determined the melting point of this ionic liquid to be *ca.* 10 °C using a thermocouple probe and visual observations.

5.1.3. *Preparation of EMITrif-PVdF(HFP) Ionic Liquid-Polymer Gel*

All ionic liquid:polymer ratio are given as a weight ratio. Under ambient conditions, 0.587 g EMITrif, 0.528 g PVdF(HFP), and 2.5 mL propylene carbonate (PC) were mixed to produce a translucent solution. This mixture was poured into an aluminum weighing pan of 5 cm diameter and placed on a Corning hot plate preheated to *ca.* 75 °C. The translucent mixture gelled to a transparent solution in 1-5 minutes and was removed from the hot plate. The gel was introduced into a 500 mL freeze drying apparatus (Kontes) and evacuated for 1-2 days at 80 °C on a vacuum line to remove the organic PC solvent. The resulting 1.1:1 EMITrif-PVdF(HFP) gel was a free-standing, rubbery film (thickness ≈ 0.05 cm; area ≈ 20 cm²) and was transparent to slightly translucent in appearance.

To facilitate solvent removal, gels were also prepared using and 4-methyl-2-pentanone (MP) with a boiling point of 118 °C *versus* 240 °C for PC. Gels prepared with MP did not require vacuum for solvent removal; instead, following heat gelling on the hot plate, the MP was allowed to evaporate overnight in ambient air. In our experience, preparation with MP is preferred over PC because it greatly facilitates complete removal of the organic liquid from the final ionic liquid-polymer gel.

5.1.4. Graphite-BMIPF₆-PVdF(HFP) Electrode Preparation

1.0 g of synthetic graphite powder (1-2 μm, Aldrich), 0.1 g of PVdF(HFP) powder, and 0.2 g of the BMIPF₆ ionic liquid were combined with 4 ml of MP in a small beaker. After thoroughly mixing the components, the resulting slurry was poured into an aluminum pan and placed on a hot plate heated to *ca.* 75 °C to initiate the gelation process. The aluminum pan was then removed from the hot plate, and the MP was allowed to evaporate overnight in a laboratory hood under ambient air. The resulting graphite-BMIPF₆-PVdF(HFP) composite gel was a free-standing, flexible film; however, it was more easily torn than the pure ionic liquid-polymer gel electrolytes. For the DIME single and bipolar cells discussed below, disk electrodes were cut from this parent composite gel film.

5.2. IONIC CONDUCTIVITIES OF IONIC LIQUID-POLYMER GELS

The ionic conductivities (κ) measured using AC-impedance of ionic liquid-PVdF(HFP) gels made with EMIBF₄, EMI(triflate), and BMIPF₆ were investigated as functions of ionic liquid:PVdF(HFP) mass ratio and of temperature. Conductivities at room temperature (22 - 23 °C) and 100 °C and pertinent information on each gel are collected in Table 5. The logarithmic plots of ionic conductivity versus T^{-1} for the gels and the corresponding neat ionic liquids approximated Arrhenius behavior (*i.e.*, κ = exp(-E_A/RT)) over the limited temperature range reported here.

Ionic Liquid	Mass Ratio	κ/mS cm⁻¹ 22 - 23°C	κ/mS cm⁻¹ 100 °C	E_A/kcal mol⁻¹
EMI(triflate)	2:1 (MP)	2.2	22.9	6.6
EMI(triflate)	1:1 (MP)	0.6	9.1	8.0
EMI(triflate)	0.4:1 (MP)	0.06	4.5	8.5
EMIBF₄	2:1 (PC)	3.2	14.4	5.0
EMIBF₄	1:1 (PC)	1.4	16.4	6.6
EMIBF₄	0.5:1 (PC)	0.1	7.3	11.7
BMIPF₆	2:1 (MP)	0.6	13.7	8.8
BMIPF₆	1:1 (MP)	0.1	7.3	11.9
EMI(triflate)	neat	9.3[c]	55.3[d]	4.2
EMIBF₄	neat	13.8[f]	59.0	4.3
BMIPF₆	neat	1.8	15.6	6.1

Table 5. Ionic Conductivities of Ionic Liquid-Polymer Gel Electrolytes and Neat Ionic Liquids

From the conductivity data several observations can be made. (1) The ionic conductivities of the gels increase as the ionic liquid:PVdF(HFP) mass ratio increases. This also tracks with the physical characteristics of the gels, which progress from a relatively stiff film at the 0.4:1 or 0.5:1 mass ratio to a flexible film at the 2:1 weight ratio. (2) The E_A values increase with decreasing ionic liquid:PVdF(HFP) mass ratios so that the enhancements in κ with increasing mass ratios are not as dramatic at 100 °C as they are at room temperature. (3) The ionic conductivities of the gels track with the ionic conductivities of the corresponding neat ionic liquids.

While the data above limits temperature to 100 °C, the ionic liquid polymer gels are stable to much higher temperatures. In fact, the conductivity of the EMITrif-PVdF(HFP) displays no unusual behavior at the melting point of the PVdF(HFP) (140 - 145 °C) and reaches 41 mS/cm at 205 °C. The conductivity declined to 7 mS/cm at 230 °C, and the gel sample decomposed to a brown, less-flexible material. This decomposition is probably a result of air oxidation, as another sample of a EMITrif-PVdF(HFP) gel exhibited no visual decomposition up to 250 °C under vacuum. It is worth noting that neat EMITrif is thermally stable to >400 °C. Therefore, because the ionic liquids are not volatile, even at elevated temperatures, and PVdF(HFP) is thermally stable to >300 °C, the ionic liquid-PVdF(HFP) gels are excellent electrolytes for high temperature operation.

5.3. DIME CELLS BASED ON IONIC LIQUID-POLYMER GEL ELECTROLYTES

Because the cation and anion of the neat ionic liquid electrolytes can be reductively and oxidatively intercalated, respectively, into graphite electrode, simple inexpensive electrochemical cells can be constructed using only two graphite electrodes as the anode and cathode and having operating voltages of 3 to 3.5 V. We have termed such systems dual intercalating molten electrolyte (DIME) cells, as already mentioned above. Because solid-state electrolytes are more desirable than liquid electrolytes for battery design and construction, we decided to examine the ionic liquid-PVdF(HFP) as electrolytes in DIME cell configurations.

For the anode and cathode graphite intercalation electrodes, graphite powder, BMIPF$_6$, and PVdF(HFP) were used to prepare a graphite-gel composite containing 70% by mass graphite, as described in section 5.1.4. Two 1.1 cm^2 graphite-gel disks were cut, placed on either side of a slightly larger disk (area ≈ 1.5 cm^2) of a 2:1 BMIPF$_6$-PVdF(HFP) gel electrolyte, and compressed between two stainless steel collector rods in a PFA Teflon® tee-cell. The DIME cell was charged and discharged for several cycles, though charging was limited to only a small percentage of full depth of discharge. Chronopotentiometric curves collected at charge and discharge currents of ±0.9 mA/cm^2 with the DIME cell exposed to ambient air gave coulombic efficiencies for the two cycles shown of only *ca.* 60%, but the open-circuit potential (E_{open}) was 3.77 V, consistent with DIME cells operating with neat ionic liquids.

One of the major advantages of solid-state electrolytes over liquid electrolytes is that solid-state electrolytes are amenable to bipolar cell construction. Therefore, components for two single DIME cells (electrode area = 0.38 cm^2) were stacked into the tee-cell test unit, with a disk of aluminum foil placed between the individual cell components. As with the single DIME cell, the components of this bipolar cell became cold welded into a laminated dimensionally stable unit upon compression. Charge and discharge curves collected at ±1.3 mA/cm^2 gave coulombic efficiency of only ca. 40%; however, the bipolar cell voltage was 7.86 V, approximately twice the single cell value. Therefore, the bipolar DIME cell configuration appears to function with the ionic liquid-PVdF(HFP) gel as the solid-state electrolyte.

6. Other Electrochemical Energy Applications of Ionic Liquids and Ionic Liquid-Polymer Systems

6.1 CAPACITOR APPLICATIONS

Ionic liquids possess nearly ideal properties for high-power electric double layer (EDL) capacitor applications: high ion concentration, nonflammable, thermal stability, and wide electrochemical window. The latter property is particularly important because the energy storage of a capacitor is given by $E = \frac{1}{2}CV^2$, where E is energy in joules/g, C is specific capacitance (F/g; F = Farads = coulombs/V) of the capacitor, and V is the maximum voltage. Therefore, the high voltage window of the ionic liquid-based could produce high specific energy capacitors if appropriate electrodes can be identified. Drawbacks to the ionic liquids are their fairly high viscosities and relatively large ions, which may limit accessibility to the highly porous electrodes used in EDL capacitors. The addition of organic co-solvents can ameliorate the high viscosity, but at the expense of increased flammability, volatility, and possibly chemical and electrochemical activity.

A number of ionic liquids and mixtures of ionic liquids and organic solvents have been considered for double layer capacitor applications [11]. The area capacitances of ionic liquids at a mercury electrode have been determined to be ca. 12 μF/cm^2, and voltage windows are typically greater than 4 V for the perfluoroanion ionic liquids. Addition of organic co-solvents to the ionic liquid tends to lower the area capacitance of the electrolyte while increasing conductivity. Furthermore, the ionic liquids-polymer gels are ideally suited for solid-state double layer capacitors because the ionic liquid will flow from the gel into a porous electrode structure as long as proper electrode wetting is achieved.

In addition to EDL capacitors, faradaic capacitors using thin-film redox active electrodes can also be developed using ion liquids and ionic liquid-polymer gel electrolytes. The high ion concentration inherent to ionic liquids is particularly attractive for p- and n-dopable conducting polymer electrodes where fast ion injection and expulsion is required for high power levels. It should be pointed out that the ruthenium oxide based faradaic capacitors function only in a protonic medium and will not operate in a typical aprotic ionic liquid electrolytes.

Given the variety of ionic liquids and mixtures available, there is significant opportunity to develop ionic liquid capacitors for energy storage and high power applications.

6.2 FUEL CELLS

The only fuel cell technology developed with molten salts remains the molten carbonate fuel cell. This technology is rather mature and is entering the commercial marketplace.

Although the room-temperature ionic liquids would appear to be well suited to fuel cell applications, this specific area of research appears to have been relatively ignored. In a closely related area, we have developed gas separation membranes and catalytic membrane reactors using ionic liquids, where the nonvolatile nature of the melts was exploited to fabricate stable liquid membranes. In addition, Noble and Koval recently described electrochemically-driven gas separation membranes at the 2001 ACS meeting in San Diego.

To our knowledge, the only reported studies attempting to apply the room-temperature ionic liquids to fuel cell development have been performed by our research group and a group at DuPont [40]. In both studies, the ionic liquid was imbibed into Nafion membrane, and the conductivity of the resulting composite was evaluated at elevated temperatures. It was found that the composite maintained ionic conductivity at temperatures in excess of 100 °C, above which unmodified Nafion dehydrates and losses its ionic conductivity. However, the nature of the conducting species in the Nafion-ionic liquid composite is still unknown. For proper proton exchange membrane (PEM) fuel cell performance, a high transport number for proton is needed.

6.3. PHOTOVOLTAIC SYSTEMS

Grätzel has employed ionic liquids as electrolytes for a photovoltaic devices based on the following photoelectrochemical energy conversion steps [41-43]:

$$Ru(bipy)_3^{2+} + h\nu \rightarrow {}^*Ru(bipy)_3^{2+} \tag{4}$$

$$^*Ru(bipy)_3^{2+} \rightarrow Ru(bipy)_3^{3+} + e^- (TiO_2) \tag{5}$$

$$Ru(bipy)_3^{3+} + 3/2\ I^- \rightarrow Ru(bipy)_3^{2+} + 1/2\ I_3^- \tag{6}$$

$$1/2\ I_3^- + e^-\ (Pt) \rightarrow 3/2\ I^- \tag{7}$$

The ionic liquids have exceptional properties for photovoltaics, most note worthy of these are (1) tolerance to high temperatures under solar illumination; (2) resistance to freezing at low ambient temperatures; (3) electrochemically and chemically inert toward a range of device fabrication materials; (4) high ionic conductivity for high flux rates; (5) nonflammable and nonvolatile for safe operation, tolerance to micro-leaks, and negligible vapor pressure upon heating; and (6) hydrophobic character to minimize uptake of water. Given these ideal characteristics of ionic liquids, it is not be surprising that photovoltaic devices based on ionic liquids appear to be near commercialization.

7. References

1. Tuck, Clive D. S. (1991) *Modern Battery Technology*, Ellis Howard, Chichester.
2. Linden, D. (1995) *Handbook of Batteries,* Second edition, McGraw-Hill, New York.
3. Scrosati, B. and Vincent, C.A (1997) *Modern Batteries*, Second edition, Arnold, London.
4. Pistoia, G., (1994) *Lithium Batteries* Elsevier, Amsterdam.
5. Winter, M., Besenhard, J. O., Spahr, M. E., and Novák, P. (1998) *Adv. Mater.* **10**, 725.
6. Idota, Y., Kubota, T., Matsufuji, Maekawa, and Miyasaka (1997) *Science* **276**, 1395.
7. Mao, O. and Dahn, J. R. (1999) *J. Electrochem. Soc.* **146**, 423.
8. Foster, D. L., Wolfenstine, Read, J. R., and Behl, W. K. (2000) *J. Electrochem. Soc.* **3**, 203.
9. Park, H.-K., Smryl, W. H., and Ward, M. D. (1995) *J. Electrochem. Soc.*, **142**, 1068.
10. Wong, J. L. and Keck, J. H. (1974) *J. Org. Chem.* **39**, 2398.
11. McEwan, AB, Ngo, H. L., LeCompte, K., and Goldman, J. L. (1999), *J. Electrochem. Soc.* **146**, 1687.
12. Suarez, P. A. Z., Selbach, V. M., Dullius, J. E. L., Einloft, S., Piatnicki, C. M. S., Azambuja, D. S., de Souza, R. F., and Dupont, J. (1997), *Electrochimica Acta* **42**, 2533.
13. Gifford, P.R. and Palmisano, J. B. (1988) *J. Electrochem. Soc.* **35**, 650.
14. Scordilis Kelley, C. and Carlin, R. T. (1993) *J. Electrochem. Soc.*, **140**, 1606.
15. Riechel, T. L. and Wilkes, J. S. (1992) *J. Electrochem. Soc.*, **139**, 977.
16. Scordilis Kelley, C. and Carlin, R. T. (1994) *J. Electrochem. Soc.*, **141**, 1606.
17. Gray, G. E., Kohl, P.A., and Winnick, J. (1995) *J. Electrochem. Soc.*, **142**, 3636.
18. B. J Piersma (1994) in *Proceedings of the Ninth International Symposium on Molten Salts* C. L. Hussey, D. S. Newman, G. Mamantov, and Y. Ito, Editors, PV 94-13, p. 415, The Electrochemical Society Proceedings Series, Pennington, NJ.
19. Fuller, J., Osteryoung, R. A., and Carlin, R. T. (1995) *J. Electrochem. Soc.*, **142**, 3632.
20. Fuller, J., Osteryoung, R. A., and Carlin, R. T. (1995) *J. Electrochem. Soc.*, **143**, L145.
21. Peled, E. (1994) *Proceedings of the Symposium on Lithium and Lithium-Ion Batteries*, S. Megahed, B. M. Barnett, and L. Xie, Editors, PV 94-28, p. 1, The Electrochemical Society Series, Pennington, NJ (1994).
22. Ma, Y., Doeff, M. M., Visco, S. J., and De Jonghe, L. C. (1993) *J. Electrochem. Soc.*, **140**, 2726.
23. Doeff, M. M., Visco, S. J., Ma, Y., Peng, M., Ding, L., and De Jonghe, L. C. (1995) *Electrochim. Acta*, **40**, 2205.
24. Fuller, J., Carlin, R. T., and Osteryoung, R. A. (1997) *J. Electrochem. Soc.*, **144**, 3881; Carlin, R. T. and Fuller, J, U.S. Patent 5,552,238, September 3, 1996.
25. Ui, K., Koura, N., and Idemoto Y. (1999) *Electrochemistry* **67**, 706.
26. Carlin, R. T., Fuller, J., Kuhn, K., Lysaght, M. J., and Trulove, P. C. (1996) *J. Applied Electrochem.*, **26**, 1147.
27. Ui, K., Koura, N., Idemoto, Y., and IIZuka, K. (1997) *Denki Kagaku* **65**, 161.
28. Carlin, R. T. and Osteryoung, R. A. (1994) *J. Electrochem. Soc.*, **141**, 1709.
29. Pickup P. G. and Osteryoung, R. A. (1995) *J. Electroanal. Chem.*, **195**, 271.
30. Pickup P. G. and Osteryoung, R. A. (1984) *J. Am. Chem. Soc.*, **106**, 2294.
31. Janiszewski, L. and Osteryoung R. A. (1987) *J. Electrochem. Soc.*, **134**, 2787.

32. Janiszewski L. and Osteryoung, R. A. (1988) *J. Electrochem. Soc.*, **135**, 116.
33. Oudard J. F., Allendorfer, R. D., and Osteryoung, R. A. (1988) *Synthetic Metals*, **22**, 407.
34. Tang J. and Osteryoung, R. A. (1991) *Synthetic Metals*, **44**, 307.
35. Tang J. and Osteryoung, R. A. (1991) *Synthetic Metals*, **45**, 1.
36. Koura, N., Ejira, H., and Takeishi, K. (1993) *J. Electrochem. Soc.*, **140**, 602.
37. Carlin, R. T., De Long, H. C., Fuller, J., and Trulove, P. C. (1994) *J. Electrochem. Soc.*, **141**, L73.
38. Fuller, J. Breda, A. C., and Carlin, R. T. (1997) *J. Electrochem. Soc.* **144**, L67.
39. Fuller, J. Breda, A. C., and Carlin, R. T. (1998) *Electroanal. Chem.*, **459**, 29.
40. Doyle, M., Choi, S., Proulx, G. (2000) *J. Electrochem. Soc.* **147**, 34.
41. Hagfeldt A. and Grätzel, M. (1995) *Chem. Rev.*, **95**, 49.
42. Papageorgiou, N., Athanassov, Y., Armand, M., Bonhôte, P., Pettersson, H., Azam, A., and Grätzel, M. (1996) *J. Electrochem. Soc.*, **143**, 3099.
43. Bonhôte, P., Dias, A.-P., Papageorgiou, N., Kalyanasundaram, K., and Grätzel, M., *Inorg. Chem.*, **35**, 1168.

8. Acknowledgements

Research reported from our laboratories was sponsored by the Air Force Office of Scientific Research and the Department of Energy. The Office of Naval Research supported the preparation and presentation of this manuscript through its Research Opportunity for Program Officers program. Opinions, interpretations, conclusions, and recommendations are those of the authors and are not necessarily endorsed by the sponsoring agencies.

SYNTHESIS AND CATALYSIS IN ROOM-TEMPERATURE IONIC LIQUIDS

P. J. SMITH, A. SETHI AND T. WELTON
*Department of Chemistry, Imperial College,
London SW7 2AY, U.K.*

Abstract

The use of ionic liquids as solvents for synthesis and catalysis is discussed. We focus on how the ionic liquids can interact (or not) to make changes in the observed reactivities.

1. Introduction

Synthetic chemistry is dominated by the study of species in solution. Although any liquid may be used as a solvent, relatively few are in general use. However, as the introduction of cleaner technologies has become a major concern throughout both industry and academia, the search for alternatives to the most damaging solvents has become a high priority. Solvents are high on the list of damaging chemicals for two simple reasons: (i) they are used in huge amounts and (ii) they are usually volatile liquids that are difficult to contain. Ionic liquids are attracting a great deal of attention as possible replacements for conventional molecular solvents in a number of processes.[1] These include reactions of the traditional stoichiometric organic synthesis arsenal and contemporary interests in catalysis for organic synthesis. Features that make ionic liquids attractive include:

- they have no significant vapour pressure, reducing solvent losses and facilitating simple containment;
- they are often composed of poorly coordinating ions, so they have the potential to be highly polar non-coordinating solvents;
- they are good solvents for a wide range of both inorganic and organic materials and unusual combinations of reagents can be brought into the same phase;
- they are immiscible with a number of organic solvents and provide a non-aqueous, polar alternative for two-phase systems, some ionic liquids can be used in biphasic systems with water.
- they are thermally robust and may be used over a wide range of temperatures.

It would be impossible to cover all of the reactions[1,2] that have been tried in ionic liquids in this paper, so we will focus on a small number of reactions that exemplify how ionic liquids can be used to change reactivities. Friedel-Crafts reactions were the first to be investigated in the chloroaluminate(III) ionic liquids and make use of the high concentrations of Lewis acid species that it is possible to achieve in these systems. For Diels-Alder reactions we describe how we have determined which interactions with the ionic liquids lead to changes in reactivities. In rhodium catalysed hydrogenation reactions it is the fact that the ionic liquid does not interact strongly with the catalyst that provides the interesting results, whereas in palladium catalysed C-C coupling reactions it is the strong interaction of the ionic liquids with the catalysts that explains the observed behaviour. Most of the ionic liquids referred to in this paper are based upon *N, N*-disubstututed imidazolium salts (Figure 1).

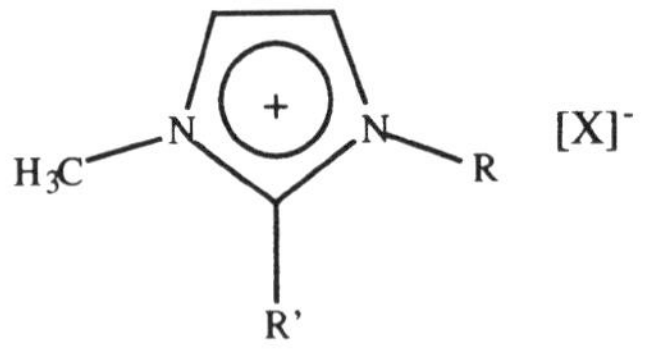

$R = C_2H_5, C_3H_7$ or C_4H_9
$R' = H$ or CH_3
$X = PF_6, BF_4, Tf_2N$ or TfO

Figure 1. Some imidazolium salts

2. Fiedel-Crafts reactions

It is unsurprising that electrophilic aromatic substitutions were the first organic reactions to be investigated in the room-temperature chloroaluminate(III) ionic liquids.[3] The high concentration of chloroaluminate(III) species coupled with the good solubility of simple arenes in the acidic ionic liquids makes them ideal solvents for these reactions, and it is possible to combine their function as a solvent and a catalyst. Of the arenes tested by reaction with 1-chloropropane, only nitrobenzene failed to react. As with conventional systems, polyalkylation was common, e.g., reaction between an excess of chloroethane and benzene led to the formation of a mixture of mono (12%), di (11%), tri (33%), tetra (24%), penta (17%), and hexa (2%) substituted products. As usual, polyalkylation can be minimized by use of a large excess of the arene but not totally eliminated. Basic ionic liquids do not provide adequate catalytic activity for alkylation reactions to occur. In order for the Friedel-Crafts reactions to occur, it is necessary to be able to form an electrophile in the ionic liquid. Luer and Bartak[4] demonstrated that even in a moderately acidic (0.52 mol % $AlCl_3$) ionic liquid, dissolution of chlorotriphenylmethane leads to the formation of the triphenylmethyl carbonium ion. Reactions with 1-chloropropane and 1-chlorobutane lead to the formation of products resulting from secondary carbonium ions, which implies that alkylation occurs *via* the dissociated carbonium ions. Mixing benzene and hexamethylbenzene in an acidic ionic liquid did not lead to the formation of toluene, xylenes, mesitylenes, etc., showing that there is no dissociation of the methyl substitutes when the ionic liquid does not contain protic impurities.

Friedel-Crafts acylation reactions of aromatic compounds have also been carried out in the [emim]Cl-$AlCl_3$ ionic liquids.[3] For the reaction of acetyl chloride with benzene, it was determined that the rate at which acetophenone was produced was dependent on the Lewis acidity of the ionic liquid, which is in turn dependent on the ionic liquid composition. The reaction between acetyl chloride and the acidic ionic liquid was followed by ^{1}H NMR. The results suggested a stoichiometric reaction between CH$_3$COCl and [Al$_2$Cl$_7$]$^-$. Indeed, it is possible to isolate solid [CH$_3$CO][AlCl$_4$] from the ionic liquids.

3. Diels-Alder cycloadditions

The Diels-Alder reaction remains one of the most useful carbon-carbon bond-forming reactions in organic chemistry. The possibility of using ionic liquids as a solvent to carry out Diels-Alder cycloaddition reactions, has been explored by several groups.[5,6,7,8,9]

The reaction of cyclopentadiene with methyl acrylate has been widely investigated in a range of molecular solvents and solvent influences on the *endo/exo* selectivity of the reaction are well known. They may be viewed as being due to the "polarity" of the solvent leading to the stabilisation of the more polar (*endo*) activated complex.[10] The effect has also been attributed to solvophobic interactions that generate an "internal pressure" and promote the association of the reagents in a "solvent cavity" during the activation process.[11] As highly ordered hydrogen-bonded solvents, ionic liquids have the potential to have dramatic effects on Diels-Alder reactions. Although not only a solvent effect, the addition of a Lewis acid is also known to have a dramatic effect on these reactions.

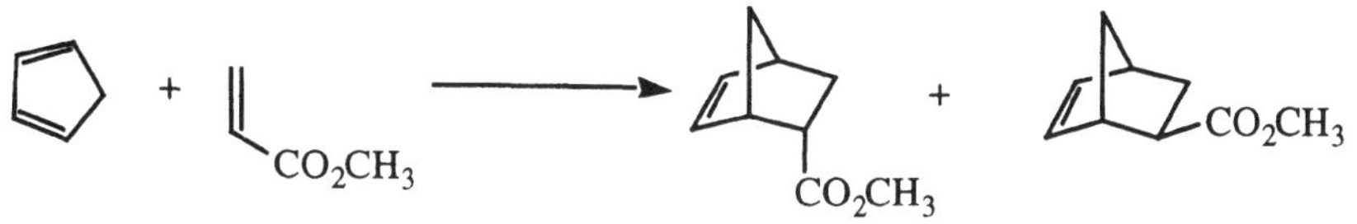

Scheme 1. The reaction of cyclopentadiene and methyl acrylate

Table 1 shows the *endo/exo* selectivity for the reaction in three ionic liquids, [EtNH$_3$][NO$_3$] (6.7:1), [bmim][BF$_4$] (4.6:1) and [bmmim][BF$_4$] (2.9:1). These compare to 6.7:1 for methanol, 5.2:1 for ethanol, 4.2:1 for acetone and 2.9:1 for diethyl ether, under similar conditions.[10] Clearly the ionic liquids show a spread of behaviours that is just as great as that seen for molecular solvents. It has previously been shown[12] that all three of the ring protons in *N,N*-disubstituted imidazolium cations can hydrogen bond to anions and that the strongest of these interactions occurs at the 2-position of the ring. With its N-H protons, ethyl ammonium nitrate is expected to have even stronger interactions. Hence, the selectivity of the reaction appears to follow degree of hydrogen bonding.

TABLE 1. The Diels-Alder addition of methyl acrylate and cyclopentadiene in three ionic liquids.

Ionic Liquid	*endo:exo* ratio
[bmmim][BF$_4$]	2.9:1
[bmim][BF$_4$]	4.6:1
[EtNH$_3$][NO$_3$]^5	6.7:1

To investigate this phenomenon further, we compared the *endo/exo* ratio for the reaction in 5 ionic liquids with a common cation but with different anions (Table 2). Since the cation remains the same in all of these liquids, its ability to hydrogen bond donate remains constant. However, the ability of the anion to hydrogen bond accept changes with the different ionic liquids. For the [bmim]$^+$ cation, the ^{1}H nmr chemical shift of the proton of the 2-position of the imidazolium ring (H^2) in a neat ionic liquid can be used as a measure of the degree of hydrogen bonding between the cation and anion.[12] The greater the chemical shift then the greater the cation-anion hydrogen bonding.

If the explanation of the difference in the *endo/exo* ratios in different ionic liquids was that hydrogen-bonding interactions between the cation and anion of the ionic liquids lead to increased solvophobic interactions and hence internal pressure, it would be expected that as the chemical shift of H^2 increased the *endo/exo* ration would increase. This is clearly not the case, indeed the opposite is true (Table 2). Hence another explanation is required.

TABLE 2. The Diels-Alder addition of methyl acrylate and cyclopentadiene *endo/exo* ratio as a function of hydrogen bonding of [bmim]$^+$ to the anions.

Ionic Liquid	*endo:exo* ratio	δ(H^2)/ppm	E_T^N
[bmim][CH$_3$SO$_3$]	3.8:1	8.54	0.62
[bmim][TfO]	4.5:1	7.86	0.64
[bmim][BF$_4$]	4.6:1	7.63	0.67
[bmim][ClO$_4$]	4.7:1	7.84	0.65
[bmim][PF$_6$]	4.8:1	7.34	0.67

Lewis acid effects on the reaction occur by the Lewis acid coordinating to the carbonyl oxygen of the methyl acrylate. It has been shown that in a Lewis basic [emim]Cl-AlCl$_3$ (48 mol % AlCl$_3$) ionic liquid the *endo/exo* ratio of the addition of methyl acrylate to cyclopentadiene is 5.25:1 under conditions similar to ours.[8] By changing to an acidic regime the (51 mol % AlCl$_3$) ratio leaps to 19:1.[8] We have ourselves used scandium(III) triflate as a Lewis acid catalyst in a [bmim][TfO] ionic liquid and achieve an *endo/exo* ratio of 16.5:1. It has been suggested that, in Diels-Alder reactions in dichloromethane with added imidazolium salts, their role is to act as Lewis acid catalysts.[13] This led us to investigate the possibility that a similar interaction was occurring in the ionic liquids themselves.

Table 2 shows the normalized Reichardt's E_T^N polarity scale values for the ionic liquids. This scale is dominated by the ability of the solvent to stabilize the ground state of the dye through hydrogen bonding to the phenoxide site of the dye, giving a measurement of the liquid's ability to hydrogen bond to a solute. It can be seen that the *endo/exo* ratio correlates well with the E_T^N value of the ionic liquid, and hence the ability of the liquid to hydrogen bond to a solute. Hence, the *endo/exo* ratio and associated acceleration of the Diels-Alder addition of cyclopentadiene and methyl acrylate in ionic liquids is controlled by the ability of the liquid to act as a hydrogen-bond donor (cation effect) moderated by its hydrogen-bond acceptor ability (anion effect). This may be described in terms of two competing equilibria. The cation can hydrogen bond to the anion:

$$[\text{bmim}]^+ + \text{A}^- \rightleftharpoons [\text{bmim}]...\text{A}$$

$$K'_{eqm} = \frac{[[\text{bmim}]...\text{A}]}{[[\text{bmim}]^+][\text{A}^-]}$$

The cation can hydrogen bond to the methyl acrylate:

$$[\text{bmim}]^+ + \text{MA} \rightleftharpoons [\text{bmim}]...\text{MA}$$

$$K'_{eqm} = \frac{[[\text{bmim}]...\text{MA}]}{[[\text{bmim}]^+][\text{MA}]}$$

It can be clearly seen that the concentration of the hydrogen bonded cation-methylacrylate adduct is inversely proportional to the equlibrium constant for the formation of the cation-anion hydrogen bonded adduct (K'_{eqm}).

4. Hetrogenization of Homogeneous Catalysts

Transition metal homogeneous catalysis is seen as one of the most promising routes to the improvement of the "atom economy" of a wide range of chemical processes, that is to maximize the number of atoms of all raw materials that end up in the products. The great advantage of homogeneous catalysis over heterogeneous catalysis is that all of the metal centres are available to the reagents, and so it is inherently more efficient. However, it does have the major drawback that it can be difficult to separate the catalyst from the products of the reaction. This leads to a waste of precious resources and to the danger of passing highly reactive chemicals into the environment.

Biphasic catalysis is a method to heterogenize a catalyst and product into two separate and immiscible phases without losing the selectivity and efficiency inherent in homogeneous catalysis.[14] The catalyst resides in solution in one of the two phases, and the substrate resides in the other phase. During reactions, the two layers are vigorously stirred, thus allowing suitable interaction of catalyst and substrate; once the reaction has reached the appropriate stage, the stirring is stopped and the mixture of phases separates into two layers, one containing the product and the other containing the catalyst. Separation is carried out by simple decantation, and in principle, the catalyst solution is available for immediate reuse.

There are three main areas of interest in biphasic catalysis. The ionic liquid-organic system is the subject of this paper. The main biphasic system used is comprised of an aqueous-organic mixture, and such processes are very effective and have been implemented in commercial processes for both oligomerization and hydroformylation reactions. This system can be augmented by the use of phase-transfer catalysts. While this system has many advantages, it precludes the use of water sensitive reagents or, often more importantly, catalysts. Also, trace amounts of organic materials are particularly difficult to remove from water. Another system uses fluorinated solvents to generate fluorous-organic biphasic reaction conditions.[15] Here, the affinity of the fluorous phase for highly fluorinated solutes is used to isolate the catalysts in the fluorous layer. This allows the use of water sensitive materials but requires the use of specially prepared catalysts (that are not always available) and expensive solvents. Also the modification of catalyst ancillary ligands that is required to achieve solubility in the fluorous phase often has a detrimental effect on the catalyst's activity.

5. Hydrogenation reactions in ionic liquids

The hydrogenation reaction of C=C bonds catalyzed by transition-metal complexes is probably one of the most widely studied reactions of homogeneous catalysis; however, the separation of products from reactants remains problematic. Initial experiments using [Rh(nbd)PPh$_3$][PF$_6$] (where nbd = norbornadiene) as the catalyst for the hydrogenation of pent-1-ene in a variety of ionic liquids showed their potential as solvents for isolating the catalyst in a biphasic protocol.[16]

TABLE 3. Hydrogenation of 1-pentene by Osborn's catalyst {[Rh(nbd)(PPh$_3$)$_2$][PF$_6$]}.

Solvent	Conv. (%) Pent-1-ene	Yield pentane	Yield pent-2-ene	TOF (min^{-1})
Acetonitrile	0	0	0	-
Acetone	99	38	61	0.55
[bmim][SbF$_6$]	96	83	13	2.54
[bmim][PF$_6$]	97	56	41	1.72
[bmim][BF$_4$]	10	5	5	0.15

For both [SbF$_6$]$^-$ and [PF$_6$]$^-$, ionic liquids hydrogenation rates are significantly greater than seen for the same catalyst in acetone, presumably due to the stabilization of the Rh(III) intermediate. When cyclohexa-1,3-diene was used as the substrate in the [emim][SbF$_6$] ionic liquid, cyclohexene was generated with 98% selectivity at 96% conversion. However, the results using a [bmim][BF$_4$] ionic liquid were disappointing. This was attributed to the presence of dissolved chloride ions in the ionic liquid that coordinate to the metal centre and deactivate the catalyst. As is common with rhodium catalysts, it was also found that isomerization of pent-1-ene to pent-2-ene occurred under the reaction conditions. Unlike traditional homogeneous systems, where it has little effect, the cis/trans selectivity of the isomerization was dependent upon the anion. In the ionic liquids, the cationic catalyst is in direct contact with neighbouring anions, whereas in a conventional solvent they will be solvent separated. It is probably this close contact that leads to the influence of the anion on the product distribution. The selectivity and the turnover frequencies of the reactions appear to be linked. These interesting results require more detailed investigation.

The hydrogenations of propene and ethane have been investigated in ionic liquids supported on membranes with Osborne's catalyst.[17] Here the short chains of the olefins lead to no possibility of isomerization. Attempts were made to relate the rates of both ethene and propene hydrogenation in a variety of ionic liquids to the coordinating ability of the anion of the ionic liquids. However, when the u.v. spectrum of [Cu(acac)(tmen)]$^+$ is used as a measure of this ability,[18] only a poor correlation is achieved. Clearly the situation is complex and requires further investigation.

TABLE 4. Hydrogenation of propene by Osborn's catalyst {[Rh(nbd)(PPh$_3$)$_2$][PF$_6$]}.

Ionic liquid	TON/min^{-1}	λ_{max} (1) / nm	Solubility /mol dm^{-3} atm^{-1}
[emim][Tf$_2$N]	1.58	546.0	0.14
[emim][TfO]	0.67	595.0	0.098
[emim][BF$_4$]	0.42	-	0.046
[bmim][PF$_6$]	0.19	516.5	0.17
[pmmim][BF$_4$]	0.17	-	0.47

6. Palladium catalysed C-C coupling reactions

One of the most successful catalytic applications of room temperature ionic liquids has been in the use of palladium catalysed reactions that lead to the formation of C-C bonds. Among these have been Heck,[19] Suzuki,[20] Stille,[21] and allylic alkylation coupling reactions.[22] Since different workers have used different reactions, direct comparison is difficult. However, it has been shown that reactivity of the haloarenes follows the same trend as seen in molecular solvents with increasing reactivity for heavier halogens.

In conventional molecular solvents these reactions suffer from a number of drawbacks such as catalyst loss into the product, catalyst decomposition and poor reagent solubilities. Many of these problems are greatly reduced or avoided completely when room-temperature ionic liquids are used as the solvents. The question becomes one of whether there is a common reason for the success of these reactions, or is it mearly a coincidence. There are two principal issues of catalyst activity and catalyst stability.

The palladium catalysed coupling reactions have been carried out in phosphonium, ammonium, N-alkylated pyridinium and disubstituted imidazolium ionic liquids. We will concentrate on those reactions conducted in disubstituted imidazolium ionic liquids.

TABLE 5. The Heck reaction of iodobenzene and ethyl acrylate to give *trans*-cinnamate in a variety of ionic liquids with 2 mol % Pd(OAc)$_2$.[19c]

Ionic liquid	additive	base	Temp/°C	Time/h	Yield/%
[bmim][PF$_6$]	PPh$_3$ (4 mol %)	Et$_3$N	100	1	99
[pmim]Cl	None	Et$_3$N	80	72	10
[pmim]Cl	None	NaHCO$_3$	100	24	19

The addition of a neutral Group 15 (usually phoshine) donor to the reaction mixture is common to the reactions described.[19-22] The absence of the donor (where it has been reported) leads to minimal reaction and extensive catalyst decomposition to palladium black (e.g. Table 5). It is well known that phoshines can act as ligands to stabilize low-valent palladium complexes and as reducing agents to generate palladium(0) species, at least one of which is probably the active catalyst. It has also been shown that the nature of the of the donor has a dramatic effect on both the reactivity and the stability of the system (e.g. Tables 6 and 7). In allylic alkylations electron-donating phoshines (e.g. PCy$_3$) gave higher rates while more elecron-withdrawing P-donors (e.g. P(OPh)$_3$) gave only low conversions.[22b] The same effect was seen in Heck reactions.[19c] This suggests that the phosphine is directly involved in the catalysis, probably as a ligand in the active catalyst complex.

TABLE 6. The Heck reaction of 4-bromoanisole and ethyl acrylate to give ethyl-4-methoxycinnamate in [bmim][PF$_6$] with a variety of Group 15 ligands.[19c]

additive	Time /h	Yield /%
none	20	7
PPh$_3$	72	65
P(tol)$_3$	24	65
P(OPh)$_3$	24	1.5
dppe	18	<1
dppf	18	<1
AsPh$_3$	12	2
SbPh$_3$	72	<1

TABLE 7. Allylation of dimethyl malonate in [bmim][PF$_6$] with a variety of phosphine additives.[19c]

Phosphine	Time /h	Conv /%
P^{t}Bu$_3$	22	75
PCy$_3$	1	100
P^{n}Bu$_3$	1	100
P(4-C$_6$H$_4$OMe)$_3$	1	100
PPh$_3$	6	100
P(4-C$_6$H$_4$CF$_3$)$_3$	20	0
P(OPh)$_3$	18	0

We have focused on the Suzuki reaction in [bmim][BF$_4$].[20] The reaction of 4-bromotoluene with phenylboronic acid in [bmim][BF$_4$] was initially investigated, affording 4-methylbiphenyl in a 69 % yield after 10 min, without catalyst decomposition (Table 8). The reaction can also be achieved with one-tenth the catalyst concentration generally required. Once the catalytic solution has been generated the reaction can be conducted under air with similar results and no catalyst decomposition.

The scope of the reaction in [bmim][BF$_4$] was investigated with electron-rich and electron-deficient halogenoarenes (Table 8). Notably, 4-methoxybiphenyl is afforded in a 40% yield in 6 h (2 TON h^{-1}) applying the original conditions. In [bmim][BF$_4$] an 81% yield is afforded in 10 min (401 TON h^{-1}, Table 8), which is in the order of 200 times the original reactivity. Despite these enhancements, chlorobenzene was still inactive even after 3 h.

During our investigations of the optimization of the Suzuki reaction in ionic liquids, we noted that the quality of the reactions and the stability of the catalyst in solution was unreliable. Further investigation revealed that this was dependent upon the batch of ionic liquid being used. After initially demonstrating that the addition of a large amount of [bmim]Cl to the reaction mixture hindered the reaction, but did not lead to obvious precipitation of palladium black we realized that the addition of a source of halide ions prevented the decomposition of the palladium catalyst and lead to higher yielding reactions (Table 9). It was determined that a ratio of 4 equivalents of halide with respect to the palladium was required to prevent visible evidence of catalyst decomposition. Hence, we chose to investigate this phenomenon.

TABLE 8. Scope of the Suzuki cross-coupling reaction in [bmim][BF$_4$]: variation of the arylhalide.

X	R	Yield (%)[a]	TON	TON h^{-1}
I	H	86 (95)	72	430
Br	H	93 (95)	78	465
Cl	H	1 (1)	1	5
Br	Cl	17	14	85
Br	COCH$_3$	67 (97)	56	335
Br	CH$_3$	69 (92)	58	345
Br	OCH$_3$	81 (92)	67	401

[a] Isolated yields of corresponding cross coupled product based on arylhalide. Purity confirmed by GC, GC-MS and ^{1}H NMR. Isolated yields after 3 h in parentheses.

^{31}P NMR showed signals at 22.6 and 21.8 ppm, respectively, and ESI-MS revealed the presence of [(PPh$_3$)$_2$Pd(bmimy)X]$^+$ {where: (bmimy) = 1-butyl-3-methylimidazolylidene, X = Cl (**3a**) or Br (**3b**)}. The assignments of these spectra were confirmed by comparison to ionic liquid solutions of independently prepared samples. The analogous [(PPh$_3$)$_2$Pd(pyimypy)Br]Br {where: (pyimypy) = 1,3-bis(2-pyridyl) imidazolylidene} complex has recently been prepared by the addition of 1:1 molar ratio of [pyimypyH]Br to Pd(OAc)$_2$ in the presence of PPh$_3$ and KBr in THF.[23] In addition, the related neutral complex, [(PR$_3$)Pd(peimy)I$_2$] {where: (peimy) = 1,3-di(1′-(R)-phenylethylimidazolylidene} and PR$_3$ = PPh$_3$ and PCy$_3$} has also been reported.[24]

TABLE 9. The effect of added [bmim]Cl on the Suzuki reaction of 4-bromotoluene and phenylboronic acid

	Yield / %	TON h^{-1}
[bmim][BF$_4$]	30	150
[bmim][BF$_4$] + [bmim]Cl (4:1 c.f. Pd)	66	330
[bmim][BF$_4$] + [bmim]Cl (1:1 c.f. 4-bromotoluene)	12.9	1.33

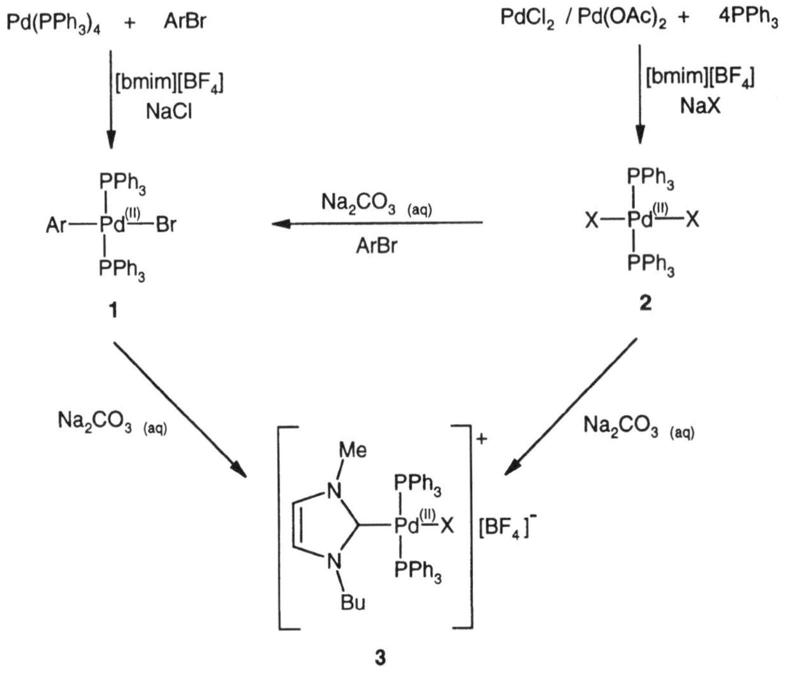

Scheme 2. The formation of [(PPh$_3$)$_2$Pd(bmimy)X]$^+$

The use of transition metal imidazolylidene complexes as catalysts has attracted much attention, particularly in palladium catalysed coupling reactions.[25] There has been some speculation that the formation of palladium imidazolylidene complexes may explain the success of many of these reactions.[19c] Hence, the *in situ* formation of **3** from [bmim][BF$_4$] has implications for a number of palladium-phosphine catalysed coupling reactions in disubstituted imidazolium based ionic liquids. While some reports suggest that the observed enhancements in such reactions can be explained in terms of the unique ionic liquid environment, the involvement of these complexes prepared *in-situ* cannot be discounted.

References

[1] (a) T. Welton, *Chem. Rev.*, 1999, **99**, 2071.
[2] P. Wasserschied and W. Keim, *Angew. Chem. Int. Ed. Engl.*, **2000**, 39, 3772
[3] J. J. Boon, J. A. Levisky, J. L. Pflug and J. S. Wilkes, *J. Org. Chem.*, 1986, **51**, 480.
[4] G. D. Luer and D. E. Bartak, *J. Org. Chem.*, 1982, **47**,1238.
[5] D. A. Jaeger and C. E. Tucker, *Tetrahedron. Lett.*, 1989, **30**, 1785.
[6] A. Sethi, T. Welton and J. Wolff, *Tetrahedron Lett.*, 1999, **40**, 793.
[7] M. J. Earle, P. B. McCormac and K. R. Seddon, *Green. Chem.*, 1999, 23.

[8] C. W. Lee, *Tetrahedron Lett.*, 1999, **40**, 2461.
[9] P.Ludley and N. Karodia, *Tetrahedron Lett.*, 2001, **42**, 2011.
[10] J. A. Berson, Z. Hamlet and W. A. Mueller, *J. Am. Chem. Soc.*, 1962, **84**, 297.
[11] R. Breslow, *Acc. Chem. Res.*, 1991, **24**, 159.
[12] A. G. Avent, P. A. Chaloner, M. P. Day, K. R. Seddon and T. Welton, *J. Chem. Soc., Dalton Trans.*, 1994, 3405-3413.
[13] J. Howarth, K. Hanlon, D. Fayne and P.McCormac, *Tetrahedron Lett.*, 1997, **38**, 3097.
[14] *Aqueous-Phase Organometallic Catalysis: Concepts and Applications*; Cornils, B.; Herrmann, W. A., Eds.; Wiley-VCH: Weinheim, 1998.
[15] Horváth, I. T.; Rabai, J. *Science* 1994, **266**, 72.
[16] Y. Chauvin, L. Mussmann, H. Olivier, *Angew. Chem., Int. Ed. Engl.*, 1995, **34**, 2698.
[17] T. H. Cho, J. Fuller and R. T. Carlin, *High Temp. Mater. Process.*, 1998, **2**, 543.
[18] M. J. Muldoon, C. M. Gordon and I. R. Dunkin, J. Chem. Soc., Perkin 2, in press.
[19] (a) D.E. Kaufmann, M. Nouroozian and H. Henze, *Synlett.*, **1996**, 1091, (b) V. P. W. Böhm and W. A. Herrmann, *J. Organomet. Chem.*, **1999**, 572, 141, (c) A. J. Carmichael, M. J. Earle, J. D. Holbrey, P. B. McCormac and K. R. Seddon, *Org. Lett.*, **1999**, *1*, 997; (d) V. P. W. Böhm and W. A. Herrmann, *Chem. Eur. J.*, **2000**, *6*, 1017; (e) L. Xu, W. Chen and J. Xiao, *Organometallics*, **2000**, *19*, 1123; (f) L. Xu, W. Chen J. Ross and J. Xiao, *Org. Lett.*, **2000**, 3, 295-297.
[20] C. Mathews, P. Smith and T. Welton, *Chem. Commun.*, 2000, 1249.
[21] S. T. Handy, X. Zhang, *Org. Lett.*, 2001, 3, 233-236.
[22] W. Chen, L. Xu, C. Chatterton and J. Xiao, *Chem. Commun.*, 1999, 1247, (b) J. Ross, W. Chen, L. Xu, and J. Xiao, *Organometallics*, 2001, 20, 138-142.
[23] J.C.C.Chen and I.J.B.Lin, *Organometallics*, 2000, 19, 5113.
[24] (a) T.Weskamp, V. P. W. Böhm and W. A. Herrmann, *J. Organomet. Chem.*, **1999**, *585*, 348; (b) W.A.Herrmann, V. P. W. Böhm, C.W.K.Gstöttmayr, M.Grosche, C.Reisinger and T.Weskamp, *J. Organomet. Chem.*, **2001**, 617, 618.
[25] (a) D. Bourissou, O. Guerret, F. P. Gabbaï and G. Bertrand, *Chem. Rev.*, **2000**, *100*, 39, (b) T. Westkamp, V. P. W. Böhm and W. A. Herrmann, *J. Organomet. Chem.*, **2000**, *600*, 12.

COORDINATION COMPOUNDS IN MELTS

S.V. VOLKOV
V.I. Vernadskii Institute of General & Inorganic Chemistry of the UNAS
32/34 Palladin avenue, 03680 Kiev 142, UKRAINE

The first information about the possibility of existence of complex compounds in molten salts was based on investigations using thermal phase analysis. The presence of well-defined maxima (distectic points) in liquidus curves in phase diagrams for binary salt systems indicates the formation of such compounds, which separate out as a solid phase. In the opinion of many authors, they must be also in the liquid (molten) phase, at least at temperatures close to liquidus. This hypothesis, however, was called in question by some authors, who stated that the presence of distectic points in phase diagrams does not suggest that there are complex compounds of definite composition in salt melts. Therefore, as early as 30-40 years ago, the possibility of existence of such complex compounds in salt melts was a point at issue.

Further investigations, first electrochemical (potentiometry, polarography, chronopotentiometry) and then spectroscopic investigations, showed definitely pronounced complex formation to be also observed in molten salt systems. Electrochemical and spectroscopic methods make it possible not only to determine the complexity function, but also to find the main parameters characterising a complex species, namely, the formation (instability) constant of the complex species, the electronic and geometrical structure, as well as some kinetic data.

The determining contribution to the establishment of the structure of complex compounds in melts was made by spectroscopic investigations (electronic and vibrational spectra). Diffraction and radiospectroscopic studies of molten salts have also promise in this respect. Thus, the question of the fact of complex formation in salt melts is no longer a debatable question. The concept "coordination compound in molten salts" is the subject of more discussion since different authors attach different meaning to this concept. It is because of this that this question is considered at first.

Now investigations in the field of the coordination chemistry of salt melts are aimed not so much at the characterisation of the complex formation phenomenon itself as at its connection with other processes occurring in such media: electrolysis, chemical reactions, catalysis, electroplating, etc. As a result of this, the number of investigations in the field of the coordination chemistry of salt melts has markedly increased in recent years, and the total number of publications dealing with complex formation in salt melts exceeded 1000 long ago.

Meanwhile, there was no monograph in literature till 1977 devoted just to this subject. To fill up this gap, the author with his colleagues took the trouble to write such a monograph and to deliver today a short lecture on this subject.

1. Definition of coordination compound (ion) in molten salts. Differences from other systems and specific features

As was pointed out in a number of papers, "the concept of complex ions in molten salts is quite different from the generally accepted notions in aqueous solutions". "When the relative concentrations of cations, anions and the solvent become comparable, it is difficult to isolate an individual form... In other words, the complexation reaction cannot be written as the hypothetical scheme: $M^{2+} + Y^- = MY^+ +$ ".

Indeed, though the coordination chemistry of salt melts is a particular case of the chemistry of coordination compounds, it has its own peculiarities. Firstly, salt melts are mostly maximally concentrated electrolytes, and the theory of strong electrolytes is known to be far from being resolved. Secondly, salt melt as a solvent gives rise mainly to the Coulomb long-range mode of interaction in contrast to weak short-range intermolecular interactions of an aqueous or some typical nonaqueous solvent. Thirdly, from the said above it follows that the structure of complexes in molten salts differs from that of "ordinary" complexes in aqueous and nonaqueous media due to the immediate outer-sphere cationic environment of the inner complex anion. Fourthly, in salt melts it is impossible in most cases to differentiate between the notions "solvent" and "ligand"; this differentiation is often conventional. Fifthly, in salt melts, conditions often exist under which the complexing-metal ion concentration greatly exceeds the ligand concentration, and this may be observed in the absence of the solvent.

In the last case, it is not clear whether we have to do with a quasiautocomplex or with a complex in which a ligand acts as the central atom, and metal ions act as ligand, or cluster (chain, network) structures are formed. In view of this, the first necessity arises: development of the concept "coordination compound" in ionic melts with the consideration of the peculiarities of such a coordination and with the establishment of the thermodynamic and kinetic boundaries of the existence of complex ions.

As early as 1975, we formulated the characteristic features of coordination compound in molten salt systems.

Discrete coordination compounds in a homogeneous molten medium include compounds (ions), which are characterised by the following:

- coordination phenomenon, namely distinct constant geometrical arrangement of ions of one sort about an ion of another sort - complexing-metal ion;

- the specificity of the composition, which manifests itself by discrepancy between the formal oxidation state of the complexing metal and coordination number and consists in a noticenable rearrangement of the electron shells of interacting ions with the formation of at least partially covalent bond;

- lifetime, which exceeds the time of contact between unbonded ions and is, in principle, a constant quantity, which is characteristic of any thermodynamic or kinetic property of disorder or "dissociation".

This formulation, which is proposed for molten salt systems, makes it possible not only to emphasize the most important and characteristic features of discrete complex ions in melts, but also to differentiate them from associates (as species which, during solution formation, either do not rearrange their electron shells at all and retain the structure of corresponding individuum or are bound by pure Coulomb forces) and from static complex groups or "preferred coordination" (as unisolated groups of atoms, to which the concepts of the constancy of their thermodynamic or kinetic heterolytic dissociation or disordering reaction "constants" cannot, in principle, be extended).

2. Homogeneous complex formation in molten salts

To investigate complexation processes in melts, various physicochemical methods are employed, which may be approximately divided into two large groups: investigations under equilibrium and nonequilibrium conditions. The objectives of these methods are to establish the fact of complex formation and the composition of complex species and to determine the stability of this species, the coordination number of the complex and its structure (electronic and geometrical), the formation (decomposition) rate constants of the complex ion or its lifetime (Table 1).

TABLE 1. Methods and parameters of investigation of complex formation in molten salts under equilibrium conditions

Methods of investigation:	
Thermal phase analysis	Electron spectroscopy
Cryometry	Raman spectroscopy
Solubility method	IR spectroscopy
Distribution method	Radiospectroscopy (ESR, NMR)
Vapour pressure method	Diffractometry
Potentiometry (EMF)	
Calorimetry	
Molar volume (density)	
Surface tension, etc	
Characteristics to be determined:	
$\Delta G = RT \ln K_{instab} = -RT \ln \beta_{stab}$	Composition of the complex
$\Delta H = \Delta G + T\Delta S$	Coordination number
ΔS	Structure of the complex:
ΔV	geometrical
$\Delta \beta$	electronic
C_p	

The main peculiarity of the methods for the investigation of complex formation under equilibrium conditions is that they are mainly based on the thermodynamic properties of the systems under investigation. Another peculiarity of these methods is that complex formation is generally investigated under thermodynamic equilibrium conditions.

The aims of investigation by this group of methods are to establish the fact of complex formation, the composition and structure of the complex and to determine the values of stability (formation) constants; and in the best case: the full set of thermodynamic characteristics (ΔG, ΔH, ΔS) of the complex.

The methods for the investigation of the nonequilibrium state of systems are usually characterised by the so-called nonequilibrium properties (Table 2). Such properties are very often referred to as transport properties. Examples of them are heat conductivity, diffusion, viscosity, electrical conductivity. The regularities characterising these properties cannot be described in terms of classical thermodynamics. Recently, they began to be described in terms of the thermodynamics of irreversible processes.

TABLE 2. Methods and parameters of investigation of complex formation in molten salts under nonequilibrium conditions

Methods of investigation:
Determination of viscosity (η_i)
Investigation of electrical conductivity (λ_i) and ion mobility (U_i)
Determination of diffusion coefficients (D_i)
Investigation of transport numbers (t_i)
Polarography (φ_i)
Chronopotentiometry ($i\tau 1/2$)
Impedance measurements (C_i)
Other kinetic measurements ($\vec{v}, \vec{k}$)

Parameters to be determined:
$\Delta G^E = \Delta G_{exp} - RT(x_1 \ln x_1 + x_2 \ln x_2 + ...)$ ⎫
$\Delta H^E = \Delta H_{exp}$ ⎪ Deviations
$\Delta V^E = \Delta V_{exp} - (x_1 V_1 + x_2 V_2 + ...)$ ⎬ from parameters
$\Delta S^E = \Delta S_{exp} + R(x_1 \ln x_1 + x_2 \ln x_2 + ...)$ ⎭ for ideal system
$\Delta \eta = \eta_{exp} - \eta_{addit.}$
$\Delta \lambda = \lambda_{exp} - \lambda_{addit.}$ ⎫
- - - - - - - - - - - ⎬ Deviations
- - - - - - - - - - - ⎭ from additive values
- - - - - - - - - - -

Electrolysis and the phenomena accompanying it also belong to irreversible and (in the general case) nonequilibrium processes. Therefore, such methods of investigation as ion electromigration, polarography (or voltammetry in general) and chronopotentiometry are also to be classed as nonequilibrium ones. If electrolysis is conducted at a constant or linearly slowly increasing voltage, the processes occur under stationary conditions, and in the case of infinitesimal time intervals they may be regarded as those approaching equilibrium processes. This relates in particular to classical polarography. However, taking into account all types of polarography, it is better to class it among nonequilibrium methods.

Of the above methods, polarography and chronopotentiometry are of the greatest interest in quantitative aspect. They enable the most complete quantitative description of complex species that are formed in salt melts. The other methods are qualitative and, in the best case, semiquantitative in character. Some of them, for example electrical conductivity method, are, as their classical variants, only of historical interest.

All these methods should be considered and analysed in terms of the idea of similarity of kinetic relations that describe the simplest transfer processes, whether it be the case of concentration gradient, heat gradient, chemical or electrochemical potential gradient (Fick equation, Fourier equation, etc). As activation processes, they should be characterized by activation energies and entropies of activation by comparing their different values (since a specific potential barrier U_i corresponds to each its property) for different properties. However, such data for molten salt systems are obviously insufficient. To judge the processes occurring in real systems, use is made mainly (though not always: for example, there are a number of comparisons of activation energies for viscosity, electrical conductivity, etc) of deviations of experimentally measured parameter values from the values for an ideal system.

The interpretation of the properties of an ideal system appears to be the most complex problem in the description of the properties of real molten salt systems. Whereas the thermodynamic characterisation of an ideal solution has been rigorously developed (the formation of such systems is not accompanied by a change in the mixing enthalpy and volume, and entropy varies, for example according to the Temkin model for molten salts, as $\Delta S_{conf} = -R \Sigma N_i N_i \ln N_i^{v+} N_i^{v-}$ (where N_i^{v+}, N_i^{v-} are the ionic fractions of cations and anions and v^+, v^- the numbers of ions into which the salts dissociate)), there are no theoretically justified equations for nonequilibrium properties for the case of the simplest (ideal) solution. Usually, additive isotherms of solution properties are simply adopted for them.

3. Effect of outer-sphere cations on complex formation in molten salts

When investigating complex formation in aqueous and nonaqueous media, outer-sphere ions are usually considered to play a minor part or even ignored. Meanwhile, a convincing proof of their participation in complexation processes is a substantial change in "complexation reaction yield" or stability constants in solutions, which is caused by these "indifferent" ions.

In aqueous and mixed solutions, the influence of outer-sphere ions is reduced for dilute electrolytes to allowance for the total ionic force.

However, already at a certain, even relatively low, ion concentration ($\mu \geq 0.1$) it is impossible to describe real processes, and to solve problems for more concentrated solutions, one has to resort to empirical equations, which are linearly dependent on μ, because it cannot be estimated theoretically. Meanwhile, this does not save the situation either since at high concentrations, the pure concept of ionic force, according to which the properties of ions depend only on the total ionic force and are quite independent of their nature, cannot be employed. It is understandable that even dilute solutions of molten salts (for example, MY_2 in AY) are concentrated electrolytes with $\mu \gg 0.1$, which cannot be characterised on the basis of this explanation of ionic force. This is their first distinction from other solutions, which requires taking into account their direct contact with each other.

It is in molten salt systems that groups of the $\{MY_m^{n-m}, YA^+\}$ type exist, which differ basically from outer-sphere complexes $\{ML_m^{n+}, Y^-\}$ in aqueous solutions in structure, the probability of the existence of a strong bond between MY_m^{n-m} and YA^+ cations, thermodynamic, spectroscopic and other characteristics (Figure 1).

To obey the principle of electroneutrality and alternation of counter-ions of molten salt system, it is to be admitted that in it, unlike ion pairs in water and nonaqueous solvents, the complex MY_m^{n-m} ion is spherically surrounded by many A^+ cations in the region of dilute solutions; this environment is to some extent symmetrical and close to that observed in A_2MY_4 crystals (where M is Co, Ni, Cu and Y is Cl, Br) in the range of comparable concentrations.

In solution:

$$MY_n + pAY + Solv \longrightarrow [ML_m]Solv + pA^+\cdot Solv + (n+p-m)(Y-L)^- Solv$$

(a)

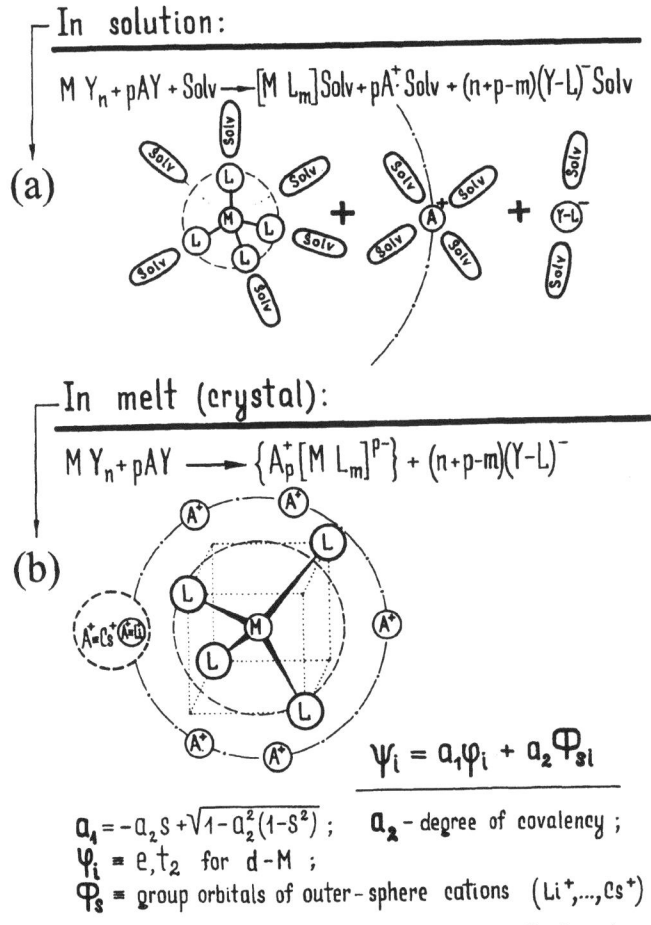

In melt (crystal):

$$MY_n + pAY \longrightarrow \{A_p^+[ML_m]^{p-}\} + (n+p-m)(Y-L)^-$$

(b)

$A = Cs^+ (A=Li)$

$$\psi_i = a_1 \varphi_i + a_2 \Phi_{sl}$$

$a_1 = -a_2 S + \sqrt{1 - a_2^2(1-S^2)}$; $\quad a_2$ - degree of covalency ;
$\varphi_i = e, t_2$ for d-M ;
Φ_s = group orbitals of outer-sphere cations $(Li^+,...,Cs^+)$

Figure 1. Effect of the outer-sphere ions of the second coordination sphere on the properties of the inner complex in solutions (a) and melts (b).

And, at last, one more distinction of a number of molten salt solutions from "ordinary" aqueous and nonaqueous solutions. Since the complexing metal is always solvated, an equilibrium between solvate formation and complex formation is established in these systems:

$$M(Solv)_m^{n+} + mL^- = ML_m^{n-m} + m(Solv),$$

which shifts to the right in the presence of outer-sphere A^+ cations with high solvation energies, that is with high ionic potentials (z/r) or small size, in the solution. The same regularity manifests itself in the investigation of complex formation in molten salt systems with a solvent.

In two-component salt systems (M^{n+}, Y^-, A^+) with a common anion, however, the determining factor is not the nature of the equilibrium between solvate formation and complex formation but the competition (with polarisation accompanying it) between

M^{n+} and A^+ ions for complex formation. As a result, A^+ cations of larger size with low ionic potentials must facilitate the formation of the MY_m^{n-m} complex.

Let us cite as an example data (Figure 2) on diffusion coefficients and on the size (r) of diffusing particles, which follows from them; it is evident from these data that in the CsCl melt with weak dissociation of complexes, "atomic" complexes are the predominant diffusing form, whereas in the NaCl melt even Th(IV), Zr(IV) and Pu(III) complexes are strongly dissociated, and the predominant diffusing species in melts are their ions.

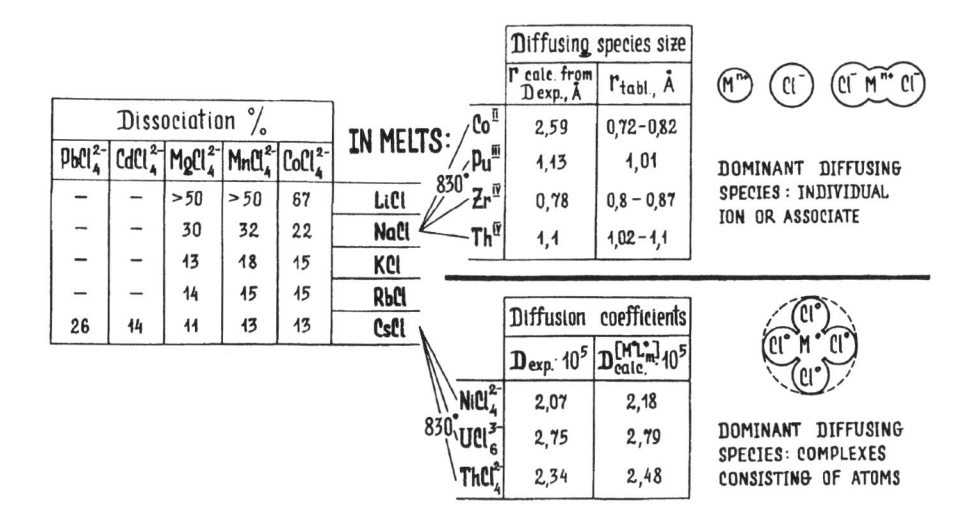

Figure 2. Degree of dissociation and size of diffusing species of metal chloride complexes in alkali chloride melts.

It is the competition effect of outer-sphere cations that substantially decreases the stability constants of various metal ion complexes in melts as compared to those in an aqueous medium (Table 3).

TABLE 3. Instability constants of complex ions in nitrate, chloride* melts and aqueous solutions

Complex ion	in melt	in water	Complex ion	in melt	in water
*CeF^{2+}	4-7.75·10^{-3}	6.3·10^{-4}	AgBr$_2^-$	3.2-7.2·10^{-3}	7.8·10^{-8}
*LaF^{2+}	2.7-4.2·10^{-3}	1.7·10^{-3}	AgBr$_3^{2-}$	1.7·10^{-5}	1.3·10^{-9}
*VO$_2$F$_4^{2-}$	3.23·10^{-9}	1.4·10^{-12}	CdBr$_4^{2-}$	1·10^{-2}	2·10^{-4}
AgCl$_2^-$	2.1·10^{-3}-6·10^{-4}	1.7·10^{-5}	PbBr$_2$	2.8-4·10^{-2}	1.2·10^{-2}
PbCl$^+$	1.6·10^{-3}	2.3·10^{-2}	ZnBr$^+$	0.04	4
PbCl$_3^-$	1.4-5.5·10^{-2}	1.4·10^{-2}	AgI$_3^{2-}$	4.3·10^{-8}	1.4·10^{-14}
ZnCl$^+$	2.8-7.3·10^{-4}	0.19	CdI$_4^{2-}$	8·10^{-9}	8.3·10^{-7}
CdCl$_3^-$	4-6·10^{-5}	3.4	AgNH$_3^+$	1·10^{-3}-6·10^{-4}	6.3·10^{-4}
CdCl$_4^{2-}$	1.5·10^{-6}	9.3·10^{-3}	Ag(NH$_3$)$_2^+$	2.17·10^{-7}	9.3·10^{-8}
CuSO$_4^-$	0.37	0.112-0.42	*Ag(CN)$_2^-$	1.34·10^{-11}	8·10^{-22}

This competition effect of outer-sphere cations (A^+) must be always taken into account in melts even in the case of alkali metal cations. When the complexing and outer-sphere cations are equivalent, the differences between them are obliterated, and we shall have to do with heteronuclear complexes. They will be dealt with later after the discussion of the complex-cluster model of the structure of molten salt systems.

4. Complex-cluster structure of homogeneous molten salt systems with complex formation

On the basis of spectroscopic, diffraction, thermodynamic and other characteristics, we proposed as early as 1975 a complex-cluster model of the structure of molten salt systems where complex formation is observed. For simplicity, let us consider it for binary AY-MY_2 systems, where M is 3d-metal ion, and A^+ is outer-sphere alkali metal cation (Figure 3).

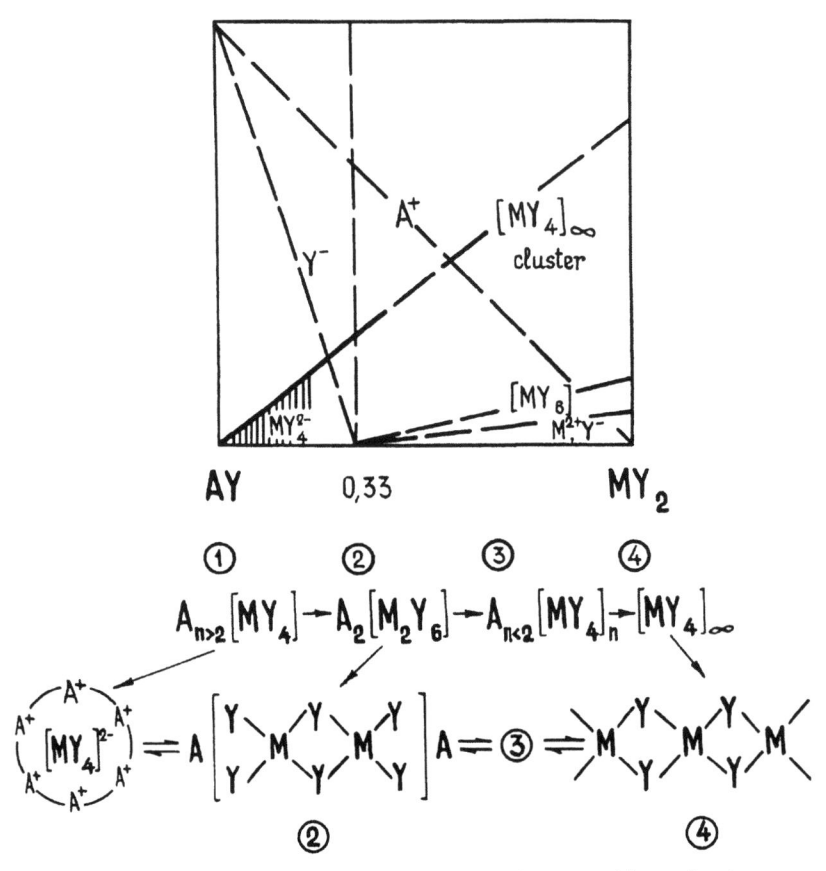

Figure 3. Complex-cluster structural model of molten salt systems with complexation.

The range $0 < X_{MY_2} < 0.1$ is regarded as containing, besides uncoordinated Y⁻ ions and A⁺ cations, discrete maximally coordinatively saturated complexes of tetrahedral type (MY_4^{2-}), which are not distorted by the spherical fields of outer-sphere A⁺ cations (Figure 4). This can be evidenced, for example, by the constancy of the ion activity coefficients, the validity of the Nernst law, obeying of absorption spectra to the Buger-Lambert-Bar law, the constancy of absorption band half-widths (indicating the invariability of the symmetry of the complex) and by other characteristics.

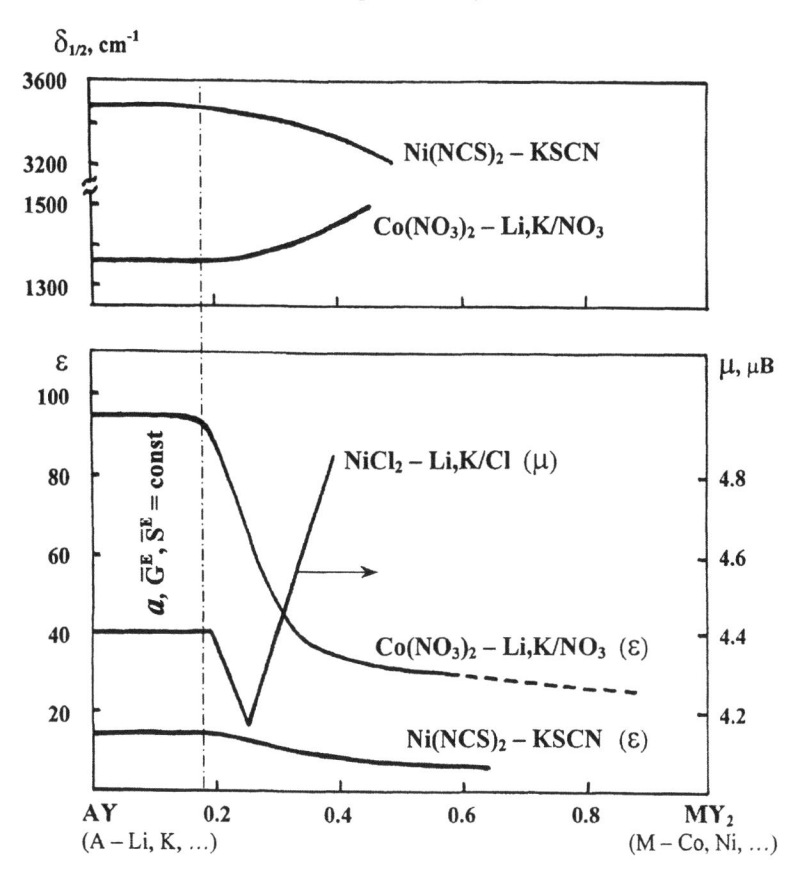

Figure 4. Thermodynamic (a, $\overline{G}^E$, $\overline{S}^E$), spectroscopic (ε, δ) and magnetic (μ) properties of complexes in molten salts as a function of their concentration.

In the range $0.1 \leq X_{MY} \leq 0.33$, electronic spectra exhibit first distorted tetrahedral discrete maximally coordinatively saturated complex MY_4^{2-} ions and then dissociation products too. The same was indicated by X-ray and displacement enthalpy data. This behaviour is due to the fact that as the concentration of complexes increases, the conditions for their complete isolation by the rather spherical field of outer-sphere cations disappear. The appearing asymmetrical outer-sphere A^+ cation environment of the complex MY_4^{2-} leads first to its distortion and then to dissociation. Only so can the dissociation process in systems without resolvation be explained.

The structural model of melts in the range $0.33 < X_{MY} < 1$ is based on the concepts of textural features preferential for these conditions: mainly statistical groups of cluster type with tetrahedral coordination of the cell and, probably, a small number of fluctuationally discrete MY_4^{2-} ions and quasicrystalline fragments.

The proposed model may be corroborated by independent measurements of some structurally sensitive property. It is seen that the value and constancy of $\mu = 4.4\mu_s$ at $0 < X_{NiCl_2} < 0.136$ indicate unambiguously the existence of the undistorted tetrahedral complex $NiCl_4^{2-}$ in this range. The decrease in μ in the range $0.136 \leq X_{NiCl_2} \leq 0.3$ is logically explained by the distortion of the $NiCl_4^{2-}$ complex due to the asymmetrical outer-sphere A^+ cation environment. In the distorted complex $NiCl_4^{2-}$, the degeneracy of the $^3T_1(F)$ ground state is removed, owing to which the orbital contribution only of the second order becomes possible, and, as a result, the magnetic moments are undervalued. At $X_{NiCl_2} > 0.3$, the inconstancy of the μ value was observed, which supports the fact that the third feature of coordination compound cannot be applied to the species that exist at this composition and indicates rather a noticeable interaction between magnetic particles combined into clusters (Figure 4).

Thus, the structural model is based on direct measurements of the above properties of melts and proceeds from concepts of preferential textural features for each specific region of composition. It by no means negates the presence in real systems of more complex ionic-molecular equilibria involving simple ions and intermediate complexes.

The complex-cluster model considered is an equilibrium model, which is based on the analysis of equilibrium properties of melts and concentrates on preferential, predominant structure forms of the melt. When such complexing systems are investigated under nonequilibrium conditions by so-called nonequilibrium methods, other species are also revealed, which are often simpler and kinetically more mobile.

5. Heteronuclear complex formation in homogeneous molten salt systems

Let us now consider the case of the presence of not one but at least two ions of complexing metals, for example zinc and cadmium, in a molten salt system. These cases are extremely interesting for the deposition, by direct electrolysis, not of an individual metal but of the desired alloy or compound from heteronuclear coordination compounds formed in the melt.

On the problem of alloy electrolysis (Figure 5). Raman spectra of molten systems $ZnCl_2$-(Li, K)Cl$_{eut}$ demonstrate a single frequency (280 cm^{-1}) in the composition range 0.03-0.33 mole fraction of $ZnCl_2$. This is Zn-Cl bond vibration in the complex anion $ZnCl_4^{2-}$. The frequencies 292 cm^{-1} and 230 cm^{-1} in melts with higher $ZnCl_2$ concentration are assigned to Zn_2Cl_6 dimer vibration and the lattice vibration of the $ZnCl_2$ salt proper.

Raman spectra of molten systems $CdCl_2$-$(Li, K)Cl_{eut}$ demonstrate similarly in the range 0.03-0.33 mole fraction of $CdCl_2$ a single Cd-Cl bond vibration frequency (252 cm^{-1}) in the $CdCl_4^{2-}$ complex (Figure 6). A concentrated melt exhibits between x = 0.5 and individual $CdCl_2$ the lattice vibration frequency 208 cm^{-1} of the $CdCl_2$ melt proper.

X_{ZnCl2}	t, ^{0}C	$\nu_1(A_1)^*$, cm^{-1}		
1.00	500	230 (st)	-	-
0.80	550	230 (st)	-	292 (w)
0.67	550	230 (w)	-	292 (w)
0.50	560	230 (vw)	-	292 (st)
0.33	570	-	280 (st)	-
0.25	590	-	280 (st)	-
0.11	580	-	280 (st)	-
0.06	580	-	280 (st)	-
0.03	590	-	280 (st)	-
$ZnCl_2$ crystal	316	224 [lit]		
melt	500	230		

X_{CdCl2}	t, ^{0}C	$\nu_1(A_1)^*$, cm^{-1}	
1.00	630	208 (st)	-
0.80	620	208 (st)	-
0.67	620	208 (w)	252 (w)
0.50	610	208 (vw)	252 (w)
0.33	600	-	252 (st)
0.23	620	-	252 (st)
0.09	570	-	252 (st)
0.06	580	-	252 (st)
0.03	590	-	252 (st)
$CdCl_2$ crystal	560	235 [lit]	
melt	630	208	

* (st – strong, w – weak, vw – very weak)

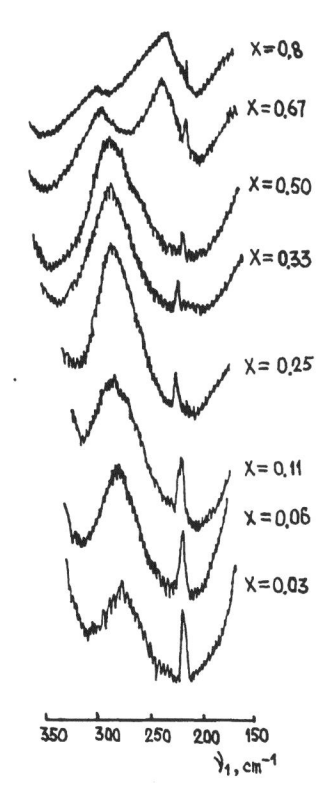

Figure 5. Raman spectra of molten salts $ZnCl_2$-$(Li, K)Cl_{eut}$.

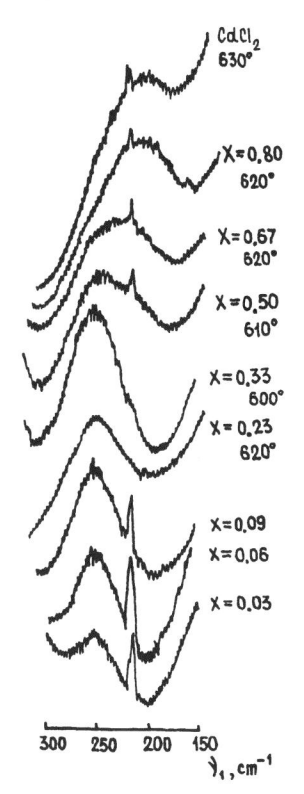

Figure 6. Raman spectra of molten salts $CdCl_2$-$(Li, K)Cl_{eut}$.

$X_{CdCl2} + X_{ZnCl2}$ (1:1)	t, °C		$\nu_1(A_1)$, cm^{-1}		
1.00	600	216 (st)	-	-	-
0.80	610	216 (st)	-	-	-
0.67	600	216 (st)	-	265 (w)	-
0.50	600	-	-	263 (st)	-
0.33	600	-	-	263 (st)	-
0.15	600	-	-	263 (st)	-
0.10	570	-	-	256 (st)	-
0.06	570	-	252 (w)	-	280 (w)
0.03	590	-	252 (st)	-	280 (st)

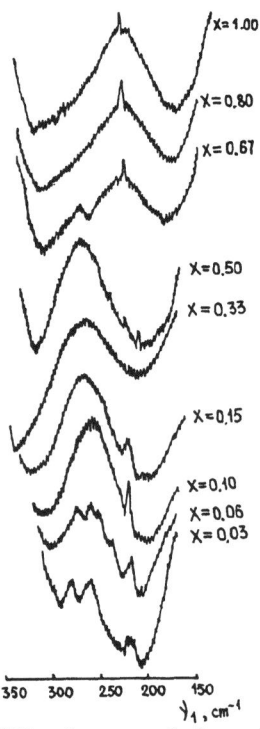

Figure 7. Raman spectra of molten salts ZnCl$_2$-CdCl$_2$-(Li, K)Cl$_{eut}$
(st – strong, w – weak).

Raman spectra of the mixed melt ZnCl$_2$-CdCl$_2$-(Li,K)Cl$_{eut}$ demonstrate at different constituent ratios both frequencies relating to the individual complex ions ZnCl$_4^{2-}$ and CdCl$_4^{2-}$ and new frequencies, whose values are between the Zn-Cl and Cd-Cl bond vibration frequencies (Figure 7).

All this spectral information can be represented as follows (Figure 8): In the low-concentration range up to 0.1 mole fraction of metal ions, they exist as individual complex anions; as their concentration increases in the interval 0.1<X≤0.5, the spectra show their heteronuclear complexes, and above X<0.5 the structure of mixed ZnCl$_2$-CdCl$_2$ salts proper is formed.

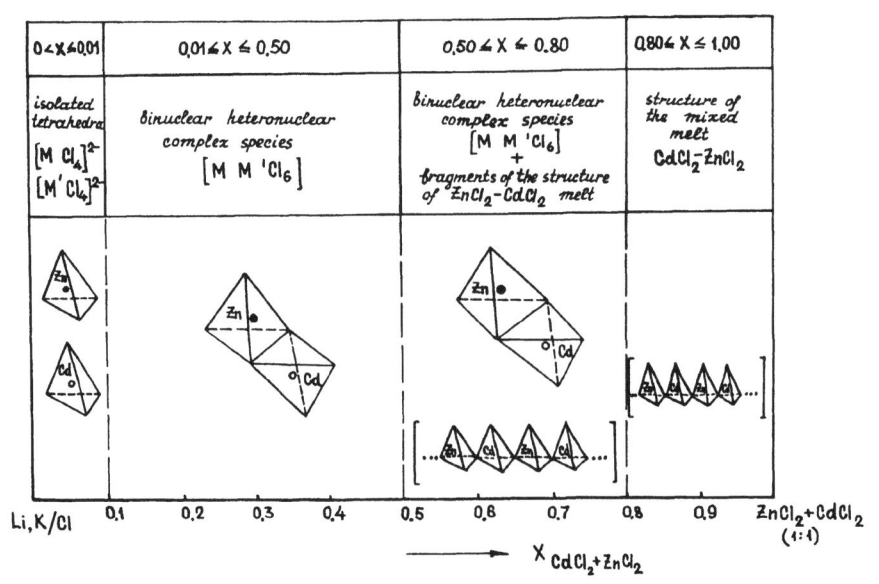

Figure 8. Predominant forms of Zn(II) and Cd(II) chloride complexes in different concentration ranges of the molten system ZnCl$_2$-CdCl$_2$-(Li, K)Cl$_{eut}$.

Polarograms of melts containing Zn(II) and Cd(II) and their mixtures corroborate unambiguously this picture at different concentrations in (Li, K)Cl$_{eut}$ (Figure 9).

The individual waves of Zn^{2+} (-1.5 V) and Cd^{2+} (-1.1 V) merge into a single potential (-1.3 V) at 400°C, as their concentration increases to 0.07 mol/l and give allows Zn-Cd. Raising the temperature (to 450°C) leads to a destruction of their heteronuclear complex and again to the observation of individual Zn^{2+} and Cd^{2+} waves. Such Raman

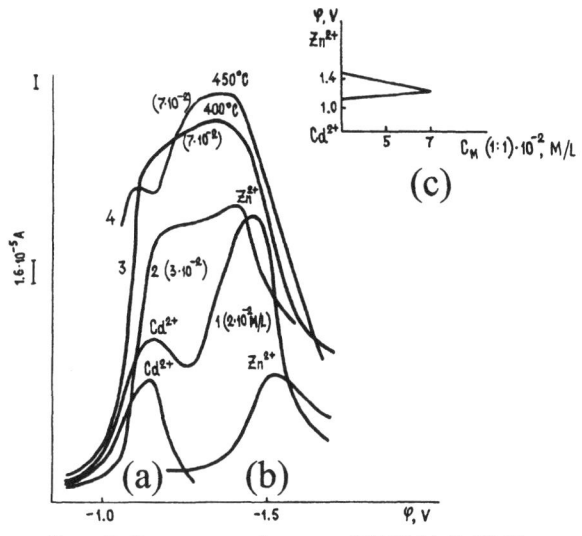

Figure 9. Square-wave polarogram of Cd(II) (a), Zn(II) (b) and Cd(II)+ Zn(II) (1:1) (curves 1-4) in the melt (Li, K)Cl$_{eut}$.

spectroscopic investigations were also carried out for other systems, for example AlCl$_3$-MgCl$_2$; and alloys, such as Al-Mg, Al-Si, etc., were obtained directly.

The next exemple (Figure 10) demonstrates a striking instance of the synthesis of titanium diboride at a single potential of –2.2 V of a heteronuclear complex being discharged in the KCl-NaCl-NaF-(TiCl$_3$-NaBF$_4$) melt.

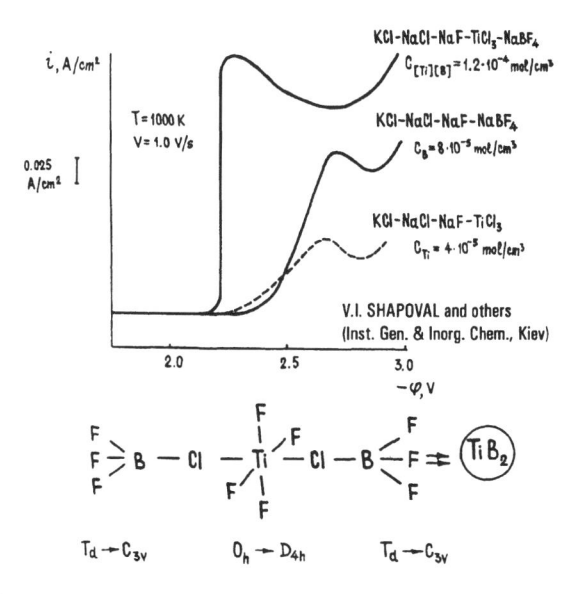

Figure 10. Voltammetric investigations of the molten systems TiCl$_3$-(KCl-NaCl-NaF), NaBF$_4$-(KCl-NaCl-NaF) and TiCl$_3$-NaBF$_4$-(KCl-NaCl-NaF) and isolation of TiBr$_2$ from heteronuclear complexes.

6. Heterogeneous-heterophase complex formation at the melt-gas (-liquid, -solid) interface

Let us now proceed to the new conceptual views which we have been developing in recent years (Table 4). Along with the well-known coordination compounds in the homogeneous (liquid, solid, gaseous) phase, heterogeneous-heterophase coordination chemistry, in which the formation and/or transformation of coordination compounds take place at the interface and which was mentioned by me in 1981, is of still greater interest (especially in technological respect). In the case under discussion they take place at the interfaces: melt/gas (1), melt/molecular liquid (2), melt/ionic anisotropic liquid (liquid crystal) (3) and melt/solid, crystal (4).

TABLE 4. Heterogeneous – heterophase high-temperature coordination chemistry

Phases		Process	Complex type
Melt	Gas	- Metal-complex catalysis - Fuel cells	Mixed-ligand complex
	Molecular liquid	Extraction	Mixed-ligand complex
	Ionic anisotropic liquid (liquid crystal)	Optical information recording	Homo-polynuclear complex
	Crystal	- Electrolysis, electrosynthesis; - Hot corrosion	Homo-, poly- and/or heteronuclear complex

The most important processes which make use of heterophase coordination compounds with electrolytes-melts are, for example, fuel cells and metal-complex catalysis in the first case, extraction in the second case, optical information recording in the third case, and electrolysis, electrosynthesis, "hot corrosion" and chemical dissolution or precipitation of solid substances in melts in the fourth case, which is perhaps the commonest one. In all of these cases, we have to deal with different-ligand, homopolynuclear, heteropolynuclear and other kinds of complexes.

Let us discuss, as an example of heterogeneous coordination chemistry, the formation of heteronuclear and heterophase complexes on electrodes (Figures 11, 12).

It is clear that metal electrodeposition from electrolytes, including molten salts, cannot be understood without studying the electrode-electrolyte interaction and double electrical layer. Conway wrote as early as 1975 that "it is useful to regard specific adsorption established experimentally as donor-acceptor interaction between ligands and metal, which is influenced by ion solvation and solvent adsorption". And this is in the total exactly heterogeneous-heterophase coordination chemistry, where the electrode metal atom acts as the coordination centre, and the ligand is, depending on electrolyte composition, either a simple anion of the background solvent or a complex anionic species in the case of the presence of p-, d- or f-metal ions in the melt. When the cathode surface charge is positive (which is generally the case in metal electrodeposition from molten salts), Cl^- anions accumulate in the double electrical layer in the case of background molten chlorides. Addition of complexing metal ions, for example Co(II) or Ni(II) ions, to such a melt leads to the coordination-electrosorption of the complexes formed by them and to the replacement of Cl^- by more complicated multicharged complexes-ligands, such as $CoCl_4^{2-}$, $NiCl_6^{4-}$, to form heterophase complexes: solid electrode metal-bridge Cl^- anion-metal ion in melt.

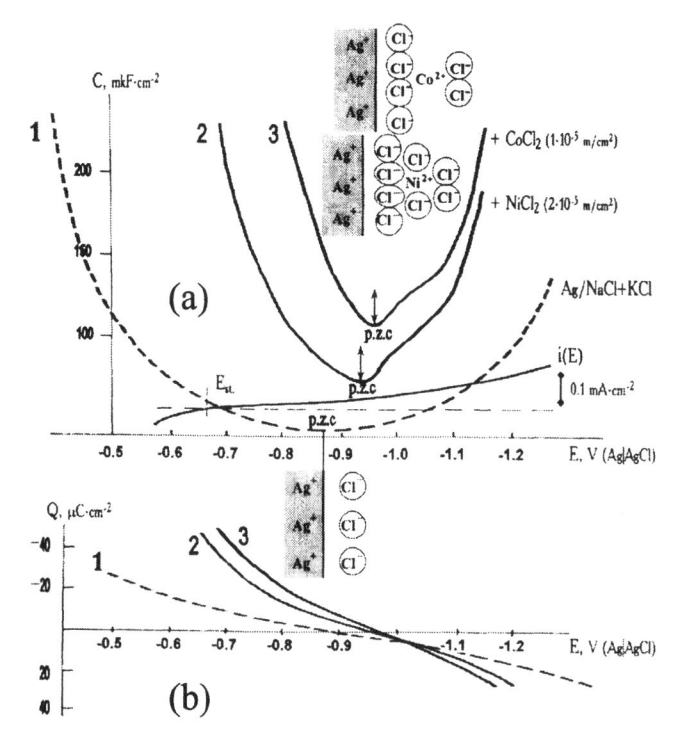

Figure 11. Plots of capacities (a) and charges (b) of a double electrical layer of silver in the melts NaCl-KCl (1), NaCl-KCl-NiCl$_2$ (2) and NaCl-KCl-CoCl$_2$ (3).

If the strength of bonds between the metal surface (Ag^0) and the existing anions-ligands Cl$^-$ and those coordinated (electrosorbed) in the double layer exceeds only slightly the strength of bonds in the complex, for example NiCl$_6^{4-}$, added to the melt and to the double layer, Cl$^-$ ligands are replaced by the NiCl$_6^{4-}$ ligand, and hence the capacity increases (though only slightly due to the "looseness" of the complex), the double-layer charge increases (steeper plots of double-layer capacity against potential), and the zero-charge point shifts slightly towards more negative values (Figure 11-a).

If the strength of bonds between the metal surface (Pt^0) and the same Cl$^-$ ligands is much higher than that of bonds in the complexes added, for example NiCl$_6^{4-}$, a limited amount of NiCl$_6^{4-}$, along with Cl$^-$, in the Pt electrode double layer leads even to a decrease in initial capacity and to the actual constancy of the sharp double layer charge (Pt electrode-Cl$^-$) due to the "looseness", that is large volume, of the complex (Figure 12-a).

The same is evidenced by the data on the effect of the formation of the more compact complex CoCl$_4^{2-}$ in the melt on the structure and capacity of the Pt electrode double layer. Though the double-layer capacity and charge increase noticeably with the formation of CoCl$_4^{2-}$ in the double layer, as could be expected, the shift of the zero-charge point towards more positive values supports the high strength of the initial Pt^0 –

Cl⁻ bond in the heterogeneous complex; as a result, Co^{2+} in this layer can only increase the positive charge.

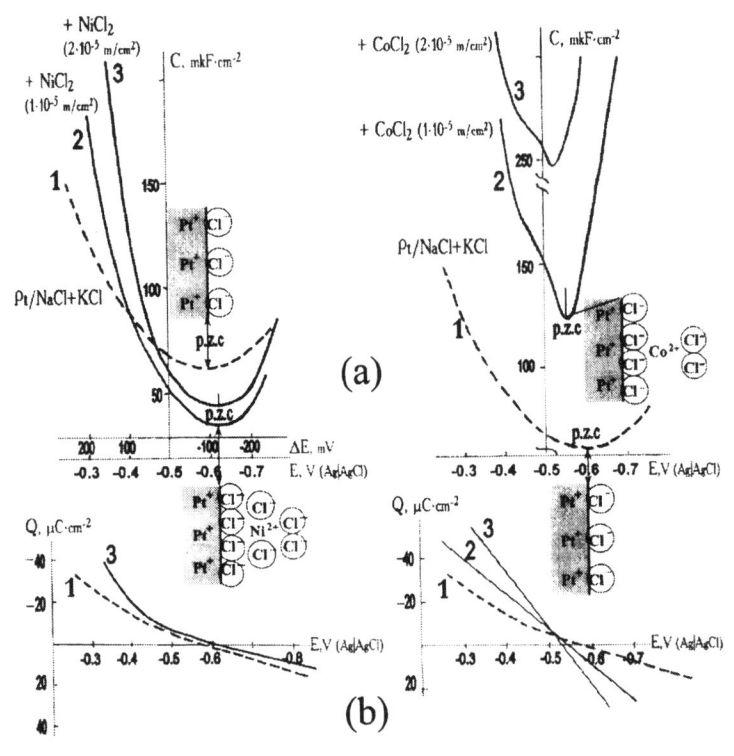

Figure 12. Plots of capacities (a) and charges (b) of a double electrical layer of platinum in the melts NaCl-KCl (1), NaCl-KCl-NiCl₂ (2) and NaCl-KCl-CoCl₂ (3).

This picture becomes still more vivid when considering the dependence of the surface charge on potential (Figures 11-b, 12-b). The presence of the "loose" complex $NiCl_6^{4-}$ in the Pt electrode double layer does not practically change its charge, which is formed through Cl⁻ ligands strongly bonded to the platinum surface, whereas the incorporation of Co^{2+} into this double layer increases greatly its charge. It is interesting to note that the nickel complex in the Ag electrode double layer with less strongly bonded Cl⁻ ligands behaves in much the same way. The example given of the formation of heterophase complexes, such as $Ag^{\delta+}- NiCl_6^{4-}$, $Pt^{\delta+}- Cl^- - CoCl_3^-$, etc., is a particular case of heterogeneous coordination chemistry.

In general, however, the approaches of heterogeneous coordination chemistry are universal for the explanation and prediction of a great variety of processes (not only for melts), such as sorption, extraction, catalysis, corrosion, electrosynthesis, etc. According to our concept, these independent branches of chemistry would flourish still better if they were considered from the general standpoints of heterogeneous coordination chemistry.

References

General References:
- Volkov S., Grischenko V., Delimarsky Yu. *Coordination Chemistry of Solt Melts*, Kiev, Naukova Dumka Publishers, 1977. - 332 p. (in Russian).
- Volkov S., Yatsymirsky K. *Spectroscopy of Molten Salts*, Kiev, Naukova Dumka Publishers, 1977. - 223 p. (in Russian).
- Volkov S., Zasukha V. *Quantum Chemistry of Coordination Condesed Systems*, Kiev, Naukova Dumka Publishers, 1985, 296 p. (Russian).
- Electron structure and high-temperature chemistry of coordination compounds // Charkin O.P., Volkov S.V., et al.-ed/ Buslaev, Nova Science Publ.- N-Y.-1996, 242 p.
- Volkov S. Spectroscopy of Molten Salts, *Rev. Chim. Miner.*, 1978, **15**, #1, p. 59-68.
- Volkov S.High-Temperature Coordination Chemistry, Abstr. 29-th IUPAC, Cologne, 1983, p.60.
- Volkov S. Spectroscopic Investgation of Molten Salt Systems - Theory and Experiment, *Pure and Appl. Chem.*, 1987, **59**, #9, p.1154-1164.
- Volkov S.Chemical Rections in Molten Salts and their Classification, *Chem.Soc.Rev.*, 1990, **19**, p.21-28.
- Volkov S., Bandur V.A., Buryak N.I. The Structurew andCatalytic properties of Coordination Polyhedre of 3d- Metals in Metaphosphate Melts , *Z. Naturforsch.*, A. - 1994, **49**, 1539 - 1543.
- Volkov S., Buryak N.I., Bandur V.A. Oxidartive metal-Complex Catalysis in Melts, Molten Salt Chem. & Techn., Trans. Tech. Publ., 2000, **7**, p.549- 558.

References of figures

[1] to Fig. 1- Volkov S. High- Temperature Coordination Chemistry, *J. Inorg. Chem* (Russian), 1986, **31**, #11, p. 2748 –2757.
[2] to Fig. 2. - Volkov S., Grischenko V., Gorodysky A. Diffusion Model for Ions in Molten Salt Systems with Complex Formation, YII Int. Conf. on Non-aqueous Solutions, Regensburg, FRG, 1980, v.1, p.A5.
[3] to Fig. 3. - Volkov S.On the Concept of Coordination Compound and the Model of Structure of of Molten Salt Systems with Complex Formation, *Chem. Zvesti* (Russian), 1976, **30**, #6, p.819-831.
[4] to Fig. 4. - Volkov S., Babushkina O. Complex-Claster model of Structure of of Molten Salt Systems with Complex Formation, *Ukr. Chem. J* (Russian), 1983, **49**, #10, p.1011-1015.
[5] to Fig. 5. - Volkov S., Babushkina O. Raman Spectra and Structure of $ZnCl_2$ —Li,K/Cl and $CdCl_2$ - Li,K/Cl melts , *Ukr. Chem. J.* (Russian), 1990, **56**, #7, p.678-683.
[5] to Fig. 6. - Volkov S., Babushkina O. Raman Spectra and Structure of $ZnCl_2$ —Li,K/Cl and $CdCl_2$ - Li,K/Cl melts , *Ukr. Chem. J.* (Russian), 1990, **56**, #7, p.678-683.
[6] to Fig. 7. - Volkov S., Babushkina O. Heteronuclear Complex Formation in the melts $ZnCl_2$ - $CdCl_2$ - Li,K/Cl on the data of Raman Spectroscopy, *J. Inorg. Chem.* (Russian), 1990, **35**, #11, p. 2881-2887.
[7] to Fig. 8. - Babushkina O.,Volkov S. Raman Spectroscopy of the Heteronuclear Complexes in the $ZnCl_2$ – $CdCl_2$ –Li,K/Cl and $AlCl_3$- $MgCl_2$ - Li,K/Cl melts, *J. Mol. Liquids*. 1999, **83**, #1-3, p.131 – 141.
[8] to Fig. 8. - Volkov S. Heteronuclear Complex Formation in Salt Melts in the Problem of the Electrolysis of Alloys and Compounds, Abstr. of the I West Pasific Electrochemistry Symp., Tokyo, Japan, 1992, S1, p. 113-114.
[9] to Fig. 9. - Volkov S.Problems of Electrodeposition of Alloys and Compounds from Heteronuclear Metal Complexes in Melts, - Abstr. of 46 ISE Meeting, China, 1995, v.2. p.6-8.
[10] to Fig. 11. - Volkov S. Panov E., Lapshin V. Specific Adsorption (Heterophase Complex Formation) of Ni and Co complexes on the Silver Electrode in Molten Chlorides, *Ukr.Chem. J.* (Russian), 1998, **64**, # 9, p. 33 –35.
[11] to Fig. 12. - Volkov S. Panov E. Specific Electrode Adsorption (or Heterophase Coordination Chemistry) of Complexes in Melts, in *Advanced Molten Salts- from Structural Aspects to Waste Processing*, Biegell H., 1999, p. 605- 611.

CALORIMETRIC METHODS

M. GAUNE-ESCARD
Institut Universitaire des Systèmes Thermiques Industriels
Technopole de Château Gombert
5 Rue Enrico Fermi
13453 Marseille Cedex 13, France

Introduction

Calorimetric techniques constitute a powerful tool to investigate materials.
The methods used for the characterization of thermodynamic properties for molten salts include temperature, enthalpy and heat capacity measurements as mixing enthalpy and phase diagram determinations for their mixtures.

We describe here these methods and their application to different kinds of melts, that also correspond to different kinds of experimental calorimetric equipment

1. Mixing Calorimetry up to 1200K

CALVET-TYPE TWIN MICROCALORIMETER

We describe in the following a Calvet-type twin microcalorimeter, operable up to 1200 K, built along the original specifications of Calvet in Marseille.

The principle of the microcalorimeter was proposed in 1923 by Tian [1], but only some 20 years later, thanks to Calvet [2], did this apparatus become easy to use over a very large temperature range (100-1200 K). This isoperibolic calorimeter has been described many times, so only essential characteristics are reiterated here.

A microcalorimeter is composed mainly of the following parts (Fig. 1a):
- An external steel enclosure surrounding a cylindrical furnace H; heating is provided by four resistors, one on the bottom, one on the top, and two on the cylindrical walls.
- A calorimeter block, made of 3 massive (B, C, and D), top, middle, and bottom parts of alumina, in which two cavities accommodate one thermopile each. This block is surrounded by three enclosures (E, F, and G) acting as thermal and electric shields.

Figure 1b shows the cylindrical thermopile with thermocouple junctions : it is constituted by 22 hollow alumina support disks with Pt/Pt-10%Rh thermocouple junctions arranged as shown in Fig. 1c. They constitute the "hot" (internal) and "cold" junctions, respectively of this sensing device. Each thermopile is located in a cylindrical bottom-closed tube, a so-called "calorimeter cell" of thin-walled alumina. This cell is 17 mm diameter and 80 mm high, and the maximum thermal flux is integrated by the 400 thermocouples..

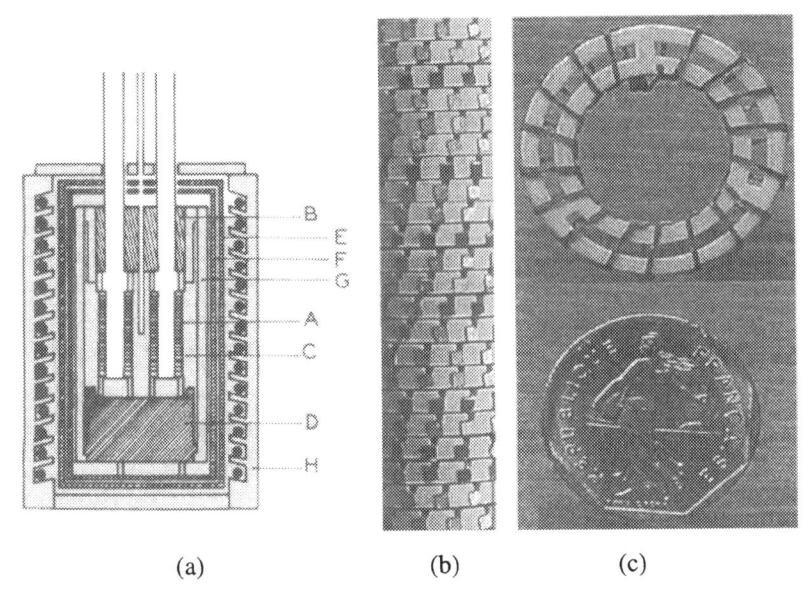

(a)   (b)   (c)

Figure 1.   (a) - Schematic diagram of a Calvet micro-calorimeter. (b) – thermocouple junctions ; (c)- Pt/Pt-Rh ribbon and hollow alumina support disk.

In order to obtain good stability of the apparatus with respect to time and temperature, the two thermopiles are connected in opposition and this twin construction in large measure eliminates most problems associated with any exterior thermal perturbation .

The furnace temperature is monitored with an electronic programmer; the sensing element is either a high-temperature resistance probe or a thermocouple, located close to the heating element. The experimental temperature is controlled by means of a thermocouple situated in the center of the calorimetric block.

1.2. EXPERIMENTAL TECHNIQUES

Several experimental techniques can be used for the determination of enthalpies of mixing. In all of them, the essential condition to be met is elimination of all the effects arising generally from material-atmosphere interactions (e.g., with oxidizing or hygroscopic substances), from solvent or solute-crucible interactions, from difficulties of mixing (A and B with very different densities), from stirring necessary to homogenize the final product, from introduction of a solute at a temperature different from that of the solvent, and so on.

The main techniques [3] for high temperature liquid systems are indicated below.

1.2.1. Solid-Liquid Mixing

The first and also simplest method to achieve the mixing of two substances is the so-called *"drop method"*, described at length by Kubaschewski and Evans[4], it consists in dropping into the liquid salt A, considered as the solvent and maintained at the experimental temperature T_E a weighed amount of salt B, previously stabilized at temperature T_O (generally room temperature). This technique, which is very easy to operate, has been applied to microcalorimetry (Fig. 2). The measured enthalpy corresponds to the enthalpy of the reaction:

$$n_B B(s,T_o) + n_A A(l,T_E) \rightarrow (n_A + n_B) \text{ AB (l or s},T_E)$$

and includes not only the enthalpy of mixing, that is to be determined, but also the enthalpy increment (enthalpy of fusion + heat capacity term) of salt B on the usually large temperature range T_o to T_E

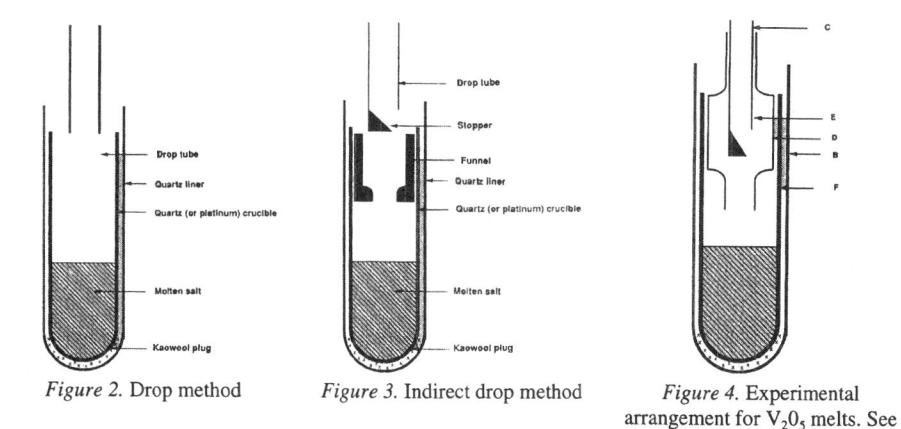

Figure 2. Drop method

Figure 3. Indirect drop method

Figure 4. Experimental arrangement for V_2O_5 melts. See text for description

Unfortunately, for many systems, the uncertainty in the enthalpy increment term is of the same order of magnitude as the enthalpy of mixing itself and can, therefore, lead to unreliable experimental enthalpy of mixing data.

In order to improve the accuracy of the previous method, an *"indirect drop"* method was developed (Fig. 3), which involves preheating of the sample B. The mixing experiment is thus carried out in two steps. First, sample B is dropped into the funnel, where preheating takes place; this drop is guided by the drop tube, the lower end of which acts as a stopper by obstructing the funnel aperture. Second, when B is thermally equilibrated, a simple vertical shift of the drop tube allows the mixing to be performed. In some cases, a stirring device was added in order to ensure homogeneity of the mixture [5]. In order to keep the samples either under vacuum or under inert atmosphere (argon U), the upper part of the system involves a set of taps and rings, and all the manipulation of the drop tube and of the stirrer is electromagnetically operated.

Mixtures of $K_2S_2O_7$, V_2O_5, and K_2SO_4 were investigated using an experimental technique [6] derived from the "indirect drop method,". All experiments concerning measurements of the heat of mixing were performed by addition of a solid (V_2O_5 or K_2SO_4) to a liquid ($K_2S_2O_7$-, or $K_2S_2O_7$-V_2O_5 mixtures). In these experiments, a modified experimental setup was constructed as shown in Fig. 4.

The outer quartz tube (B) has an outer diameter of 18 mm and is 600 mm long; the lower 100-mm part fits very closely to the alumina walls of the calorimetric cavity in order to secure an optimized horizontal heat conductivity. A standard-taper joint of borosilicate glass at the top is connected to the quartz tube outside the calorimeter by a taper joint. A slow gas flow is maintained through the system before and during the experiment. Through the taper joint, two Pyrex or quartz tubes lead into the calorimetric cavity. The inner tube (C) (outer diameter, 6 mm) is equipped with a small funnel at the top outside the calorimeter and a hole in the wall just above the smoothly ground end of the glass rod inside the calorimeter.

The end of the glass rod also acts as a stopper (E) in the reservoir of the outer tube, which is well tightened to the taper joint by conventional fittings consisting of a screw and a silicon packing. Just below the reservoir and in close contact with it, the crucible (F), containing the melt, is located. This quartz or borosilicate crucible has outer dimensions close to the inner dimensions of the tube B to optimize the horizontal heat flow. The experiments concerning the heat of mixing were performed as follows. The crystalline V_2O_5, or K_2SO_4 was weighed with an accuracy of 10^{-2} mg. The final weight of the crystals was measured after the dust had been removed from their surface by several drops from about 50 cm. The crystals were accepted as dust-free when the deviation between the last two preliminary weighings was less than a few hundredths of a milligram. A crystal, typically with a mass of 10-30 mg, was dropped through the funnel of the tube C into the reservoir D, and the thermal disturbance was registered on a recorder and fed to a system for automatic integration. After the baseline was reestablished—usually within 20 min—the tube C was raised about 2 cm by manipulation from the outside, allowing the crystal to drop from the reservoir into the melt below.

The drop methods described above allow several successive additions of B to be made during the same experimental run, while investigating the composition dependence of the enthalpy of mixing. The main advantages of the drop methods are therefore their simplicity and rapidity; this contrasts with those more direct and accurate liquid-liquid mixing techniques, which make it possible to measure the mixing of liquid components directly and, in principle, with a greater accuracy.

1.2.2. Liquid-Liquid Mixing

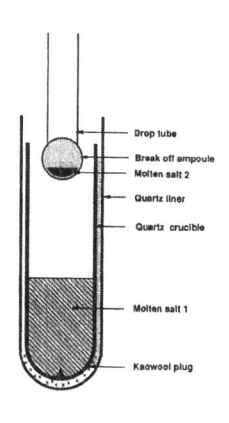

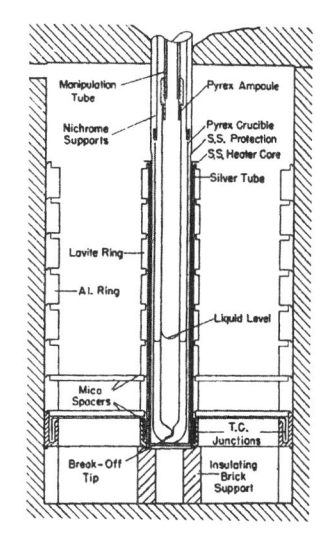

Figure 5. Break-off bubble method.

Figure 6. Break-off ampoule method.

The break-off bubble method [7] (Fig. 5) has often been used with molten salt mixtures. Inside the Pyrex (or quartz) liner, a cylindrical crucible contains the liquid salt A while salt B is contained in a spherical thin-walled ampoule. This Pyrex (or quartz) ampoule has to be thin enough to be broken with a single stroke. The mixing of the two liquids is initiated by crushing the bubble against the bottom of the crucible. The thermal change arising from the ampoule break-off was checked and found very small and reproducible through blank experiments.

This method is similar in its principle to that used by Kleppa [8] in his pioneering calorimetric investigation of molten salts. In those mixing experiments, salt B was contained into an ampoule consisting in a Pyrex tube fitted with a break-off tip (Fig. 6). However, because of surface tension effects, both molten salts probably wet the crucible and ampoule walls before mixing, possibly resulting in a final composition slightly deviating from that calculated from the initial amounts of components. Indeed, in the above break-off bubble method, the mixing homogeneity is preserved since the melt is contained in a single space (crucible) instead of two containers (crucible + ampoule) with a risk of concentration gradient.

The above break-off bubble method was used recently for the investigation of lanthanide halide melts [9-17] and Figure 7 shows as an example the enthalpy of mixing of neodymium iodide with the alkali halides [16].

The methods of mixing described above are general, but the choice of a suitable technique of mixing for a particular system should always be made in accordance with

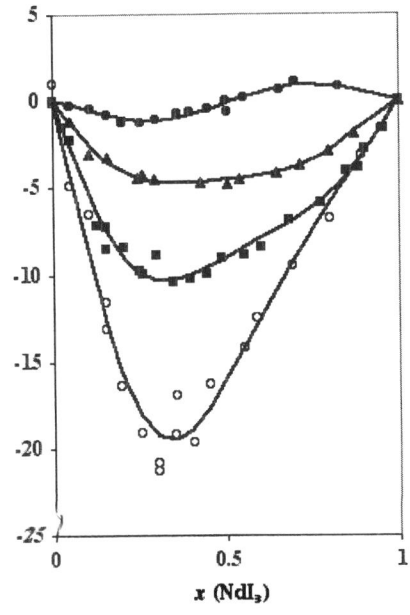

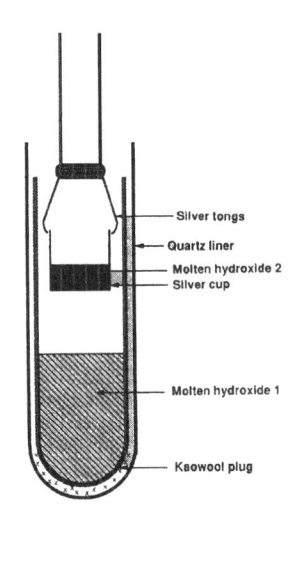

Figure 7 Enthalpy of mixing of NdI$_3$-MI systems [hh]
● - LiI ; ▲- NaI ; ■ - KI ; ❑ - CsI

Figure 8. Suspended cup method

When the melt reacts with Pyrex (or quartz), as for instance molten alkali metal hydroxide mixtures, a "suspended cup" method (Fig. 8) was developed [18]. Inside the Pyrex (or quartz) liner, a cylindrical silver crucible contained the liquid hydroxide A while the salt B was contained in a small silver cup. This silver cup was held by tongs and could be released by external operation of the tongs. The mixing of the two liquids was initiated by dropping the silver cup into the crucible. The thermal excursion arising from this drop was very small and reproducible.

However, when the salts to be mixed have very high vapor pressures, it is impossible to use a classical mixing device. Suitable cells had to be constructed for this purpose. For instance, mixtures of BiCl$_3$ and KCl were [19] Measurements were carried out in a high-temperature Calvet-type microcalorimeter at 690 K, 22K below the normal boiling point of BiCl$_3$ [20]. At this temperature, the vapor pressure of BiCl$_3$ is 69.4 kPa. Pyrex glass was used to prepare the break-off spherical ampoules (Fig. 9). The bottom part was a thin-walled membrane of strictly controlled mechanical properties. The thickness of the membrane was selected so that it was possible to break the ampoule in the calorimeter without any measurable thermal effect but, on the other hand, so that the membrane was strong enough to stand atmospheric pressure after evacuation of the ampoule. The ampoules were filled in a glove box with weighed portions of BiCl$_3$, evacuated, sealed and stored in vacuum to avoid diffusion of air through the membrane. The other parts of the calorimetric cell were also made of Pyrex.

The crucible, filled with a weighed portion of KCl prior to measurement, included at the top a ground-glass ball joint that ensured tightness after the ampoule was broken. The crucible lid, constituted by the male part of the ball joint, moved the ampoule down for break-off. Before experiment the vertical distance between the lid joint and crucible joint was adjusted by means of spacers so that it would be from 0.4 to 0.6 mm. This device was placed in a protective quartz tube filled with argon.

The time required for thermal equilibration into the calorimeter before experiment was 4 h. By breaking off the ampoule, half of its spherical bottom was crushed so that it was possible to mix thoroughly liquid bismuth chloride with solid potassium chloride.

Measurements in which the ampoules were not completely broken were disregarded. No endothermic effects related to evaporation were found.

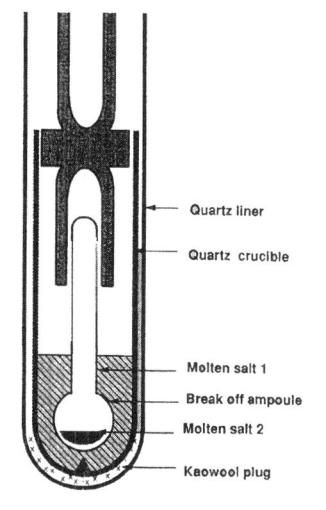

Figure 9. Experimental arrangement for the investigation of bismuth chloride melts.

Any baseline shift after mixing was very slight; it was never larger than 1 mm, which is less than 1% of the peak height. In most cases, no detectable change of the thermogram baseline was observed The calibration constant k of the calorimeter was determined on the basis of the thermal change for zinc solidification. Calibration measurements were carried out in the same cell in which enthalpy of mixing was measured. The molar enthalpy of fusion of zinc, $\Delta_{fus}H_m = 7.343 \pm 0.083$ kJ/mol was taken from Chiotti *et al* [21]. The standard deviation $s(k)$ of the calibration constant of the calorimeter was $0.021k$.

The same method was used for the investigation of mixtures of BiCl$_3$ with the other alkali metal chlorides [22].

The experimental constraints are most severe when the mixture under consideration has a sublimable component like AlCl$_3$. Ternary mixtures of AlCl$_3$, KCl, and AlCl$_3$NH$_3$ were investigated [23]. Both reactants, the AlCl$_3$ + KCl liquid mixture and AlCl$_3$NH$_3$(l), respectively, were contained in closed Pyrex ampoules, both kept inside the calorimeter itself during measurements. The inner ampoule containing AlCl$_3$NH$_3$ had a very thin base which could easily be broken on the sharp edge of the outer ampoule with a rapid movement of the manipulation rod sticking out of the calorimeter.

Calibration was performed using α-Al$_2$O$_3$ from the National Bureau of Standards. Because the experimental cell was closed, the calibration could not be carried out in the cell itself, but had to be performed directly in the steel protection tube situated in the calorimeter proper. The calibration constant obtained in this manner is somewhat uncertain. Tests were made [24] by varying the rest position of the α-Al_2O_3 piece relative to the calorimeter itself and also the cell arrangement by letting the α-Al_2O_3 piece fall into a Pyrex tube situated inside the steel protection tube. Variations of the order of ±5% were observed.

The procedure employed in the experiments was as follows. The inner cell was first filled with the desired amount of $AlCl_3NH_3$ in a glovebox, and transferred to a vacuum line evacuated to 10^{-5} torr. The cell was then back-filled with Ar to adjust the pressure so that the pressure in the inner and the outer cell were approximately equal at the working temperature of the calorimeter. This was done to ensure stable cell behavior when the inner cell wall was broken during the mixing experiment. The cell was then sealed and transferred back to the glovebox, where it was mounted into the outer cell after this cell had been filled with the necessary amounts of $AlCl_3$ and KCl. The outer cell with its contents was then transferred to the vacuum line, evacuated to 10^{-5} torr, sealed, and fused onto the Pyrex manipulation rod.

Before the cell was introduced into the calorimeter, it was preheated at a somewhat higher temperature than the calorimeter temperature to speed up the melting of the $AlCl_3$-KCl mixture. Complete dissolution of KCl(s) was observed visually.
Complete mixing of the three salts after breaking the inner ampoule was ensured by rapid stirring of the cell. These manipulations had a minor influence on the measured heat change and no baseline shifts were observed.

Some other examples are given below that were used for the investigation of other very reactive melts. These calorimetric devices are rather sophisticated and great care was taken in order to obtain reliable enthalpy of mixing data.

For the investigation of zinc halide-containing melts, [25] a novel experimental arrangement and procedure was adopted in order to control the weight losses due to evaporation of the zinc halides at 665°C. Inside the fused-silica liner, a cylindrical crucible contains the low-vapor-pressure salt in the liquid state (LiX, CsX, AgX). The zinc halides are contained in an evacuated "double break-off" fused-silica bubble. The pressure of the zinc halide liquids inside this bulb was approximatively 0.5 to 0.7 atm, while the argon pressure in the liner was maintained at 1 atm.

The double break-off tube eliminated the problem of volatilization of the zinc halide from the time when the liner was introduced into the calorimeter until the mixing was initiated. This period ranged from 45 min to 2 h.

The mixing of the two liquids was started by crushing the lower break-off tip against the bottom of the crucible. As the pressure in the bulb was lower than 1 atm, a portion of the melt contained in the crucible was sucked up into the bulband was mixed there with the zinc halide. The mixing process was then completed by crushing the upper break-off tip. This allowed the zinc halide to be released into the crucible and mixed with the remainder of the melt. However,after the mixing operation was started, zinc halide was lost by evaporation and an endothermic baseline shift was observed. In order to correct for this baseline shift, a series of blank experiments was performed.

For the investigation of lead oxide-containing melts, the following experimental procedure was adopted [26-27]. About 60 g of oxide melt was contained in an 80% gold-20% palladium crucible of about 17-mm diameter and 75-mm height. The lead oxide to be dissolved in the melt was kept in a very shallow platinum cup of about 10-mm diameter. This cup was attached, by means of three platinum wires, to a fused-silica tube which could be manipulated from outside the furnace system. The solution reaction was initiated by lowering the platinum cup into the melt. Stirring was accomplished by means of a platinum-covered graphite plunger. The liquid PbO was displaced and brought into reaction by inserting the plunger into the platinum cup. The other oxide component was added to the solvent in a similar manner, but in this case adequate stirring was achieved simply by moving the platinum cup up and down in the gold-palladium crucible. Corrections were made for the heat change associated with each stage. This heat change is endothermic in character and is largely due to mass displacement in the vertical temperature gradient of the calorimeter. When a plunger was used, the stirring correction was 10-50% of the total heat of reaction; without a plunger, the correction was 10-20%.

Enthalpies of mixing in binary liquid carbonate mixtures have been measured [28]. Due to the corrosive nature of molten alkali carbonates, these cannot, unlike most other salts, be contained in fused-silica containers and be introduced by the usual "break-off" technique. Experience has shown that attack on Palau (20% Pd-80% Au) in "acid" alkali carbonate melts, that is, in melts kept under a relatively high CO_2, pressure, may be considered negligible.

The experimental arrangement used to perform these calorimetric measurements included plunger as well as dipper crucible that could be manipulated from the outside the calorimeter itself.

Immediately before insertion into the calorimeter, the fused-silica liner with its contents was preheated for a period of 15 min at about 50 K above the operating temperature of the calorimeter (~ 1150 K). The weight loss of the most volatile carbonate (Rb_2CO_3) was about 0.3%; in spite of the relatively small vaporization losses, attack by the vapors on the fused-silica liner and shield was considerable, and the lower parts of the device had to be rebuilt after 10 to 15 experiments.

The mixing of the two salts was achieved by vertical manipulation of the plunger and dipper. After the initial mixing process, three additional stirring operations were carried out at 1-min intervals to ensure complete mixing.

2. Mixing Calorimetry above 1200 K

Beyond 1200K, other calorimeters should be used because of the intrinsic limitation of the building materials.

For the investigation of high-temperature melts, a commercial Setaram enthalpimeter was used.

The whole calorimetric assembly includes the following parts:
(1) The vertical cylindrical furnace consists of a graphite resistor surrounding a gastight alumina tube, 23 mm i.d., 600 mm long, in which the calorimetric detector

is located and localizing the experimental chamber. The geometry of the resistor is such as to provide a constant-temperature zone, 140 mm long, in the central part of the tube. This furnace has an external water-cooled jacket and can be heated up to about 2000°C.

(2) The furnace is supplied with low-voltage current. Its temperature is controlled by an electronic system monitored by a Pt/Pt-13% Rh thermocouple within the central part of the furnace.

(3) A set of valves and flow meters enables evacuation of the furnace and the experimental chamber, and the flow or maintenance under pressure of a purified gas.

(4) The samples are introduced into the calorimeter and maintained under experimental conditions, at ambient temperature and after a preliminary evacuation, with a very simple charging device very similar to that previously designed for Calvet calorimeters.

(5) The features of the calorimetric detector, and of the mixing and stirring systems, located within the experimental chamber, vary according to the nature of the experiment.

(6) Acquisition and processing of enthalpy data treatment is computer-operated.

Figure 10 is a vertical section of the very high temperature calorimeter: C_l and C_t are the laboratory and reference crucibles, respectively. These alumina crucibles are placed vertically and coaxially in the center of the experimental chamber. The gastight alumina tube T constitutes the external shield of the experimental chamber.

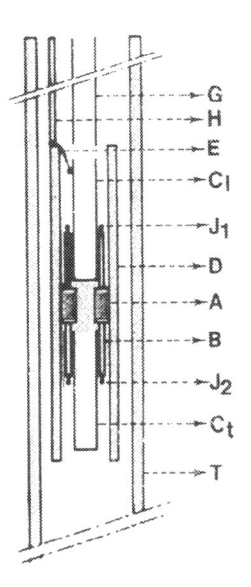

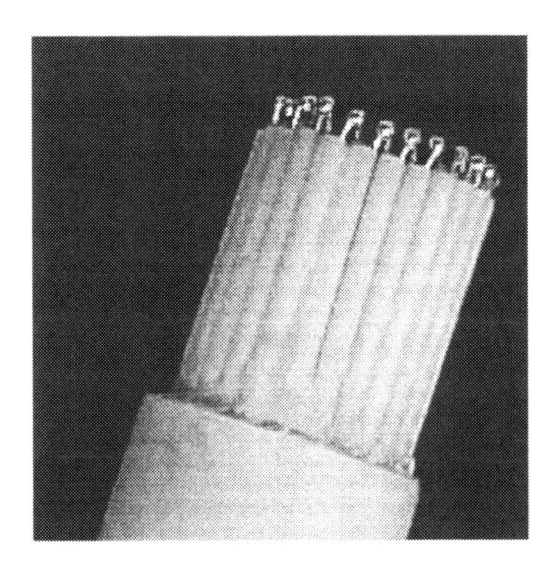

Figure 10. Vertical section of the very high

Figure 11 : détails of the sensing device.

temperature calorimeter. See text for description

The middle part of each crucible is in contact with the junctions of the sensing Pt-6% Rh/Pt-30% Rh thermocouples, J_1 and J_2 (Fig. 11).

The tube D acts both to maintain in the proper place the ring supporting the thermocouples and as to constitute a thermal screen to the assembly. The whole assembly is suspended in the central part of the furnace by three hollow alumina tubes (H) which protect the junction wires of the thermocouples.

A systematic study showed that, with this kind of calorimeter, the two following conditions are very important in order to obtain reproducible and accurate values of the thermal changes upon mixing: (1) a high sensitivity of the detector, and (2) an appropriate integration of the thermal flux produced within the laboratory cell by the heat change upon mixing. The former condition is best met by using a sufficient number of thermocouples, and the latter by having a regular arrangement of the thermopile junctions around the laboratory crucible. Therefore it was necessary to make some modifications to the commercial design and several calorimetric detectors were built in accordance with these conditions [29-30].

In one detector version, the 16 upper thermocouples junctions were located alternately on two horizontal levels about 10 mm apart. For technical reasons, and since no thermal excursion arises within the reference crucible (C_t), the lower junctions were on the same horizontal level. Another kind of detector had more thermocouples (25-66).

2. 1. METHODS OF MIXING

Because of its easy operation, the so-called "drop method," already described, is usually used to obtain the enthalpy of formation of a liquid mixture AB. However, it should be stressed once more that the enthalpy change during liquid-liquid mixing is calculated as the difference between two rather large quantities, thus introducing large errors limits. For instance, numerically, the enthalpy of fusion of silicon ($\Delta_{fus}H_{,Si}$ = 50.54 kJ/mol) is significantly higher than the maximum value of the enthalpy of mixing of most silicon-based alloys. The situation is the same for molten ionic mixtures such as K_2SO_4-NaF [31] for which the enthalpy of mixing is found to be 6.7 kJ/mol at x_{NaF} =0.6 and the correction term about 250 kJ/mol.

In order to improve the results, it is therefore necessary to eliminate or severely limit the enthalpy increment term (enthalpy of fusion and heat capacity term) of the sample B added to the liquid bath A. This can be achieved if the sample B is introduced into the liquid bath A at a temperature equal or close to the experimental temperature T_E.

The "indirect drop method" was adapted this kind of calorimeter, as shown schematically in Figure 12. Sample B, contained in the charging system at temperature T_O, is dropped into the calorimeter and remains in the funnel (F) thermostated at temperature T_F. This temperature T_F is measured with a Pt-6% Rh /Pt-30% Rh thermocouple located within the small alumina sphere (S) (6 mm diameter) which

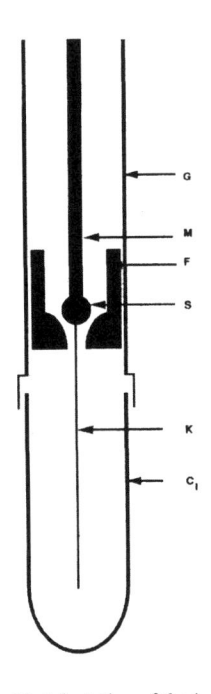

Figure 12. Adaptation of the indirect drop method. G, Alumina drop tube; M, alumina rod; F, funnel; S, alumina sphere; K, stirrer, C_1, laboratory crucible

closes the aperture of F. When sample B reaches thermal equilibrium, the funnel (F) is opened by lifting the thin alumina rod (M). The mixing process takes place in C_1 The temperature T_F, and therefore the magnitude of the correction term, depends on the position of the funnel in the tube (T). The funnel (F) may be attached with a small alumina wedge at a distance from crucible C_1 which can be adjusted, for example, at temperatures equal to or less than T_E.

When the temperature of the funnel, T_F, is less than the fusion temperature of sample B, T_{fusB}, the experiment is generally quite facile. On the other hand, many factors have to be taken into account to obtain reliable results when $T_F > T_{fusB}$. The liquid B should have suitable physicochemical properties (vapor pressure, viscosity, chemical reactivity with respect to the container, surface tension, etc.) to remain in the funnel F during thermal stabilization and to completely flow out when F is open. This can be achieved with molten metals, but these conditions are very difficult to fulfill with molten salts since they generally completely creep out before F is open.

For some mixing experiments, a large difference between the densities of the components A and B restricts obtaining a homogeneous mixture AB. In these cases, a stirring device completes the previous system of mixing. A thin alumina rod (K), sufficiently long to dip into the liquid bath, is added to S and acts as a stirrer. The vertical movement of M homogenizes the liquid AB. Such a device was used to investigate the ternary alloys Al-Ge-Si [32] (Si is added to the liquid Al + Ge) and the ionic mixture LiF-K_2SO_4[33].

A number of metallic and ionic mixtures such as Au-Sn [34] Ag-Au-Pd [35] Al-Ge-Si [32] NaF-Na_2SO_4 and NaF-K_2SO_4 [36] NaF-Rb_2SO_4 [37] ZrF_4-MF (M = Li, Na, K, Rb) [38] ternary ZrF_4-based melts [39], AlF_3- based mixtures [40], KF-NdF_3 [41] were investigated in this way over a temperature range between 1000 and 1500°C. Whenever possible, we compared our results with those obtained from other calorimetric methods.

In the course of a critical analysis [42] of all the calorimetric data published on the enthalpy of mixing of the Au-Sn system, we showed that the difference is less than 2% between two sets of measurements at 1000 K obtained either with a Calvet microcalorimeter or with the apparatus described above. Although at higher temperatures the few data available for other binary alloys do not make possible such a comparison, this good agreement evidenced the reliability of the method.

Also, an automatic sample charger was designed and developed which enables a completely automated operation of the very high temperature enthalpimeter [43]. It allows a complete experimental run with successive addition of 30 samples; each individual mixing experiment is computer operated and calorimetric thermograms are also automatically integrated.

It should be noted, however, that no apparatus or method can be considered universal in the domain of high-temperature mixing calorimetry, and adaptations should always be made in accordance with the particular requirements of the system under investigation.

3. Enthalpy and Related Temperature Changes

For enthalpy and related temperature change determinations, different techniques such as differential thermal analysis (DTA) and differential scanning calorimetry (DSC) can be used. The purpose of these differential techniques is to record the difference between the enthalpy change which occurs in a sample and that in some inert reference material when they are both heated.

However the results obtained strongly depend on the performance of the calorimetric system used.

Generally DTA techniques should only be considered as semiquantitative (see, for instance. Fig. 13a) : the sensing element in each differential cell is generally of modest sensitivity (1-2 thermocouple or a resistance probe) and more importantly, heat detection, restricted to the surface of the experimental cells, does not not fully integrate the heat flow in the bulk sample . This results in baseline drift of thermograms that induces large uncertainties in the evaluation of thermal effects.

Therefore differential enthalpic analysis (DEA; see, for instance. Figure 13b) or DSC should be preferred.

A Calvet microcalorimeter, described above for isothermal operation, can also be operated as a differential enthalpic analyzer if a linear temperature programmer is added to the temperature regulator; although its thermal inertia is such that the heating rate does not exceed a few degrees per hour, this inconvenience is largely compensated by the great sensitivity and the large experimental volume.

This microcalorimetric technique is very reliable and most efficient when the sample under investigation undergoes several phase transitions at temperatures very close altogether: it is then possible to separate the resulting thermal effects (Fig. 13b), while conventional DTA would only give overlapping peaks (Fig. 13a). Several uranium(IV) halides and compounds of alkali metals and uranium(IV) halides were successfully investigated [24,44-48] using such a technique.

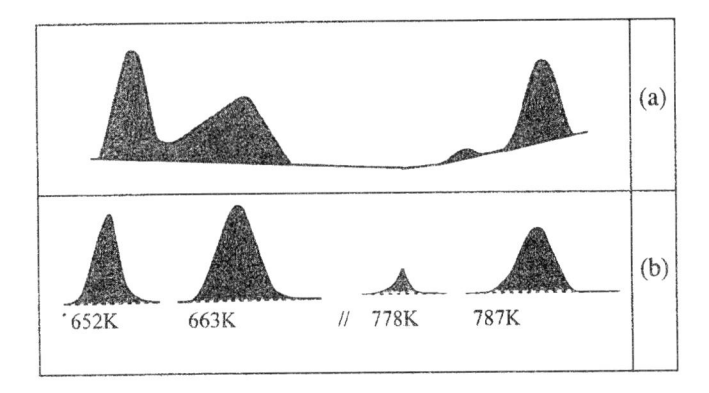

Figure 13. Phase transitions in Na_2UBr_6: (a) DTA thermogram; (b) DEA thermogram

The same features exist in the DSC 121, commercialised by Setaram, since this differential scanning calorimeter is a miniaturized version of the Calvet microcalorimeter (Fig. 14). In addition to a great sensitivity and efficient integration of heat flux, its small inertia allows fast heating and cooling runs.

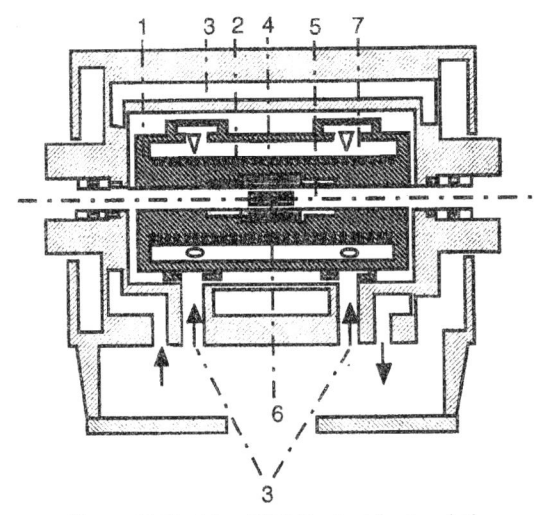

Figure 14 : Heat flux DSC. See text for description

It consists of a calorimetric block made of silver (1) with embedded heating elements (2) and a water- (3) or gas- (7) cooling system.
The heat flux detector, consisting of two cylindrical thermopiles (4) connected in opposition and built along the same principle as the Calvet microcalorimeter (several hundreds of thermocouples)

The calorimetric block accommodates two alumina tubes (5), each with an internal protecting liner. These Inconel tubes protect against any accidental molten salt leak that could damage the DSC equipment. The (reference and experimental) sealed cells (6), made of quartz or Inconel, fit snugly into these tubes, in the constant temperature zone of the block. An Argon stream prevents any high temperature oxidation.

The maximum operation temperature is 1100 K.

Several examples regarding the temperatures and enthalpies of solid-solid and solid-liquid phase transitions in the series of lanthanide halides LnX_3 and of stoichiometric compounds they form with alkali halides MX can be found in recent papers [49-52].

4. Heat Capacity Determinations

Heat capacity at constant pressure, C_p is the derivative with respect to temperature:of the enthalpy change induced by a temperature variation :

$$C_p = dH/dT$$

At high temperature, the methods used for C_p determination, are based on the simultaneous measurement of the enthalpy temperature variation *vs.* time at a programmed rate of temperature increase.

In indirect methods, *heat content* measurements are performed on a large temperature range, for instance by drop calorimetry, and C_p is derived by analytic derivation of heat content plots *vs.* temperature.

In direct methods, the sample is *heated* over a large temperature range either *continuously* or by *successive small temperature increments* (Fig. 17), with a linear dependence of temperature on time

4.1. HEAT CONTENT

Heat content measurements at high temperature are generally easier than direct heat capacity determinations. They were used widely to get early Cp data on molten salts. The enthalpy increment of a substance between the temperatures T_1 and T_2 (with $T_1 < T_2$), $[H(T_2) - H(T_1)]$, is generally measured by drop calorimetry. Two general techniques are employed, depending on whether measurements are carried out at high temperature (T_2) or at low temperature (T_2):

1. The sample is heated at high temperature (T_2), and the actual heat content measurement is performed in a calorimeter at the experimental temperature T_1.
2. The sample is at low temperature (T_1), and the actual heat content measurement is performed in a high-temperature calorimeter at the experimental temperature T_2; this is the so-called "reverse drop method."This method should be preferred in principle for melts since non-equilibrium final states can be obtained on cooling.

Figure 15 shows the enthalpy increment $[H(T) - H(298)]$ measured for a hypothetical solid and liquid sample. The enthalpy of fusion can be deduced as the difference between the enthalpy of the liquid and of the solid at the fusion temperature T_F. The heat capacities of the solid and the liquid, $C_{p,s}$ and $C_{p,l}$ is obtained from the temperature dependence of $[H(T) - H(298)]$.

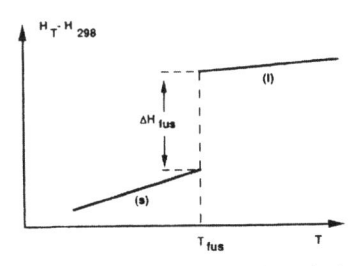

Figure 15. Determination of the enthalpy increment for a hypothetical liquid and solid sample.

For instance, using a high-precision adiabatic-shield drop calorimeter, Holm et al [53] measured the enthalpy increments $[H(T) - H(298 \text{ K})]$ of the congruently melting 2:1 and 1:1 compounds of alkali chlorides and magnesium chloride; from the results obtained at several experimental temperatures corresponding to solid and liquid samples, they determined the enthalpies of fusion of these compounds. Values of the heat capacities for the molten salt mixtures were also derived.

Two samples of each of the compounds were loaded into platinum containers of known mass. The containers were evacuated carefully inside a glovebox to remove the air. The glovebox was filled with purified nitrogen. After evacuation, the containers were filled with purified argon. They were then sealed by arc-welding a cup-shaped platinum lid to the rim of the container.

The sample was equilibrated in a vertical laboratory furnace and lifted into the silver calorimeter, which was placed above the furnace. The calorimeter was surrounded by silver shields, electrically heated to maintain quasi-adiabatic conditions. The furnace temperature was measured by a Pt/Pt-10% Rh thermocouple and the calorimeter temperature by a quartz thermometer.

Steady-state conditions were usually obtained after 10 to 20 min, depending on the furnace temperature. The calorimeter temperature during the period of experiments ranged from 300 to 330 K with a mean of 315 K. The heat capacity values for the compounds were estimated from those of the binary compounds by the relation:

$$C_p = nC_p(\text{AlkCl}) + C_p(\text{MgCl}_2), \quad n = 1 \text{ or } 2$$

4.2. THE RATIO METHOD

When a sample material is subjected to a linear temperature increase, the rate of heat flow into the sample is proportional to its instantaneous heat capacity. By regarding this rate of heat flow as a function of temperature and comparing it with that for a standard material under the same conditions, we can obtain the heat capacity, C_p as a function of temperature. The procedure has been described in detail and a precision for specific heat determinations of 0.3% claimed in some cases [44].

Figure 16 shows the principal method of heat capacity determination.

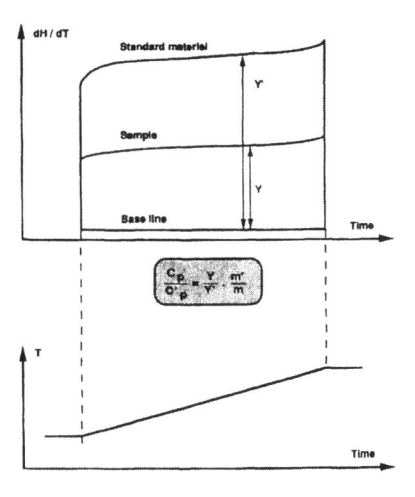

Figure 16. Heat capacity determinations by the ratio method.

Empty cells are placed in the sample and reference holders. An isothermal base-line is recorded at the lower temperature, and the temperature is then programmed to increase over a range. An isothermal baseline is then recorded at the higher temperature as indicated in the lower part of Figure 16. The two baselines are used to interpolate a baseline over the scanning section, as shown in the upper part of Figure 16. The procedure is repeated with a known mass of sample in the sample cell, and a trace of *dH/dT* versus time is recorded. The deviation from the base line is due to the absorption of heat by the sample, and we may write:

$$dH/dT = mC_p(dT_p/dt)$$

where m is the mass of the sample, C_p the heat capacity, and (dT_p/dt) the programmed rate of temperature increase

While the above equation would yield values of C_p directly, in order to minimize experimental errors, the procedure is repeated with a known mass of standard sample, the heat capacity of which is well established. Thus, only two ordinate deflections at the same temperature *(Y* and *Y'* in Fig. 16) are required to yield a ratio of the C_p values of the sample and standard. This global method has a modest accuracy.

4.3. THE "STEP" METHOD

This "step method" consists in *successive small temperature increments* (see Fig. 17), with a linear dependence of temperature on time. Each small temperature step is followed by an isothermal delay, which ensures thermal re-equilibration of the sample.

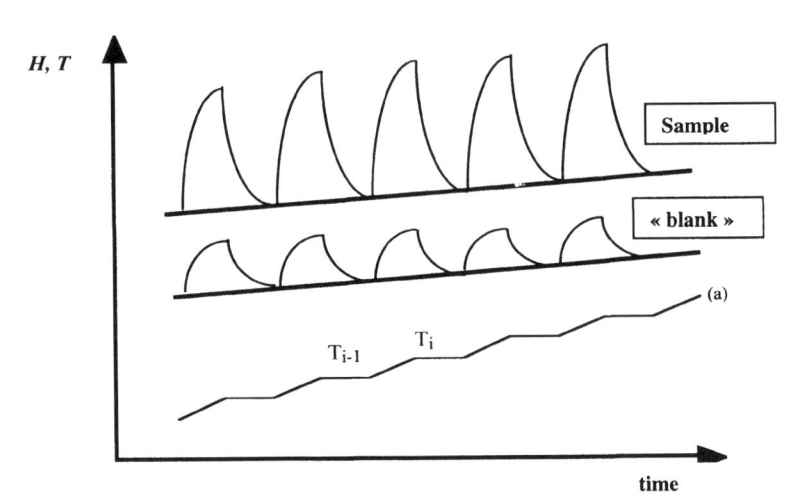

Figure 17 : Heat capacity measurement by the step method

From the observation of the thermal disequilibrium between the two cells during a heat pulse, the heat capacity of the sample located in the working cell can be obtained as a function of the temperature. The two experimental crucibles contained in the cells are chosen in order to have as similar a mass as possible.

Since the experimental parameters are the time dependence of enthalpy and of temperature, the heat capacity equation can be written :

$$C_p = \left(\frac{\delta H}{\delta T}\right)_p = \left(\frac{\delta H}{\delta t}\right)_p / \left(\frac{\delta T}{\delta t}\right)_p$$

Since temperature T varies linearly against time t, the integration between times t_1 and t_2, that correspond to temperatures T_1 and T_2, yields an average heat capacity value $\overline{C_p}$ on the small temperature interval $[T_1 - T_2]$:

$$\int_{t_1}^{t_2}\left(\frac{\delta H}{\delta t}\right)_p dt = \int_{t_2}^{t_1} C_p \left(\frac{\delta T}{\delta t}\right)_p dt = \int_{T_2}^{T_1}\left(\frac{\delta H}{\delta T}\right)_p dt = \int_{T_2}^{T_1} C_p dt = \overline{C_p}\left[T\right]_{T_1}^{T_2} = \overline{C_p}\left[T_2 - T_1\right].$$

Therefore this method provides nearly "true" heat capacity values of materials, except in the vicinity of characteristic temperatures (phase transition, etc.) where the corresponding enthalpy increments superimpose to those induced by the temperature increments of the "step method" and invalidate the $\overline{C_p}$ evaluation.

During the experiment, the cells are maintained in a purified argon flow. In the temperature range from 300 to 1100 K, Cp measurements are carried out step by step, each heating step, generally 5K at the heating rate 1.5 K mn^{-1}, is followed by an constant temperature plateau for 400 s.

The same heating procedure ("blank experiment") should be repeated with two empty cells identical to those used for the expérimental sample run. The heat capacity of sample is obtained at each temperature from the difference between the enthalpy increments obtained, at each temperature step, in the two experimental series.

This method has been applied to several lanthanide halides LnX3 and stoiechiometric compounds they form with alkali halides MX [51-52,54-57]. Indeed, most MX-LnX3 systems exhibit phase diagrams characterized by stoichiometric compounds, the formation of which and the domain of existence vary depending on the nature of the alkali metal and of the lanthanide.

This can be exemplified with the KCl-PrCl3 phase diagram [58] reported in Figure 18.

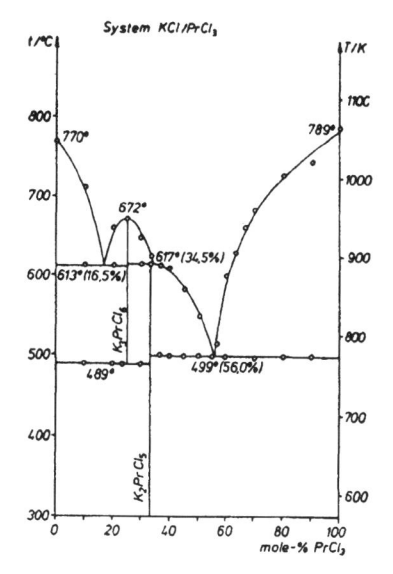

Figure 18. Equilibrium phase diagram of the KCl- PrCl3 system

The congruently melting K3PrCl6 does not exist at room temperature and forms peritectically from K$_2$PrCl$_5$ according to the reaction :

$$KCl + K_2PrCl_5 \rightarrow K_3PrCl_6$$

Figure 19 reports as an example, the heat capacity dependence on temperature of the K$_3$PrCl$_6$ compound [56].

The formation of the stoichiometric compound can be clearly seen at T = 768 K while fusion occurs at T = 944 K, in excellent agreement with our previous DSC investigations [13].

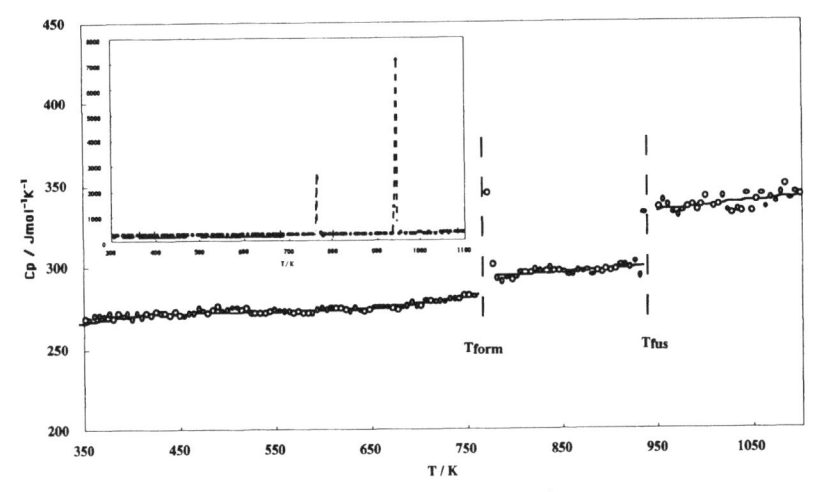

Figure 19. Molar heat capacity of K$_3$PrCl$_6$ against temperature :
o - experimental values, solid line – polynomial fitting

5. Calorimetric Determination of Phase Diagrams

Several physicochemical methods make it possible to determine the equilibrium lines of a phase diagram. The most extensively used so far is thermal analysis [3].
Isobaric determination of phase equilibria in binary or multicomponent systems is based on either the temperature dependence or the concentration dependence of a physicochemical parameter.

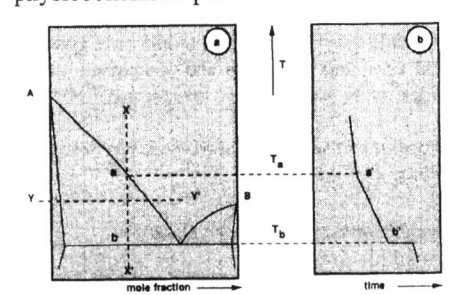

Figure 20. Phase diagram determination by thermal analysis: (a) hypothetical binary phase diagram; *XX'*, isopleth; *YY'*, isotherm; (b) corresponding thermogram

Figure 20a shows a very simple type of condensed binary system; the characteristic shape [59] of the thermal analysis thermogram (Fig. 20b) is obtained by following the thermal effects arising from a linear temperature variation against time (isopleth *XX'*); points *a'* and *b* correspond to the intersections of the equilibrium lines *AE* and *ES* of the phase diagram. In the same way, the equilibrium points *c* and *c'* can be detected by following the change in a physical parameter such as electrical conductivity or viscosity.

The choice of a suitable method for the determination of equilibrium lines depends on many factors such as the physical and chemical properties of components (reactivity, volatility, etc.) and the nature of the equilibrium phase diagram.The determination by thermal analysis, for instance, of a liquid miscibility gap or of a steep liquidus line is rather critical, and other methods seem more suitable.

In many cases, however, calorimetry, and microcalorimetry particularly, is able to provide valuable information; for example, it can be used either to obtain directly equilibrium points or to calculate indirectly some equilibrium lines of a phase diagram. The principle of the method is as follows. For the sake of clarity, the present work has been limited to binary mixtures, but generalization to multicomponent systems can be readily accomplished. At constant temperature and pressure, it is relatively easy to measure the enthalpy of formation ΔH_M of a liquid, single-phase mixture A-B. [4-8, 18, 60-63]. Measurements are performed for mixtures of different compositions, and the plot of ΔH_M versus the mole fraction x_B of component B has a nearly parabolic shape. When, at the experimental temperature, more than one phase exists at the mole fraction x_B, then the enthalpy curve exhibits a characteristic shape which can easily be explained by the "lever rule."

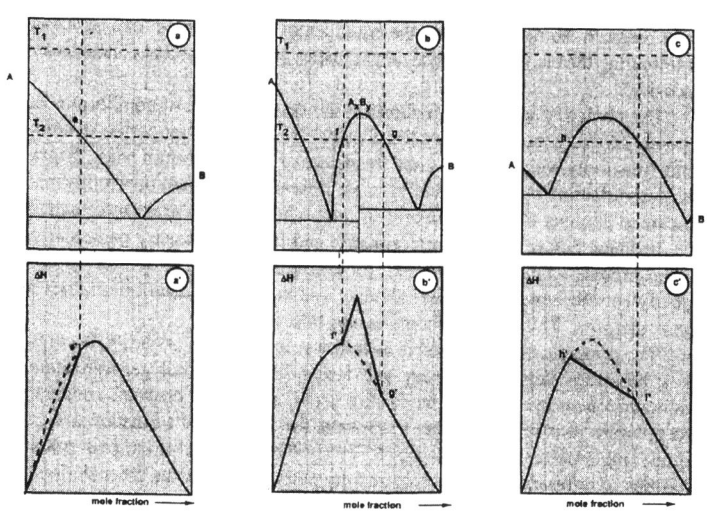

Figure 21. Schematic representation of the composition dependence of the enthalpy of formation of multiphase mixtures: (a) simple binary system; (b) binary system with a definite compound; (c) binary system with a liquid miscibility gap

Figure 21 gives an idea of the shape of the $\Delta H_M = f(x_B)$ curve obtained at several temperatures and for different kinds of liquidus. Only three kinds of fairly simple equilibrium diagrams have been chosen for this theoretical example (diagrams a-c in Fig. 21) The dashed lines on the figures correspond to the experimental temperatures T_1 and T_2 for which, in the whole concentration range, the mixtures are either single-phase (T_1) or two-phase (T_2). Diagrams a'-c' in show the shape of ΔH_M curves obtained at T_1 and at T_2. We used the simplifying hypothesis, which is very often accepted, that enthalpy of mixing does not change with temperature. Each liquidus crossing corresponds to an angular point *(g-g', h-h',* etc.) on the $\Delta H_M = f(x_A)$ curve. The break points *h'* and *i'* refer to the intersection of $\Delta H_M = f(x_A)$—a quasiparabolic curve for a liquid single-phase region—with the straight line *h'i'*. The segment *h'i'* arises from the linear variation (lever rule) of the amounts of the conjugated liquids L_1 and L_2, the enthalpies of formation of which are $\Delta H_M L1$ and $\Delta H_M L2$. For the second diagram, the linear parts correspond to the existence of the compound A_xB_y. This microcalorimetric determination can therefore provide not only the coordinates of equilibrium points but also the thermodynamic quantities of formation of these mixtures, quantities which will be useful in the calculation of equilibrium phase diagrams. This technique is generally completed by differential thermal analysis performed with the same apparatus and with the same sample.

Isothermal calorimetry allows the enthalpy variations corresponding to the formation of single-phase liquid mixtures of many-phase liquid-liquid and solid-liquid mixtures at the same time to be measured directly and also, in some cases, the equilibrium temperatures of the liquidus to be determined. We showed that these measurements are obtainable over a wide temperature range for two- or three-component ionic and metallic mixtures; obviously, this process can be extended to n-component mixtures, with the usual difficulties of graphical representation.

Moreover, in the absence of a good knowledge of Gibbs enthalpies of formation and postulating some simplifying hypothesis, the liquidus lines or surfaces of the equilibrium phase diagram can often be calculated from such experimental results. If the temperatures calculated in this way agree with those obtained experimentally, it is then possible to propose a set of consistent values for the mixture considered.

Differential scanning calorimetry can also be used to determine phase diagrams. This is illustrated in Fig. 22. The A-B binary system is a hypothetical mixture exhibiting the following features:

- Component B exists in two allotropic forms.
- The A_xB_y compounds melt incongruently.
- Solid solubility occurs in the A-rich region.

Seven thermograms (a-g) are given as examples. They correspond to:
- fusion of pure A (a_1)
- solid-state transition and fusion of pure B (g_1 and g_2)
- fusion of the eutectic mixture E (d_1)
- In the non-eutectic mixtures corresponding to thermograms c, e, and f, the peaks c_1, e_1, and f_1, occurring at the eutectic temperature, have a smaller area than the corresponding eutectic peak d_1.
- In the mixture corresponding to thermogram f, the first melt forms isothermally at the eutectic temperature and its amount increases on heating until the liquidus is reached at f_3. At the transformation temperature, however, the amount of pure B still present as a solid undergoes the isothermal solid-state transition indicated by peak f_2.

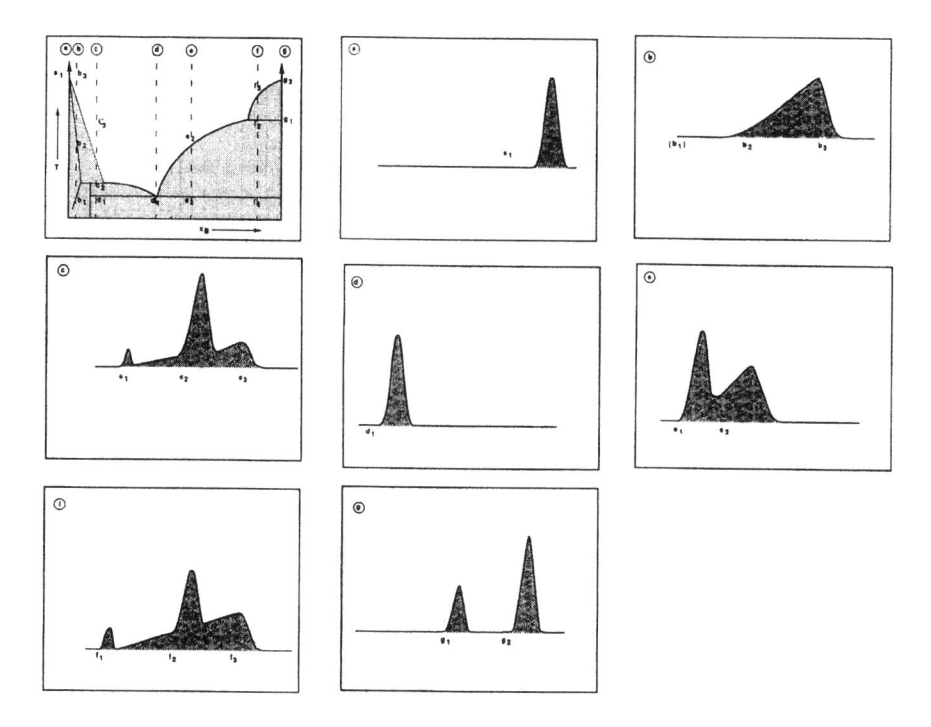

Figure 22. Phase diagram determination by differential scanning calorimetry. Thermograms (a) to (g) correspond to different compositions of the hypothetical A-B phase diagram.

- Thermogram e is self-explanatory.
- The features of thermogram c are similar to those of thermogram f; in this case, however, peak c_2 corresponds to the isothermal peritectic transformation into liquid and mixed crystals of the amount of A_xB_y still present as a solid.
- Finally, in thermogram b, the melting process begins at b_2 and ends at b_3, while the solid-state reaction denoted by b_1 can hardly be detected by DSC or, generally speaking, by a thermal method (other physical methods such as, for instance, electrical conductivity measurements are more sensitive for this purpose)

This technique was used very recently for the determination of the $NaCl$-$EuCl_2$ phase diagram [64]. Indeed, several processes are under development at the international level dealing with reprocessing of nuclear wastes and recycling of spent fuel [see for instance pp 249-261 and 263-282]. The aim is to remove the actinides and lanthanides from spent fuel. Lanthanides are extracted by molten salt electrorefining and it is important to characterize phase equilibria in the process. Very little thermodynamic data were available on $EuCl_2$ and, concerning the $NaCl$-$EuCl_2$ phase diagram, the only information available in recent literature seemed unreliable [65].

Figure 23 and Figure 24 show the phase diagram détermined by DSC and the determination of the eutectic composition from a Tamann construction [59], respectively.

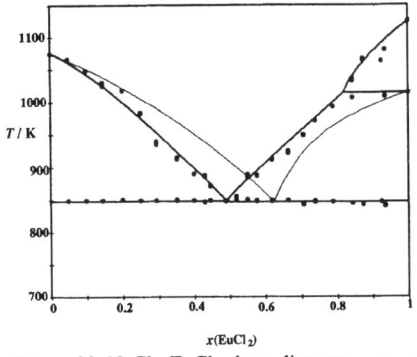

Figure 23. NaCl - EuCl$_2$ phase diagram. —●— : this work, [64] — : Koyama *et al.* [65].

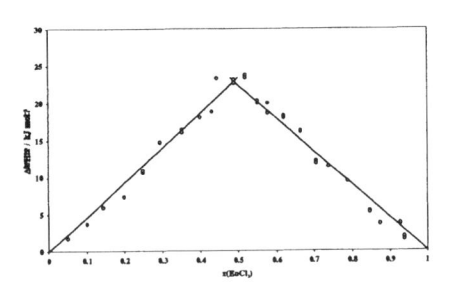

Figure 24. Tamman diagram of NaCl - EuCl$_2$ system.

6. Conclusion

The present chapter reports the main calorimetric methods which can be used to investigate molten salts and their mixtures. The essential purification of melts has not been discussed in any detail here, since each melt system demands individual consideration; this was provided in detail in appropriate chapters of previous volumes of this series. Of course, it was not possible to cover all aspects of high-temperature calorimetry and all fields of applications. As in any research field, calorimetric experimentation with molten salts can yield results of both fundamental and applied interest. The former aspect covers modeling and theoretical developments of molten salts as high-temperature liquids; the latter deals with the many applications of molten salts as materials. These aspects are not independent, and molten salt technology should not ignore the important theoretical advances that have been or are being made.

References

1. A.Tian, *J. Chim. Phys.* **20,** 132 (1923)
2. E. Calvet and H. Prat, *Microcalorimetrie,* Masson, Paris (1955); E. Calvet and H. Prat, *Recents progrès en microcalorimetrie,* Dunod, Paris (1958)
3. M. Gaune-Escard and J. P. Bros, *Thermochim. Acta* **31,** 323 (1979)
4. 0. Kubaschewski and E. L. L. Evans, *La thermochimie en métallurgie,* Gauthier-Villard, Paris (1964)
5. H. Eslami, Thesis, Universite de Provence, Marseille (1976)
6. R. Fehrmann, M. Gaune-Escard, and N. J. Bjerrum, *Inorg. Chem.* 25, 1132 (1986
7. M. Gaune-Escard, Thesis, Universite de Provence, Marseille (1972)
8. 0. J. Kleppa, *J. Phys. Chem.* **64,** 1937 (1960)
9. M. Gaune-Escard, L. Rycerz, A. Bogacz. Enthalpies of mixing in the $DyCl_3$-NaCl, $DyCl_3$-KCl and $DyCl_3$-$PrCl_3$ liquid systems. *J. Alloys and Compounds,* **204.** 185-188, (1994).
10. R. Takagi, L. Rycerz, M. Gaune-Escard. Mixing enthalpy and structure of the molten NaCl-$DyCl_3$ system. *Denki Kagaku,* **62,** 3, 240, (1994).
11. M. Gaune-Escard, L. Rycerz, W. Szczepaniak, A. Bogacz. Enthalpies of mixing in the $PrCl_3$-$CaCl_2$ and $NdCl_3$-$CaCl_2$ liquid systems, *Thermochimica Acta,* **236,** 51-58, (1994).
12. M. Gaune-Escard, L. Rycerz, W. Szczepaniak, A. Bogacz, Calorimetric investigation of the $NdCl_3$-MCl (M = Na, K, Rb, Cs). *Thermochimica Acta,* **236,** 67-80, (1994).
13. M. Gaune-Escard, L. Rycerz, W. Szczepaniak, A. Bogacz, Calorimetric investigation of the $PrCl_3$-NaCl and $PrCl_3$-KCl liquid mixtures. *Thermochimica Acta,* **236,** 59-66, (1994).
14. M. Gaune-Escard, A. Bogacz, L. Rycerz, W. Szczepaniak. Formation enthalpies of the MBr-$LaBr_3$ liquid mixtures (M = Li, Na, K, Rb, Cs). *Thermochimica Acta,* **279,** 1-10 , (1996).
15. M. Gaune-Escard, L. Rycerz. Mixing enthalpy of $TbCl_3$-MCl liquid mixtures . M = Li, Na, K, Rb, Cs. *High Temp. Material Processes,* **2,** n°4, 483-496 (1998).
16. M. Gaune-Escard, L. Rycerz. Calorimetric investigation of the NdI_3-MI systems ((M = Li, Na, K, Cs). *Molten Salt Forum, 5-6, 217 (1998)*
17. F. da Silva, L. Rycerz , M. Gaune-Escard, *Z. Naturforsch 2001 (under press)*
18. *H.* Aghai-Khafri, J.P. Bros, M. Gaune-Escard, *Chem. Thermodynamics, 8, 331-338, (1976)*
19. W. Lukas, M. Gaune-Escard, and J. P. Bros, *J. Chem. Thermodyn.* **19,** 717 (1987)
20. J. W. Johnson, W. J. Silva, and D. J. Cubicciotti, *J. Chem. Phys.* **69,** 3916 (1965)
21. D. Chiotti, G. Gartner, E. Stevens, and Y. Saito, *J. Chem. Eng. Data* **11,** 571 (1966)
22. Z. Benkhaldoun, Thesis, Universite de Provence, Marseille (1985)
23. G. Hatem, M. Gaune-Escard, J. P. Bros, and T. 0stvold, *Ber. Bunsenses Phys Chem* **92,** 751 (1988).
24. Y. Fouque, M. Gaune-Escard, W. Szczepaniak, and A. Bogacz, *J. Chim. Phys.* **75,** 360 (1978)
25. G. N. Papatheodorou and 0. J. Kleppa, *Z. Anorg. Allg. Chem.* **401,** 132 (1973)
26. J. L. Holm and 0. J. Kleppa, *Inorg. Chem.* **6,** 645 (1967)
27. T. 0stvold and 0. J. Kleppa, *Inorg. Chem.* **8,** 78 (1969)
28. B. K. Andersen and 0. J. Kleppa, *Acta Chem. Scand. Ser. A* **30,** 751 (1976)
29. M. Gaune-Escard and J. P. Bros, *Can. Met. Q.* 13(2), 335 (1974)
30. G. Hatem, P. Gaune, J. P. Bros, F. Gehringer, and E. Hayer, *Rev. Sci. Instrum.* 52, 585 (1981)
31. G. Hatem and M. Gaune-Escard, *J. Chem. Thermodyn.* **11,** 927 (1979)
32. J. P. Bros, H. Eslami, and P. Gaune, *Ber. Bunsenges. Phys. Chem.* **85,** 333 (1981)
33. G. Hatem and M. Gaune Escard, 9th Experimental Thermodynamic Conference, London April 16-18, 1980
34. E. Hayer, K. L. Komarek, J. P. Bros, and M. Gaune-Escard, *Z. Metallkde.* 72, 109 (1981)
35. J. M. Miane, M. Gaune-Escard, and J. P. Bros, *High Temp.-High Pressures* **9,** 465 (1977)
36. G. Hatem, M. Gaune-Escard, and A. Pelton, *J. Phys. Chem.* **86,** 3039 (1982)
37. G. Hatem and M. Gaune-Escard, *J. Chem. Thermodyn.* **16,** 897 (1984)
38. G. Hatem, F. Tabaries, and M. Gaune-Escard, *Thermochim. Acta* **149,** 15, (1989)
39. K. Mahmoud, Thesis, Universite de Provence, Marseille (1989)
40. P. Peretz, G. Hatem, M. Gaune-Escard, M. Hoch, *Thermochimica Acta* **262,** *45-54 (1995)*

41 G. Hatem, M. Gaune-Escard, *J. Chem. Thermodynamics*, **25**, 219-228 (1993).
42 E. Hayer, K. Komarek, J. P. Bros, and M. Gaune-Escard, *4th International Conference on Liquid and Amorphous Metals, Grenoble*, July 7-11, 1980
43 D. El Allam, Thesis, Universite de Provence, Marseille (1989)
44 M. J. O'Neil, *Anal. Chem.* **38**, 1331 (1966)
45 A. Bogacz, W. Wisniowski, Y. Fouque, J. P. Bros, and M. Gaune-Escard, *J. Cat. Anal. Therm* **XIV**, 339 (1983)
46 A. Bogacz, J. P. Bros, Y. Fouque, M. Gaune-Escard, and W. Szczepaniak, *J. Chem. Soc., Faraday Trans. 1* **80**, 2935 (1984)
47 Y. Fouque, J. P. Bros, M. Gaune-Escard, M. Wisniowski, and A. Bogacz, *Ber. Bunsenges Phys. Chem.* **89**, 777 (1985)
48 J. P. Bros, M. Gaune-Escard, W. Szczepaniak, A. Bogacz, and A. W. Hewat, *Acta Crystallogr. Sect. B*, 43, 113 (1987)
49 M. Gaune-Escard, L. Rycerz, W. Szczepaniak, A. Bogacz, Entropies of phase transitions in the the M_3LnCl_6 compounds (M = K, Rb, Cs ; Ln = La, Ce, Pr, Nd) and K_2LaCl_5, *J. Alloys and compounds*, **204**, *189-192, (1994).*
50 M. Gaune-Escard, L. Rycerz, W. Szczepaniak, A. Bogacz. Enthalpies of phase transition of the lanthanide chlorides $LaCl_3$, $CeCl_3$, $PrCl_3$, $NdCl_3$, $GdCl_3$, $DyCl_3$, $ErCl_3$ and $TmCl_3$. *J. Alloys and Compounds*, **204,** 193-196, (1994).
51 L. Rycerz, M. Gaune-Escard. Enthalpy of phase transition and heat capacity of stoichiometric compounds in $LaBr_3$-MBr systems (M = K, Rb, Cs). *J. Thermal Analysis and Calorimetry*, **56,** 355-363 (1999).
52 L. Rycerz, M. Gaune-Escard. Enthalpies of phase transitions and heat capacity of $SmCl_3$, $EuCl_3$, $TbCl_3$, $ErCl_3$ and $TmCl_3$ *Z. Naturforsch 2001 (sous presse)*
53 J. L. Holm, B. J. Holm, B. Rinnan, and F. Gr0nvold, *J. Chem. Thermodyn.* 5, 97 (1973)
54 M. Gaune-Escard, A. Bogacz, L. Rycerz, W. Szczepaniak, Heat capacity of $LaCl_3$, $CeCl_3$, $PrCl_3$, $NdCl_3$,$GdCl_3$,$DyCl_3$. *J. Alloys Compounds*, **235**, 176-181, (1996).
55 M. GAUNE-ESCARD, A. BOGACZ, L. RYCERZ, W. SZCZEPANIAK Heat capacity of LaCl3, CeCl3 , PrCl3, NdCl3 ,GdCl3 ,DyCl3 *J. Alloys Compounds, 235, 176-181, (1996).*
56 k. M. GAUNE-ESCARD, L. RYCERZ Heat capacity of the K3LnCl6 compounds - Ln = La, Ce, Pr, Nd. *Z. Naturforsch.54 a, 229-235, (1999).*
57 l. M. GAUNE-ESCARD, L. RYCERZ Heat capacity of the Rb3LnCl6 compounds - Ln = La, Ce, Pr, Nd. *Z. Naturforsch.54 a, 397-403, (1999).*
58 H.J. Seifert, J. Sandrock and J. Uebach, *Z. anorg. Allg. Chem.* 555, 143 (1987)]
59 G. Tamman, *Z. Anorg. Chem.* **37**, 303 (1903); *Z. Anorg. Chem.* **45,** 24 (1905); *Z. Anorg. Chem.* **47**, 289 (1905); see also J. E. Ricci, *The Phase Rule and Heterogeneous Equilibrium*, Dover, New York (1966)
60 0. J. Kleppa, *J. Phys. Chem.* **61,** 1120 (1957)
61 0. J. Kleppa, *J. Phys. Chem.* **65,** 843 (1961)
62 J. P. Bros, Thesis, Universite de Provence, Marseille (1968)
63 G. Hatem, Thesis, Universite de Provence, Marseille, (1980)
64 F. da Silva, M; Gaune-Escard, *Z. Naturforsch 2001 (under press)*
65 Y. Koyama, R. Takagi, Y. Iwadate and K. Fukushima, *J. Alloys Comp.* 260, 75 (1997)

SUBJECT INDEX

Accelerator-driven system	251	CaO-MgO-SiO_2-MnO-Al_2O_3	233
Actinide fluorides	279	CaO-TiO_2	232
Actinide recycling	251	CaO-TiO_2-SiO_2	228
Actinides	249, 279	Capacitor	341
Activity	183, 216	Carbonate melts	383
Ag deposition	168	Catalysis	345
$AlCl_3$ + LiSCN	311	Cathode	322
$AlCl_3$ clusters	12	Chemical short-range order	4
$AlCl_3$-EMIC	335	Chloroaluminates	329
$AlCl_3$-EMIC-LiCl	333	Coherent	118
$AlCl_3$-KCl-$AlCl_3$-NH3	381	Complex	185
$AlCl_3$-MBIC	168	Complex compounds in molten	
$AlCl_3$-NaCl-KCl	206	salts	357
AlF_3 melts	386	Complex-cluster structure	364
AlF_3-NaF clusters	15	Computer simulation	136
Alkali halides	6, 10, 62	Container	271
Alkali hydroxides	380	Container materials	285
AlkCl-$AlCl_3$	196	Correlation	246
Alloy formation	288	Cr deposition	298
Analysis (DTA)	387	Cs_2FeCl_4	58
Anode	322	$Cs_2NaFeCl_6$	60
As_2O_3	83	$CsCl$-$BeCl_2$	70
B_2O_3-Na_2O	237	$CsCl$-$FeCl_3$	73
Battery	322	$CsFeCl_4$	60
$BeCl_2$	68	CuCl	138
$BiCl_3$- Alkali chlorides	381	CuCl-KCl	188
$BiCl_3$- KCl	381	Currentless transfer	289
BMICH$_3$SO	348	Data mining	242
BMIClO$_4$	348	Database	242
BMIM BF$_4$	347	Definition	358
BMIPF$_6$	338, 348	Descriptor	248
BMISbF$_6$	350	Diels-Alder	347
BMITfO	348	Differential enthalpic Analysis	
Borate	235	(DEA)	387
Boson peak	84	Differential Scanning	
Break-off ampoule	379	Calorimeter (DSC)	388
Break-off bubble	379	Differential Thermal Analysis	
$Ca(NO_3)_2$-KNO_3	311	(DTA)	387
$CaCl_2$	140	Divalent halide	137
CaO-Al_2O_3-SiO_2	227	Divalent systems	65
CaO-FeO-SiO_2	224	Double strata system	254
CaO-MgO-SiO_2	222	Drop method	377

DyCl$_3$	143	Immiscibility	317, 349
Dynamic properties	96	Incoherent	118
Dynamics	19	Indirect drop method	377
Effect of outer-sphere cations	61	Inelastic	122
Elastic	22	Informatics	242
Electrochemical processes	292	Interfacial phase transition	151
Electrochemical scanning probe	164	Interionic forces	6
		Ionic binding	8
		Ionic liquid	312, 327, 345
Electrochemical techniques	283	Ionic transport	17
Electrocrystallisation	162	Isotopic substitution	124
Electrodeposition	162	K$_2$O-B$_2$O$_3$	238
Electrodes	287	K$_2$PrCl$_5$	393
Electroreduction	292	K$_2$S$_2$O$_7$	378
Electrorefining	251	K$_2$SO$_4$	378
EMlBF$_4$	338	K$_2$SO$_4$-NaF	385
EMlBF$_4$-LiBF$_4$	333	K$_2$ZnF$_4$	67
EMlTf$_2$N	351	K$_3$PrCl$_6$	393
EMlTfO	351	KCl-MgCl$_2$	186
EmlTrif-PVdF(HFP)	338	KCl-PrCl$_3$	393
Enthalpimeter	383	KF-NdF$_3$	386
Enthalpy of fusion	387	Lanthanide fluorides	279
Enthalpy of transition	387	Lanthanide halide	379
EuCl$_2$	398	Lanthanides	249, 279
Fast breeder reactor	251	Lead oxide melts	382
F-Center	161	Lewis acid	315, 347
Fission products cleanup	278	Li$_2$O-B$_2$O$_3$	238
Fragility	87	LiCl-KCl/Bi	256
Fuel cell	342	LiCl-KCl/Cd	256
Fuel coolant	265	LiF-AlF$_3$	192
Fuel Cycle	253, 265	LiF-BeF$_2$	270
Fuel salt	265	LiF-BeF$_2$-NaF	270
Ge deposition	170	LiF-BeF$_2$-ThF$_4$	270
Gel	337	LiF-K$_2$SO$_4$	386
Glass forming	213, 306	Liquid-liquid mixing	378
Glass transition	84	LiNO$_3$	144
Glass-forming salts	83	Lithium battery	314, 323
Green chemistry	305	Low melting	306
Heat Capacity	389	Macroscopic	145
Heterogeneous-heterophase complex formation	370	Materials compatibility	272
		Metal fuel	251
Heteronuclear complex formation	366	Metal halide	135
		Metal-molten salt	159
Heteronuclear complexes	299	Microcalorimeter	375
Hf deposition	290	Microscopic	134
Homogeneous complex formation	359	Modelling	213
		Modelling (unmeasurable)	206
Imidazolium	346	Modelling (binary)	180

Modelling (ternary)	204	Primitive model	4
Molecular dynamics	136	Principal Component Analysis	243
Molten salt	134	Pulsed neutron source	111
Molten salt reactor	265	Pyrochemistry	249
Monatomic system	121	Radical distribution function	124
Monovalent halide	136	Radwaste	263
Multi-electron transfer	299	Raman spectroscopy	49
$NaBF_4$-NaF	270	Rare-earth halides	73
NaCl	137	Ratio method	390
$NaCl$-$AlCl_3$	190	Rayleigh-Brillouin spectroscopy	97
$NaCl$-$EuCl_2$	398	Re deposition	289
NaF-AlF_3	189, 192	Recycling of spent fuel	398
NaF-K_2SO_4	386	Refractory metal complexes	292
NaF-Na_2SO_4	386	Reversibility-irreversibility	297
NaF-Rb_2SO_4	386	Room temperature molten salts	168
Na-Na_3AlF_6 clusters	16	Room-temperature	345
Nb deposition	290	$SaCl_3$	55
NdI_3-Alkali iodides	379	Safety	263
Nd-$NdCl_3$-LiCl-KCl	195	$Sc_2NaScCl_6$	58
Neutron diffraction	120	Scanning tunneling Microscopy (STMI)	164
Neutron scattering	108	Scattering cross-section	116
Neutron source	111	Screening	4
Ni deposition	171	Shell model	8
NiAl deposition	171	Silicate	213
Nitrates	144, 311	Solid-liquid mixing	377
Nuclear Energy	249	Solvent	345
Nuclear reactor	111	Spallation source	111
Nuclear Waste	398	$SrCl_2$	139
Optical properties	158	Static approximation	120
Organic cation	316	Step method	391
Organic electrolytes	326	Structural	134
Oxide	213	Structure	1, 57
Oxide fuel	253	Structure factor	121
Pair correlation function	120	Structure of cathodic deposits	297
Pair distribution function	122	Suspended cup method	380
Partial structure factor	123	Synchrotron	111
Pd catalyst	351	Synthesis	345
Perfluorinated anion	312, 327	Ta deposition	290
Phase diagram	10,155,186,220,310, 394	Temperature of fusion	387
Photon correlation spectroscopy	101	Temperature of transition	387
Photovoltaics	342	Tetravalent halides	78
Polarization effect	318	$ThCl_4$	81
Polyatomic System	123		
Polymer	337		
Polymerization degree	220		
Potential	314		
Preparation	283, 338		

ThCl$_4$-CsCl	81
Theory	116
Thermodynamics	151, 179, 213, 256
Ti deposition	286
Transport properties	269
Trivalent halides	12, 72, 142
V$_2$O$_5$	378
Wetting phenomena	159
X-Ray diffraction	108
Xylene	309
ZnCl$_2$	382
Zn halide melts	140, 382
ZnCl$_2$-AlCl$_3$	84
ZrF$_4$-alkali fluorides	386
ZrF$_4$-KF	80

9 781402 004599

Made in the USA
Lexington, KY
20 January 2010